Flexible Energy Conversion
and Storage Devices

Flexible Energy Conversion and Storage Devices

Edited by Chunyi Zhi and Liming Dai

WILEY-VCH

Editors

Prof. Chunyi Zhi
City University of Hong Kong
Department of Materials Science
and Engineering
83 Tat Chee Avenue
Kowloon
Hong Kong

Dr. Liming Dai
Case Western Reserve University
Macromolecular Science and Engineering
10900 Euclid Avenue
Cleveland, OH 44406
United States

Cover Image: © FGorgun/iStockphoto

Library of Congress Card No.: applied for

British Library Cataloguing-in-Publication Data
A catalogue record for this book is available from the British Library.

Bibliographic information published by the Deutsche Nationalbibliothek
The Deutsche Nationalbibliothek lists this publication in the Deutsche Nationalbibliografie; detailed bibliographic data are available on the Internet at <http://dnb.d-nb.de>.

Print ISBN: 978-3-527-34253-2
ePDF ISBN: 978-3-527-34260-0
ePub ISBN: 978-3-527-34262-4
oBook ISBN: 978-3-527-34263-1

Typesetting SPi Global, Chennai, India
Printing and Binding C.O.S. Printers Pte Ltd, Singapore

Printed on acid-free paper

10 9 8 7 6 5 4 3 2 1

Contents

Preface

Flexible devices, including flexible and wearable optoelectronics for real-time health monitoring and live sport data collection, play an increasing role in our daily life. However, most recently developed flexible electronics are still powered by batteries with a rigid structure, and can hence be referred as pseudo-flexible devices. Therefore, it is more important than ever to develop flexible and wearable energy conversion and storage devices. Along with rapid progresses in the solid-state flexible devices based on two-dimensional and polymer materials, there are also recent research activities on the development of flexible energy conversion and storage devices.

On one hand, the development of flexible energy conversion and storage devices opens up promising new applications for energy. On the other hand, many recent developments in flexible current collectors, electrodes and solid-state or quasi-solid-state electrolytes, and new electrochemical mechanisms, have paved the way toward flexible energy conversion and storage devices . So far, many new materials and structures have been reported to make traditional energy conversion and storage systems into flexible forms while various prototypes of new flexible energy devices have been demonstrated. This is a truly interdisciplinary field, which will unveil a variety of new challenging research and development opportunities for chemists, physicists, materials science researchers, engineers, and medical practitioners. While significant progress has been achieved, many fundamental and technological obstacles are impeding the practical applications of flexible energy conversion and storage devices. A timely comprehensive overview in an edited book form on such a rapidly growing field of such significance is highly desirable. The aim of this book is to provide timely comprehensive overviews by leading experts on various critical issues in this emerging field. Through such a critical overview, our understanding of flexible energy devices and the associated materials as well as the fundamental principles for energy conversion and storage will significantly increase, as will the interest to a broad range of scientists and insights for future development. This book contains chapters, which are all written by top researchers who have been actively working in this field and have been delivering the recent advances in areas from materials syntheses, through fundamental principles, to device applications. This approach fosters an understanding of the scientific basis of flexible energy devices, and then extends this knowledge to the development, construction, and application of functional energy systems. As such, this book will further stimulate and advance the research and development of this intriguing field.

We would like to thank our colleagues who have amiably dedicated their time, energy, and expertise in contributing to this book. Without their contributions, it would not have been possible to collect information on so many state-of-the-art processes and techniques in this expeditiously developing field. We would also like to thank the employees of the Wiley Group for their very kind and patient cooperation throughout the preparation of this book. We have enjoyed our communication with all the contributors and publishing editors. We wish that this book will serve the flexible device community for many years to come.

November 9, 2017

Chunyi Zhi
Professor
City University of Hong Kong
Kowloon Tong, Hong Kong

Liming Dai
Kent Hale Smith Professor
Case Western Reserve University
Cleveland, OH, USA

1

Flexible All-Solid-State Supercapacitors and Micro-Pattern Supercapacitors

Yuqing Liu[1], *Chen Zhao*[2], *Shayan Seyedin*[3], *Joselito Razal*[3], *and Jun Chen*[1]

[1] *University of Wollongong, ARC Centre of Excellence for Electromaterials Science, Intelligent Polymer Research Institute, Australian Institute of Innovative Materials, Innovation Campus, Wollongong, NSW 2522, Australia*
[2] *Guangdong University of Technology, School of Materials and Energy, No. 100 Waihuan Xi Road, Guangzhou, Guangdong 510006, China*
[3] *Deakin University, Institute for Frontier Materials, 75 Pigdons Road, Geelong, VIC 3216, Australia*

1.1 Introduction

Research on flexible energy storage devices has received significant attention and has led to the development of novel power sources for various types of flexible and wearable electronics. Currently available energy storage devices are too rigid and bulky for the next generation of flexible electronics. Therefore, the development of high-performance and reliable power sources that need to be light, thin, and flexible becomes critical. Furthermore, these energy storage devices need to be functional under various mechanical deformations, such as bending, twisting, and even stretching.

To date, progress has been made in developing energy storage systems such as lithium ion batteries and supercapacitors (SCs) for diverse applications in personal consumer electronics, electric vehicles, and uninterruptable power supplies. Lithium ion batteries were commercialized by SONY in 1990 [1]. As illustrated in Figure 1.1, they can achieve an energy density as high as 180 Wh kg^{-1}. However, lithium ion batteries suffer from slow power delivery or uptake, which inhibits their applications in energy storage systems where fast and high power is needed [2]. Conversely, conventional capacitors have a much higher power density than batteries, but their energy density lower. Supercapacitors have the potential to bridge the gap between batteries and conventional capacitors. A typical supercapacitor exhibits an energy density of ~5 Wh kg^{-1}, which is lower than that of batteries, but a much higher power density of ~10 kW kg^{-1} can be achieved in a few seconds. Significant efforts have been made recently to increase the energy density of the supercapacitors to be close to or beyond that of the batteries without sacrificing their high power density.

Most studies on supercapacitors have primarily focused on using liquid electrolytes, such as aqueous solutions, organic solutions, and ionic liquids. Since most of these liquid electrolytes are toxic and corrosive, precise packaging

Flexible Energy Conversion and Storage Devices, First Edition. Edited by Chunyi Zhi and Liming Dai.

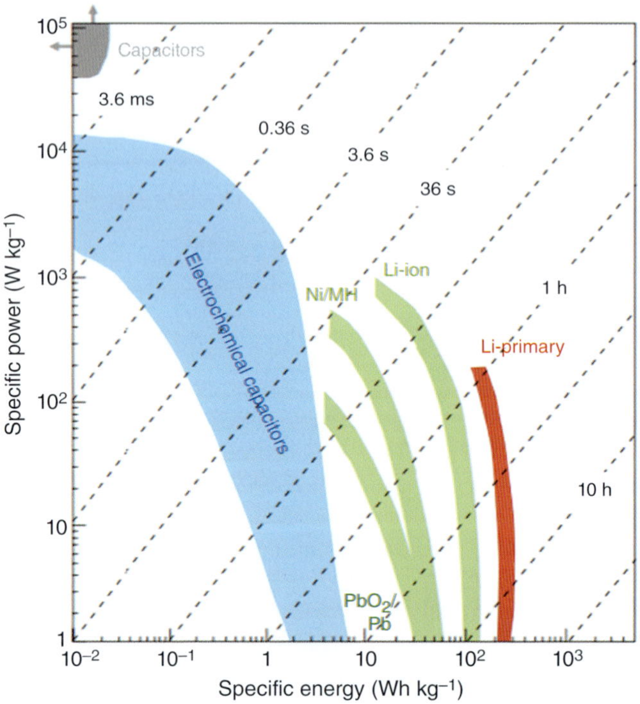

Figure 1.1 Ragone plot for various types of energy storage devices. *Source:* Simon and Gogotsi 2008 [1]. Copyright 2008. Reproduced with permission from Nature Publishing Group.

is required to avoid any possible leakage of these electrolytes. An alternative solution to the leakage problem, which can also reduce the packaging cost, is the use of a solid-state electrolyte. Flexible solid-state supercapacitors present important advantages over traditional ones, which include miniaturization and flexibility, decreased weight, improved energy and power density, and ease of handling. These advantages make them a promising new class of energy storage devices for flexible and wearable electronics.

Supercapacitors are generally classified into two types based on their energy storage mechanisms [1]: (i) electrochemical double-layer capacitor (EDLC), which stores energy by the adsorption of anions and cations on the surface of electrodes and (ii) pseudocapacitor, which derives the capacitance from the storage of charge in redox materials in response to a redox reaction.

The first EDLC was demonstrated and patented by General Electric in 1957 [3]. Generally, an EDLC is composed of two carbon-based electrodes with high surface area, an electrolyte, and a separator (Figure 1.2a). When the supercapacitor is charged, cations in the electrolyte move to the negatively polarized electrode, and anions immigrate to the positively polarized electrode. The concept of a double layer was first described by Helmholtz in 1853, who stated that two layers of opposite charge form at the electrode–electrolyte interface. This simple model was then refined by Gouy and Chapman: they introduced a diffuse layer, which arose from the accumulation of ions close to

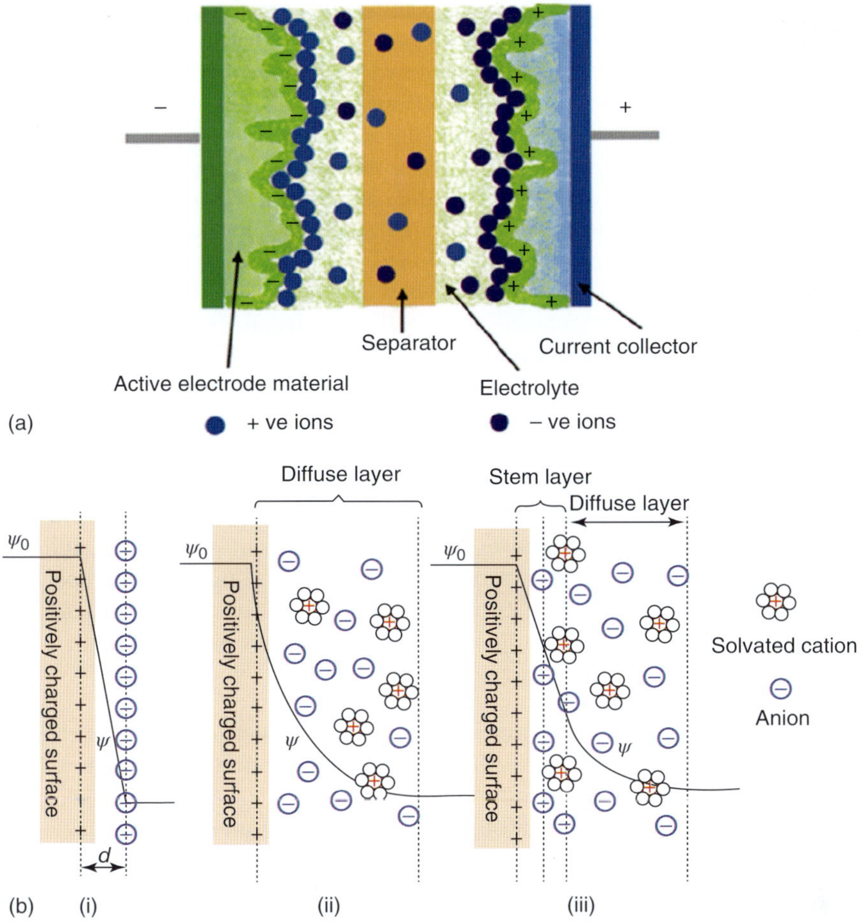

Figure 1.2 (a) Schematic structure of an EDLC. *Source:* Zhang and Zhao 2012 [4]. Copyright 2012. Reproduced with permission from John Wiley & Sons. (b) (i) Helmholtz model, (ii) Gouy–Chapman model, and (iii) Stern model of double layer. *Source:* Zhang and Zhao 2009 [5]. Copyright 2009. Reproduced with permission from Royal Society of Chemistry.

the electrode surface. However, the Gouy–Chapman model overestimated the EDL capacitance since the capacitance increases inversely with the separation distance. Later, Stern combined these two models to define the two parts of ion distribution; the compact layer and the diffuse layer. In the compact layer, the hydrated ions are adsorbed strongly by the electrode, and the compact layer is composed of specifically adsorbed ions and nonspecifically adsorbed counter ions. The diffuse layer is what the Gouy–Chapman model has defined. The capacitance in the EDL can be considered as a combination of the capacitances from the compact and diffuse layers (Figure 1.2b) [5]. The capacitance of an EDLC is generally given by the following Equation (1.1) for a plate capacitor:

$$C = \frac{\varepsilon_r \varepsilon_0 A}{d} \quad (1.1)$$

where ε_r is the dielectric constant of the electrolyte, ε_0 is the dielectric constant of vacuum, A is the surface area of the electrode which is accessible to the electrolyte ions, and d is the effective thickness of the double layer. However, some work revealed that the specific capacitance and the surface area do not show a linear relationship [6, 7]. This nonlinearity is ascribed to the presence of some micropores that are inaccessible to large solvated ions. Huang et al. accounted for this pore distribution and separated the capacitance according to pore size [8]. Their models fit well with the experimental results, no matter what type of carbon materials and electrolytes are employed. To date, various carbon materials such as activated carbon, carbon nanotubes (CNTs), and graphene are the most widely used active electrode materials due to their high specific surface area, high conductivity, easy processing, and high chemical stability.

The capacitance of EDLCs strongly depends on the surface area and pore size distribution of the electrode materials. In contrast, a pseudocapacitor stores energy through fast and reversible redox reactions between the electrolyte and the electrode materials. When a potential is applied to a pseudocapacitor, fast and reversible redox reactions occur on the electrode materials and the charge passes across the double layer, resulting in a current passing through the supercapacitor [9]. The typical electrode materials for this kind of supercapacitor include conducting polymers (CPs) [10] and metal oxides or hydroxides [11]. The charging process of conducting polymers can be expressed as follows:

$$\mathrm{CP} \rightarrow \mathrm{CP}^{n+}(A^-)_n + n\mathrm{e}^-$$

The anions in the electrolytes transfer into the backbone of polymer chains. During discharging, the anions are ejected from the backbone and immigrate into the electrolyte.

Among metal oxides with pseudocapacitance, Ruthenium oxide (RuO_2) is the most promising material. This is because it is conductive and has a very high theoretical specific capacitance of ~2000 F g^{-1} [2]. In acidic electrolytes, the oxidation states of Ru can change from Ru(II) to Ru(IV), which involves rapid reversible electron transfer together with the electron adsorption of protons on the surface of RuO_2. This process can be described by the following reaction [9]:

$$\mathrm{RuO_2} + x\mathrm{H}^+ + x\mathrm{e}^- \leftrightarrow \mathrm{RuO}_{2-x}(\mathrm{OH})_x \quad (1.2)$$

where $0 \leq x \leq 2$.

1.2 Potential Components and Device Architecture for Flexible Supercapacitors

A flexible solid-state supercapacitor is generally composed of flexible electrodes, a solid-state electrolyte, and a flexible packaging material. The flexible electrode should have high capacitance, low self-discharge, long cycling stability, and high mechanical stability. The solid-state electrolyte needs to possess high ionic conductivity and good thermal and mechanical stability. The solid-state electrolyte can also serve as the separator eliminating the need for an additional separator [12].

1.2.1 Flexible Electrode Materials

1.2.1.1 Carbon Materials

Carbon nanomaterials such as 1D CNTs and 2D graphene, with structures composed of conjugated sp^2 carbons, have been widely studied for flexible energy storage applications.

CNTs are tubular allotropes of carbon with graphitic structures (Figure 1.3). They can be classified into two types: single-walled carbon nanotubes (SWCNTs) and multiwalled carbon nanotubes (MWCNTs) [14]. CNTs have attracted significant interest in recent years for developing high-performance flexible supercapacitors due to their high electrical conductivity, unique pore structure, and good mechanical stability [15]. CNTs possess moderate specific surface area compared to activated carbons; however, comparable specific capacitance has been reported for CNTs. Niu et al. reported a MWCNT-based electrode that exhibited a maximum specific capacitance of ~110 $F\,g^{-1}$ [16]. SWCNT-based electrodes achieved a specific capacitance as high as 180 $F\,g^{-1}$ [17]. Recent studies have shown that aligned CNTs are more efficient in ion transportation compared to the randomly entangled CNTs. A high specific capacitance of 365 $F\,g^{-1}$ has been obtained for an MWCNT array based electrode [18].

Graphene, a one-atom thick single-layer graphitic carbon (Figure 1.4), has also attracted significant attention as electrode materials for flexible supercapacitors because of its high electrical properties, large surface area, and high mechanical strength [20–23]. The theoretical specific surface area of graphene is about 2630 $m^2\,g^{-1}$ [24], leading to the theoretical double-layer capacitance of 550 $F\,g^{-1}$ [25]. However, this high level of capacitance is difficult to achieve as the result of the restacking of the graphene sheets during the electrode preparation process. Stoller et al. investigated reduced graphene oxide (rGO) as electrode materials [24]. The individual graphene sheets partially agglomerated during the reduction process, leading to moderate specific capacitances of 135 and 99 $F\,g^{-1}$ in aqueous and organic electrolytes, respectively. Zhu et al. reported that KOH-activated

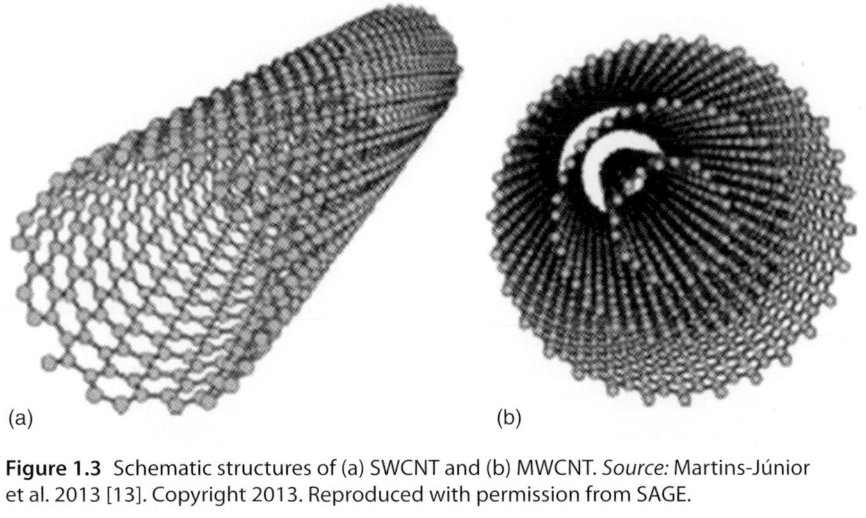

Figure 1.3 Schematic structures of (a) SWCNT and (b) MWCNT. *Source:* Martins-Júnior et al. 2013 [13]. Copyright 2013. Reproduced with permission from SAGE.

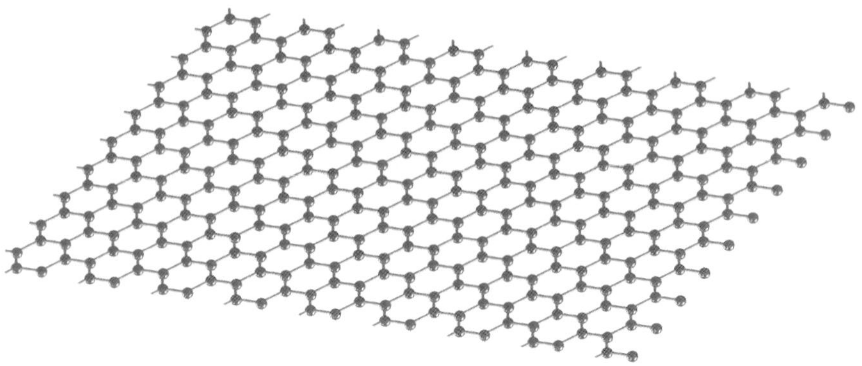

Figure 1.4 Schematic structure of one single-layer graphene. *Source:* Yan et al. 2012 [19]. Copyright 2012. Reproduced with permission from Royal Society of Chemistry.

graphene could achieve a specific surface area of up to 3100 $m^2 g^{-1}$, resulting in a specific capacitance of 166 $F g^{-1}$ at a current density of 5.7 $A g^{-1}$ [26].

1.2.1.2 Conducting Polymers

Conducting polymers are organic polymers that can conduct electricity. In 2000, the Nobel Prize in Chemistry was awarded to Alan J. Heeger, Alan MacDiarmid, and Hideki Shirakawa for their outstanding work on conducting polymer polyacetylene [27]. During the past decades, numerous CPs have been developed, and have found widespread application in polymer light-emitting diodes (LEDs) [28], sensors [29], artificial muscles [30], anticorrosion coatings [31], and in energy conversion and storage devices [10, 32]. CPs are suitable as electrode materials for flexible supercapacitors as they possess many advantages: low cost, environmental friendliness, high conductivity in the doped state, high capacitance, ease of synthesis, and the possibility of manufacturing them as self-supporting and flexible thin films [10, 33–36]. The most commonly studied CPs in supercapacitors are polypyrrole (PPy) [37], polyaniline (PANi) [35], and derivatives of polythiophene (PTh), such as poly(ethylenedioxythiophene) (PEDOT) [38]. The structures of the common CPs are shown in Figure 1.5.

CPs can be p-doped (with anions) or n-doped (with cations). Electrochemical p-doping of conducting polymer occurs by removing electrons from the polymer chain and the addition of anions to the polymer backbone to balance the positive charge. The reverse mechanism (i.e. the addition of electrons to the backbone and balancing the negative charge with cations) accounts for the electrochemical n-doping of conducting polymers. PPy and PANi can only be p-doped in supercapacitor systems due to the highly negative potentials required for n-doping, which is beyond the reduction potential limit of the electrolyte [39]. PTh and its derivatives can be both p-doped and n-doped [40, 41].

Electrodes made from CPs showed much higher specific capacitances compared to carbon materials. Fan and Maier prepared a highly porous PPy electrode on Ti foil using the cyclic voltammetry (CV) method [42]. The resultant PPy showed a specific capacitance of about 450 $F g^{-1}$. PPy doped with Nafion ions or perchlorate exhibited a specific capacitance of 344 or 355 $F g^{-1}$, respectively [43].

Figure 1.5 Chemical structures of PPy, PANi, PTh, and PEDOT (undoped form).

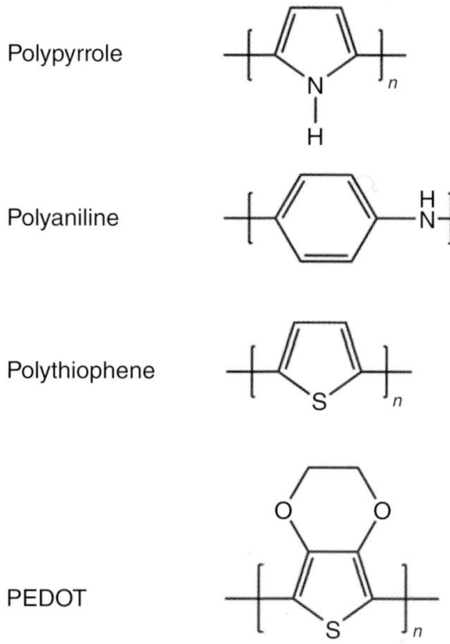

PANi was reported to show a specific capacitance higher than 500 F g^{-1} [44–54]. A derivative of PTh, poly(tris(4-(thiophen-2-yl)phenyl)amine), can even achieve a specific capacitance greater than 990 F g^{-1} [55]. The drawback of the CPs-based electrodes is that they suffer from structural degradation caused by swelling and shrinking of CPs during long-term cycling, leading to the decay of their electrochemical performance. It is suggested that electrodes prepared from nanostructured CPs [56] or composites of CPs with carbon-based materials [57–59] can show superior cyclic stabilities.

1.2.1.3 Composite Materials

Carbon materials, such as CNTs and graphene, can be employed as a conducting substrate for the integration of pseudocapacitive materials to boost the electrode performance [60]. The combination of carbon materials with conducting polymers has been proven to be effective for preparing electrodes for supercapacitors due to the synergistic effect of electrochemical double-layer capacitance and pseudocapacitance [61]. This strategy endows the CPs with improved cyclic stabilities as carbon materials such as CNTs and graphene can act as frameworks to stabilize the CPs, limiting their swelling and shrinking during the cycles [62]. Apart from CPs such as PPy [63, 64], PANi [65, 66], and PEDOT [67], various metal oxides, including MnO_x [68, 69], RuO_2 [70], V_2O_5 [71], and Co_3O_4 [72] have also been explored for the fabrication of flexible carbon-based composite electrodes.

1.2.2 Solid-State Electrolytes

The use of solid-state electrolytes addresses the electrolyte leakage issue, and also facilitates the device-packaging process. The most widely used solid-state electrolytes in supercapacitors are polymer electrolytes. The ideal polymer electrolytes should be of high ionic conductivities at room temperature, low electronic conductivities, and good mechanical stabilities. The polymer electrolytes typically achieve their ionic conductivities through the movement of protons, lithium ions, or the ionic species in ionic liquids [73].

Proton-conducting polymer electrolytes, composed of acid/polymer blends are most widely used solid electrolytes for flexible supercapacitors. The main polymer used in this system is poly(vinyl alcohol) (PVA), and H_2SO_4 and H_3PO_4 are often chosen as the proton conductors [64, 74–77]. This type of polymer electrolytes is typically prepared by the addition of an acid to the aqueous polymer solution. Freestanding electrolyte films, which exhibit conductivity in the range of 10^{-6}–10^{-3} S cm^{-1}, can then be obtained by evaporating the water. During the cell-assembling process, the electrodes are immersed in the hot polymer electrolyte solution or the electrolyte solution is cast on the electrode surface. The polymer electrolyte solution can penetrate the active layer of the electrodes, increasing the electrochemically active surface area [73]. After drying, two electrodes with polymer electrolyte are pressed face-to-face to form an integrated cell. Proton-conducting polymer electrolytes exhibit the highest conductivity among all polymer electrolytes. They can be prepared and used under ambient conditions and the presence of water helps to maintain high conductivity. However, their main drawback is the relatively narrow voltage window (usually limited to 1 V) [73].

Lithium-ion-conducting polymer electrolytes are generally prepared by mixing a polymer, such as poly(ethylene oxide) (PEO), poly(acrylonitrile) (PAN), poly(methyl methacrylate) (PMMA), and poly(vinylidene fluoride) (PVDF) with a lithium salt dissolved in an organic solvent. Solvents such as ethylene carbonate (EC), propylene carbonate (PC), ethyl methyl carbonate (EMC), dimethyl carbonate (DMC), diethyl carbonate (DEC), dimethyl formamide (DMF), and tetrahydrofuran (THF) were found to also act as a plasticizer giving rise to a higher conductivity [78]. Lithium-ion-conducting polymer electrolytes are prepared by dissolving various lithium salts in organic solvents and immobilizing in a polymer matrix. The lithium-ions-based solid electrolytes can operate in a wide electrochemical window. However, they are moisture sensitive and require oxygen and moisture-free environment for cell assembling.

Ionic liquids are a class of liquid slats. Their unique properties, such as low volatility and flammability, high thermal stability, and wide potential window, make them suitable electrolytes for supercapacitor application [79]. They can be trapped in a polymer matrix or a silica network, form an ion gel that can be used as solid electrolytes [80–83].

1.2.3 Device Architecture of Flexible Supercapacitor

The electrochemical performance of a supercapacitor, is dependent not only on the properties of the materials used in each component, but also on how these components are designed, matched, and combined to form a device that can satisfy the performance requirements. Here, we discuss two widely studied architectures used for flexible supercapacitor (fSC) devices: sandwiched stacked structure and interdigitated planar structure [84–86] (Figure 1.6).

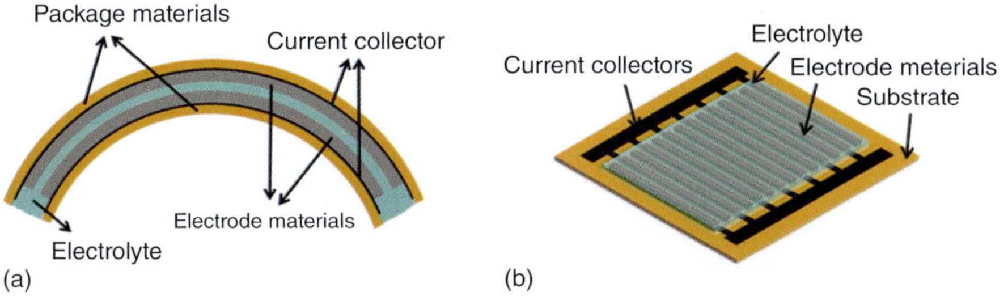

Figure 1.6 Schematic diagrams of flexible supercapacitors with (a) conventional sandwiched structure and (b) interdigitated structure.

The early all-solid-state fSCs adopted the conventional sandwiched structure design with a stacked configuration of current collector/film electrode/solid electrolyte/film electrode/current collector (Figure 1.6a). This design attracted the attention of researchers due to its extremely simple structure and relatively easy fabrication process. However, when the thicker electrodes are required, this design often suffers from serious ion transport limitations due to the consequently increased ion transport path [84, 87]. Moreover, portable electronic devices should not only be flexible, but also need to have small size and thickness. Consequently, the bulky sandwiched architecture is not compatible with the planar geometry of the miniaturized microelectronics and not suitable for electronic circuitry [84].

To overcome this problem, a planar design with several microelectrodes (as each electrode) interdigitally arranged on a flexible substrate such as terephthalate (PET) or polyimide (PI) films or papers (interdigitated design), was developed as a new emerging device structure (Figure 1.6b) [84]. The fabrication of interdigitated microelectrodes often involves an additional micro-patterning step before or after film electrode deposition. Compared with the stacked sandwiched structure, the plannar interdigitated structure possesses the following advantages:

- *Space-saving device.* Without a layer of solid electrolyte used as separator in the sandwiched design, the fSC device with interdigitated design is thinner and smaller and can store and provide more power in a limited space without the volume being wasted on the separator layer.
- *Shorter ion transport paths.* The advances in micro-patterning techniques make it possible to accurately control very small distances between adjacent electrodes thereby shortening the ion transport path. This results in decreasing the ion transport resistance, increasing the rate capability and power density of the device.
- *Better penetration of electrolyte ions.* The increased exposure of the microelectrodes to the electrolyte in the interdigitated design can increase the ion accessibility of the inner surface of the electrode materials resulting in an improved energy storage.
- *Facile interconnection and integration within the electronic systems.* Most applications require a current or voltage that is higher than what can be delivered by a single SC device. Hence, the connection of the SC devices in series, parallel or a combination of those are needed. Having the electrodes in

the same plane would facilitate these connections on one chip, avoiding the intricate wired interconnection between sandwiched SCs [88].

The drawbacks of the interdigitated design are as follows:

- *Less areal energy.* The footprint area of the interdigitated SCs includes the active surface areas of both electrodes and an inactive gap surface in between, while the footprint area of the sandwiched SCs is only defined by the active materials. Hence, a higher areal energy can be achieved for SCs in the sandwiched configuration. The adverse impact of the interdigitated SCs can be minimized by decreasing the distance between the adjacent electrodes. However, it normally involves more expensive micro-fabrication techniques with fine control on resolution.
- *Additional consumption of expenses and labor in the patterning step.* This is the main challenge for the scale-up of the interdigitated design for commercial applications. Efforts are being devoted to develop scalable micro-fabrication techniques for various electrode materials, which will be discussed in Section 1.3.2.

Sandwiched SCs are still the industrially favorable device design due to their facile and low-cost fabrication process, although the interdigitated structures show many advantages. However, with the development of the miniaturized wearable electronics and the continuous progress in micro-patterning techniques, the interdigitated SCs may find an increased application.

1.3 Flexible Supercapacitor Devices with Sandwiched Structures

1.3.1 Freestanding Films Based Flexible Devices

Flexible supercapacitor devices can be fabricated from freestanding electrode films. One of the active materials used in preparing flexible and freestanding electrodes is CNT. CNTs can be assembled into freestanding films through vacuum filtration-induced self-assembly. In a typical process, the stable CNT dispersion is first formed by breaking up the large bundles of CNTs in water (with the aid of a surfactant) or in organic solvents using ultrasonic energy. During the filtration process, CNTs are trapped inside or on the surface of the membrane pores, forming an interconnected and entangled film. The CNT film can then be peeled off after sufficient film thickness has been achieved [89]. Aligned CNT films can also obtained by drawing the vertically oriented CNT arrays [90]. Such aligned CNT films are ultrathin, transparent, stretchable, and conductive. Chen et al. [91] developed a transparent and stretchable supercapacitor electrode by continuously drawing a transparent CNT sheet out from a CNT forest onto a poly(dimethylsiloxane) (PDMS) substrate and then coating a PVA/H_3PO_4 polymer electrolyte solution onto the CNT sheet. The flexible supercapacitor device was made by assembling two electrodes in either a parallel or cross configuration (Figure 1.7a). The transparent supercapacitor showed a specific capacitance of

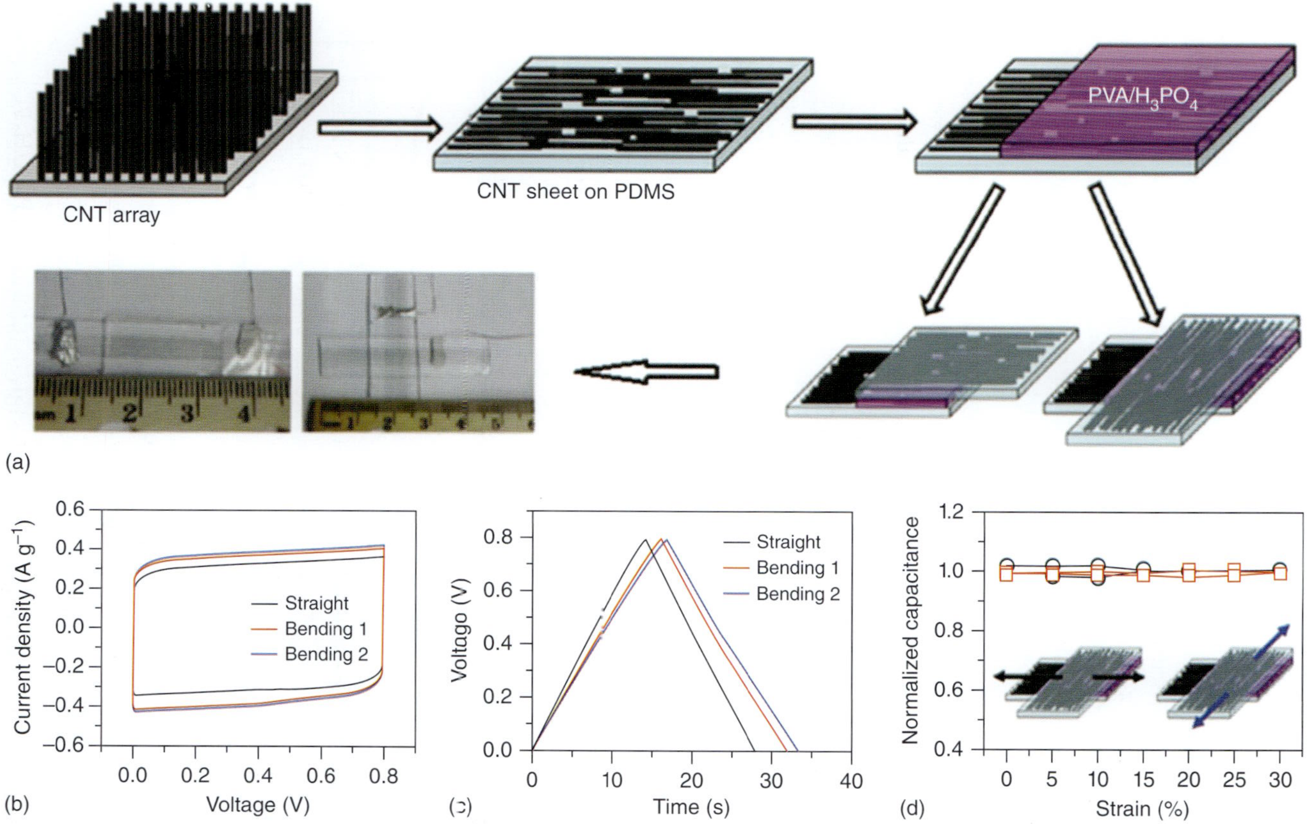

Figure 1.7 (a) Schematic illustration of the process for fabricating the transparent and stretchable supercapacitor. (b) CV curves (at scan rate of 0.1 V s^{-1}), and (c) charge–discharge curves (at constant current density of 0.2 A g^{-1}) of the supercapacitor under different states. (d) Normalized specific capacitance of the supercapacitor as a function of tensile strain. *Source:* Chen et al. 2014 [91]. Copyright 2014. Reproduced with permission from Nature Publishing Group.

7.3 F g^{-1} and can be bent or biaxially stretched to 30% without noticeable changes in its electrochemical performance (Figure 1.7b–d).

Freestanding graphene films or papers, have also shown great promise as flexible electrodes for supercapacitors because they are ultrathin, flexible, and lightweight. Graphene films have been fabricated through various solution processing methods, such as vacuum filtration, spin-coating, Langmuir–Blodgett, layer-by-layer deposition, and interfacial self-assembly [92]. However, during the fabrication process, the graphene sheets tend to aggregate and restack due to the strong $\pi - \pi$ interactions and van der Waals forces, limiting the available surface area and the diffusion of electrolyte ions, resulting in deteriorating electrochemical performance. To prevent the restacking of graphene sheets, two main strategies have been proposed: (i) to add spaces such as CNTs and polymers [93–95] and (ii) to produce 3D porous graphene networks [96, 97].

Shi and coworkers [94] fabricated a flexible electrodes from rGO /poly(vinyl pyrrolidone) (PVP) composite film (Figure 1.8a,b). The PVP chains were found to effectively intercalate between the graphene sheets, acting as spacers to

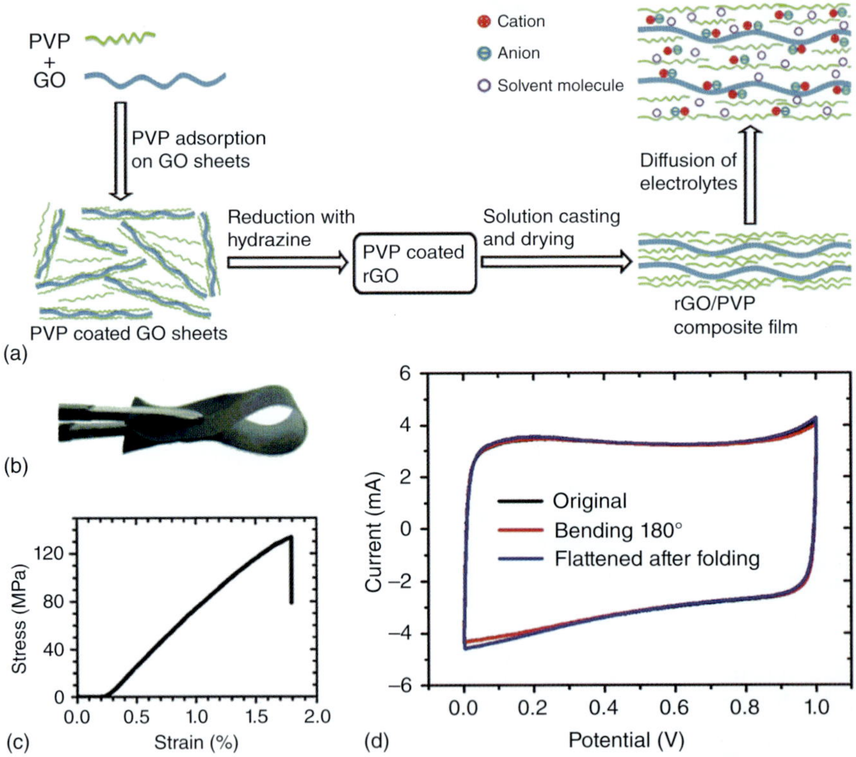

Figure 1.8 (a) Schematic illustration of fabrication of rGO/PVP composite film and the diffusion of the electrolyte onto the surfaces of rGO sheets. (b) Photograph and (c) tensile stress–strain property of the flexible rGO/PVP film. (d) CV curves of the solid-state device based on the rGO_{55}/PVP_{45} composite film with a thickness of 2.4 μm in various bending states (scan rate = 100 mV s^{-1}). *Source:* Huang et al. 2014 [94]. Copyright 2014. Reproduced with permission from Royal Society of Chemistry.

prevent the graphene sheets restacking. The composite films showed a high mechanical strength of ~121.5 MPa (Figure 1.8c). The solid-state supercapacitors based on the rGO/PVP composite containing 55% rGO (by weight) showed high volumetric specific capacitances (67.4 $F\,cm^{-3}$ at 1 $A\,g^{-1}$), good rate-capability (51.4 $F\,cm^{-3}$ at 100 $A\,g^{-1}$), and excellent flexibility (Figure 1.8d). Park and coworkers [95] prepared Nafion functionalized rGO films using a supramolecular assembly approach and vacuum filtration. The integration of Nafion not only prevented the restacking of graphene sheets, but also improved the interfacial wettability between the electrodes and electrolyte. All-solid-state supercapacitors were fabricated by sandwiching Nafion membranes between the functionalized rGO thin films. The specific capacitance of functionalized rGO was 118.5 $F\,g^{-1}$, which is about two times higher than the reported value for the pure rGO (62.3 $F\,g^{-1}$). After bending at a radius of 2.2 mm, the CV curves of the supercapacitor exhibited almost the same shape as those in the flat state, indicating high mechanical flexibility.

Tremendous attention has also been paid to 3D porous graphene network to prevent the aggregation of the graphene sheets. The general approaches involved include freeze drying [98], hydrothermal [96], and organic sol–gel [99]. Shu et al. [97] prepared a flexible freestanding 3D graphene with interconnected porous structure by freeze drying a wet GO gel, followed by thermal and chemical reduction (Figure 1.9a). The water in the precursor GO gel helped to prevent the restacking of the graphene sheets. The porous graphene paper was highly flexible (Figure 1.9b). A flexible solid-state supercapacitor was fabricated using those graphene papers and PVA/H_2SO_4 polymer electrolyte (Figure 1.9c,d). The graphene paper in the device delivered a specific capacitance of 137 $F\,g^{-1}$ at 1 $A\,g^{-1}$. And it can maintain 94% of its capacitance under bending (Figure 1.9e). 3D graphene macrostructures such as graphene hydrogels can also be prepared by a one-step hydrothermal process. Xu et al. [96] pressed the hydrothermally reduced graphene hydrogel onto gold coated polyimide substrate with a PVA/H_2SO_4 polymer electrolyte to assemble flexible solid-state supercapacitors The graphene gel film in the solid-state device achieved a high specific capacitance of 186 $F\,g^{-1}$ at 1 $A\,g^{-1}$, and the device showed stable electrochemical performance under different bending angles.

Apart from the wet process to create 3D graphene structures, Kaner's group used a standard LightScribe CD/DVD drive and developed a solid-state laser scribed strategy to produce 3D porous graphene electrodes (Figure 1.10) [100]. The laser irradiation reduced the GO to laser-scribed graphene (LSG), and the initially stacked GO sheets were converted to well-exfoliated graphene (EG) sheets. The obtained LSG films showed excellent conductivity (1738 $S\,m^{-1}$) and mechanical flexibility (1% change in the electrical resistance after 1000 bending cycles). Flexible solid-state supercapacitor based on the LSG films and PVA/H_3PO_4 electrolyte showed comparable performance with those used H_3PO_4 as the liquid electrolyte. Bending had almost no effect on the capacitive behavior of the device, and only ~5% change in the device capacitance was observed after 1000 bending cycles.

Compared to pure carbon based materials, pseudocapacitive materials exhibit higher specific capacitance. They have been incorporated into the carbon

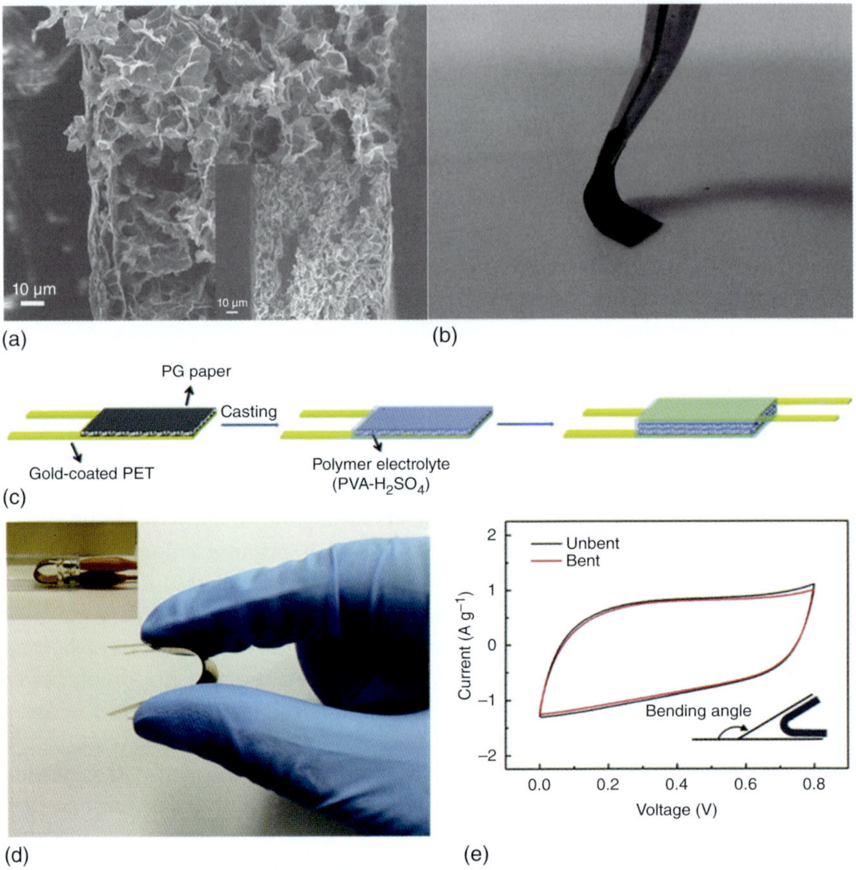

Figure 1.9 (a) Cross-sectional SEM image of a porous graphene paper. (b) Photograph of the porous graphene paper demonstrating its flexibility. (c) Schematic procedure to fabricate a solid-state supercapacitor from porous graphene paper. (d) Photograph of the flexible device at a bending state. (e) CV curves of the flexible device at bending and relaxation state (scan rate = 10 mV s^{-1}). *Source:* Shu et al. 2015 [97]. Copyright 2015. Reproduced with permission from Royal Society of Chemistry.

based flexible electrodes to fabricate solid-state supercapacitors with a higher electrochemical performance. For example, Meng et al. [77] developed a solid-state paper-like supercapacitor based on PANi coated free-standing CNT film and PVA/H_2SO_4 gel electrolyte. This device showed a specific capacitance of 31.4 F g^{-1} and excellent cycling stability.

1.3.2 Flexible Substrate Supported Electrodes Based Devices

Another approach to fabricate flexible solid-state supercapacitor is to use flexible substrate to support the electrodes. Mechanically flexible, thin, and lightweight plastics such as polyethylene terephthalate (PET) have been widely used as the supporting substrates for flexible supercapacitors. Kaempgen et al. [101] demonstrated a thin film supercapacitor based on spray coated SWCNT networks

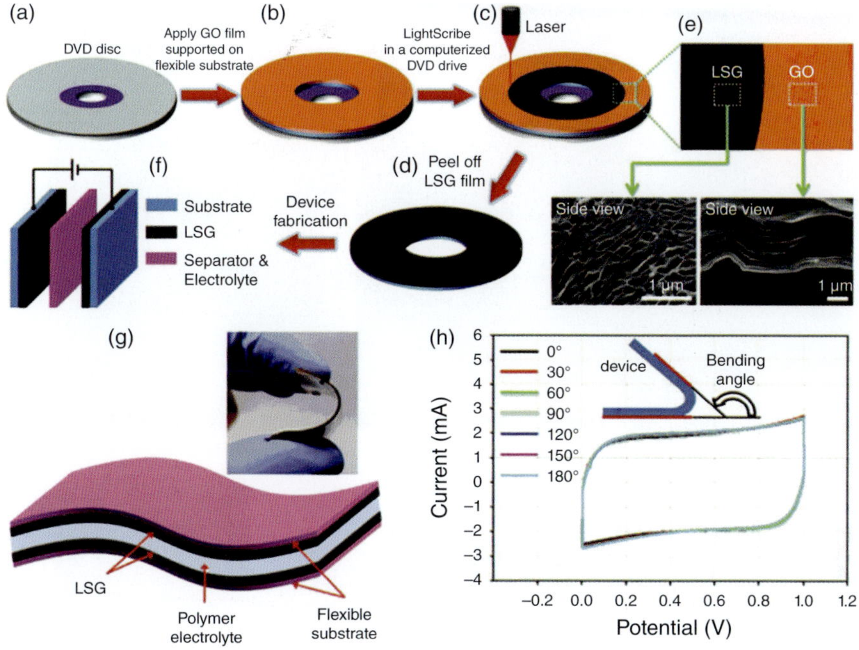

Figure 1.10 (a–d) Schematic illustration of the fabrication of LSG. (e) SEM images of GO and LSG. Schematic of (f) a conventional structured device and (g) a solid-state device based on LSG (Inset in (g) indicates the flexibility of the device). (h) CV curves of the devices at different bending angles. *Source:* El-Kady et al. 2012 [100]. Copyright 2012. Reproduced with permission from American Association for the Advancement of Science.

(served as both electrode and current collector) on a PET substrate using a PVA/H_3PO_4 gel electrolyte (Figure 1.11a,b). The SWCNT network showed a specific capacitance of ~36 $F\,g^{-1}$ in the solid-state device. Chen et al. [102] reported a transparent solid-state stretchable supercapacitor based on wrinkled graphene electrodes that can sustain a strain of up to 40% (Figure 1.11c,d). The wrinkled graphene was synthesized by chemical vapor deposition (CVD) on a wrinkled copper film. After PDMS coating and etching, the wrinkled graphene was well transferred onto PDMS. The solid-state supercapacitor showed a stable performance over hundreds of stretching cycles.

Although solid polymeric substrates have been widely used as flexible substrates, they suffer from limited active materials loading and diffusion of electrolyte. Alternatively, porous substrates, such as paper and textile, have been used for flexible supercapacitors. In some cases, only the surface of the papers is made conductive, and the bulk of papers act as a flexible support. Yao et al. [103] introduced a pencil-drawing and a subsequent PANi electrodeposition to fabricate graphite/PANi hybrid electrodes on a paper. The hybrid electrode was made by first drawing a conductive graphite layer on a paper using a pencil. Electrochemical deposition of PANi nanowire networks was then carried out on the graphite layer. The solid-state supercapacitors was assembled by two graphite/PANi electrodes sandwiching a PVA/H_2SO_4 electrolyte layer and achieved a high energy density of 0.31 $mWh\,cm^{-3}$ at a power density of 0.054 $W\,cm^{-1}$. Liu et al. [104]

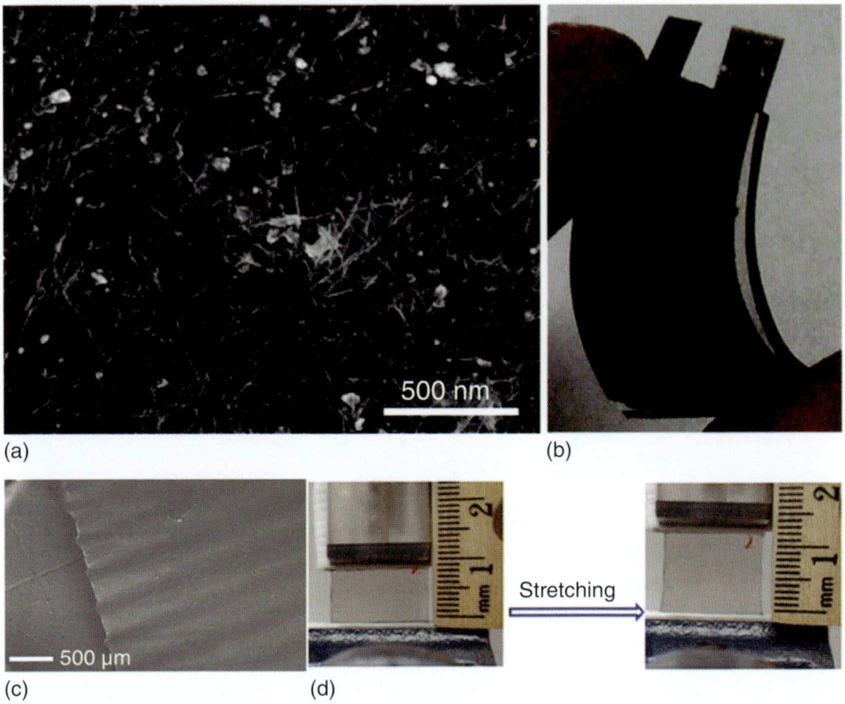

Figure 1.11 (a) SEM image of the SWCNT network. (b) Photograph of an SWCNT network based flexible supercapacitor. *Source:* Kaempgen et al. 2009 [101]. Copyright 2009. Reproduced with permission from American Chemical Society. (c) SEM images of the wrinkled graphene film. (d) Photographs of the wrinkled graphene-based flexible supercapacitor device at different stretching states. *Source:* Chen et al. 2014 [102]. Copyright 2014. Reproduced with permission from American Chemical Society.

demonstrated PANi-rGO-cellulose paper nanocomposite electrodes. The fabrication of nanocomposite electrodes started with coating a GO layer on the cellulose fibers in the paper, and then the GO-coated cellulose fiber paper was used as a template to assemble porous rGO networks in the pores of the paper by a hydrothermal process. PANi was then chemically polymerized on the rGO paper to obtain PANi-rGO/paper composite paper (Figure 1.12a). The solid-state supercapacitor was assembled by sandwiching two such composite papers with PVA/H_2SO_4 polymer electrolyte (Figure 1.12b). The calculated specific capacitance of PANi-rGO/paper electrode was about 224 $F\,g^{-1}$, and the solid-state device showed stable performance at folding and bending states (Figure 1.12c–e).

Textiles have hierarchical porosity in the fibers and yarns, and there is also free space between fibers and yarns. Such inherently porous structures facilitate the accommodation of active materials. Yun et al. [105] designed a solid-state stretchable textile supercapacitor using PPy-MnO_2/CNT coated textile electrodes and PEO-based gel electrolyte. A piece of textile made of cotton was dipped into SWCNT ink to obtain a conductive textile, MnO_2, and PPy was then electrodeposited on the SWCNT-coated textile. The textile supercapacitor exhibited an energy density of 31.1 $Wh\,kg^{-1}$ and a power density of 22.1 $kW\,kg^{-1}$.

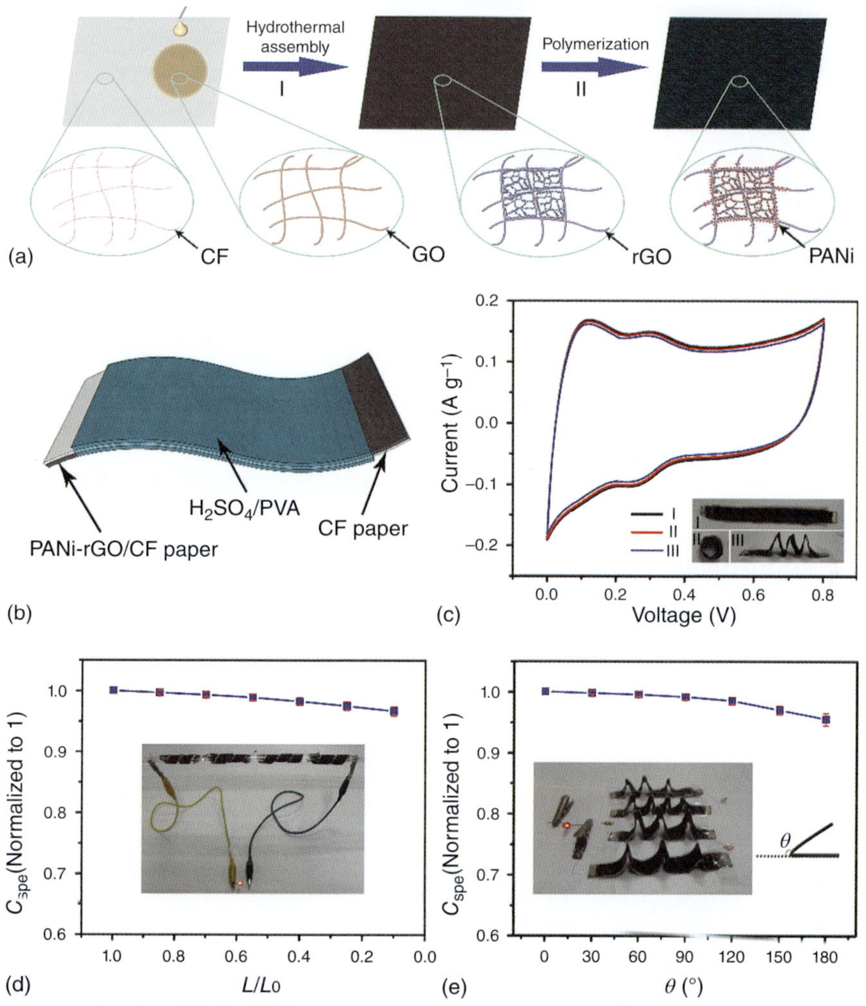

Figure 1.12 (a) Schematic diagram of the preparation of the PANi-rGO/cellulose fiber composite paper. (b) Schematic diagram of an all-solid-state integrated supercapacitor device. (c) CV curves of the all-solid-state supercapacitor under different deformation conditions (scan rate: 2 mV s^{-1}). Normalized specific capacitance of the all-solid-state supercapacitor under (d) bending and (e) folding states. *Source:* Liu et al. 2014 [104]. Copyright 2014. Reproduced with permission from John Wiley & Sons.

The textile device showed a stable electrochemical performance under bending and stretching deformations. Highly conductive carbon cloths were also employed as substrates as an alternative for flexible textile supercapacitor fabrication. For instance, Yuan et al. [106] fabricated a solid-state supercapacitor using a carbon cloth as a current collector and mechanical support for the carbon nanoparticles/MnO_2 nanorods hybrid electrode and PVA/H_3PO_4 electrolyte (Figure 1.13a). The device showed an excellent electrochemical performance with an energy density of 4.8 Wh kg^{-1} at a power density of 14 kW kg^{-1}. The device could be bended without scarifying its performance (Figure 1.13b).

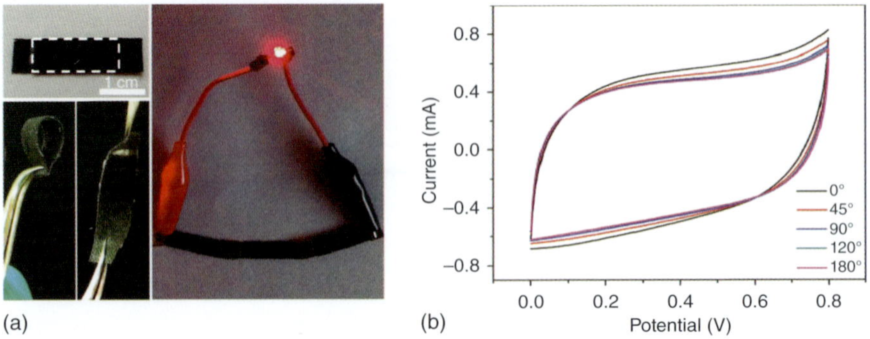

Figure 1.13 (a) Photographs of the solid-state supercapacitor made of carbon nanoparticles/MnO_2 nanorods hybrid electrode and a carbon cloth at normal, bent, and twisted state. The image on the right shows a red LED turned on by three supercapacitors connected in series. (b) CV curves for the solid-state supercapacitor at different bending angles. *Source:* Yuan et al. 2012 [106]. Copyright 2012. Reproduced with permission from American Chemical Society.

1.4 Flexible Micro-Supercapacitor Devices with Interdigitated Architecture

The fabrication of flexible microsupercapacitors with interdigitated design involves a micro-patterning step before or after the flexible film electrode formation. Recent developments in micro-manufacturing technology have opened pathways for the patterning step of the microsupercapacitor fabrication that suit the type of active materials and the required performance (either high energy density or high power density). The fabrication methods can be mainly classified into three categories: in situ synthesis of active materials on a pre-patterned surface, directly printing of existing active materials, and patterning the well-developed film electrode.

1.4.1 In situ Synthesis of Active Materials on Pre-Patterned Surfaces

Active materials can be synthesized on the pre-patterned surface via various synthesis techniques (such as electrochemical polymerization [107, 108], electrolytic deposition [109, 110], electrophoretic deposition [88], chemical deposition [111], etc.). Here, a pre-patterned surface is firstly prepared by coating and patterning a thin layer of photoresist (PR) or removable inks on a flexible substrate. In the next step, the current collectors are coated on the flexible substrate using the patterned PR or ink as the mask followed by the deposition of the active material. The masks are then removed chemically either before or after the active materials deposition, whenever they are no longer needed.

Sung et al. [107] reported a PR patterning using the conventional photolithography technique which was the earliest and the most common method of mask preparation for the current collector coating. In this approach, a patterned interdigitated gold microelectrode was first prepared on a silicon wafer by photolithography. PPy was then electrochemically polymerized on the pre-patterned gold layer. In the next step, a layer of solidified PVA-H_3PO_4 gel electrolyte was firmly attached to the PPy microelectrode arrays. Upon detaching both PPy and the electrolyte layer from the silicon substrate, a small, lightweight, freestanding, and flexible microsupercapacitor (fMSC) device was achieved. A relatively poor electrochemical performance (with areal capacitance of the device <0.5 mF cm^{-2}) was observed for this fMSC device emanating from the primitive device assembly technique such as the lack of current collectors and inappropriate coating of the electrolyte (simply attaching the dried gel electrolyte film to the electrode materials). Nevertheless, this study demonstrated the feasibility of fabricating flexible all-solid-state microsupercapacitor devices.

Using similar techniques (photolithography and in situ chemical polymerization), Wang et al. [111] further modified and improved the fabrication steps of fMSCs (Figure 1.14). They directly prepared PANi-based microelectrodes with Au/Cr as the current collector on a flexible PET film as the substrate compared to the silicon wafer. Moreover, instead of attaching a layer of the solidified electrolyte on active materials, the PVA-H_2SO_4 gel electrolyte was drop-cast on the surface of the microelectrodes and then solidified. Taking the advantage of the advanced nanostructure of PANi electrodes, existence of current collectors, and better penetration of the electrolyte, the as-prepared fMSCs exhibited an excellent device capacitance of 1.17 mF cm^{-2} and 25.4 F cm^{-3} (23.52 mF cm^{-2} and 588 F cm^{-3} for electrode materials), a high scan rate capability (>10 V s^{-1}), and a low leakage current. In addition, the feasibility of increasing the operating voltage and/or current by simply making series and/or parallel connecting circuits of MSC units on one PET film was firstly demonstrated. This is important in industrial applications where the required voltage or current is typically higher than what one single fMSC can supply. Following this approach, not only conducting polymers but carbon materials such as rGO [88], CNT [82], and metal oxides [112] were also fabricated. Despite the high resolution of the prepared interdigitated microelectrodes (minimum width of array ~50 μm) and the excellent fMSCs device performance, the high cost and the sophisticated process of photolithography limited its further development toward commercialization.

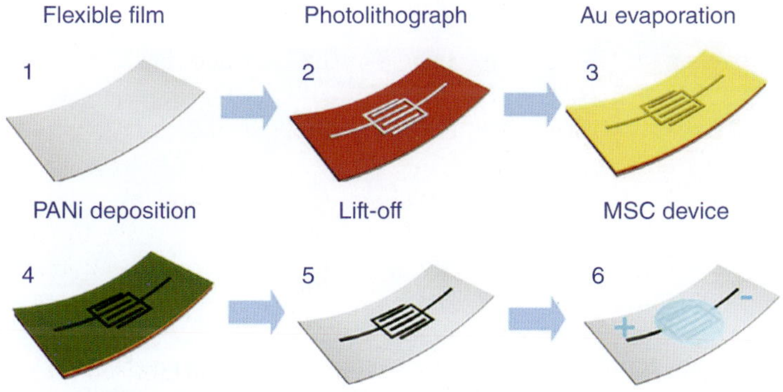

Figure 1.14 A schematic of fabrication process for interdigitated microelectrodes of PANi nanowire arrays on a flexible film. *Source:* Wang et al. 2011 [111]. Copyright 2011. Reproduced with permission from John Wiley & Sons.

Commercial and printable inks can also be used as the mask material instead of PR. Using a common office laser printer, Ye's group printed a circuit template as a mask on a PET film [108, 110]. They deposited Au/PANi and Au/MnO_2/Au (AMA) stacked microelectrodes, respectively, on the mask by beam evaporation of Au current collectors and the electrodeposition of the active materials (Figure 1.15a–h). The fMSC devices were achieved by lifting off the printed circuit template using THF. As the circuit pattern can be designed on a personal computer using software, the dimension parameters and the number of patterns could be easily adjusted. Hundreds of patterns can be printed on a flexible PET film (Figure 1.15j,k) indicating the great potential of this approach for scalable fMSCs fabrication. Jiang et al. used a Marker Pen Lithography technique, in which writing sacrificial ink patterns were written using commercial marker pens on the substrate and were used as the mask for the deposition of current collectors and active materials (PEDOT and PANi) [113]. Despite the relatively low resolution (minimum printed width >250 μm for laser printer, ~500 μm for

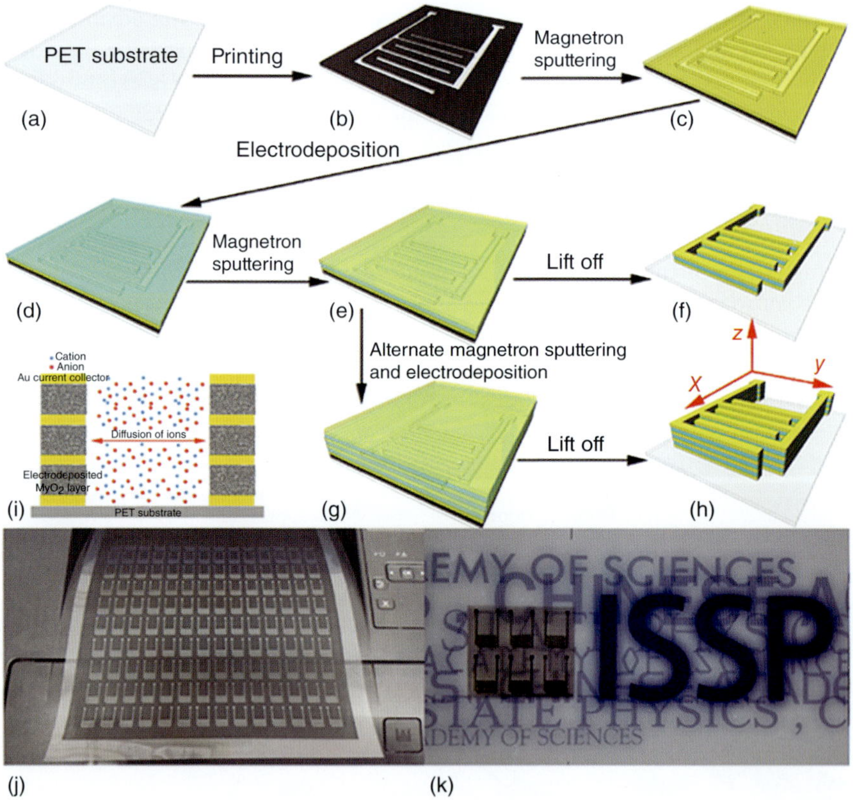

Figure 1.15 (a–h) Schematic illustrations of the fabrication procedure for the on-chip fMSCs with AMA stacked hybrid electrodes and single AMA sandwiched hybrid electrode units. (i) Schematic diagram of the ion diffusion pathway in the AMA stacked hybrid electrodes. (j) A photograph of a large area of printed interdigitated patterns on a flexible PET film. (k) A photograph of the as-prepared on-chip fMSCs with AMA stacked hybrid electrodes. *Source:* Hu et al. 2016 [110]. Copyright 2016. Reproduced with permission from John Wiley & Sons.

maker pen) that needs to be improved using a more precise printer to achieve higher device performance, the simplicity of these innovative strategies opened up new avenues for easy and scalable fabrication of fMSCs.

The fMSCs prepared from pseudocapacitive electrode materials showed superior areal and volumetric capacitances (9–41 $mF\,cm^{-2}$ and 50–147 $F\,cm^{-3}$ respectively) compared to the relatively low capacitance of the carbon-based fMSCs (<1 $mF\,cm^{-2}$ and <20 $F\,cm^{-3}$ respectively) normalized by the area and volume of the whole device. The poor performance of carbon-based devices can be explained by the ultrathin thickness (~25 nm to ~1 μm) of the in situ synthesized active materials, as carbon-based materials have poorer adhesion to the current collectors and substrates than pseudocapacitive materials. Meanwhile, the relatively low intrinsic volumetric capacitance exacerbates the poor areal capacitance of the ultrathin carbon-based device [87]. Also, some of the techniques used for the in situ deposition of carbon materials (such as electrostatic spray deposition of GO/CNT [114], CVD of graphene [115] and CNT [116], etc.) require stringent conditions, e.g. relatively high temperature, that may not be compatible with the commonly used low-cost PET and paper substrates used in fMSCs fabrication. In addition, the post-processing of carefully and fully removing PR residual or ink masks is labor-intensive and may result in losing some of the active materials. These limitations motivated the researchers to seek alternative ways for the preparation of carbon-based fMSCs.

1.4.2 Direct Printing of Active Materials

Direct printing of inks containing active materials onto a flexible substrate is another approach by which microelectrodes can be fabricated for use in fMSCs. Starting from the carbon materials powders, a stable suspension or a viscous slurry is obtained first. A binder, surfactant, and/or conductive additives may also be added to the suspension to improve the printability and/or conductivity of the active materials. Compared to the sophisticated steps required in the preparation of microelectrodes using the pre-patterning approach, the direct printing of the active materials can realize the deposition and patterning simultaneously in one step, thus reducing materials usage and process complexities. Different printing techniques (such as inkjet printing [15–17, 117–119], extrusion printing [18, 120], screen printing [13, 20, 121, 122], roll-to-roll printing [21, 22, 123, 124], etc.) have been exploited to prepare patterned microelectrodes from carbonaceous materials to make fMSCs. Highly dispersed and stable active materials with suitable particle sizes and appropriate rheological properties are required for successful patterning using printing techniques.

In screen printing, the ink is forced through a screen with pre-patterned open meshes and then contact with the substrate (Figure 1.16a). Screen printing can be carried out in a large scale with uniform thickness and within a short period. It also allows for patterning on a variety of substrates, not only on a smooth plastic film, but also on porous cloth and papers. Screen printing has enabled the integration of various composite active materials with superior nanostructures into fMSCs, which were incompatible with other fabrication strategies. For instance, using screen printing techniques, Wang et al.

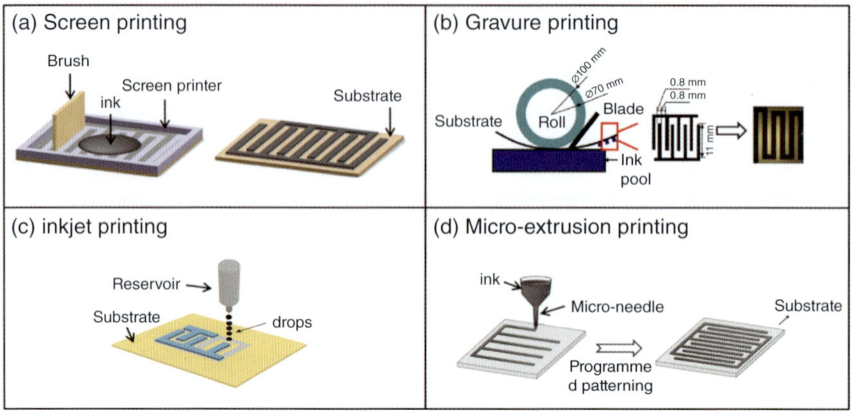

Figure 1.16 Schematic diagrams of (a) screen printing, (b) gravure printing, (c) inkjet printing, and (d) micro-extrusion printing. *Source:* Reproduced from References [87, 120, 123, 125] with permission from IOP Publishing Ltd, AIP Publishing LLC, Nature Publishing Group, and Elsevier.

fabricated MnO_2/onion-like carbon (MnO_2/OLC)-based fMSCs, in which the MnO_2/OLC nanocomposites have unique nanourchin shapes with the diameter of 200–500 nm [125]. The optimized device exhibited high areal and volumetric capacitances of ~7.04 mF cm^{-2} and ~7.04 F cm^{-3}. Similarly, Zhu et al. screen printed interdigitated electrodes using the ink of novel 3D nanocomposites consisting cobalt oxide (CoO) nanoflowers woven with CNTs networks [121]. High volumetric and areal capacitances of ~17.4 F cm^{-3} and ~17.4 mF cm^{-2}, respectively, and a high energy density of ~3.48 mWh cm^{-3} were achieved for the fMSC device. In another study, nitrogen-doped graphene was screen printed at a relatively high electrode thickness of ~10 μm achieving a high areal capacitance of ~3.4 mF cm^{-2}, which is much higher than the areal capacitance of carbon-based fMSCs prepared by the pre-patterning approach.

Gravure printing is another fast printing technique for fMSC patterning, which is used traditionally for the roll-to-roll printing of newspapers, magazines, and packages (Figure 1.16b). In this approach, an image or a pattern is carved into the surface of a cylinder, and then the inks are transferred from the carved cells onto the substrate as a result of the high printing pressure and adhesive forces between the ink and substrate. Carbon materials (e.g., crumpled graphene) and nanostructured composite (e.g., hybrid MoS_2@S-rGO consisting of sulfonated rGO and MoS_2 nanoflowers) were gravure printed into microelectrodes with the thickness of ~10 μm [123, 124]. High areal capacitances of 6.65 and 6.56 mF cm^{-2} have been achieved by these graphene and MoS_2@S-rGO-based fMSCs, respectively. However, both screen printing and gravure printing use pre-patterned screens or cylinders making the adjustment of the patterning parameters difficult. Moreover, the resolution of ~300 μm for screen printing and ~700 μm for gravure printing are relatively poor, limiting the size and the performance of the final fMSC device. To circumvent these problems, digital printing technology, such as inkjet printing and extrusion printing that have high pattern precision were used.

Inkjet printing is a digital, noncontact printing technology that provides precise control over patterns position through a moving *XY* printing head and over the thickness of the printing (Figure 1.16c). Inkjet printing allows printing on a variety of substrates using simple drawn patterns on a computer using software. Using a GO ink and inkjet Lee and coworkers printed a pair of 4 interdigitated patterns with dimensions of 1.75 mm × 1.890 mm with 80 μm spacings between the adjacent interdigitated fingers on a flexible PET film [118]. After the in-depth reduction of GO to graphene using a photo-thermal technique with a xenon flash camera, a porous graphene-based fMSC with a maximum volumetric capacitance of 0.82 $F\,cm^{-3}$ was obtained. Hersam and coworkers also used an inkjet printer for the fabrication of graphene-based fMSCs from printable and highly conductive graphene inks and achieved an improved volumetric capacitance (~9.3 $F\,cm^{-3}$) in the PVA/H_3PO_4 solid-electrolyte [119]. They demonstrated the scalable printing of well-defined patterns on a flexible polyimide film. Asymmetric patterns can also be achieved by inkjet printing different metal-based and carbon-materials-based inks and can be used for fabricating asymmetric fMSC devices. Pang et al. made an asymmetric fMSC device for the first time by inkjet printing lamellar potassium cobalt phosphate hydrate [$K_2Co_3(P_2O_7)_2{\cdot}2H_2O$)] nanocrystal whiskers dispersion as the ink for cathode and graphene as the ink for anode [117]. The device with high pattern resolution (70 ~ 80 μm) delivered a volumetric capacitance of ~6.0 $F\,cm^{-3}$ with a maximum volumetric energy density of ~0.96 $mWh\,cm^{-3}$ in PVA/KOH electrolyte. They also printed silver inks as current collectors prior to the deposition of active materials and were able to improve the rate capability of fMSCs. In another study, an all-inkjet-printed solid-state fMSC was fabricated using printable CNT active materials and UV curable gel electrolytes [126]. This work opened up a new way for the fabrication of fMSCs and their application in next-generation fully printed wearable electronics. The main challenge in inkjet printing is that only limited materials have been inkjet printed since it is difficult to prepare highly dispersed materials with small particle sizes that are able to produce fluid jets. In addition, only ultrathin films (normally less than 1 μm, maximum of ~10 μm) have been printed, which restricts the areal capacitance of fMSCs. Micro-extrusion printing is promising in overcoming this problem.

Micro-extrusion printing is considered to be a more scalable patterning technique than the inkjet printing to obtain both 2D and 3D architectures (Figure 1.16d) [120, 127]. This technique involves extruding a viscous ink through a nozzle mounted on an *XYZ* motor with controlled speed and programmable trajectory. Sun et al. fabricated graphene-based interdigitated microelectrodes by extrusion printing of a concentrated GO aqueous solution (20 $mg\,ml^{-1}$) [120]. By tuning the needle diameter, extrusion speed, and scan rate, they were able to print identical microelectrodes with a width of ~600 μm and an inter-spacing of ~500 μm on a flexible PET substrate. After chemically reducing GO into rGO and coating a layer of PVA-H_2SO_4, an all-solid-state fMSC with a high areal capacitance of ~19.8 $mF\,cm^{-2}$ and a volumetric capacitance of ~41.8 $F\,cm^{-3}$ was achieved. Despite limited use of extrusion printing in the fabrication of fMSCs, it has been widely applied in various other fields involving microbatteries, biosensors, and solar cells. For example, 3D interdigitated microbattery (3D-IMA) architectures have been fabricated by printing concentrated lithium-oxide-based inks [127]. The 16-layer 3D-IMA possesses electrode width of ~60 μm and spacing of ~50 μm, indicating the great potential of the extrusion printing technique in the fabrication of fMSCs with tailored 3D architectures.

In general, printing techniques require carefully synthesized and optimized ink formulation of active materials. Parameters such as ink viscosity, printing speed, and surface treatment of the substrates are important factors that affect the efficacy of printing techniques. Nevertheless, the high compatibility of active materials, facile control of resolution, and thickness of the patterns, operating at normal atmospheric temperature, low cost, and the availability of a wide range of printing techniques that can cater to different inks, make the printing techniques promising in the development of fMSCs.

1.4.3 Patterning of Well-Developed Film Electrodes

Patterning the pre-prepared film electrodes using techniques such as selectively etching or converting is another approach to developing fSC devices. Active materials used in this category can be carbon materials, conducting polymers, and metal oxides, or a combination of these. Methods for patterning of film electrodes are classified into mask/template patterning and mask-free patterning processes.

Plasma etching is a conventionally used dry-etching method for mask patterning. Electrode films are first prepared directly or transferred to a desired flexible substrate via one of the film formation methods (such as spin coating and vacuum filtration). A thin layer of a current collector (e.g. Au) is then coated on the film with a customized mask. Then the current collector acts as a protection mask against oxygen plasma etching and leads to the designed patterned microelectrodes. This method has been used to develop high-performance electrode films (e.g. graphene [128], graphene/polyaniline [129], graphene/MnO_2/Ag nanowires [130], and $NiFe_2O_4$ nanofibers [131]) for fMSCs. In another study, an alternating stacked graphene/PANi films were prepared by layer-by-layer vacuum filtration [129]. As shown in Figure 1.17, the 2D pseudocapacitive graphene-conducting polymer nanosheets (PANi-G) and electrochemically exfoliated graphene were alternately and densely stacked layer by layer, in which the mesoporous PANi-G with easily accessible ion-transporting channels acts as a pseudocapacitance and highly conductive EG provides an electron-conducting network for rapid electron transport. Coupled with an interdigitated architecture, the fabricated fMSC exhibited an extremely high device areal capacitance of ~52.5 mF cm^{-2} and volumetric capacitance of ~109 F cm^{-3}.

Xue et al. used a microfluidic etching method, in which a patterned PDMS elastomer stamp acted as the template for etching [132]. They pressured the PDMS stamp on an electrospun MnO_2 film, and then different aqueous etching solutions were passed from the microfluidic channels to etch the MnO_2 resulting in interdigitated microelectrodes. The PDMS stamp is cheap and stable against most etchants that react with a variety of active materials, making the microfluidic etching a versatile and low-cost approach for electrode patterning. Both of the above etching methods rely on masks or templates for the development of the desired patterns and are labor-intensive and are not very congenial to adjust

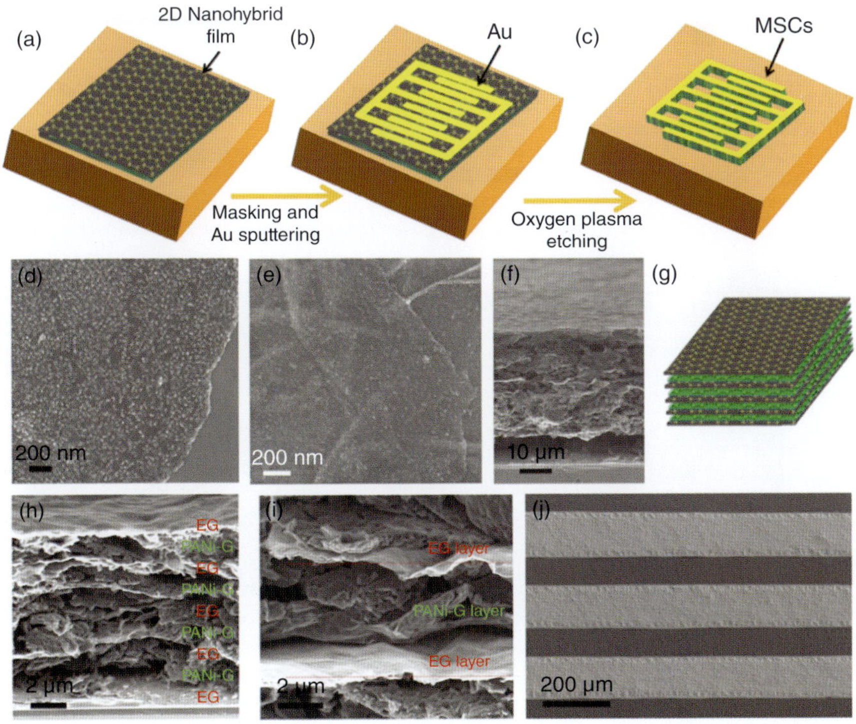

Figure 1.17 (a–c) Illustration of the fabrication procedure for an in-plane fMSC with interdigital fingers. SEM images of (d) a 46 nm thick PANi-G nanosheet and (e) PPy-G nanosheet. (f) SEM image and (g) schematic illustration of micrometer-thick 2D nanohybrid film, showing a layer-stacked structure. (h,i) Cross-sectional SEM images of a 2D nanohybrid film. (j) SEM image of the 2D nanohybrid film with a tilt angle of 30°. *Source:* Reproduced from Reference [129] with permission. Copyright 2014. Wiley-VCH Verlag GmbH & Co. KGaA, Weinheim.

the patterning parameters for commercial applications. To solve these problems, mask-free patterning methods were gradually developed.

The application of laser for directly writing of graphene-based fMSCs was reported by several groups and has shown great potential for commercialization [80, 133, 134]. In this method, thin films of graphene precursors (graphene oxide and commercialized polyimide sheets) are selectively converted and patterned to graphene interdigitated microelectrodes using a laser system [80, 133, 134]. For example, El-Kady and Kaner described a scalable fabrication of graphene fMSCs by direct laser writing on GO films using a standard LightScribe DVD burner (Figure 1.18) [80]. By this means, more than 100 MSCs could be readily generated on a single disc in ~30 min. Dimensions of the interdigitated were adjusted to control the energy and power handling. Considering lasers' wide application in industry for metal cutting and welding, and the wide range of wavelengths and powers that are available, laser patterning is a promising tool to fabricate fMSCs for real applications. However, the areal capacitances of those laser-patterned graphene fMSCs were still relatively low, only 1–16 mF cm^{-2}. The reason for the low capacitance of the fMSCs may be because the conversion of graphene

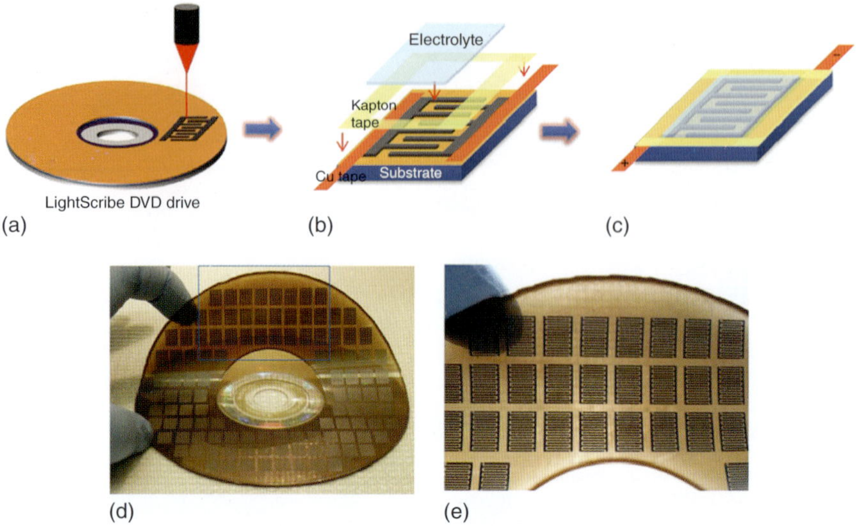

Figure 1.18 Fabrication of graphene based MSC (a–c) Schematic diagrams showing the fabrication process for an LSG microsupercapacitor. (d,e) Photographs of more than 100 micro-devices produced on a single run on a flexible substrate. *Source:* El-Kady and Kaner 2013 [80]. Copyright 2013. Reproduced with permission from Nature Publishing Group.

can only be performed on the surface of graphene precursor thin-films, limiting its applicability for thick films. Also, the residual precursor of graphene left between the fingers would not only reduce the ion diffusion rate but also affect the penetration of electrolyte ions from the lateral side of the microelectrodes upon increasing the film thickness. In addition, this technique is limited for graphene-based fMSCs and may not be applicable to other active materials.

Recently, the laser lithography technology (known as laser-cutting) has been used as an etching tool to create patterned channels on various kinds of film electrodes such as MoS_2 [135], activated mesophase pitch (aMP) [136], mushroom-derived carbon (MDC) [137], rGO-PEDOT/PSS [138], and MXene [139]. This laser-cutting approach "etches" unwanted material away from the electrode as opposed to "writing" the electrode to generate patterned electrode design on the substrate (Figure 1.19a). Yadav et al. applied laser-etching on an MDC electrode film (~7 μm thickness) to generate interdigitated patterns (electrode width of ~300 μm and interspace between electrodes ~100 μm) [137]. The fabricated fMSCs exhibited an areal capacitance of ~9.05 mF cm^{-2} and had a volumetric capacitance of ~12.92 mF cm^{-3} based on the area and volume of the whole device. Liu et al. used laser lithography and fabricated an rGO-PEDOT/PSS fMSC with a higher film thickness of ~57.9 μm and achieved a higher areal capacitance of ~84.7 mF cm^{-2} and volumetric capacitance of ~14.5 F cm^{-3} [138]. They demonstrated that 10 fMSCs with the desired microelectrode widths can be readily prepared in just 90 s (Figure 1.19b). A minimum distance between adjacent fingers of ~100 μm was achieved (Figure 1.19c). Series and/or parallel interconnection could also be easily established prior to the device fabrication, exemplifying the simplicity and convenience of the fabrication

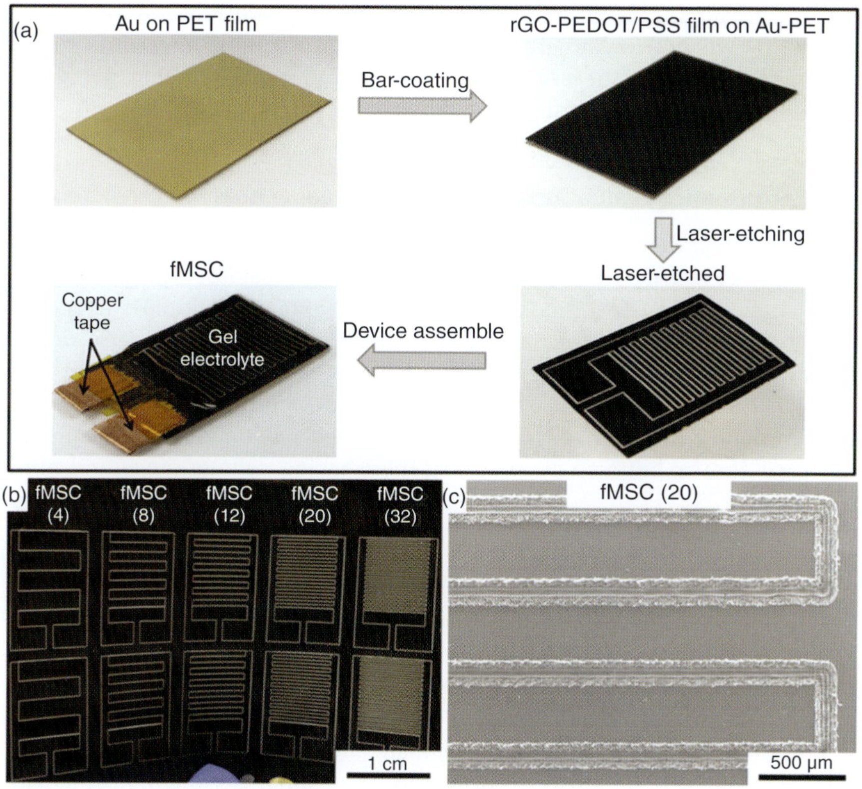

Figure 1.19 (a) Schematic diagram of the fabrication of all-solid-state rGO-PEDOT/PSS based planar fMSCs on a flexible PET film using laser-cutting technique. (b) Photos of laser-etched fMSCs with different configurations on one rGO-PEDOT/PSS film. (c) SEM image of an fMSC (20) pattern (top view). *Source:* Liu et al. 2016 [138]. Copyright 2016. Reproduced with permission from John Wiley & Sons.

process. It is believed that this technology can be applied to a wide range of electrode materials and film thicknesses provided that residual precursors (or any unwanted materials) can be completely removed to create well-separated microelectrodes. The simplicity of this technology (i.e. computer-generated intricate patterns that can be "written" on any material such as polymers, nanomaterials, and many others on the desired substrates with complex shapes and structures) can pave the way for rational engineering of novel materials suitable for a wide range of wearable electronic devices not limited to energy storage applications.

1.5 Performance Evaluation and Potential Application of Flexible Supercapacitors

Supercapacitors act as a bridge between the energy source and energy outlet and are often combined with other energy units for practical applications. However, the lack of a standardized criterion to evaluate the performance of an SC device makes it difficult for developers and manufactures to design and choose the applicable fSC device for integrating into a specific wearable electronic circuit. Hence, we will discuss the performance evaluation of fSCs before going into more detail on their integration in real application.

1.5.1 Performance Evaluation of Flexible Supercapacitors

Energy and power densities are two key parameters to evaluate the performance of an SC device. The maximum energy E (J) stored in a capacitor is proportional to the capacitance C (F) and the maximum operation voltage window V (V) according to Equation (1.3) [140]. The maximum power (P_{max}) supplied by the SC device, i.e. the rate of energy delivery per unit time, can be calculated by Equation (1.4), where R_S is the equivalent series resistance of the device [140]. Hence, the excellent performance of an SC device can be achieved not only by increasing the device capacitance, but also by widening the working voltage range and minimizing the R_S [78].

$$E = \frac{1}{2}CV^2 \tag{1.3}$$

$$P_{max} = \frac{V^2}{4R_S} \tag{1.4}$$

Many all-solid-state flexible supercapacitors with excellent energy and/or power properties have been reported by researchers. However, there is still a lack of standardized criteria to accurately evaluate and compare the performance of the SC devices. It is gradually realized that the traditional method of normalizing the device properties (e.g. capacitance, energy, power) by the mass of active materials is unreliable. This is because the mass of active materials is negligible compared with that of the device. In the past 5 years, an increasing number of reports evaluated the device performance in terms of the volume of the whole device. This approach can better reflect the realistic performance of the SC device in practical applications, especially when the miniaturized wearable electronic systems allow limited space for the energy storage unit. However, some of these devices are based on ultrathin electrodes with densely packed structures, which cannot be scaled up when electrode thickness increases due to the consequently increased electrode resistance and affected ion accessible surface area [88, 138, 141, 142]. The high volumetric performance of these electrodes cannot be translated to the high device performance when certain thicknesses of the electrode are required. Hence, to have meaningful evaluation of the device performance, the properties of the fSC devices should be normalized to both the foot-print area and the volume of the entire device that includes the electrodes, solid-state electrolyte, the separator, and current collectors. Furthermore, the thickness of the device and each component should also be provided for comparison.

The other key parameters to evaluate the performance of the fSC device are the flexibility and the mechanical properties of the device, which are necessary for the wearable energy storage devices to perform under various types of deformations such as stretching and bending. There is also a lack of standardized methods to evaluate the mechanical properties of the electrodes, electrolyte, or even the entire device. However, researchers have tried various ways to evaluate the mechanical properties of the fSC device and its components. For instance, the mechanical performance of the electrode film has been evaluated by repeatedly kinking, bending, and winding it, by crumpling it into a ball and by folding it into a paper crane and it was found that the electrode film did not fail under these destructive attempts [143, 144]. The electrode film has also been agitated, ultrasonicated, and bent in water and ethanol and was observed to maintain its integrity indicating the robustness of the electrode film [143, 144]. Tensile testing is also an effective approach to evaluate the mechanical properties of the fSC electrode or device under stretching conditions [143, 144]. The investigation of the integrity and the capacitance performance of the fSC device under bending or twisting conditions has become a common method to evaluate the mechanical properties of the device. No changes in the electrochemical performance of the fSC device between flat, different bending angles, and various twisting conditions are indicative of a functional flexible device [142]. In addition, fSCs will need to be robust enough to provide reliable operation under mechanical impact. Li et al. also evaluated the electrochemical performance of the fSC device when the device was subject to repeated impact force (i.e. applying a swinging hammer with an energy of ~2 J) [119]. The slight changes in the electrochemical performance of the device under bending and impact tests indicate the suitability of the device for flexible and portable energy storage applications.

1.5.2 Integration of Flexible Supercapacitors

Initially, most of the high-performance fSCs were used to directly power simple electronic components like LEDs or a timer [138, 142, 145]. To realize the current or voltage requirement of electronics, the fSCs were often connected in series and/or parallel combinations. Researchers continued exploring the applications of fSCs in functional electronics for real world applications [122, 125, 131, 146–150]. Lu et al. used three asymmetric fSCs made from MnO_2/Fe_2O_3 electrodes and connected them in series to charge a mobile phone (HTC, A320e) [146]. Yun et al. integrated a series of fMSC arrays (based on PANi-wrapped MWCNT electrode material) to a patterned-graphene NO_2 gas sensor [150]. The sensor could detect NO_2 gas for more than 50 min after fully charging the MSCs. Wang et al. demonstrated that a $GeSe_2$-based in-plane supercapacitor could power a photodetector based on CdSe nanowire film [148]. The SC-powered photodetector had a current response similar to the white light irradiation (100 mW cm^{-2}) achieved using an external power source.

Supercapacitors can also store energy from sustainable and renewable energy harvesting units (e.g. solar cells [151, 152], nanogenerators [153]). The charged SCs could power electronic devices continuously without using an external power supply or a battery, thus forming a self-powered system. Flexible all-solid-state supercapacitors have been integrated with piezo-electric nanogenerators, triboelectric generators, and various solar cells (e.g. dye-sensitized solar cells, perovskite hybrid solar cells) as energy source to power wearable LEDs, UV/NO_2 gas sensors, strain sensors, and pressure sensors [151, 153–155]. Jung et al. reported a wearable self-powered energy system sewn into a knitted shirt (Figure 1.20). In this system, CNT-RuO_2-based flexible SCs work as

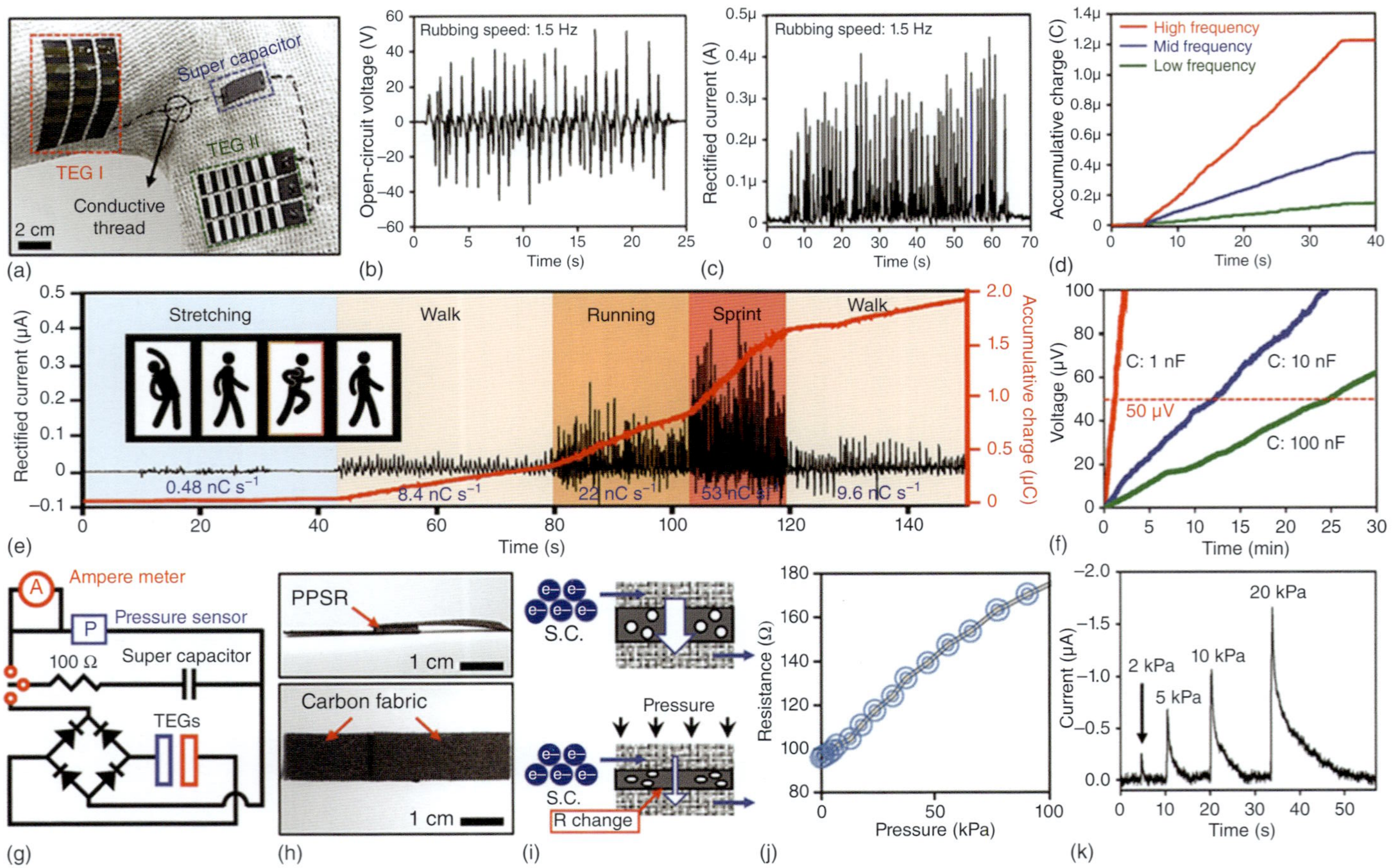

TEG I
Super capacitor
TEG II
Conductive thread
2 cm
(a)
Rubbing speed: 1.5 Hz
Open-circuit voltage (V)
Time (s)
(b)
Rubbing speed: 1.5 Hz
Rectified current (A)
Time (s)
(c)
High frequency
Mid frequency
Low frequency
Accumulative charge (C)
Time (s)
(d)
Stretching
Walk
Running
Sprint
Walk
0.48 nC s^{-1}
8.4 nC s^{-1}
22 nC s^{-1}
53 nC s^{-1}
9.6 nC s^{-1}
Rectified current (μA)
Accumulative charge (μC)
Time (s)
(e)
C: 1 nF
C: 10 nF
C: 100 nF
50 μV
Voltage (μV)
Time (min)
(f)
Ampere meter
Pressure sensor
100 Ω
Super capacitor
TEGs
(g)
PPSR
1 cm
Carbon fabric
1 cm
(h)
S.C.
Pressure
S.C.
R change
(i)
Resistance (Ω)
Pressure (kPa)
(j)
20 kPa
10 kPa
5 kPa
2 kPa
Current (μA)
Time (s)
(k)

Figure 1.20 Demonstration and measurement of electrical signals of the energy supply devices. (a) Image of the devices applied to a knit shirt and connected by conductive threads (TEGs size: 1.5 cm × 6 cm). (b) Open-circuit voltage, and (c) rectified current generated by the TEG from arm swings. (d) The generated charge accumulation from TEG at different frequencies. (e) Demonstration of the human activity sensor. The generated electricity was recorded during the illustrated jogging activity. (f) Charging of various capacitors with TEG. (g) Electric circuit diagram for the pressure sensor. (h) Digital photo of the pressure sensor. Side view (top), top view (bottom). (i) Schematic diagram of the pressure sensing mechanism. Original state (top), under pressure (bottom). (j) Resistance change of the pressure sensor. (k) Change of current in the SC as a function of the pressure applied to the pressure sensor. *Source:* Jung et al. 2014 [154]. Copyright 2014. Reproduced with permission from John Wiley & Sons.

an energy storage unit to store the charges generated from a triboelectric generator (TEG) through human activities like running or walking and provide power for a pressure sensor [154]. By monitoring the slope of the charge accumulation generated from the subject performing normal jogging, the activity of the subject can be tracked. These units were based on carbon fabrics, allowing them to be woven onto the designated locations in conventional clothing and interconnected by conductive threads to form a truly wearable self-powered system. Besides wearable electronics, fSCs were also integrated into a body-attachable multisensor system by Kim et al. (Figure 1.21) [155]. In this work, individual sensors (a fragmentized graphene foam (FGF) strain sensor and an MWCNT/SnO_2 based NO_2/UV sensor), an array of nine CNT-based fMSCs, and a wireless radio frequency (RF) power receiver were transferred onto a flexible and biocompatible silicon Ecoflex substrate. Interconnections were achieved by the embedded liquid metal lines. The fMSCs charged by a wireless RF power receiver enable the operation of multisensors. The fabricated sensor system could be attached to the skin and used to successfully detect biosignals, such as a neck pulse, saliva swallowing, voice, and body movements and could provide a stable sensing signal upon exposure to NO_2 gas and UV light. Although their fMSC array (power supply <10 mW at 1.5 V) is not suitable for the commercially available RF transceiver chips (require power of ~100 mW at 3 V), this study shows that fMSCs have great potential for applications in the next-generation body-attachable healthcare and environmental sensor systems.

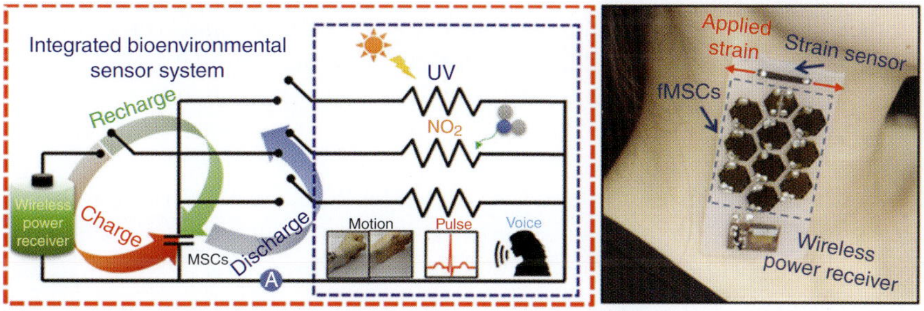

(a) (b)

Figure 1.21 (a) Circuit diagram of an integrated system. (b) Photograph of a fragmentized graphene foam (FGF) sensor attached on the skin of a neck. *Source:* Kim et al. 2016 [155]. Copyright 2016. Reproduced with permission from John Wiley & Sons.

1.6 Conclusions and Perspectives

Although great progress has been achieved in the fabrication of all-solid-state flexible supercapacitors for both sandwiched and interdigitated architecture, much work remains to be done to realize its full commercialization. Further work is needed for developing scalable fMSCs in electrodes, fabrication techniques, and device integration aspect.

Firstly, efforts are needed to develop flexible electrode materials that have the possibility to scale up their performance with increased electrode thickness. Hence, exploring flexible electrode materials that combine the merits of high electrical conductivity, large ion-accessible surface area, densely packed structure, and excellent electrochemical stability, is still a main direction.

Secondly, laser-etching and 3D micro-extrusion printing are the two scalable techniques for fMSCs patterning. However, laser-etching is limited to only symmetric fMSCs and electrode materials with simple architecture. Fully 3D printing of flexible energy-storage devices (including electrode, electrolyte, and package materials) with layer-by-layered 3D electrode architecture, either symmetric or asymmetric design, will be a promising field in the near future.

Thirdly, the combination of fSCs and other electronics (such as solar cells, sensors, and photodetectors) in smart flexible devices is also an important development direction. Besides simplifying connecting wires or circuit lines, developing highly integrated multifunctional wearable devices (like supercapacitor-sensor, photo-supercapacitor, and thermal supercapacitors) is worth exploring.

References

1 Simon, P. and Gogotsi, Y. (2008). *Nat. Mater.* 7: 845.
2 Yan, J., Wang, Q., Wei, T., and Fan, Z. (2014). *Adv. Energy Mater.* 4: 1300816.
3 Kötz, R. and Carlen, M. (2000). *Electrochim. Acta* 45: 2483.
4 Zhang, J. and Zhao, X.S. (2012). *ChemSusChem* 5: 818.
5 Zhang, L.L. and Zhao, X.S. (2009). *Chem. Soc. Rev.* 38: 2520.
6 Qu, D. and Shi, H. (1998). *J. Power Sources* 74: 99.
7 Endo, M., Maeda, T., Takeda, T. et al. (2001). *J. Electrochem. Soc.* 148: A910.
8 Huang, J., Sumpter, B.G., and Meunier, V. (2008). *Chem. Eur. J.* 14: 6614.
9 Wang, G., Zhang, L., and Zhang, J. (2012). *Chem. Soc. Rev.* 41: 797.
10 Snook, G.A., Kao, P., and Best, A.S. (2011). *J. Power Sources* 196: 1.
11 Lokhande, C.D., Dubal, D.P., and Joo, O.-S. (2011). *Curr. Appl. Phys.* 11: 255.
12 Wang, C. and Wallace, G.G. (2015). *Electrochim. Acta* 175: 87.
13 Martins-Júnior, P.A., Alcântara, C.E., Resende, R.R., and Ferreira, A.J. (2013). *J. Dent. Res.* 92: 575.
14 Baughman, R.H., Zakhidov, A.A., and de Heer, W.A. (2002). *Science* 297: 787.
15 Wen, L., Li, F., and Cheng, H.-M. (2016). *Adv. Mater.* 28: 4306.
16 Niu, C., Sichel, E.K., Hoch, R. et al. (1997). *Appl. Phys. Lett.* 75: 1480.
17 An, K.H., Kim, W.S., Park, Y.S. et al. (2001). *Adv. Funt. Mater.* 11: 387.
18 Chen, Q.-L., Xue, K.-H., Shen, W. et al. (2004). *Electrochim. Acta* 49: 4157.

19 Yan, L., Zheng, Y.B., Zhao, F. et al. (2012). *Chem. Soc. Rev.* 41: 97.
20 Pumera, M. (2010). *Chem. Soc. Rev.* 39: 4146.
21 Pumera, M. (2011). *Energy Environ. Sci.* 4: 668.
22 Brownson, D.A.C., Kampouris, D.K., and Banks, C.E. (2011). *J. Power Sources* 196: 4873.
23 Sun, Y., Wu, Q., and Shi, G. (2011). *Energy Environ. Sci.* 4: 1113.
24 Stoller, M.D., Park, S., Zhu, Y. et al. (2008). *Nano Lett.* 8: 3498.
25 Dai, L., Chang, D.W., Baek, J.-B., and Lu, W. (2012). *Small* 8: 1130.
26 Zhu, Y., Murali, S., Stoller, M.D. et al. (2011). *Science* 332: 1537.
27 Shirakawa, H., Louis, E.J., MacDiarmid, A.G. et al. (1977). *J. Chem. Soc., Chem. Commun.* 578.
28 Friend, R.H., Gymer, R.W., Holmes, A.B. et al. (1999). *Nature* 397: 121.
29 Wan, P., Wen, X., Sun, C. et al. (2015). *Small* 11: 5409.
30 Spinks, G.M., Mottaghitalab, V., Bahrami-Samani, M. et al. (2006). *Adv. Mater.* 18: 637.
31 Mirmohseni, A. and Oladegaragoze, A. (2000). *Synth. Met.* 114: 105.
32 Peng, T., Sun, W., Huang, C. et al. (2014). *ACS Appl. Mater. Interfaces* 6: 14.
33 Arbizzani, C., Mastragostino, M., and Meneghello, L. (1996). *Electrochim. Acta* 41: 21.
34 Ghosh, S. and Inganas, O. (1999). *Adv. Mater.* 11: 1214.
35 Ryu, K.S., Kim, K.M., Park, N.-G. et al. (2002). *J. Power Sources* 103: 305.
36 Ramya, R., Sivasubramanian, R., and Sangaranarayanan, M.V. (2013). *Electrochim. Acta* 101: 109.
37 Muthulakshmi, B., Kalpana, D., Pitchumani, S., and Renganathan, N.G. (2006). *J. Power Sources* 158: 1533.
38 Laforgue, A., Simon, P., Sarrazin, C., and Fauvarque, J.-F. (1999). *J. Power Sources* 80: 142.
39 Naoi, K., Suematsu, S., and Manago, A. (2000). *J. Electrochem. Soc.* 147: 420.
40 Levi, M.D., Gofer, Y., Aurbach, D. et al. (2000). *J. Electrochem. Soc.* 147: 1096.
41 Skompska, M., Mieczkowski, J., Holze, R., and Heinze, J. (2005). *J. Electroanal. Chem.* 577: 9.
42 Fan, L.-Z. and Maier, J. (2006). *Electrochem. Commun.* 8: 937.
43 Kim, B.C., Ko, J.M., and Wallace, G.G. (2008). *J. Power Sources* 177: 665.
44 Dhawale, D.S., Vinu, A., and Lokhande, C.D. (2011). *Electrochim. Acta* 56: 9482.
45 Guan, H., Fan, L.Z., Zhang, H., and Qu, X. (2010). *Electrochim. Acta* 56: 964.
46 Liu, J., Zhou, M., Fan, L.Z. et al. (2010). *Electrochim. Acta* 55: 5819.
47 Wang, Y.G., Li, H.Q., and Xia, Y.Y. (2006). *Adv. Mater.* 18: 2619.
48 Ismail, Y.A., Chang, J., Shin, S.R. et al. (2009). *J. Electrochem. Soc.* 156: A313.
49 Kuila, B.K., Nandan, B., Böhme, M. et al. (2009). *Chem. Commun.* 5749.
50 Cheng, Q., Tang, J., Ma, J. et al. (2011). *J. Phys. Chem. C* 115: 23584.
51 Wang, K., Huang, J., and Wei, Z. (2010). *J. Phys. Chem. C* 114: 8062.
52 Zhou, H., Chen, H., Luo, S. et al. (2005). *J. Solid State Electrochem.* 9: 574.
53 Gupta, V. and Miura, N. (2005). *Electrochem. Solid-State Lett.* 8: A630.
54 Li, G.R., Feng, Z.P., Zhong, J.H. et al. (2010). *Macromolecules* 43: 2178.

55 Roberts, M.E., Wheeler, D.R., McKenzie, B.B., and Bunker, B.C. (2009). *J. Mater. Chem.* 19: 6977.
56 Huang, J., Wang, K., and Wei, Z. (2010). *J. Mater. Chem.* 20: 1117.
57 Li, X. and Zhitomirsky, I. (2013). *J. Power Sources* 221: 49.
58 Biswas, S. and Drzal, L.T. (2010). *Chem. Mater.* 22: 5667.
59 Xu, C., Sun, J., and Gao, L. (2011). *J. Mater. Chem.* 21: 11253.
60 Reddy, A.L.M., Gowda, S.R., Shaijumon, M.M., and Ajayan, P.M. (2012). *Adv. Mater.* 24: 5045.
61 Frackowiak, E., Khomenko, V., Jurewicz, K. et al. (2006). *J. Power Sources* 153: 413.
62 Zhao, C., Shu, K., Wang, C. et al. (2015). *Electrochim. Acta* 172: 12.
63 Lu, X., Dou, H., Yuan, C. et al. (2012). *J. Power Sources* 197: 319.
64 Li, S., Zhao, C., Shu, K. et al. (2014). *Carbon* 79: 554.
65 Wang, K., Zhao, P., Zhou, X. et al. (2011). *J. Mater. Chem.* 21: 16373.
66 Wu, Q., Xu, Y., Yao, Z. et al. (2010). *ACS Nano* 4: 1963.
67 Alvi, F., Ram, M.K., Basnayaka, P.A. et al. (2011). *Electrochim. Acta* 56: 9406.
68 He, Y., Chen, W., Li, X. et al. (2013). *ACS Nano* 7: 174.
69 Cheng, Y., Lu, S., Zhang, H. et al. (2012). *Nano Lett.* 12: 4206.
70 Chen, P., Chen, H., Qiu, J., and Zhou, C. (2010). *Nano Res.* 3: 594.
71 Boukhalfa, S., Evanoff, K., and Yushin, G. (2012). *Energy Environ. Sci.* 5: 6872.
72 Yuan, C., Yang, L., Hou, L. et al. (2012). *Adv. Funct. Mater.* 22: 2560.
73 Gao, H. and Lian, K. (2014). *RSC Adv.* 4: 33091.
74 Zhao, C., Wang, C., Gorkin, R. et al. (2014). *Electrochem. Commun.* 41: 20.
75 Zhao, C., Wang, C., Yue, Z. et al. (2013). *ACS Appl. Mater. Interfaces* 5: 9008.
76 Niu, Z., Dong, H., Zhu, B. et al. (2013). *Adv. Mater.* 25: 1058.
77 Meng, C., Liu, C., Chen, L. et al. (2010). *Nano Lett.* 10: 4025.
78 Lu, X., Yu, M., Wang, G. et al. (2014). *Energy Environ. Sci.* 7: 2160.
79 Macfarlane, D.R., Forsyth, M., Howlett, P.C. et al. (2008). *ChemInform* 40: 1165.
80 El-Kady, M.F. and Kaner, R.B. (2013). *Nat. Commun.* 4: 1475.
81 Hong, S.Y., Yoon, J., Jin, S.W. et al. (2014). *ACS Nano* 8: 8844.
82 Lim, B.Y., Yoon, J., Yun, J. et al. (2014). *ACS Nano* 8: 11639.
83 Saricilar, S., Antiohos, D., Shu, K. et al. (2013). *Electrochem. Commun.* 32: 47.
84 Beidaghi, M. and Gogotsi, Y. (2014). *Energy Environ. Sci.* 7: 867.
85 Peng, X., Peng, L., Wu, C., and Xie, Y. (2014). *Chem. Soc. Rev.* 43: 3303.
86 Wu, Z.-S., Feng, X., and Cheng, H.-M. (2014). *Nat. Sci. Rev.* 1: 277.
87 Kyeremateng, N.A., Brousse, T., and Pech, D. (2016). *Nat. Nanotechnol.* 12: 7.
88 Niu, Z., Zhang, L., Liu, L. et al. (2013). *Adv. Mater.* 25: 4035.
89 Liu, L., Niu, Z., and Chen, J. (2016). *Chem. Soc. Rev.* 45: 4340.
90 Zhang, M., Fang, S., Zakhidov, A.A. et al. (2005). *Science* 309: 1215.
91 Chen, T., Peng, H.S., Durstock, M., and Dai, L.M. (2014). *Sci. Rep.* 4: 3612.
92 Shao, Y., El-Kady, M.F., Wang, L.J. et al. (2015). *Chem. Soc. Rev.* 44: 3639.
93 Lu, X., Dou, H., Gao, B. et al. (2011). *Electrochim. Acta* 56: 5115.

94 Huang, L., Li, C., and Shi, G. (2014). *J. Mater. Chem. A* 2: 968.
95 Choi, B.G., Hong, J., Hong, W.H. et al. (2011). *ACS Nano* 5: 7205.
96 Xu, Y., Lin, Z., Huang, X. et al. (2013). *ACS Nano* 7: 4042.
97 Shu, K., Wang, C., Li, S. et al. (2015). *J. Mater. Chem. A* 3: 4428.
98 Sun, H., Xu, Z., and Gao, C. (2013). *Adv. Mater.* 25: 2554.
99 Worsley, M.A., Pauzauskie, P.J., Olson, T.Y. et al. (2010). *J. Am. Chem. Soc.* 132: 14067.
100 El-Kady, M.F., Strong, V., Dubin, S., and Kaner, R.B. (2012). *Science* 335: 1326.
101 Kaempgen, M., Chan, C.K., Ma, J. et al. (2009). *Nano Lett.* 9: 1872.
102 Chen, T., Xue, Y.H., Roy, A.K., and Dai, L.M. (2014). *ACS Nano* 8: 1039.
103 Yao, B., Yuan, L., Xiao, X. et al. (2013). *Nano Energy* 2: 1071.
104 Liu, L., Niu, Z., Zhang, L. et al. (2014). *Adv. Mater.* 26: 4855.
105 Yun, T.G., Hwang, B.I., Kim, D. et al. (2015). *ACS Appl. Mater. Interfaces* 7: 9228.
106 Yuan, L., Chen, J., Hu, C. et al. (2012). *ACS Nano* 6: 656.
107 Sung, J.-H., Kim, S.-J., Jeong, S.-H. et al. (2006). *J. Power Sources* 162: 1467.
108 Hu, H., Zhang, K., Li, S. et al. (2014). *J. Mater. Chem. A* 2: 20916.
109 Wang, X., Myers, B.D., Yan, J. et al. (2013). *Nanoscale* 5: 4119.
110 Hu, H., Pei, Z., Fan, H., and Ye, C. (2016). *Small* 12: 3059.
111 Wang, K., Zou, W.J., Quan, B.G. et al. (2011). *Adv. Energy Mater.* 1: 1068.
112 Long, X., Zeng, Z., Guo, E. et al. (2016). *J. Power Sources* 325: 264.
113 Jiang, Q., Kurra, N., and Alshareef, H.N. (2015). *Adv. Funct. Mater.* 25: 4976.
114 Beidaghi, M. and Wang, C. (2012). *Adv. Funct. Mater.* 22: 4501.
115 Lin, J., Zhang, C., Yan, Z. et al. (2013). *Nano Lett.* 13: 72.
116 Chen, W., Beidaghi, M., Penmatsa, V. et al. (2010). *IEEE Trans. Nanotechnol.* 9: 734.
117 Pang, H., Zhang, Y., Lai, W.-Y. et al. (2015). *Nano Energy* 15: 303.
118 Jung, H., Ve Cheah, C., Jeong, N., and Lee, J. (2014). *Appl. Phys. Lett.* 105: 053902.
119 Li, L., Secor, E.B., Chen, K.-S. et al. (2016). *Adv. Energy Mater.* 6: 1600909.
120 Sun, G., An, J., Chua, C.K. et al. (2015). *Electrochem. Commun.* 51: 33.
121 Zhu, Y.G., Wang, Y., Shi, Y.M. et al. (2014). *Nano Energy* 3: 46.
122 Liu, S., Xie, J., Li, H. et al. (2014). *J. Mater. Chem. A* 2: 18125.
123 Xiao, Y., Huang, L., Zhang, Q. et al. (2015). *Appl. Phys. Lett.* 107: 013906.
124 Zhang, Q., Huang, L., Chang, Q. et al. (2016). *Nanotechnology* 27: 105401.
125 Wang, Y., Shi, Y., Zhao, C.X. et al. (2014). *Nanotechnology* 25: 094010.
126 Choi, K.-H., Yoo, J., Lee, C.K., and Lee, S.-Y. (2016). *Energy Environ. Sci.* 9: 2812.
127 Sun, K., Wei, T.-S., Ahn, B.Y. et al. (2013). *Adv. Mater.* 25: 4539.
128 Wu, Z.-S., Parvez, K., Feng, X., and Müllen, K. (2013). *Nat. Commun.* 4: 2487.
129 Wu, Z.-S., Parvez, K., Li, S. et al. (2015). *Adv. Mater.* 27: 4054.
130 Liu, W., Lu, C., Wang, X. et al. (2015). *ACS Nano* 9: 1528.
131 Li, L., Lou, Z., Han, W., and Shen, G. (2016). *Nanoscale* 8: 14986.
132 Xue, M.Q., Xie, Z., Zhang, L.S. et al. (2011). *Nanoscale* 3: 2703.
133 Lin, J., Peng, Z., Liu, Y. et al. (2014). *Nat. Commun.* 5: 5714.

134 Peng, Z., Ye, R., Mann, J.A. et al. (2015). *ACS Nano* 9: 5868.
135 Cao, L., Yang, S., Gao, W. et al. (2013). *Small* 9: 2905.
136 Huang, H.-C., Chung, C.-J., Hsieh, C.-T. et al. (2016). *Nano Energy* 21: 90.
137 Yadav, P., Basu, A., Suryawanshi, A. et al. (2016). *Adv. Mater. Interfaces* 3: 1600057.
138 Liu, Y., Weng, B., Xu, Q. et al. (2016). *Adv. Mater. Technol.* 1: 1600166-1.
139 Peng, Y.-Y., Akuzum, B., Kurra, N. et al. (2016). *Energy Environ. Sci.* 9: 2847.
140 Lu, M., Beguin, F., and Frackowiak, E. (2013). *Supercapacitors: Materials, Systems and Applications*. Germany: Weinheim.
141 Su, Z., Yang, C., Xu, C. et al. (2013). *J. Mater. Chem. A* 1: 12432.
142 Liu, Y., Weng, B., Razal, J.M. et al. (2015). *Sci. Rep.* 5: 17045.
143 Chen, Y., Du, L., Yang, P. et al. (2015). *J. Power Sources* 287: 68.
144 Dong, L., Xu, C., Li, Y. et al. (2016). *Adv. Mater.* 28: 9313.
145 Wang, G., Wang, H., Lu, X. et al. (2014). *Adv. Mater.* 26: 2676.
146 Lu, X., Zeng, Y., Yu, M. et al. (2014). *Adv. Mater.* 26: 3148.
147 Tao, J., Liu, N., Ma, W. et al. (2013). *Sci. Rep.* 3: 1.
148 Wang, X., Liu, B., Wang, Q. et al. (2013). *Adv. Mater.* 25: 1479.
149 Xu, J. and Shen, G. (2015). *Nano Energy* 13: 131.
150 Yun, J., Lim, Y., Jang, G.N. et al. (2016). *Nano Energy* 19: 401.
151 Du, P., Hu, X., Yi, C. et al. (2015). *Adv. Funct. Mater.* 25: 2420.
152 Yang, P., Xiao, X., Li, Y. et al. (2013). *ACS Nano* 7: 2617.
153 Yuan, L., Xiao, X., Ding, T. et al. (2012). *Angew. Chem.* 124: 5018.
154 Jung, S., Lee, J., Hyeon, T. et al. (2014). *Adv. Mater.* 26: 6329.
155 Kim, D., Kim, D., Lee, H. et al. (2016). *Adv. Mater.* 28: 748.

2

Fiber/Yarn-Based Flexible Supercapacitor

Yang Huang[1] *and Chunyi Zhi*[2]

[1] *Shenzhen University, College of Materials Science and Engineering, Xueyuan Road, Nanshan District, Shenzhen, 518071, China*
[2] *City University of Hong Kong, Department of Physics and Materials Science, Tat Chee Avenue, Kowloon, Hong Kong, 999077, China*

2.1 Introduction

The revolution in information technology has led to the fast development of miniaturized hardware devices that are integrating into every corner of our lives and providing us great convenience; devices that have hitherto been frequently described in science fiction books and movies. For example, Oculus Rift, one typical device in Virtual Reality (VR) developed by Facebook, can bring incredible immersive and interactive experience from a simulated three-dimensional environment only by wearing it. The blooming of such highly integrated portable electronic devices requires diverse sources of energy to power them and realize their powerful functionalities. Batteries and capacitors are the two most commonly seen energy storage (ES) devices that are applied in these specific scenarios. Compared with batteries and traditional capacitors, supercapacitors (SCs), or specifically ultracapacitors, can be a better option, since they provide relatively high power densities and energy densities, together with fast charge–discharge rate and long operation life, which are regarded as promising energy storage devices that can bridge the energy density gap between batteries and capacitors as shown in Figure 2.1 [3].

The general trend in developing a portable electronic device is to make it flexible and ensure wearable comfortability, because of which such devices will have wider applicability in different conditions and environments by accommodating to various deformations caused by intense motions when carried by the users in certain situations, for instance, bending or twisting in sport activities. As expected, the power source should also be flexible when embedded in portable devices as an indispensable part. In accordance with such requirements, numerous flexible SCs with outstanding energy storage capabilities and different geometric shapes have emerged recently. The most common configuration is the planar shape, which is relatively simple, but its flexibility is still unable to meet the ultimate goal of marketed products, namely, wearability, one of the research

Flexible Energy Conversion and Storage Devices, First Edition. Edited by Chunyi Zhi and Liming Dai.

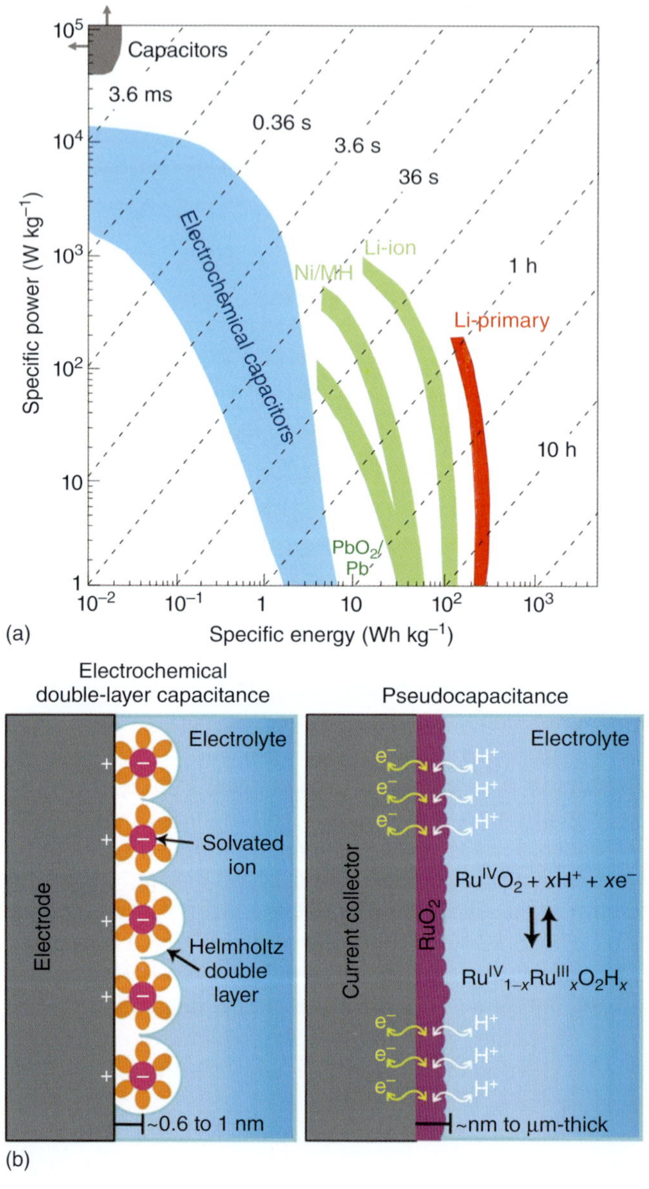

Figure 2.1 (a) Specific power against specific energy, namely Ragone plot, for different electrical energy storage devices. *Source:* Reproduced with permission [1]. Copyright, 2008, Macmillan Publishers Limited. (b) Two categories of electrode materials: electrochemical double-layer capacitor (EDLC) and pseudocapacitor, with different charge storage process. *Source:* Long et al. 2011 [2]. Copyright 2011, Reproduced with permission from Cambridge University Press.

hot spots and frontiers lately [4, 5]. In order to solve the problem of insufficient flexibility, another kind of SCs have been developed based on fibers/yarns with distinctive one-dimensional (1D) format. When compared with traditional two-dimensional (2D) planar configuration, the fiber/yarn-based SCs have several obvious advantages: (i) compact size and light weight, allowing them to meet the rigorous standards for portable devices; (ii) omni-directional flexibility, ensuring wearable comfortability and performance stability under different deformations, thus broadening the application scenarios; and (iii) weaving and knitting feasibility, the most distinguished features that enable them to be integrated into fabrics/textiles or with other wearable electronic devices to form multifunctional systems.

Because fiber/yarn-based SCs can provide maximum flexibility and consequently have great potential for fabricating wearable electronics, there are a variety of novel prototypes that have been reported over the past decade by using all kinds of fibers/yarns as substrates/current collector, ranging from organic (e.g. polyester fiber) to inorganic (e.g. carbon fibers) ones, or from conductive (e.g. metal wire) to nonconductive (e.g. rubber wire) ones [6–8]. In order to achieve high-performance energy storage, two major categories of active/electrode materials are loaded on these 1D fibers/yarns via varied approaches (e.g. electrodeposition): electrochemical double-layer capacitor (EDLC) materials and pseudocapacitor materials, similarly applied in traditional 2D planar SCs (Figure 2.1b) [2]. The EDLC realizes electrical storage by separating charges at the interface between electrodes and electrolyte without faradaic reactions that are normally determined by specific surface area, pore structure, and electrical conductivity, whereas energy storage of pseudocapacitance is achieved through fast reversible surface redox reactions, which largely depends on theoretical capacitance, electrical behavior, and geometric configuration, etc. [3, 9]. Up to now, carbon materials including activated carbon, carbon nanotube (CNT), graphene, carbon fibers, etc. are commonly used for high-performance EDLCs [1, 10, 11]. Generally, EDLCs have higher power density and better cycling stability than pseudocapacitors, but they suffer from low energy density due to a limited specific capacitance of carbon materials. Instead, pseudocapacitive materials such as metal oxides/hydroxides (RuO_2, MnO_2, Co_3O_4, $Ni(OH)_2$, etc.), and conductive polymers (polyaniline (PANi), polypyrrole (PPy), poly(3,4-ethylenedioxythiophene) (PEDOT), polythiophene, etc.) have much higher capacitance and energy density than EDLC materials while inevitably subject to low electrical conductivity and poor rate capability owing to their intrinsic electrical behavior and way of energy storage [1, 12]. A common approach to optimize the performance of fiber/yarn SCs is to fabricate hybrid composites that can take advantage of both groups of electrode materials simultaneously.

Despite the fact that significant achievements in researches have been made, the development of fiber/yarn-based SCs is still in its infancy and there are still lots of obstructions in its popularization and application in the market. First, as a promising energy storage device, further optimization of capacitance is still worth pursuing, especially considering the increasing power requirement of highly integrated powerful personal electronics. Next, some practical properties beyond energy storage should be taken into consideration, since these properties will largely affect the overall user experience on a day to day basis. For example, washability is still a challenging topic that remains to be conquered. Thanks to sustainable flexibility, fiber/yarn-based SCs can be woven or knitted into fabrics/textiles that allow sweat and air to pass through freely. After repeated use, the functional fabrics/textile would be smudged by different things (e.g. dirt or grease) which can be cleaned by simple washing. However, nearly all of the SCs are not waterproof, and therefore their mechanical strength and/or energy storage will deteriorate if soaked in water for a long time. Such similar practical features have triggered certain concerns in this field. Last, but not least, scalable production of fiber/yarn-based SCs is still far from satisfactory. Most of the reported prototypes are realized by family-style, manual workshop processes which is hard to be extended to industrialization production. The core problem is continuous production of highly conductive fibers/yarns with certain mechanical strength that can be used as substrate/current collectors. As long as it is perfectly solved, the fiber/yarn-based flexible SCs with arbitrary length will be achieved and further be woven/knitted into fabrics/textiles of any sizes since they can withstand tensile stress (2–4 MPa) generated by the commercial weaving/knitting machine [13], which could fully develop their potential in wearable electronics.

There is no doubt that fiber/yarn is the core and soul of 1D SCs that directly determine flexibility as discussed earlier. A number of fibers/yarns, including conductive and nonconductive ones, have been employed in SCs. Although there will be less barriers when intrinsic conductive fibers/yarns are chosen as substrates/current collectors, they only account for a small portion of commercial available products, whereas nonconductive fibers/yarns can be tuned into conductive fibers/yarns through some effective strategies, for example, electroless plating, without compromising their flexibility. Thus, both types of fibers/yarns have their own advantages in the fabrication of 1D flexible SCs. In this chapter, the recently emerged fiber/yarn-based flexible SCs are reviewed by using a special categorization method based on the intrinsic characteristic of fiber/yarn: electrical conductivity. According to the classification, the advantages and disadvantages in the fabrication of various intrinsic conductive and nonconductive fibers/yarns to 1D SCs along with their resultant prototypes are discussed in detail at first. Then, the current state of integrated wearable systems based on flexible fiber/yarn SCs and other 1D devices are summarized with an emphasis on design and fabrication. Lastly, future development and key technical challenges ahead are presented. It is believed that such hierarchical detailed discussion on fiber/yarn-based flexible SCs is beneficial for the researchers who work in this area, and will excite more young talent to pay attention to this fast-developing field.

2.2 Supercapacitor with Intrinsic Conductive Fiber/Yarn

Structurally, a complete SC device comprises three parts: two separate electrodes, a separator, and an electrolyte. The electrodes, composed of substrate/current collector and active material, greatly affect the electrochemical performance of SCs in terms of their electrical conductivity, microstructure, and electrochemical behavior. While using conductive fibers/yarns as substrate/current collectors, various active materials can be directly grown on them via various methods without being concerned about deterioration of electrical conductivity, since the main skeleton of the whole device is highly conductive. Nevertheless, choices of intrinsic conductive fibers/yarns, including products from commercial market and from a laboratory concept, are not that many, most of which are based on two major categories of materials: metals and carbon materials, resulting in some restrictions in designing 1D SCs. For example, metal fibers/yarns (e.g. stainless wire) might be too heavy and rigid for wearable electronics, while carbon materials (e.g. CNT fiber) based ones might be too expensive for large-scale production of wearable consumer electronics. Fortunately, a number of attempts have been made recently by researchers to enhance the practicability of these conductive fibers/yarns in fabricating flexible and wearable high-performance SCs. In the following sub-sections, a few intrinsic conductive fibers/yarns will be discussed according to their advantages and disadvantages in fabricating flexible SCs, with special emphasis on their potential in wearable electronics.

2.2.1 Carbolic Fiber/Yarn-Based Supercapacitor

As one category of highly conductive materials, carbon materials, such as active carbon, graphite, CNT, graphene, etc. are widely used in a variety of fields, of which, batteries and SCs are undoubtedly one of the most popular applications. The structures of carbon materials vary from 0D (e.g. quantum dots) to 3D (e.g. microsphere), and they can be further transformed into different geometric configurations such as 1D fiber/yarn [14], 2D film/paper [15], and 3D foam [16] when processed by different methods. Thanks to its great flexibility and extremely high electrical conductivity, the carbon-based fiber/yarn is preferred in the design and fabrication of 1D SCs. One obvious problem of these promising fibers/yarns is that some of them are still prototypes of the laboratory concept and are unsuitable for mass industrial production, while other newly developed commercial products are highly priced that hampers large-scale application. Still, many efforts have been made to develop high-performance 1D SCs targeted at solving different problems by using carbon-based fibers/yarns with new techniques. In the following text, a number of representative works will be discussed according to their novelties in both design and fabrication, revealing the potential of carbon-based fibers/yarns in the fabrication of flexible SCs.

Carbon fibers, consisting of graphite filaments about 5–10 μm in diameter, have high stiffness and tensile strength, together with high electronic conductivity and light weight, which are ideal candidates for the fabrication of flexible SC. However, pristine carbon fiber has two fatal flaws: low surface area ($<10\ m^2\ g^{-1}$) and poor porosity, and therefore it can only act as current collector or substrate to load other active materials resulting in a limited role in SCs for most of the time [17]. Recently, a facile strategy has been developed to introduce large surface area and high mesoporosity to pristine carbon fiber tows by chemical exfoliation, annealing, and reduction [18], which can be implemented for 1D flexible SC as shown in Figure 2.2a. After improvements, these activated carbon fiber tows (A-CFTs) can preserve their mechanical strength and electrical conductivity. Thanks to increased surface area, created mesopores, and residual

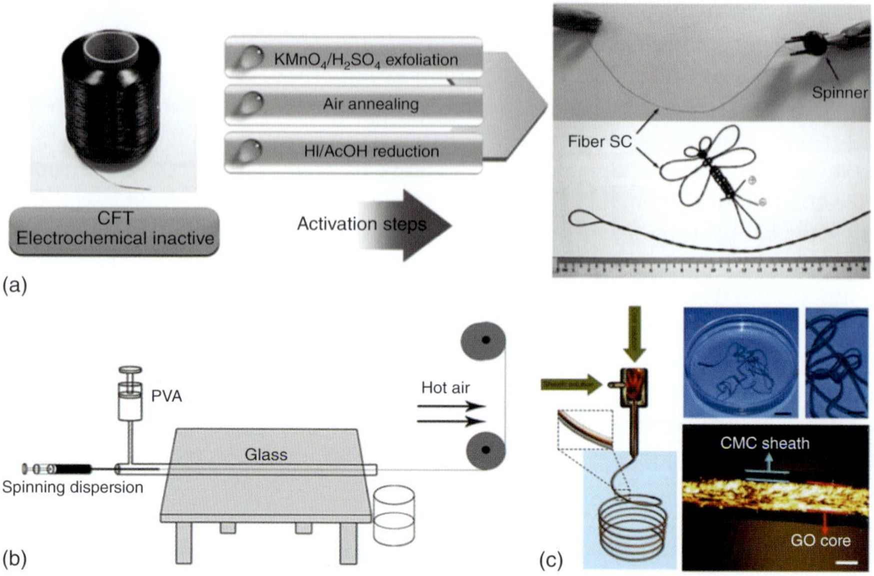

Figure 2.2 (a) Illustration of activation steps for CFTs, which can be assembled into a long-fiber SC and woven into a dragonfly knot. *Source:* Yu et al. 2015 [18]. Copyright 2015. Reproduced with permission from John Wiley & Sons. (b) Spinning process for continuous CNTs/PANi yarn. *Source:* Meng et al. 2014 [19]. Copyright 2014. Reproduced with permission from John Wiley & Sons. (c) Schematic demonstration of the coaxial spinning process and images of GO@CMC fiber. *Source:* Reproduced with permission [20]. Copyright 2014, Macmillan Publishers Limited.

functional groups, A-CFTs show a high specific capacitance of 14.2 $F\,cm^{-3}$ in aqueous H_2SO_4 electrolyte along with a high discharge rate up to 10 000 $mV\,s^{-1}$. Subsequently, they are used to assemble high-performance all solid-state fiber SCs with outstanding flexibility and weavability: their capacitance is nearly unchanged when bending from 0° to 180° after up to 2000 cycles; moreover, they can be woven into a dragonfly knot demonstrating application potential in smart textiles (Figure 2.2a). Another advantage of this method is that the production of A-CFTs can be easily scaled up while using industrial-scale manufacture process, which is favorable for realizing efficient energy storage in flexible electronics with low cost.

CNTs are also one type of attractive active materials for high-performance SCs owing to their electrical conductivity, charge transport capability, mesoporosity, and electrolyte accessibility. In earlier researches, a number of studies have applied planar networks of CNTs such as sheets, thin films, and buckypapers in constructing 2D SCs [11, 21]. With the upsurge of development in fiber/yarn-based SCs and wearable electronics, it is desirable to transform CNTs into fiber/yarn that is strong and flexible enough for weaving and knitting into fabrics. The parallel arrays of CNT forests can be drawn into continuous length webs that can be further converted into CNT yarns by twist-insertion, mechanical rubbing, and liquid densification [22–24]. Compared with carbon fiber, CNT yarn has a larger surface area and a porous structure that is beneficial for higher energy storage in constructing flexible SCs. However, capacitance and energy density of CNT is still relatively low limiting its practical application. This problem can be improved by using pseudocapacitive materials, for example, metal oxides or conductive polymers. In some cases, CNT yarns are prepared in advance, then coated or deposited with active materials on their surface, and finally assembled into a fiber-/yarn-shaped SC. Such a two-step method cannot ensure massive load of active materials, considering that most of them only accumulate on the outer surface of the fiber/yarn; in other words, there are no additional pseudocapacitance materials in core, leading to limited improvements in energy storage. In addition, the operational process is too complex to extend it to large-scale production. Therefore, developing a one-step strategy to fabricate integrated composite yarn based on CNTs and pseudocapacitance materials has unique advantages. In a recent work, a composite yarn made of single-walled CNTs and PANi nanowires was spun using a one-step process that can be directly used as high-performance integrated electrodes for flexible SCs [19]. Wet-spinning was chosen as the preliminary method for producing this composite yarn considering its compatibility to CNTs and potential in scaling up for industrial applications. As shown in Figure 2.2b, the spinning apparatus is built by using a three-branch glass tube: one branch connects a syringe with the dispersion and the other branch connects to a syringe with polyvinyl alcohol (PVA) solution, which are driven by a dual-channel micro-injection pump, and the last branch serves as outlet for the CNT-based yarn. The dispersion of CNTs and PANi coagulates when injected into coflowing PVA, subsequently forming continuous, meter-long gel yarn after being dried in hot air. The diameter of these yarns can be tuned by changing the syringe needles with different inner diameters, or by optimizing the spinning speeds and the concentration of the coagulation solutions. In order to achieve better electrical conductivity, the gel yarns were placed in 95 °C deionized water to remove the surfactants. After removing large amounts of PVA, the conductive yarn could still maintain its good mechanical properties thanks to the entangling of CNTs and PANi nanowires. As expected, these composite yarns can be readily assembled into high-performance flexible SCs after coating with gel electrolyte and twisted together. Owing to the synergistic effects of CNTs and PANi, the as-prepared yarn-based supercapacitor achieved a high capacitance of 6.23 $mF\,cm^{-2}$ at a current density of 0.2 $A\,g^{-1}$, which could be woven into textiles with traditional cotton yarns, showing a potential in building wearable electronics.

Graphene, one amazing 2D material, has been widely used in a variety of fields such as sensors, batteries, actuators, SCs, to name a few, due to its high surface area, outstanding thermal conductivity, excellent mechanical properties, and electronic conductivity. Similarly to other carbonic materials, graphene and its derivatives have been used as electrode materials for SCs in different appearances. Among them, graphene-based fibers/yarns are of particular interest due to their tiny volume and high flexibility, which are promising for flexible and even wearable SCs as discussed previously. However, both graphene-based and other carbon-based fibers/yarns are easily short-circuited since they are close to each other when assembled into a compact fiber-/yarn-shaped SC. Although the coated solid electrolyte can act as separator and partly reduces chances of short-circuiting, the coating process is too complex and time-consuming;

besides, it is difficult to ensure even coating of the electrolyte layer, resulting in low capacitance and energy density. Thus, it is desirable to develop a simple and effective method that can prepare sheath-protected fiber/yarn electrodes with outstanding energy storage capability. A coaxial wet-spinning assembly strategy is proposed recently to prepare polyelectrolyte-wrapped graphene core-sheath fibers that can be used directly as contactable and interweaving yarn electrodes while avoiding the short circuit effectively [20]. As shown in Figure 2.2c, the spinneret has two inlets that connect the inner and outer channels: aqueous graphene oxide (GO) solution was used as the inner spinning dope, while aqueous solution of ionically conductive and electrically insulative sodium carboxymethyl cellulose (CMC) was chosen as the outer spinning dope. Then, GO from the inner channel and CMC from the outer channel were injected simultaneously into the coagulating bath that shaped rapidly forming continuous microfibers. After chemical reduction of hydroiodic acid, conductivity of GO@CMC fibers were greatly improved achieving a high value of ~7000 $S\,m^{-1}$ while the insulative polyelectrolyte sheath maintained well. Such continuous well-defined core-sheath fibers can be directly used as electrodes and assembled into flexible SCs when coated with PVA gel electrolyte. The as-prepared GO@CMC yarn SC exhibited an extremely high areal capacitance of 127 $mF\,cm^{-2}$, corresponding to a volumetric capacitance of 114 $F\,cm^{-3}$ and a length capacitance of 3.8 $mF\,cm^{-1}$. Like the wet-spinning method for CNT fibers, this technique also has the potential of large-scale productivity and can prepare ultralong fibers, up to 100 m. As a demonstration, two individual 40-cm-long fibers were used as electrodes to interweave a cloth SC, which showed a capacitance of 28 mF at a current density of 10 μA. Owing to the protection of the CMC sheath, the cloth SC could work safely under the bending angle of 180° along three directions without concerns related to short-circuiting, revealing outstanding mechanical flexibility.

2.2.2 Metallic Fiber/Yarn-Based Supercapacitor

Metallic fiber/yarn, for instance, gold and silver wire, has been used as decoration in clothing and textiles since ancient times, and many of these exquisite antique textiles can be found in museums all over the world. Nowadays, there are two major processes that are applied in the fabrication of metallic fibers/yarns, namely, laminating process and metalizing process. Or, they may also be bundles drawn from larger diameter wires producing smaller fibers/yarns with micrometer sizes. After years of development, a variety of metallic fibers/yarns with different ingredients and sizes have been manufactured and are available commercially. Because of high electronic conductivity and outstanding mechanical flexibility, they can be readily utilized as ideal current collectors/substrate for high-performance flexible SCs; besides, their capability to withstand the input tensile stress generated by commercial weaving and knitting machines allows them to be assembled into wearable systems with other fiber-/yarn-shaped electronics in large-scale industrial production. Moreover, in order to improve the practicability, some intrinsic characteristics of metallic fiber/yarn can be exploited to enhance energy storage capability and/or add new functions to the flexible SCs. However, as compared with carbon-based fiber/yarn, relative heavy weight is an inevitable issue which might hinder its popularization and application in wearable electronics to a certain degree. One effective solution is to make metallic fibers/yarns composed of numerous wires of tiny volume (e.g. bundle drawing and twisting technique) that can realize increased surface area with less weight. In this sub-section, a few recently reported representative SCs based on metallic fiber/yarn are discussed in terms of their design and fabrication, as well as on their potential applications in wearable electronics.

Although conventional metallic fibers/yarns demonstrate the potential of weavability, knittability, and wearability, it will be more desirable when they are scaled down to micrometer size considering the requirements of tiny volume and light weight in flexible SCs. To address this challenge, continuous conductive yarns with micrometer size (180–250 μm) were fabricated by twist-bundle-drawing multiple high-temperature spun stainless 316 l thin fibers together, which have high conductivity, outstanding tensile strength (over 700 MPa), excellent corrosion resistance, and thermal stability while retaining great flexibility equivalent to common cellulose yarns [13]. These obtained yarns can be used as high performance electrodes after the deposition of reduced graphene oxide (rGO), MnO_2 and PPy as active materials as shown in Figure 2.3a. Synergistic effects of hierarchical structure on conductive stainless yarn leads to higher electrochemical performance and better mechanical flexibility: the inner rGO layer can improve charge transfer of pseudocapacitive MnO_2 interlayer, whereas the outermost PPy layer serves as a flexible stress buffer and protects integrity of SC during different deformations besides of its contribution to capacitance. The fiber-shaped SC assembled by these modified yarn electrodes show high capacitance and energy densities without compromising cycling stability and deformation stability. The specific capacitance and energy density was calculated to be 31 $mF\,cm^{-1}$, 411 $mF\,cm^{-2}$, and 0.037 $mWh\,cm^{-2}$ using galvanostatic discharge curves. Its capacitance maintained over 92% of its original value after 4950 charging/discharging cycles at a current density of 80 $mA\,cm^{-3}$. In practical application, deformation stability is also important, since the flexible SC is subjected to various deformations with the repeat use. Thanks to great flexibility of stainless yarn and protection of PPy, the yarn SC retains over 80%, 91%, and 103% of origin capacitance after 1000 cycles of bending, knotting and twisting, respectively, showing the outstanding deformation stability. Then, as a demonstration of taking advantage of their flexibility and weavability, a 15 cm × 10 cm cloth was woven by using seven yarn SCs in series, which could power 30 light-emitting diodes (LEDs) after fully charged (Figure 2.3a). This work offers some quotable experience for general process in modern textile production industry, for instance, forming a hierarchical structure on a twist-bundle-drawing yarn, inspiring the design and fabrication of various wearable electronics.

Generally, metallic fibers/yarns are simply used as current collectors/substrates for SCs in most cases making a negligible contribution to capacitance. It will be much better if these conductive fibers/yarns can enhance their roles in the energy storage process, especially considering the fact that their weight and volume account for a large proportion of the total weight and volume of SCs. Recently,

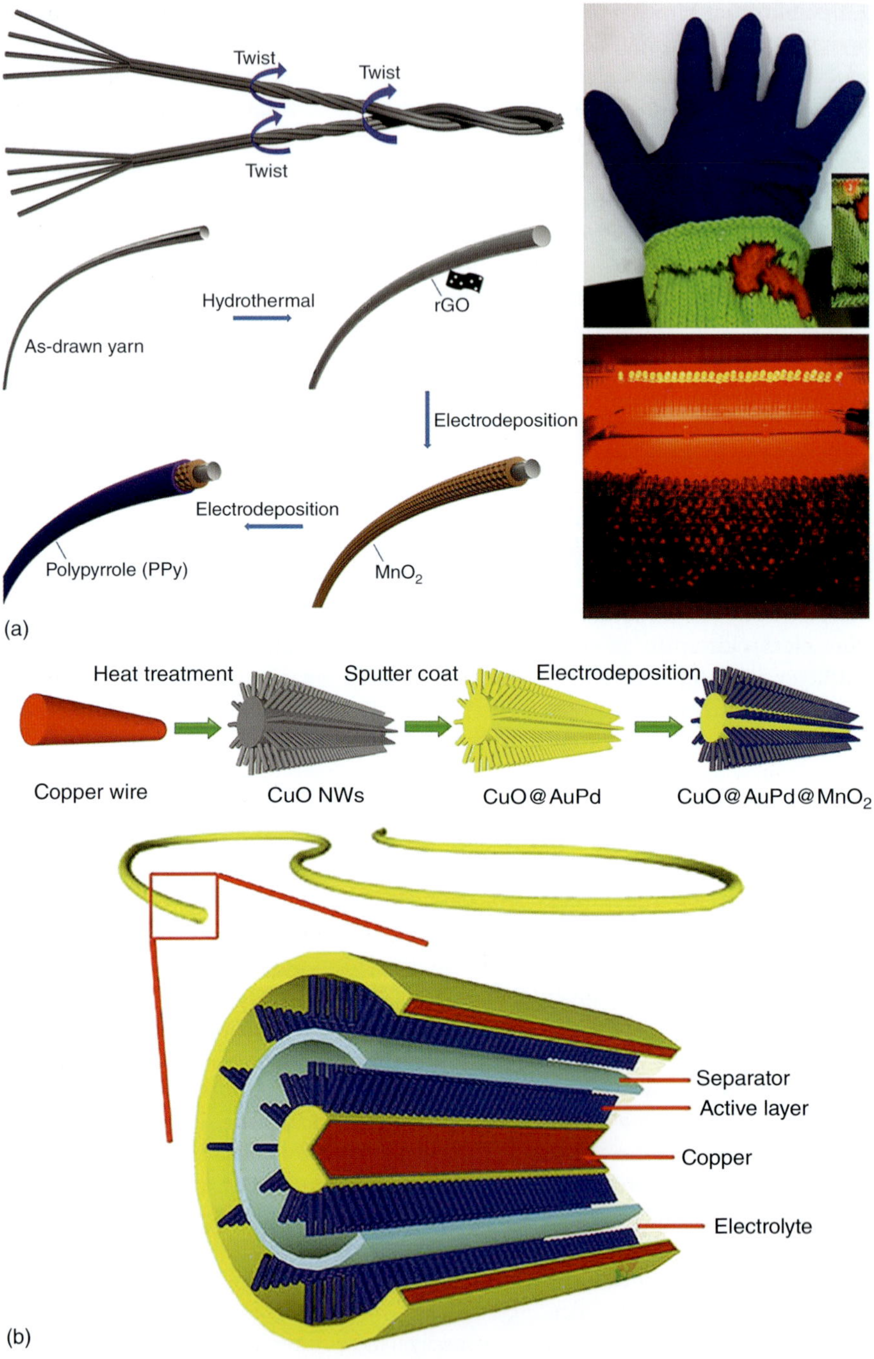

Figure 2.3 (a) Schematics of stainless yarn fabrication and modification, photographs of the energy storage textiles made of SC yarns. *Source:* Huang et al. 2015 [13]. Copyright 2015. Reproduced with permission from American Chemical Society. (b) Schematic illustration of CuO@AuPd@MnO_2 NWs fabrication process and assembly of coaxial SC. *Source:* Yu and Thomas 2014 [25]. Copyright 2014. Reproduced with permission from John Wiley & Sons.

a flexible coaxial supercapacitor (CSC) cable has been developed by using a copper wire with large area, template-free, high aspect ratio, and freestanding $CuO@AuPd@MnO_2$ core–shell nanowhiskers (NWs) design, as illustrated in Figure 2.3b [25]. The NWs are grown on the outer surface of the copper wire with high aspect ratio and thus avoid structural collapse effectively, resulting in relatively large surface area that can improve capacitance. In addition, the modification of pristine copper core wire was accomplished by mild temperature heating at 500 °C for 4 h which is useful to scale up the process to prepare any length. Despite their role in limiting current leakage from the inner core, the CuO NWs exhibit poor conductivity and limited capacitance. Thus, it is preferable to load other active materials with high theoretical value to further improve the capacitance of electrodes. MnO_2 was chosen and conformably electrodeposited on the CuO NWs. However, before electrodeposition, a thin layer of gold–palladium (AuPd) was deposited onto the surface of the CuO NWs, which served as an anode material for MnO_2 and as additional current collector between the active materials. An integrated CSC was built by simply inserting the inner electrode into the outer tubular electrode both of which were modified with similar $CuO@AuPd@MnO_2$ core–shell NWs structure (Figure 2.3b). The CSC exhibited a volumetric energy density of 0.55 mWh cm^{-3} with a volumetric power density of 413 mW cm^{-3}. After folding at 180° for 100 cycles, the device retained 93.4% of its original capacitance showing outstanding flexibility and bendability; besides, it maintained 99% of the initial value over 5000 charging/discharging cycles, demonstrating good cycling stability. There is another advantage of this CSC, and that is dual functionality, i.e. it can work simultaneously as an energy storage device and as an electrical cable. As shown in a demonstration, the electrical cable section did not interfere with the energy storage section, since cyclic voltammetry (CV) curves showed no differences before and after the connection of the external power source.

In a previous example, the concept of dual functionality of the SC has been referred to, which can realize additional functions beyond energy storage and hence extend the applicative fields. This kind of multifunctional SCs is getting more and more attention, because the embedded multifunction might be a perfect solution to certain pain points that users are familiar with. For example, the reliability of fiber/yarn-based SCs during actual applications is always a vital issue that arouses concern in research and the industrial field, especially given various deformations they will inevitably face when used in wearable electronics. With the increase of serve life, accumulation of these deformations will become irreversible, leading to malfunction of the whole device. Thus, there is an urgent need to solve this problem and improve the reliability. Lately, by using traditional nickel–titanium (NiTi) wires, a novel shape memory supercapacitor (SMSC) has been developed in order to improve the reliability after severe plastic deformation takes place [26]. NiTi wire, one conductive alloy, can recover back to its original shape very shortly even after serious plastic deformation when heated to phase change temperature. The automatic shape recovery characteristic makes it a perfect solution for deformation accumulation. With high conductivity and great flexibility, MnO_2 and PPy can be electrodeposited onto the surface of the NiTi wire easily to form a hierarchical flexible electrode, as shown in Figure 2.4a.

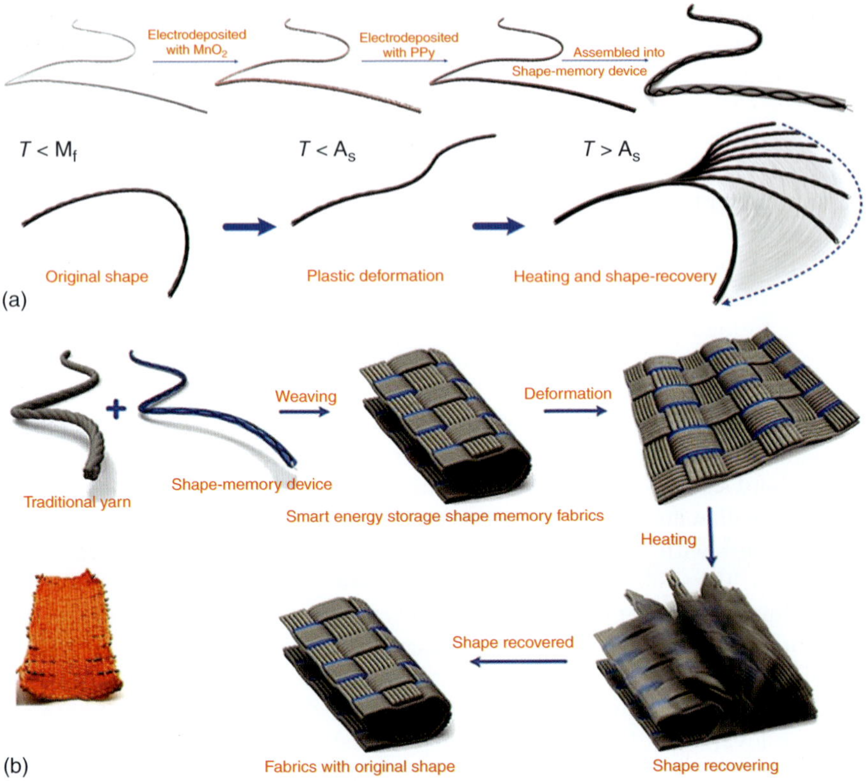

Figure 2.4 (a) Schematics of SMSC fabrication and its shape memory effect. (b) Schematic illustration of SMSCs woven with traditional yarns to form a textile. *Source:* Huang et al. 2016 [26]. Copyright 2016. Reproduced with permission from Royal Society of Chemistry.

After coated with solid electrolyte, a flexible SMSC with amazing shape memory effect is achieved, which exhibits a high specific capacitance of 198.2 $F\,g^{-1}$ at a current density of 1 $A\,g^{-1}$. Moreover, even after 15 cycles of plastic deformation and restoration, shape recovery capability and capacitance did not deteriorate obviously, reaching a high value of 85% and 86%, respectively. Afterward, several SMSCs were knitted with traditional yarns to fabricate a novel shape memory energy storage textile (Figure 2.4b). By connecting the embedded SMSCs in series or parallel, different voltage and current needs can be met, showing the possible scalability. Furthermore, this shape memory textile will recover back to its original shape after heating to the triggering temperature of SMSC, which can be utilized as a temperature alarm. This study opens a new path in developing functional SC relying on the intrinsic characteristics of metallic fiber besides its high conductivity, which will inspire to design other multifunctional flexible SCs.

2.2.3 Hybrid Conductive Fiber/Yarn-Based Supercapacitor

Since both carbolic fibers/yarns and metallic fibers/yarns have their own advantages and disadvantages in building flexible 1D SCs, it is rational to design and fabricate a device with hybrid fibers/yarns, and thus make best use of their advantages and bypass the disadvantages. However, such SCs are not commonly reported when compared with devices developed with a single type of fiber/yarn, owing to the difficulties in coordinating the properties of two different materials. For example, there can be considerable differences in wettability of carbon-based fibers/yarns and metal-based fibers/yarns, which leads to less processing methods in fabricating high-performance flexible SCs. Nevertheless, a lot of effort has been made to achieve such a kind of device with hybrid fibers/yarns, and the typical ones will be discussed in detailed in terms of their characteristics.

Integrated nanodevices are becoming more and more popular recently, because they can provide a diversity of functions with broad applicabilities when compared with conventional single function ones. As shown in Figure 2.5a, a novel fiber-based flexible asymmetric SC with integrated photodetecting system has been developed by using nickel fibers with Co_3O_4 nanowires as the positive electrode and carbon fibers coated with graphene as the negative electrode, which can realize energy storage and optoelectronic detection simultaneously [27]. First, porous Co_3O_4 nanowires were grown regularly on the nickel fibers via a facile solvothermal method, which had a length of 1–2 μm and a diameter of 50–100 nm. Then, a bundle of carbon fibers/graphene were carefully entangled around this nickel fiber which was previously immersed in the electrolyte, and thus a microscale fiber-based flexible asymmetric SC was obtained. The as-prepared flexible integrated devices showed an increased capacitance from 0.234 to 0.695 $F\,cm^{-3}$ by enlarging the potential voltage window from 0–0.6 to 0–1.5 V. In addition, this flexible SC can act as a photodetector owing to light-sensitive carbon fibers/graphene negative electrodes that can absorb a wide wavelength range between visible and infrared [29]. Under illumination, electron–hole pairs generated in graphene will recombine in tens of picoseconds, which depends on the quality and carrier concentration of graphene. After applying an external field that is generated by the fully charged SC, electron–hole pairs can be separated: electrons move to a positive Co_3O_4 electrode, while holes move to negative carbon fiber/graphene electrode, leading to improved leakage current of the SC. Thus, photodetection can be realized by monitoring the changes of leakage current. Prior to white light illumination (with an intensity of 25 and 40 $mW\,cm^{-2}$), the leakage currents of SC were around 200 nA while after illumination, it increased to about 400 nA, indicating photodetection behavior of the integrated device (Figure 2.5b). The concept of miniaturized integrated flexible SC was successfully demonstrated in this work based on the use of hybrid conductive fibers.

Fiber/yarn-based flexible SCs have high potential in constructing power electronics textiles as discussed earlier. However, capacitance and cycling electrochemical performance of these 1D SCs will deteriorate obviously when they are scaled up to a long length, owing to an increased resistance. In order to solve this significant problem, a special metal/CNT core-spun-yarn electrode for 1D flexible SCs have been developed via using six types of metal filaments (Pt, Au, Ag, AuAg, PtCu, and Cu) as current collectors [28]. The CNT yarn was prepared by twisting a continuous CNT web drawn from the multiwalled

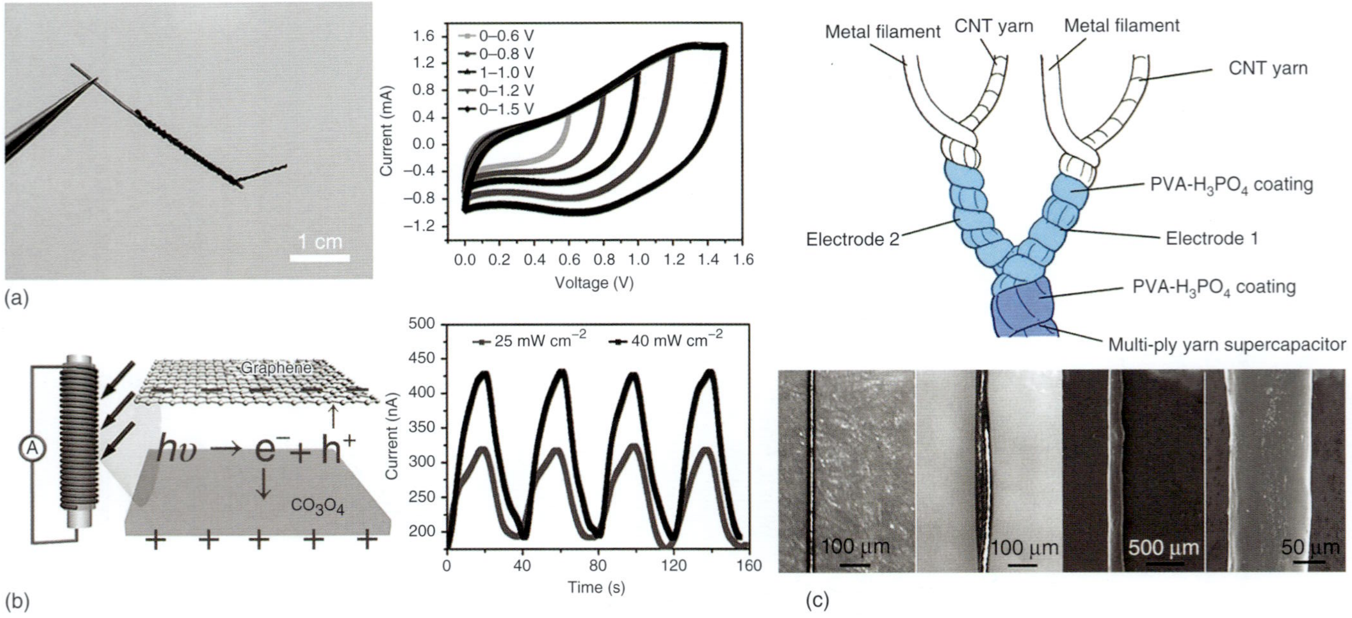

Figure 2.5 (a) Photograph of fiber based asymmetric SC and its CV curves. (b) Schematic illustration of the integrated system and incident light intensities under illumination. *Source:* Wang et al. 2014 [27]. Copyright 2014. Reproduced with permission from John Wiley & Sons. (c) Architecture of the linear SC based on metal filament and CNT yarn, corresponding SEM images of bare filament Cu, Cu + CNT electrode, and final SC device (from left to right). *Source:* Zhang et al. 2015 [28]. Copyright 2015. Reproduced with permission from American Chemical Society.

CNT forest synthesized by the chemical vapor deposition (CVD) method. Then, this CNT yarn was twisted together with a metal filament and thus formed a complete electrode. After coated with solid electrolyte, two identical electrodes were plied together to form a 1D flexible SC as shown in Figure 2.5c. All of the metal filaments could significantly improve electrochemical performance of the SC, but among them, Cu and PtCu filaments could enhance electrochemical performance much better than that by other filaments. This is because the surface of Cu-containing filaments is oxidized by acid solid electrolyte and changed into CuO, which could contribute capacitance as an additional pseudocapacitive material, while the core still remained highly conductive since it was not oxidized and perfectly maintained its role as current collector. The gravimetric capacitance increased by 20% due to an extended potential window (from 1.0 to 1.4 V) resulting from dual active materials of CNT and CuO. Meanwhile, the energy density and the power density increased by 134% and 729%, respectively, suggesting dramatic improvements. In addition, the SC maintained 91% and 97.5% of its original capacitance after 3000 charging/discharging cycles and 1000 folding/unfolding cycles, respectively. This work demonstrates that metal fiber/yarn can also contribute to the capacitance of SC through suitable treatment, instead of simply acting as a current collector/substrate, an easily overlooked potential that is worth exploring.

2.3 Supercapacitors with Intrinsic Nonconductive Fiber/Yarn

Apart from intrinsic conductive fibers/yarns, most commercial products are nonconductive ones and are made from a number of natural or synthetic materials such as cotton, silk, nylon, and Kevlar with matured technologies and processes. There are significant limitations to designing and fabricating 1D flexible SCs if these readily available nonconductive fibers/yarns are excluded from constructing devices as current collectors/substrates, particularly considering their outstanding adaptability to weaving and knitting machines that is indispensable for large-scale production of electronic textiles. Nevertheless, it is a considerable challenge to transform them into conductive fibers/yarns through appropriate strategies in view of their diverse properties (e.g. wettability) and structures (e.g. porosity). Toward this, several techniques have been demonstrated as effective methods to convert the conductivity; for example, repetitive dip coating and atomic layer deposition (ALD) [30, 31]. In addition, there are a wide range of conductive materials that can be applied to reverse the electrical conductivity of insulating fibers/yarns, for example, carbon materials and metals, and some among them can enhance capacitance by serving as electrochemical active materials at the same time, resulting in another challenge in choosing suitable candidates with better performance for capacitance [32]. In the following sub-sections, a number of representative flexible 1D SC samples based on intrinsic nonconductive fibers/yarns are introduced according to their material selections and processing methods, which are very meaningful for developing high-performance flexible devices with low cost and convenient fabrication procedures.

2.3.1 Fiber/Yarn Modified by Carbon Materials

Carbon-based materials are widely used as EDLC electrodes in SCs as discussed in previous sub-sections, owing to their high conductivity, large surface area, controllable porosity, and easy processability. Therefore, they are considered as improving agents for nonconductive fibers/yarns, while providing additional capacitance. In recent years, there are a number of studies by using different carbon-based materials, such as activated carbon, CNTs, and graphene, to modify these yarns and fabricate flexible 1D SCs. But low theoretical capacitance is inevitable for all of carbon-based SCs; thus, pseudocapacitive materials are usually incorporated with them as they suffer from intrinsic small surface area and low electrical conductivity. In the following text, modified fibers/yarns as current collectors/substrates for the fabrication of several high-performance flexible carbon-based SCs are discussed in terms of their design and process methods.

Many of the reported 1D SCs are not designed with stretchability that is a very practical function for integrating with other fiber/yarn electronics to form wearable devices particularly given the challenging states (e.g. twisting and stretching) they will encounter in daily use. Even among the reported stretchable fiber/yarn-based SCs, there are still shortcomings in the compatibility of wearability and large-scale production with low cost. In textile industry, different types of fibers/yarns with various physical and mechanical properties are readily available, which are suitable for knitting, weaving, and sewing, and among them, urethane fiber core spun yarns (UYs) are with intrinsic high stretchability that is formed by entwining cotton yarns around elastic urethane filaments as shown in Figure 2.6a. Thus, there is much possibility to develop stretchable and wearable SCs on a large scale, if UYs can be converted into conductive yarns and then serve as current collectors/substrates for SC. Inspired by this, continuous stretchable and conductive UY based fiber was successfully prepared by a scalable dipping and drying process and subsequently assembled into a stretchable and wearable 1D SC (Figure 2.6b) [33]. Compared with common nonstretchable yarns, pristine UYs can be stretched to 150% strain without obvious deterioration. Moreover, when they are woven into a fabric, the UY-based one can be stretched up to 300% strain, whereas a cotton yarn-based one can achieve only a 20% strain, suggesting that UYs have great potential in fabricating stretchable and wearable electronics. Thanks to the large contact area of outer surrounding cotton yarns, CNTs can be readily coated on the UYs through a simple repetitive dipping and drying method. In order to achieve higher capacitance, pseudocapacitive PPy was uniformly electrodeposited on the surface of UY yarns to form an integrated electrode. When stretched to 100%, the increase in the electrical resistance of PPy@CNT@UY electrode was less than 18%. After coated with solid electrolyte, two of these electrodes were further assembled into a stretchable SC, which showed a high areal capacitance of 69 $mF\,cm^{-2}$ at 5 $mV\,s^{-1}$. Additionally, more than 80% of the capacitances can be maintained when subjected to a strain of 80%. Even after 1000 cycles of repetitive stretching test (from 0% to 40%), more than 85% of its original capacitance can be retained, indicating outstanding stability under serious deformation. As expected, these stretchable SCs were

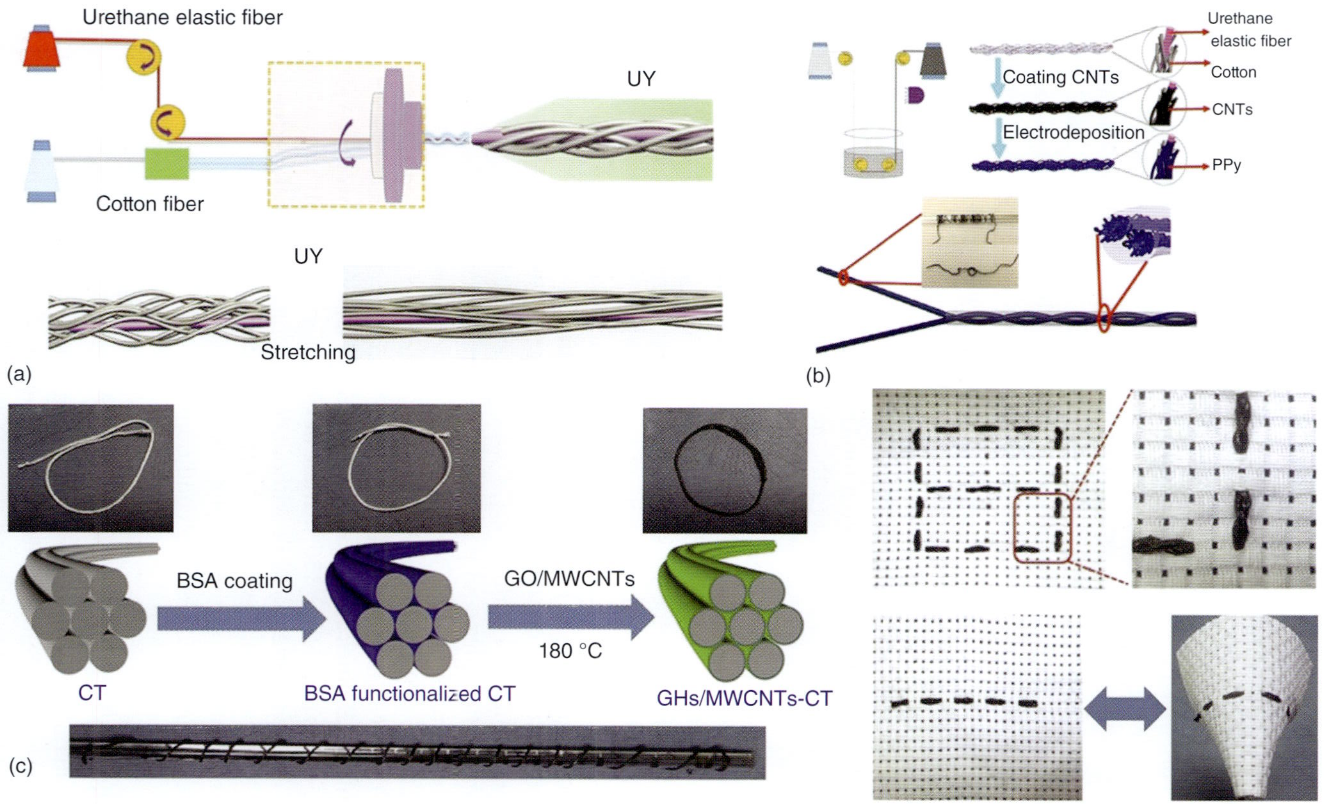

Figure 2.6 (a) Schematic of spinning process of UY and its original/stretched state. (b) Schematic of modification process and a complete all-solid-state SC, insets are image of stretchable UY based electrodes. *Source:* Sun et al. 2016 [33]. Copyright 2016. Reproduced with permission from Elsevier. (c) Schematic of CT modified by GHs/MWCNTs, a 50 cm long GHs/MWCNTs-CT, and photographs of an all-solid-state SC embedded in a fabric that is bendable. *Source:* Zhou et al. 2016 [34]. Copyright 2016. Reproduced with permission from Elsevier.

readily woven with traditional fabrics to form a large-scale energy storage textile, demonstrating a potential role in powering wearable electronics.

Graphene is a one-atom-thick layer of carbon atoms arranged in a hexagonal format with a number of amazing properties, such as exceptional electronic and thermal conductivity, extreme mechanical strength, and extraordinary elasticity and stiffness [35]. Compared with CNTs, graphene has the advantage of relatively low-cost production, since it can be obtained from natural graphite via chemical methods. As discussed previously, pure graphene fibers/yarns can be fabricated by a variety of methods, for example, wet-spinning, which are ideal current collectors/substrates for flexible SCs due to their outstanding electrochemical properties. However, considering their complex preparation process and high price, it is more desirable to exploit a technique to fabricate graphene-based conductive fibers/yarns in a more convenient process with low cost. As reported recently, natural cotton tread (CT) was modified by using a mixture suspension of graphene oxide/multiwalled carbon nanotubes (GO/MWCNTs) via a hydrothermal method, which can be directly utilized as fiber-shaped electrodes for 1D flexible SC as shown in Figure 2.6c [34]. Before hydrothermal process, CT was dip-coated with bovine serum albumin (BSA) powder solution that served as an effective adhesive and improved the adhesion of GO sheets on the CT surface. In order to prevent their aggregation and to further enhance electronic conductivity, a small amount of MWCNTs (GO:MWCNTs = 5 : 1, mass ratio) were added in the GO sheet suspension. After the immersion of BSA-pretreated CT in a mix solution of GO/MWCNTs, GO sheets (negatively charged) will spontaneously attach onto the CT surface (positively charged) firmly, because of the intense electrostatic attractions between them. When hydrothermal process was carried out, these GO sheets were gradually reduced to graphene hydrogels, resulting in the tightening of loose cotton fibers in the CT and hence improved their contact resistances. By optimizing the fabrication process, a 50 cm long GO/MWCNT-modified conductive CT was prepared and wrapped along a rod, showing the flexibility and potential large-scale production (Figure 2.6c). Subsequently, these electrodes were assembled into fiber-shaped SCs with a length capacitance of 97.73 $\mu F\ cm^{-1}$. Like most of the other reported 1D flexible SCs, the GO/MWCNTs-modified CT-based SC exhibited good mechanical and electrochemical stabilities and maintained 90.75% of its initial capacitance after 500 bending cycles (from 0° to 180°). These flexible SCs were readily knitted into fabrics with different patterns, while without compromising original mechanical and electrochemical stabilities (Figure 2.6c). The CV curves showed no obvious differences even after repeated rubbing, indicating the stability of capacitive performance under various bending conditions.

2.3.2 Fiber/Yarn Modified by Metallic Materials

Metal is a material with outstanding electrical and thermal conductivity, which can be made into fibers/yarns via different strategies as mentioned earlier. However, their relative heavy weight and stiffness might be barriers for applications in flexible and wearable SCs, when compared with other fibers/yarns that are commonly used in the textile industry. Therefore, it will be a satisfactory solution to coat a suitable thickness of metals on the surface of lightweight fiber/yarn substrates through some practical processes that can not only take advantage of high conductivity from metals, but also preserve the softness and light weight from the original fibers/yarns. To achieve this concept, high-performance flexible SCs have been designed and fabricated based on metallic modified fibers/yarns as reported lately, which will be discussed in the following text according to process route, capacitive performance, and potential application in the manufacture of electronic textile.

A significant challenge of flexible 1D SCs is to fabricate lightweight fiber/yarn electrodes with excellent electrochemical performance, namely, high capacitance, high power density, and energy density via cost-effective and scalable processes. Instead of using readily available metal yarns or by fabricating complex carbon based yarns, a "low-cost graphene/metallic textile" composite yarn has been developed and applied in a high-performance all-solid-state SC for the first time as shown in Figure 2.7a–c [36]. In this design, Ni-coated cotton fibers are used as current collectors of the electrodes due to their great flexibility, high conductivity, and good durability, which are realized by the electroless deposition method (ELD). Compared with other metallic coating methods, for example, ALD, ELD is very attractive since it can be carried out under ambient conditions on a large scale which does not need sophisticated equipment. First, readily available cotton yarns were immersed into a mixed solution of ethanoic acid and (2-(methacryloyloxy)ethyl trimethylammonium choloride-*co*-3-(trimethoxysilyl)propyl methacrylate) (P(METAC-*co*-MPTS)), which could be covalently grafted onto the cotton surface after hydrolysis and curing processes [37]. Then, they were dipped into an aqueous solution of $(NH_4)_2PdCl_4$, and $PdCl_4^{2-}$ would attach onto the copolymer layer via the strong interactions with the ammonium groups. Lastly, through immersing into an ELD bath of Ni, a thin layer of Ni was uniformly and densely deposited on the surface of cotton yarns, which converted them into highly conductive metallic yarns successfully without sacrificing flexibility and durability. In order to achieve decent capacitive performance, graphene was deposited on the pretreated metallic yarns via electrochemically electrolyzing GO, in which the yarns were used as working electrodes [38]. These GO sheets were reduced into conductive rGO that penetrated into the interval space among the fibers of yarns, after treated with hydrazine vapor. Thereafter, a wearable all-solid-state SC was fabricated by using the rGO/Ni cotton composite electrodes (Figure 2.7c), which exhibited a linear capacitance of 0.11 $F\,cm^{-1}$ and a volumetric energy density of 6.1 $mWh\,cm^{-3}$. As a proof-of-concept, these flexible SCs, acting as weft and warp yarns, were woven into an energy storage fabric with other pristine cotton yarns (Figure 2.7d), which could meet the operational voltage and power requirements by connecting the SC devices either in series or in parallel. Besides, this fabric showed good capacitive stability under different bending states. Even after up to 4000 bending cycles (from 0° to 180°), there were almost no defects in the surface of rGO/Ni cotton yarn electrodes and the volumetric capacitance only decreased by 5% together with a high columbic efficiency of 98–100%. The method of fabricating hierarchical structured composite electrodes in this study should be applicable to other textile fabrics and active materials.

can be charged by the TENG fabrics to 2.1 V in 2009s with a vibration motor at about 5 Hz, and further galvanostatically discharged at 1 μA for 811 s, indicating the validity of self-charging energy storage textile (Figure 2.8d). This work successfully demonstrates scalable production of conductive yarns, capacitive yarn SC, and energy harvesting fabrics and their integrated self-power system, which can inspire other distinctive wearable electronics.

Besides harvesting of mechanical energy, photovoltaic devices can also realize energy conversion effectively by converting irradiation of sun into usable electricity. It is of interest to fabricate them in a fiber/yarn format, which can be potentially integrated with 1D SCs to form an integrated wearable system owing to shape dominance. In pursuance of this, a novel coaxial all-solid-state "energy fiber" with outstanding flexibility has been developed by integrated functions of photovoltaic conversion (PC) and ES as shown in Figure 2.9 [41]. In this design, as for PC, the generated charges can be separated and transported to generate photocurrents effectively because of a fine interface contact of coaxial structure in the radial direction; at the same time, as for ES, this structure is favorable for rapid charge transportation because of larger effective contact area. Moreover, in order to further improve photoelectric conversion and energy storage, conductive MWCNT sheets were applied in this distinctive device. A flexible Ti wire, modified with perpendicularly aligned titania nanotubes by electrochemical anodization,

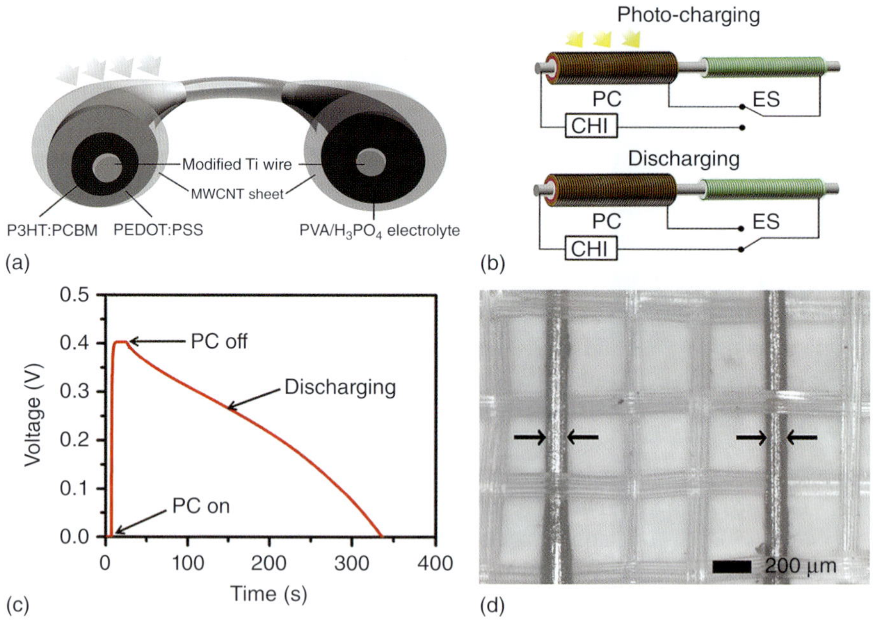

Figure 2.9 (a) Schematic of all-solid-state, coaxial and integrated "energy fiber." (b) Schematic of the circuit connection in charging and discharging. (c) Charging/discharging curve with a current during the discharging process. (d) Photograph of two "energy fibers" woven into a flexible textile. *Source:* Zhang et al. 2014 [41]. Copyright 2014. Reproduced with permission from John Wiley & Sons.

and aligned MWCNT sheets were used as two electrodes of "energy fiber." In the PC section, poly(3-hexyl thiophene):-phenyl-C_{61}-butyric acid methyl ester (P3HT:PCBM) and poly(3,4-ethylenedioxythiophene):poly(styrene sulfonate) (PEDOT:PSS) layers were first coated on modified Ti wire, and then MWCNT sheet was wrapped on as the outermost layer; whereas in the ES section, solid electrolyte was coated on the Ti wire and followed by attaching a layer of MWCNT (Figure 2.9a). On absorption of sunlight, excitons will be generated and separated into electrons and holes in the heterojunction layer between P3HT and PCBM during PC. The maximum photoelectric conversion efficiency is 1.01%, which can be maintained by more than 90% for a period of 10 days. As connected in the circuit as shown in Figure 2.9b, the "energy fiber" can realize a self-powering process: after self-charging by PC, the ES process can discharge the storage electricity very fast at a current of 0.1 μA, achieving an energy storage efficiency of 65.6% (Figure 2.9c). Owing to the coaxial structure, ensuring close contact of multiple layers, the "energy fibers" showed outstanding stability under various bending states. In addition, they could be woven with each other or with other fibers/yarns to form a flexible self-powering energy storage textile (Figure 2.9d). The combination of PC and ES devices in this study may open new avenues for a well-developed textile industry in consideration of their practical application in wearable electronics.

2.5 Conclusion and Outlook

In this chapter, a number of 1D flexible SCs have been introduced on the basis of their current collectors/substrates, namely, fibers and yarns, which are the main components in the device. According to their conductivities, these fibers and yarns can be divided into two broad categories: intrinsic conductive ones and intrinsic nonconductive ones, both of which are playing important roles in fabricating 1D flexible SCs owing to shape dominance. The intrinsic conductive fibers/yarns, can be directly used as current collectors/substrates for a variety of active materials, such as transition metal oxides and conductive polymers, without complex pretreatments. However, a narrow range of choices available in conductive fibers/yarns and the associated relatively high cost are two major obstacles that limit their applications in large-scale production of continuous fiber/yarn-based flexible SCs. While comparing with conductive ones, intrinsic nonconductive fibers/yarns, for example, cotton yarn and nylon fiber, are readily available in the market because of their massive application in traditional textile industry. A number of attempts have been made to fabricate high-performance 1D flexible SCs based on these low-cost nonconductive fibers/yarns as discussed, which intend to take full advantage of their great flexibility. Nevertheless, it is a big challenge to convert nonconductive fibers/yarns into highly conductive ones, especially considering that the existing modification processes are either too complex (e.g. ELD process) or require expensive equipment (e.g. ALD technique), hampering their large-scale production. Therefore, there are still a lot of opportunities as well as challenges in developing ideal fiber/yarn-based high-performance flexible SCs with practical methods for scaleing up production.

Great flexibility gives fiber/yarn-based SCs one big competitive advantage, which allows them to be compatible with traditional textiles via weaving/knitting with common fibers/yarns, which has become the main focus for the development of wearable electronics. By discussing plenty of distinctive 1D SC prototypes in the above sub-sections, it can be observed that many of them exhibit potential for weaving/knitting with each other and/or with other traditional fibers/yarns, while forming energy storage textiles that are flexible, deformable, wearable, and comfortable. However, it is challenging to continuously weave these fiber/yarn SCs into a textile with arbitrary large area, since most of them cannot be produced continuously with arbitrary length due to the complex fabrication technologies as aforementioned, let alone their possibilities to withstand the tensile stress generated by using commercial weaving/knitting machines. Thus, it still requires a lot of effort to make fiber/yarn-based SCs more compatible with well-developed textile technologies, in order to improve their practicability in wearable electronics.

By building integrated wearable systems based on fiber/yarn SCs and other fiber/yarn-based devices, promising multifunctional smart textiles, such as self-powering fabrics, can be achieved with extended application fields. Although combination of other functionalities and energy storage is a good concept and has been studied to a certain degree, there would be barriers of mismatches in additional functions and fiber/yarn-based SCs while designing a really harmony multifunctional system in the practical work. For example, the SCs can be fully charged within seconds by the solar cell owing to their low energy density. Hence, exquisite design of an integrated system should be made considering the compatibility between different functional parts instead of simply combining together.

In summary, fiber/yarn-based SCs are one of the promising energy storage devices that can be applied in a variety of applications according to the advantages they offer with regard to mechanical flexibility. It is expected that the integrated electronic textiles based on these SCs with enhanced performance will change our lives in the near future.

References

1 Simon, P. and Gogotsi, Y. (2008). Materials for electrochemical capacitors. *Nat. Mater.* 7: 845–854.

2 Long, J.W., Bélanger, D., Brousse, T. et al. (2011). Asymmetric electrochemical capacitors – stretching the limits of aqueous electrolytes. *MRS Bull.* 36: 513–522.

3 Simon, P., Gogotsi, Y., and Dunn, B. (2014). Where do batteries end and supercapacitors begin? *Science* 343: 1210–1211.

4 Stoppa, M. and Chiolerio, A. (2014). Wearable electronics and smart textiles: a critical review. *Sensors* 14: 11957.

5 Zeng, W., Shu, L., Li, Q. et al. (2014). Fiber-based wearable electronics: a review of materials, fabrication, devices, and applications. *Adv. Mater.* 26: 5310–5336.

6 Gulzar, U., Goriparti, S., Miele, E. et al. (2016). Next-generation textiles: from embedded supercapacitors to lithium ion batteries. *J. Mater. Chem. A* 4: 16771–16800.

7 Huang, Y., Zhu, M., Huang, Y. et al. (2016). A modularization approach for linear-shaped functional supercapacitors. *J. Mater. Chem. A* 4: 4580–4586.

8 Senthilkumar, S.T., Wang, Y., and Huang, H. (2015). Advances and prospects of fiber supercapacitors. *J. Mater. Chem. A* 3: 20863–20879.

9 Yu, D., Qian, Q., Wei, L. et al. (2015). Emergence of fiber supercapacitors. *Chem. Soc. Rev.* 44: 647–662.

10 Frackowiak, E. (2007). Carbon materials for supercapacitor application. *Phys. Chem. Chem. Phys.* 9: 1774–1785.

11 Zhang, L.L. and Zhao, X.S. (2009). Carbon-based materials as supercapacitor electrodes. *Chem. Soc. Rev.* 38: 2520–2531.

12 Wang, G., Zhang, L., and Zhang, J. (2012). A review of electrode materials for electrochemical supercapacitors. *Chem. Soc. Rev.* 41: 797–828.

13 Huang, Y., Hu, H., Huang, Y. et al. (2015). From industrially weavable and knittable highly conductive yarns to large wearable energy storage textiles. *ACS Nano* 9: 4766–4775.

14 Zhang, Y., Bai, W., Cheng, X. et al. (2014). Flexible and stretchable lithium–ion batteries and supercapacitors based on electrically conducting carbon nanotube fiber springs. *Angew. Chem. Int. Ed.* 53: 14564–14568.

15 Huang, Y., Zhu, M., Meng, W. et al. (2015). Robust reduced graphene oxide paper fabricated with a household non-stick frying pan: a large-area freestanding flexible substrate for supercapacitors. *RSC Adv.* 5: 33981–33989.

16 Zhao, Y., Liu, J., Hu, Y. et al. (2013). Highly compression-tolerant supercapacitor based on polypyrrole-mediated graphene foam electrodes. *Adv. Mater.* 25: 591–595.

17 Wang, G., Wang, H., Lu, X. et al. (2014). Solid-state supercapacitor based on activated carbon cloths exhibits excellent rate capability. *Adv. Mater.* 26: 2676–2682.

18 Yu, D., Zhai, S., Jiang, W. et al. (2015). Transforming pristine carbon fiber tows into high performance solid-state fiber supercapacitors. *Adv. Mater.* 27: 4895–4901.

19 Meng, Q., Wang, K., Guo, W. et al. (2014). Thread-like supercapacitors based on one-step spun nanocomposite yarns. *Small* 10: 3187–3193.

20 Kou, L., Huang, T., Zheng, B. et al. (2014). Coaxial wet-spun yarn supercapacitors for high-energy density and safe wearable electronics. *Nat. Commun.* 5: 3754.

21 Huang, Y., Zhong, M., Huang, Y. et al. (2015). A self-healable and highly stretchable supercapacitor based on a dual crosslinked polyelectrolyte. *Nat. Commun.* 6: 10310.

22 Dalton, A.B., Collins, S., Munoz, E. et al. (2003). Super-tough carbon-nanotube fibres. *Nature* 423: 703–703.

23 Zhang, M., Atkinson, K.R., and Baughman, R.H. (2004). Multifunctional carbon nanotube yarns by downsizing an ancient technology. *Science* 306: 1358–1361.

24 Zhang, X., Li, Q., Holesinger, T.G. et al. (2007). Ultrastrong, stiff, and lightweight carbon-nanotube fibers. *Adv. Mater.* 19: 4198–4201.

25 Yu, Z. and Thomas, J. (2014). Energy storing electrical cables: integrating energy storage and electrical conduction. *Adv. Mater.* 26: 4279–4285.

26 Huang, Y., Zhu, M., Pei, Z. et al. (2016). A shape memory supercapacitor and its application in smart energy storage textiles. *J. Mater. Chem.. A* 4: 1290–1297.

27 Wang, X., Liu, B., Liu, R. et al. (2014). Fiber-based flexible all-solid-state asymmetric supercapacitors for integrated photodetecting system. *Angew. Chem. Int. Ed.* 53: 1849–1853.

28 Zhang, D., Wu, Y., Li, T. et al. (2015). High performance carbon nanotube yarn supercapacitors with a surface-oxidized copper current collector. *ACS Appl. Mater. Interfaces* 7: 25835–25842.

29 Chitara, B., Panchakarla, L.S., Krupanidhi, S.B., and Rao, C.N.R. (2011). Infrared photodetectors based on reduced graphene oxide and graphene nanoribbons. *Adv. Mater.* 23: 5419–5424.

30 Daubert, J.S., Mundy, J.Z., and Parsons, G.N. (2016). Kevlar-based supercapacitor fibers with conformal pseudocapacitive metal oxide and metal formed by ALD. *Adv. Mater. Interfaces* 3: doi: 10.1002/admi.201600355.

31 Wang, Z., Huang, Y., Sun, J. et al. (2016). Polyurethane/cotton/carbon nanotubes core-spun yarn as high reliability stretchable strain sensor for human motion detection. *ACS Appl. Mater. Interfaces* 8: 24837–24843.

32 Fu, Y., Cai, X., Wu, H. et al. (2012). Fiber supercapacitors utilizing pen ink for flexible/wearable energy storage. *Adv. Mater.* 24: 5713–5718.

33 Sun, J., Huang, Y., Fu, C. et al. (2016). High-performance stretchable yarn supercapacitor based on PPy@CNTs@urethane elastic fiber core spun yarn. *Nano Energy* 27: 230–237.

34 Zhou, Q., Jia, C., Ye, X. et al. (2016). A knittable fiber-shaped supercapacitor based on natural cotton thread for wearable electronics. *J. Power Sources* 327: 365–373.

35 Tan, Y.B., and Lee, J.-M. (2013). Graphene for supercapacitor applications. *J. Mater. Chem. A* 1, 14814–14843.

36 Liu, L., Yu, Y., Yan, C. et al. (2015). Wearable energy-dense and power-dense supercapacitor yarns enabled by scalable graphene–metallic textile composite electrodes. *Nat. Commun.* 6: 7260.

37 Yu, Y., Zeng, J., Chen, C. et al. (2014). Three-dimensional compressible and stretchable conductive composites. *Adv. Mater.* 26: 810–815.

38 Meng, Y., Zhao, Y., Hu, C. et al. (2013). All-graphene core-sheath microfibers for all-solid-state, stretchable fibriform supercapacitors and wearable electronic textiles. *Adv. Mater.* 25: 2326–2331.

39 Napari, M., Malm, J., Lehto, R. et al. (2015). Nucleation and growth of ZnO on PMMA by low-temperature atomic layer deposition. *J. Vac. Sci. Technol. A* 33: 01A128.

40 Pu, X., Li, L., Liu, M. et al. (2016). Wearable self-charging power textile based on flexible yarn supercapacitors and fabric nanogenerators. *Adv. Mater.* 28: 98–105.

41 Zhang, Z., Chen, X., Chen, P. et al. (2014). Integrated polymer solar cell and electrochemical supercapacitor in a flexible and stable fiber format. *Adv. Mater.* 26: 466–470.

3

Flexible Lithium Ion Batteries

Xuli Chen and Yingying Ma

Hunan University, Department of Materials Science, College of Materials Science and Engineering, 2 South Lushan Road, Changsha 410082, China

3.1 Overview of Lithium Ion Battery

Batteries, which store energy in the form of chemical energy and convert the stored chemical energy into electrical energy to power other devices, have been known for hundreds of years. Rechargeable lithium ion battery, due to its ability to sustain repeated charge and discharge, has been widely used as an environmental-friendly and resource-saving power source, especially in portable devices, such as computers, cell phones, and vehicles [1]. A lithium ion battery contains a cathode (positive electrode), an anode (negative electrode), and an ionic conductive electrolyte. It powers electronic devices by providing voltage that is determined by the potential difference between the cathode and anode. Besides voltage, capacity, energy density, power density, and cyclic performance are the most critical indexes to evaluate the performance of a power source. Compared with other rechargeable batteries currently available in the market, lithium ion battery has higher voltage (about 3.7 V) and longer cyclic life (more than 1000 cycles) [2]. Figure 3.1 further compares the energy density and power density of various rechargeable batteries, and shows that lithium ion battery exhibits superiority with higher energy density, power density, and higher safety [3].

3.1.1 General Principle

In 1990, Sony claimed a novel rechargeable battery with high performance. In the battery, $LiCoO_2$ and a tailor-made carbonaceous material were used as the cathode and anode active material, respectively. During the repeated charge–discharge processes, lithium ions are inserted and/or extracted from electrodes and migrate in the electrolyte, and thus, the battery stores and delivers energy. Soon, the lithium ion battery was successfully commercialized worldwide because of its high energy density, outstanding cyclic performance, and no memory effect, properties that seriously impede the development of its

Flexible Energy Conversion and Storage Devices, First Edition. Edited by Chunyi Zhi and Liming Dai.

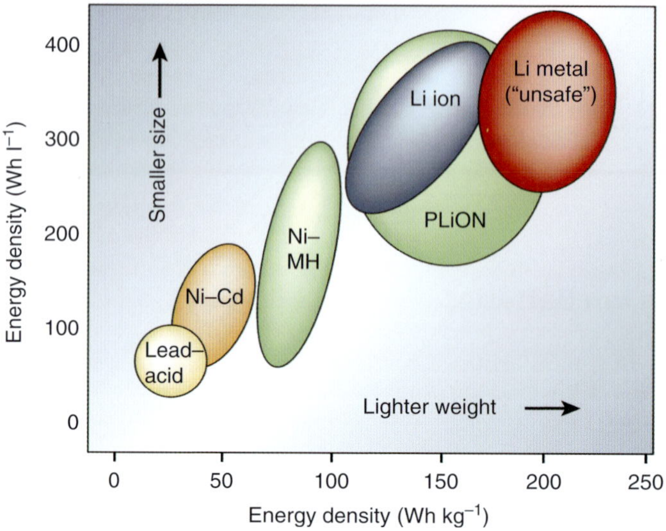

Figure 3.1 Specific energy densities of the different rechargeable batteries. *Source:* Tarascon and Armand 2001 [3]. Copyright 2001. Reprinted with permission from Nature Publishing Group.

rivals, lead–acid, nickel–cadmium (Ni–Cd), and nickel–metal hydride (Ni–MH) batteries.

At the beginning, lithium metal was used as the anode because of its high negative standard redox potential (−3.04 V vs normal hydrogen electrode (NHE)), high electrochemical capacity (3860 mAh g^{-1}), and low density (0.53 g cm^{-3}) [4]. However, it was soon found that parasitic reactions that occurred on the lithium surface heavily limited its cyclic longevity. Moreover, lithium dendrite formed on the lithium metal causes serious safety risk in the application of the battery. In view of these negative effects, graphitic carbon was researched to replace lithium metal and act as the anode. Graphitic carbon reserves lithium during the charge–discharge process by forming an intercalation compound (LiC_6). The voltage profile exhibits a stable plateau at ~0.2 V vs Li/Li^+ during lithiated and delithiated processes. The theoretical lithium storage capacity is calculated to be 372 mAh g^{-1} by the lithiated compound, LiC_6 [5].

Cathode materials are generally layered compounds, such as metal sulfides and $LiCoO_2$ in which lithium ions intercalated between the layers. For instance, titanium disulfide (TiS_2) has layered structures and can be lithiated into Li_xTiS_2, and can thus be used as energy storage materials. TiS_2 possesses remarkable reversibility because when TiS_2 is transferred into Li_xTiS_2, there is no phase transition, and therefore no rearrangement and reconstruction of lattice. $LiCoO_2$ has a similar layered structure but involves several phases during its lithiation process. As a result, because only half of the lithium in the lithiated compound can be exploited to deliver energy, the capacity of $LiCoO_2$ is only 140 mAh g^{-1}, which is comparatively low [5]. Besides TiS_2 and $LiCoO_2$, various cathodes have been investigated, such as $LiNiO_2$, $LiMn_2O_4$, $LiFePO_4$, and so on [6].

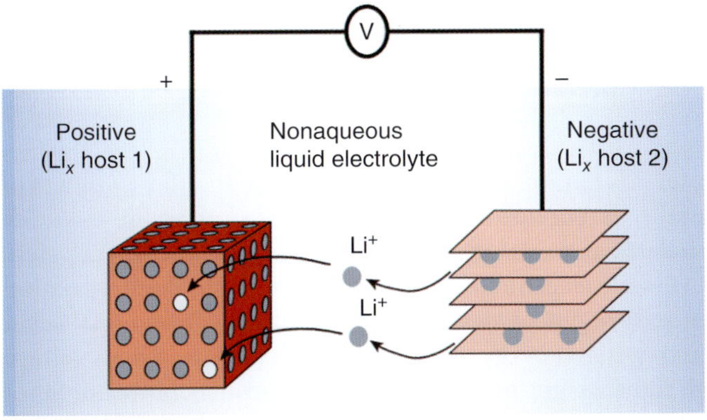

Figure 3.2 Schematic representation and operating principles of Li ion batteries during discharge process. *Source:* Tarascon and Armand 2001 [3]. Copyright 2001. Reprinted with permission from Nature Publishing Group.

In lithium ion batteries, energy delivery and storage are carried out by the migration of lithium ions between the cathode and anode during the charge–discharge process. Figure 3.2 schematically illustrates the migration of lithium ions during the discharge process and charge process [3]. For example, in the $LiCoO_2$-graphite system, the lithium ions are inserted into and extracted from the cathode and anode. The reaction in the cathode is expressed as:

$$LiCoO_2 \rightleftarrows \frac{1}{2}Li^+ + \frac{1}{2}e^- + Li_{0.5}CoO_2 \tag{3.1}$$

For $LiCoO_2$, the upper voltage during the charge process is limited to 4.2 V vs Li/Li^+, which means only half of the lithium stored in $LiCoO_2$ can be extracted [7]. If it is overcharged, electrolyte decomposition will result in parasitic reactions to prevent further delithiation. When the battery is fabricated, it is under a discharged state and the battery's work needs to start from a charge process in which the lithium is extracted from the cathode and intercalated into the anode.

For the anode, the lithiation and delithiation of graphite can be expressed as:

$$C_6 + Li^+ + e^- \rightleftharpoons LiC_6 \tag{3.2}$$

With intercalation of lithium ions, graphite experiences several stages of phase transition spanning from LiC_{24}, LiC_{18}, and LiC_{12} to the final compound LiC_6 [8]. In the first charge cycle, decomposition of electrolyte occurs and thus a passivation film is formed on the surface of the anode. On one hand, this phenomenon causes initial irreversible capacity, and on the other hand, the passivation film also protects the graphite from deterioration. Considering these effects, it is necessary to keep an excess storage of lithium in the cathode, in order to satisfy the consumption of lithium for the forming of the passivation films on the anode.

For the electrolyte used in a lithium ion battery, aqueous solutions are excluded due to the high reactivity of water with lithium metal, the low decomposition potential (1.23 V vs NHE), and also the narrow temperature window for liquid state (0–100 °C). Electrolytes of lithium salts in aprotic solvents with wide electrochemical window and good compatibility are preferred in lithium ion batteries. We will detail our discussions on cathode, anode, and electrolyte in the following parts.

3.1.2 Cathode

As a stable power source, lithium ion battery should be able to offer a voltage of more than 3.5 V. Generally, carbonaceous materials are used as anode in which lithium ion reversibly inserts and extracts at a potential of ~0 V vs Li/Li^+ electrode. Therefore, the reversible electrochemical intercalation and deintercalation of lithium ion in the cathode should proceed at a potential of 4 V vs Li/Li^+ electrode. As a critical property, high energy density, which depends on high voltage and capacity along with light weight of the battery, is vigorously pursued. Since the energy stored in the battery is determined by its output voltage and capacity, and also considering factors of density, safety, cost, extensive research has been conducted on cathode materials, including $LiCoO_2$, $LiMn_2O_4$, and $LiFePO_4$, which is a kind of recently discovered polyanionic compounds with multiple advantages, as potential candidates that can play important roles in the future.

3.1.2.1 $LiCoO_2$ with Layered Structure

$LiCoO_2$, the earliest and most commonly used cathode material, has a theoretical capacity of 274 mAh g^{-1} provided that the lithium is all extracted. But under safe voltage of 4.2 V, only half of the lithium stored in $LiCoO_2$ can be deintercalated. And thus, the reversible capacity is generally restricted to around 140 mAh g^{-1}. To circumvent this issue, strategies have been designed to partially replace cobalt in this phase with other similar metals, such as nickel or manganese, so as to protect the structure from decomposition at higher potential. The product with stoichiometric formula of $LiCo_{1/3}Ni_{1/3}Mn_{1/3}O_2$ can exhibit a reversible capacity of 158 mAh g^{-1} [9]. The capacity is then further increased by synthesizing lithium-rich layered oxides, such as $x Li_2MnO_3 \cdot (1-x) LiMn_{0.5}Ni_{0.5}O_2$ composites with a capacity of 250 mAh g^{-1} [10].

3.1.2.2 $LiMn_2O_4$ with a Spinel Structure

Well known as an economical material, $LiMn_2O_4$ possesses a spinel structure and is much more robust against the removal of lithium. Hence, 95% of the theoretical capacity is capable of being reversibly carried out. The combination of a chemically stable Mn^{+3}/Mn^{+4} couple and the spinel 3D lattice ensures that the battery's cathode material has good safety and high power capability. The spinel structure of $LiMn_2O_4$ delivers a unique 3D network to allow the Li ions diffuse multidirectionally and thus a capacity of around 80 mAh g^{-1} is realized at high drain rates [1, 11]. However, this material also suffers stability problems due to its self-discharge when used in conventional liquid electrolytes wherein the produced trace amounts of hydrogen fluoride will corrode the electrode, especially at elevated temperatures.

3.1.2.3 $LiFePO_4$ with an Olivine Structure

$LiFePO_4$ with an olivine structure possesses lots of merits, such as high capacity, chemical stability, thermal stability, no toxicity, environmental friendliness, and low cost. With acceptable voltage plateau of around 3.45 V (vs Li/Li^+) and theoretical capacity of 170 mAh g^{-1}, the main disadvantages of $LiFePO_4$ is the insulating nature (10^{-9} S cm^{-1}) that leads to slow lithium ion diffusion ($<10^{-14}$ cm^2 s^{-1}), high impedance, and inferior rate capability [12]. Therefore, various efforts have been focused on decreasing the particle size and increasing electrical conductivity so as to improve the electrochemical performance of $LiFePO_4$ cathode. It has been proved that the electrochemical performances of $LiFePO_4$ can be effectively improved by structural or morphological modifications, including coating with conducting additives, mechano-chemical activation, and supervalent metal ion doping [13, 14].

3.1.3 Anode

Lithium metal was previously used as the anode of lithium ion battery. It was soon abandoned because it could produce lithium dendrite that can pierce the separator and then cause internal short circuit and hazardous explosion. Regarding the safety problem of using lithium metal as anode, finding a safe anode substitute becomes a vital issue. In 1971, Dey demonstrated that it is possible for alloys of lithium with aluminum, silicon, zinc, etc. to be electrochemically formed in organic electrolytes which could help in the investigation of these alloys as alternative anode materials [15]. However, the cyclic stabilities of lithium alloys are quite poor due to the large volume change which brings internal stress inside the alloy crystal lattice, causing cracking and pulverization of alloy anodes. On the other hand, Sony Corporation discovered that the carbonaceous material with low voltage, excellent cyclic stability can be used as anode, and realized full battery with the C-$LiCoO_2$ rocking chair in 1991. Replacing the lithium metal with lithiated carbonaceous anode prevents the growth of irregular dendritic lithium and ensures that the battery has better cyclic longevity and safety stability.

3.1.3.1 Carbonaceous Anodes

Carbonaceous layered materials, with higher reversibility and low voltage Li intercalation–deintercalation process, were firstly used as anode materials while Sony creatively researched and commercialized the C-$LiCoO_2$ rocking-chair cell in 1991 [3]. Graphite is the most commonly used carbonaceous layered material as ion host in rechargeable batteries. Graphite intercalation compounds have several stages ranging from LiC_{24}, LiC_{18}, and LiC_{12} to the final compound LiC_6 which produces the theoretical specific capacity of 372 mAh g^{-1} for graphite under normal pressure [7, 16].

3.1.3.2 Metal Alloy Anodes

In the definition of $C = Q/U$, the charges (Q) of the anode (Q_A), the cathode (Q_C), and the battery ($Q_{battery}$) are equal ($Q_A = Q_C = Q_{battery}$), and the voltage of the battery ($U_{battery}$) is the sum of the anode (U_A) and cathode (U_C). Therefore, there is a simple relationship between the capacities of the anode (C_A), the cathode (C_C), and the battery ($C_{battery}$, based on all active materials of the battery):

$$\frac{1}{C_{battery}} = \frac{1}{C_A} + \frac{1}{C_C} \tag{3.3}$$

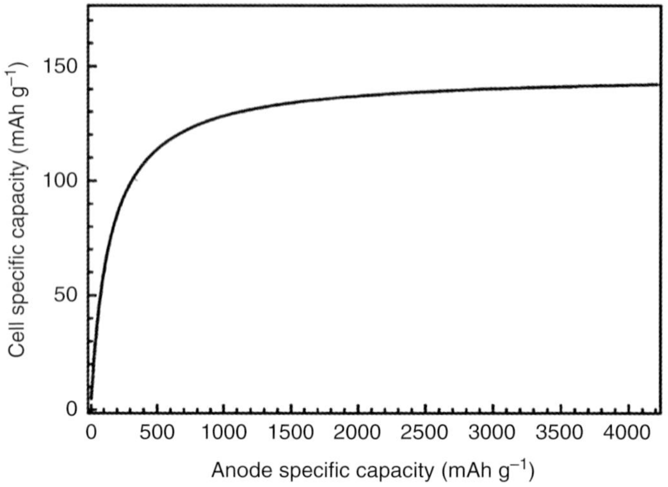

Figure 3.3 The dependence of the total capacity of a lithium ion battery ($C_{battery}$) on the anode capacity (C_A). In the battery, $LiCoO_2$ with the theoretical capacity of 147 mAh g^{-1} is taken as cathode. *Source:* Szczech and Jin 2011 [17]. Copyright 2011. Reprinted with permission from Royal Society of Chemistry.

As a result, $C_{battery}$ can be expressed as

$$C_{battery} = \frac{C_C}{1 + \frac{C_C}{C_A}} = \frac{C_A C_C}{C_A + C_C}\ (\text{mAh g}^{-1}) \tag{3.4}$$

Equation (3.4) reveals that higher values of C_A and C_C are always beneficial to get higher $C_{battery}$. Generally, C_C ranges between 140 and 200 mAh g^{-1}, and thus higher $C_{battery}$ largely depends on the improvement of C_A. For a given C_C, the dependence of $C_{battery}$ on C_A can be shown as in Figure 3.3 (using $LiCoO_2$ as the cathode). Obviously, $C_{battery}$ can be greatly improved by increasing C_A to the order of 1000 mAh g^{-1} [17].

In a lithium ion battery, the anode is considered to possess a potential vs Li/Li^+ close to 0 V. For carbonaceous layered materials that store lithium ions through the lithium intercalation process, the capacity is restricted by their crystalline structure. However, metals can store lithium ions by forming an alloy (Li_xM), and thus can exhibit higher capacities. Being polarized to a sufficiently negative potential in the electrolyte, numerous metals, such as Si, Al, Sn, Sb, Ge, Pb, and Ag have been researched and proved to be able to alloy electrochemically with lithium. These alloys possess high theoretical capacities with the formation and decomposition of Li_xM. However, most of these high-capacity alloy anodes often suffer short cycle life due to the large volume expansion and the electrolyte depletion [18].

3.1.4 Electrolyte

In lithium ion battery, electrolytes play the role of an ionic conductor to deliver lithium ions back and forth between cathodes and anodes during charge and discharge processes. Generally, liquid electrolytes are employed in a lithium ion battery because the electrodes possess porous structure and the electrolyte in liquid state can infiltrate into the electrodes better. Moreover, liquid electrolyte also facilitates the smooth transfer of lithium ions at the interface between electrolyte and solid electrodes comprising active materials, conducting additives, and polymer binders [19]. Most of the liquid electrolytes used for lithium ion batteries are nonaqueous solutions because the commonly used lithium ionic conductor, $LiPF_6$ salt, is very easy to be hydrolyzed in aqueous solution. Therefore, $LiPF_6$ salt is often dissolved in the mixture of carbonate solvents selected from cyclic carbonates–ethylene carbonate (EC), and propylene carbonate (PC), and linear carbonates–dimethyl carbonate (DMC), ethyl methyl carbonate (EMC), and diethyl carbonate (DEC) at the density of about 1 $mol\,l^{-1}$. Besides liquid electrolyte, polymer electrolyte was launched in 1973 driven by its ability to prevent internal short circuiting and electrolyte leakage. Furthermore, polymer electrolytes are more adaptable to the environment and noncombustible reaction products make the battery safer than a liquid electrolyte system.

3.2 Planar-Shaped Flexible Lithium Ion Batteries

To keep up with the development of soft portable electronics, such as rollup displays and wearable devices, it is essential to develop flexible lithium ion batteries, including planar-shaped flexible lithium ion battery and fiber-shaped lithium ion battery. In this section, the development and the advancement of flexible lithium ion batteries in planar shape will be first presented.

3.2.1 Bendable Planar Lithium Ion Batteries

The flexibility of the lithium ion battery is often determined by the electrode's flexibility. Generally, flexible substrates are employed to support the electrodes so as to endow the lithium ion batteries with flexibility and/or stretchability. According to the substrates, unstretchable planar flexible lithium ion battery can be divided into carbon-based lithium ion battery, thin metal material-based lithium ion battery, polymer lithium ion battery, and lithium ion battery with special structural design.

3.2.1.1 Bendable Carbon-Based Planar Lithium Ion Battery

Paper electronics has promoted the development of paper-based lithium ion battery. With a rough surface and porous structure, paper is convenient for ion transportation. Plus, with domestic flexibility, paper is suitable for developing bendable lithium ion battery electrode. However, conventional paper is insulating for electrons. To solve this crucial problem, Cui and coworkers have endowed paper with excellent conductivity by coating their surface with single-walled carbon nanotubes (SWCNTs) [20]. Coating conductive nanomaterials, such as carbon nanotubes (CNTs), graphene, and metal nanowire, onto a flexible substrate which is traditionally insulating, has been proved to be an effective way to fabricate flexible electrodes.

CNTs-Coated Paper-Based Bendable Half Lithium Ion Battery CNTs are ideal flexible electrode material on account of their excellent electrical conductivity, huge specific surface area, and outstanding mechanical properties [21]. Flexible lithium ion batteries based on CNTs or CNT composites have been widely researched. Ajayan and coworkers prepared CNTs/cellulose composite electrode by paving aligned CNTs on cellulose paper [22]. The resulting composite electrode can be recoverably bent to any degree, twisted, and rolled up and the properties can be well maintained during these deformations. Taking the CNTs/cellulose composite paper and a thin evaporated lithium layer as cathode and anode respectively, a flexible half battery was assembled and the capacity was observed to reach 110 mAh g^{-1} which can be repeated for several tens of cycles; therefore, a flexible cathode was successively fabricated though it was just used in half battery.

CNTs Film-Based Bendable Full Lithium Ion Battery To realize flexible full battery, it is imperative that both cathodes and anodes are all flexible. Introducing conductive paper without any rigid metal, as current collectors for both cathode and anode turned out to be an effective method. Generally, metal-free conductive papers are based on carbon materials, such as CNTs, graphene, and carbon fiber. Cui and coworkers developed a highly conductive paper using CNT film [23]. CNT films were formed on stainless steel substrates through a solution-based process. CNTs were dispersed in water using sodium dodecylbenzenesulfonate as surfactant. The resulting ink was paved onto stainless steel substrates through the doctor blade method and then dried to CNTs film with a thickness of ~2 μm. $Li_4Ti_5O_{12}$ (LTO) and $LiCoO_2$ (LCO) were used as anode and cathode active materials, respectively. These active materials were separately mixed with super P carbon and polyvinylidene fluoride (PVDF) binder. The resulting LTO and LCO slurries were applied onto the as-prepared CNTs films. As shown in Figure 3.4a, dried anodic LTO/CNTs and cathodic LCO/CNTs hybrid electrode films are peeled off from stainless steel substrates by immersing the substrates in deionized (DI) water because the interaction between CNTs and stainless steel is pretty weak. It can be seen in Figure 3.4b that the hybrid electrode films were totally flexible. In both anode and cathode, CNTs functioned as mechanical substrates and current collectors simultaneously. Figure 3.4c schematically illustrated that the peeled LTO/CNTs and LCO/CNTs films were laminated onto a paper with paper sandwiched between them. As shown in Figure 3.4d, the assembled battery was pretty flexible and no obvious degradation of properties during deformation was observed. Moreover, the assembled flexible lithium ion battery can power a light-emitting diode (LED) as shown in Figure 3.4e. The first-cycle charge–discharge profile of this paper lithium ion battery with a voltage platform of 2.0–2.2 V is given in Figure 3.4f. Using paper as both mechanical support and separator, the paper lithium ion battery exhibits excellent self-discharge performance of only 0.0054 V voltage drop after resting for 350 h (as shown in Figure 3.4g). Moreover, the insert in Figure 3.4g indicates the cyclic performance of the paper flexible lithium ion battery with 93% retention after 20 cycles and coulombic efficiency of 94–97% except the first cycle. All of these excellent mechanical and energy storage performances render this paper flexible lithium ion battery as a power source for flexible electronic devices.

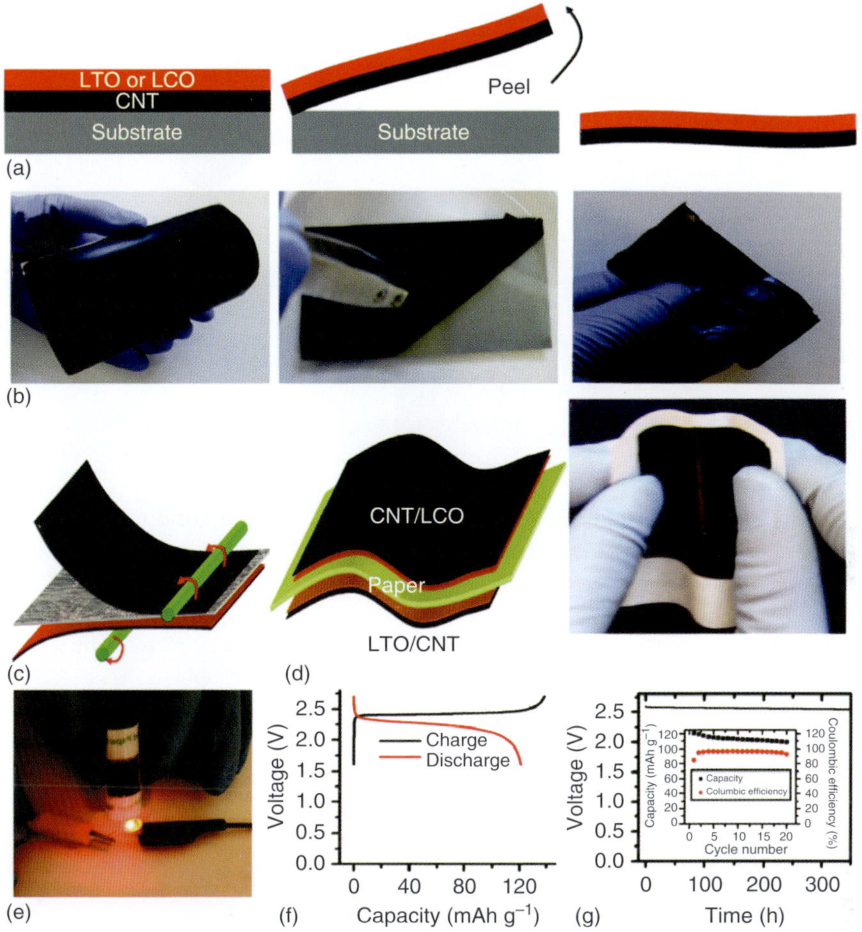

Figure 3.4 (a) Schematic illustration of fabricating process for LTO/CNTs or LCO/CNTs hybrid films. (b) (left) LTO/CNTs film coated on stainless steel substrate; (middle) the double layer film being separated from the stainless steel substrate in DI water; and (right) the final dried flexible hybrid film. (c) Schematic illustration of the lamination process. (d) Schematic structure of the final paper Li-ion battery device and the picture of the un-encapsulated Li-ion paper battery. (e) Picture of the paper battery lighting a red LED. (f) Galvanostatic charging/discharging profile of the assembled paper battery in the first cycle. (g) Self-discharge performance of the LTO–LCO full battery being charged to 2.6 V. The inset is the cyclic performance of the paper full battery. *Source:* Hu et al. 2010 [23]. Copyright 2010. Reprinted with permission from American Chemical Society.

Carbon Cloth-Based Bendable Lithium Ion Battery Unlike traditional insulating papers, carbon paper and carbon cloth can be directly used as current collectors. Liu et al. developed three-dimensional (3D) $ZnCo_2O_4$ nanowire arrays/carbon cloth possessing excellent energy storage performance [24]. As illustrated in Figure 3.5a, 3D $ZnCo_2O_4$ nanowire array was grown on the carbon cloth through a facile hydrothermal route. The hierarchical structure of the as-prepared $ZnCo_2O_4$ nanowire arrays/carbon cloth obtained by scanning

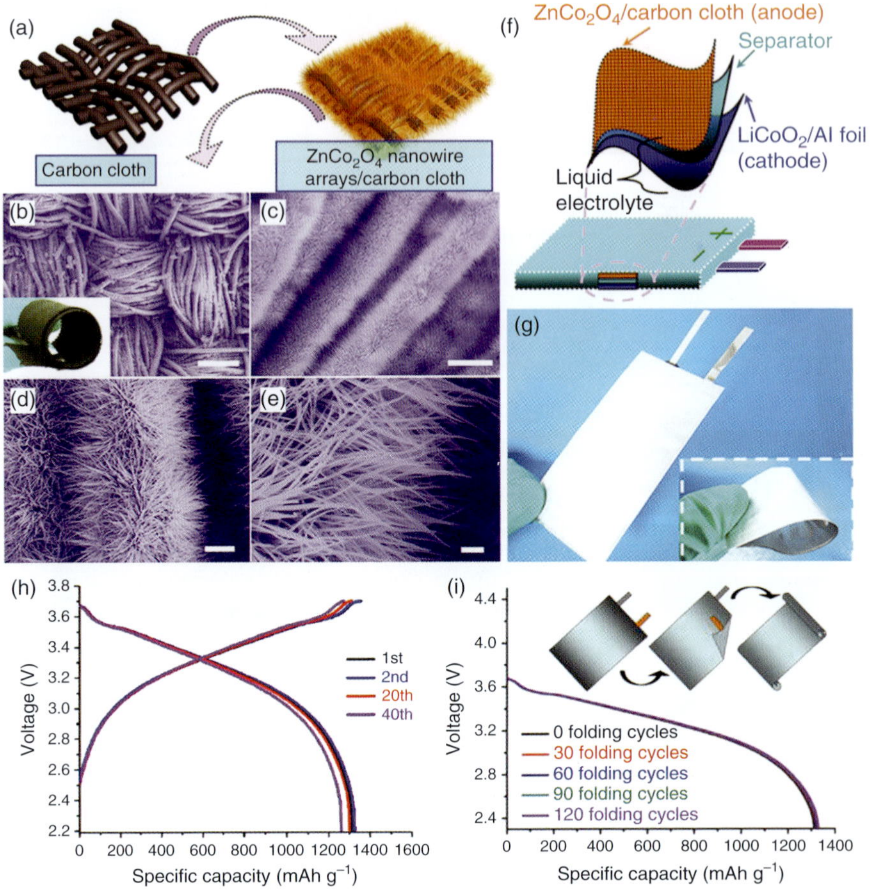

Figure 3.5 (a) Schematic illustration of the synthesis process of 3D $ZnCo_2O_4$ nanowire arrays/carbon cloth. (b–e) SEM images of the $ZnCo_2O_4$ nanowire arrays grown on carbon cloth at different magnifications. Scale bars in (b–e) are 200, 20, 5, and 1 μm, respectively. The inset in (b) is the digital image of the produced flexible electrode at rolled state. (f) Schematic illustration for the structure of the flexible full lithium ion battery based on 3D $ZnCo_2O_4$nanowire arrays/carbon cloth, $LiCoO_2$ and liquid electrolyte. (g) Photographic images of the assembled flexible full lithium ion battery. (h) Charge–discharge profiles of the 1st, 2nd, 20th, and 40th cycles of the flexible full battery. (i) The voltage versus specific capacity profiles of the flexible full battery before and after 30, 60, 90, and 120 cycles of bending. *Source:* Liu et al. 2012 [24]. Copyright 2012. Reprinted with permission from American Chemical Society.

electron microscope (SEM) images is depicted in Figure 3.5b–e. It can be obviously seen that $ZnCo_2O_4$ nanowire arrays grown on the carbon fiber cloth are well vertically aligned, which endows it with a large surface area. The uniform $ZnCo_2O_4$ nanowire arrays/carbon fiber with a diameter of 20 μm as well as the $ZnCo_2O_4$ nanowires with uniform diameter and length of 80–100 nm and 5 μm of the homogeneous structure make the integrated $ZnCo_2O_4$ nanowire arrays/carbon cloth electrode material convenient for the access of electrolyte. It should be noted that the electrode can still keep excellent flexibility and can be recoverably rolled up as shown in the insert of Figure 3.5b. By using the produced $ZnCo_2O_4$ nanowire arrays/carbon cloth as the anode and $LiCoO_2$/Al foil as the cathode, full lithium ion battery was assembled as shown in Figure 3.5f–g. The resulting battery exhibited stable capacity of 1314 $mAh\,g^{-1}$ which can be maintained with 96% after 40 cycles as shown in Figure 3.5h. Moreover, the full battery is also flexible and no obvious performance degradation was observed during bending for hundreds of cycles as shown in Figure 3.5i. And this work also demonstrated that the flexible full battery based on carbon cloth current collector can easily light an LED even at its bent state. Besides $ZnCo_2O_4$, carbon cloth is also integrated with SnO_2, $SrGe_4O_9$, $BaGe_4O_9$, Zn_2GeO_4, and Si nanowires through simple hydrothermal and spray-coating processes to fabricate flexible full lithium ion batteries [25–27].

Graphene Foam-Based Bendable Lithium Ion Battery 3D graphene foam has been extensively researched in the energy storage field due to its high conductivity, large specific surface area, and light-weight property [28, 29]. Li et al. demonstrated that the graphene foam synthesized through the chemical vapor deposition method possesses electrical conductivity as high as ~1000 $S\,m^{-1}$ and light weight with low area density of ~0.1 $mg\,cm^{-2}$ at a thickness of ~100 μm. The 3D interconnected network structures with high porosity of ~99.7% provide the graphene foam with not only a large specific surface area but also fast transport channel for the charge carriers during the charge–discharge process. Moreover, due to the high quality and carbonaceous nature of graphene building blocks, the graphene foam networks are pretty stable during electrochemical characterization as well as bending to arbitrary shapes process. All of these features render the graphene foam an ideal candidate for flexible lithium ion battery electrode [30]. Using the graphene foam as both a highly conductive and a 3D interconnected current collector, which provides fast pathway for both electrons and ions, they reported a thin, lightweight, and flexible full lithium ion battery by loading $Li_4Ti_5O_{12}$ and $LiFePO_4$ on graphene foam for use as the anode and cathode, respectively (as shown in Figure 3.6a). Owing to the high electrical conductivity and polyporous structure of the $Li_4Ti_5O_{12}$/graphene foam (as shown in the SEM and transmission electron microscope (TEM) images of Figure 3.6b–d) and $LiFePO_4$/graphene foam hybrid electrodes, the transportation of electrons and ions are highly facilitated. Therefore, the fabricated battery can be charged and discharged at high rates of up to 200 C, which means the battery can be fully discharged in 18 s. Moreover, the flexible full battery can power a red LED even under bent state as shown in Figure 3.6e. The almost indistinguishable charge–discharge profiles of as-fabricated flat battery and bent battery after bending with a radius of <5 mm for 20 times as shown in Figure 3.6f also proved the excellent flexibility of the full battery. Figure 3.6g shows the cyclic stability under both flat and bent states.

3.2.1.2 Thin Metal Material-Based Lithium Ion Battery

Considering the fact that commonly used carbon substrates, such as carbon cloths and carbon nanomaterials coated traditional papers, are much heavier than thin copper or aluminum foils with the same area, the specific energy

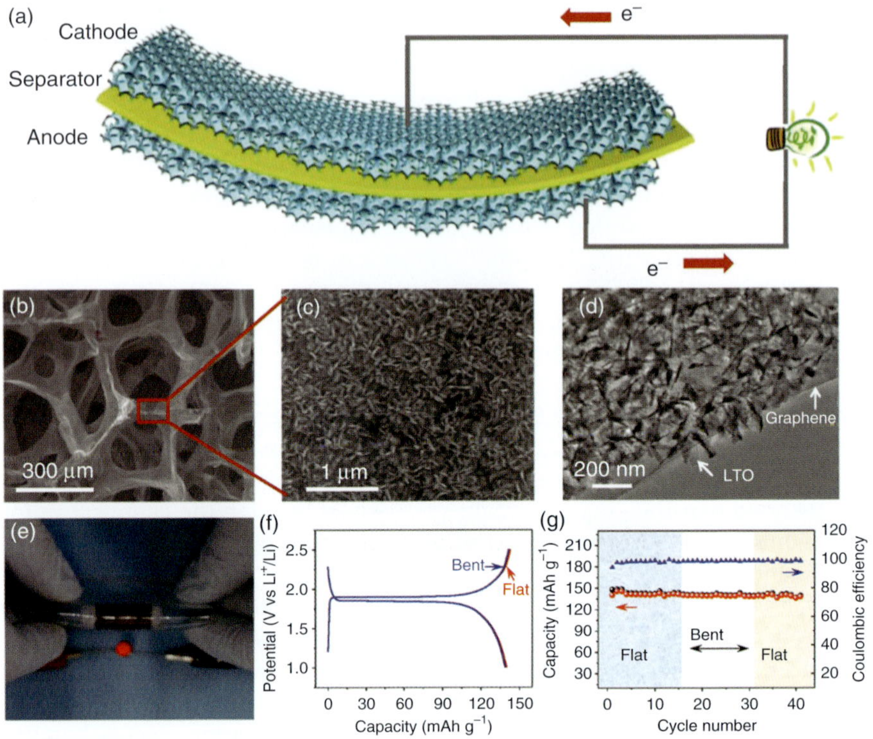

Figure 3.6 Schematic and characterization of thin, lightweight, and flexible full battery based on $Li_4Ti_5O_{12}$/graphene foam and $LiFePO_4$/graphene foam hybrid electrodes. (a) Schematic illustration of the full battery containing a cathode and an anode. (b, c) SEM images of $Li_4Ti_5O_{12}$/graphene foam at low and high magnifications. (d) TEM image of $Li_4Ti_5O_{12}$ sheet on graphene foam. (e) Photograph of the poly(dimethylsiloxane) (PDMS) encapsulated flexible full battery lighting a red LED device under bending state. (f) Galvanostatic charge/discharge profiles of the as-fabricated flat flexible full battery (red lines) and bent battery after repeatedly bending to a radius of 5 mm 20 times (blue lines), respectively. (g) Cyclic performance of the battery under flat and bent states. *Source:* Li et al. 2012 [30]. Copyright 2012. Reprinted with permission from National Academy of Sciences.

density may be sufficiently decreased by the heavy current collectors. To solve this problem, using ultrathin titanium foil as current collector, Wang et al. fabricated 3D web-like $Li_4Ti_5O_{12}$ anode which was assembled from numerous 1D nanowires (as shown in Figure 3.7) [31]. Based on this web-like $Li_4Ti_5O_{12}$ anode and $LiMn_2O_4$ cathode, a flexible full battery was assembled. The combination of $Li_4Ti_5O_{12}$ anode and $LiMn_2O_4$ cathode resulted in an operating voltage of about 2.4 V which is stable during charge–discharge cycling. Moreover, the flexible full battery exhibited both high capacities of 125 mAh g^{-1} at high current rate of 20 C and excellent flexibility with little performance degradation even at seriously bent states.

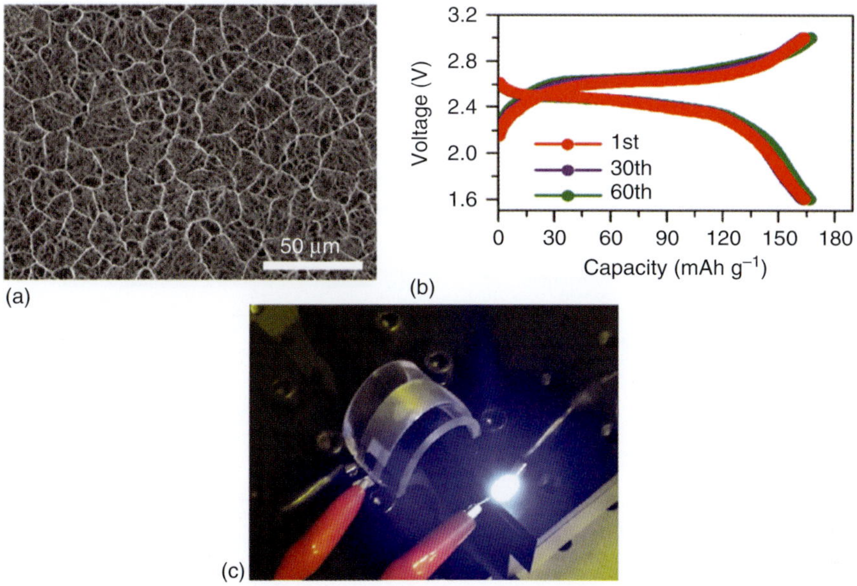

Figure 3.7 (a) SEM image of web-like $Li_4Ti_5O_{12}$ anode. (b) Charge–discharge profiles at 2 C of the full battery. (c) Photograph of full battery at bent state lighting an LED. *Source:* Wang et al. 2014. [31]. Copyright 2014. Reprinted with permission from Tsinghua University Press and Springer-Verlag Berlin Heidelberg.

3.2.1.3 Polymer-Based Lithium Ion Battery

Generally, liquid electrolytes are mostly used in lithium ion battery to transport ions. Even in flexible lithium ion batteries, electrolytes commonly used are still in liquid state which calls for proper packaging techniques that can protect liquid electrolytes and simultaneously maintain the flexibility. What is more important is that the unfavorable thermal stability of liquid electrolyte seriously limits the development of flexible lithium ion battery with respect to safety problems [32–35]. Therefore, deformable solid electrolytes have been designed and synthesized to realize fully flexible lithium ion batteries with not only high energy storage properties but also excellent thermal and mechanical stabilities. In the past several years, gel polymer electrolyte (GPE), plastic crystal electrolyte (PCE), and PCE/polymer matrix-based gel electrolyte (PCPE) have been widely explored and substantial progresses have been achieved, though the flexibility and mechanical stability are still required to be further improved [34–37].

GPE is a kind of widely used solid-state electrolyte. Being composed of polymer matrix and liquid electrolyte, GPE possesses not only excellent ionic conductivity but also good mechanical flexibility and low rates of safety failure. Kil et al. prepared a kind of imprintable, flexible, and shape-conformable composite gel polymer electrolyte (called c-GPE) composed of a UV-curved ethoxylated trimethylolpropane triacrylate (ETPTA) polymer matrix, high boiling point liquid electrolyte (1 M $LiPF_6$ in EC/PC = 1/1 (v/v)) and Al_2O_3 nanoparticles (as

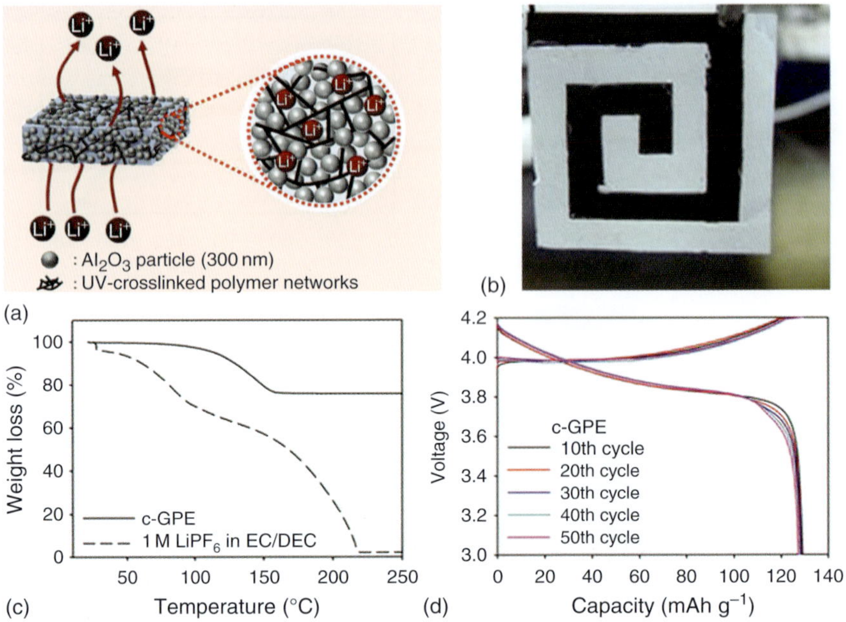

Figure 3.8 (a) Schematic illustration of the produced c-GPE composed of ETPTA, liquid electrolyte and Al_2O_3 nanoparticles. (b) c-GPE mixture before UV-crosslinking reaction. (c) Thermal gravity analysis (TGA) profiles showing difference in thermal stability compared to a conventional liquid electrolyte (1 M $LiPF_6$ in EC/DEC = 1/1 v/v) and c-GPE. (d) Cyclic stability containing capacity retention with cycling as well as coulombic efficiency of a cell based on lithium metal/flat-shaped c-GPE/$LiCoO_2$ cathode. *Source:* Kil et al. 2013 [34]. Copyright 2013. Reprinted with permission from John Wiley & Sons.

shown in Figure 3.8a) [34]. In this work, UV-crosslinked ETPTA monomer containing trivalent vinyl groups plays the role of a framework and Al_2O_3 nanoparticles are introduced as a functional filler to control the rheological properties of the GPE. As shown in Figure 3.8b, the produced GPE is also shape-conformable through UV-assisted nanoimprint lithography technology. Figure 3.8c,d indicates better thermal stability of the c-GPE compared with liquid electrolyte and the cyclic stability of lithium ion battery based on the produced c-GPE.

As compared with GPE-containing liquid electrolyte, PCE is composed of lithium salts and plastic crystals. The thermal stability of lithium ion battery can be efficiently improved using PCE as electrolyte with acceptable ion mobility [36, 37]. For example, Armand and coworkers reported PCE based on succinonitrile ($N{\equiv}C{-}CH_2{-}CH_2{-}C{\equiv}N$) with a plastic-crystal form that can be used as a highly polar medium to dissolve a large variety of salts and to achieve high ionic conductivity. Owing to high diffusivity, plasticity, and solvating power of plastic-crystal succinonitrile, a large number of materials with high ionic conductivity can be prepared and the ionic conductivity at 25 °C reached as high a value as 3 $mS\,cm^{-1}$ which is 2 orders of magnitude above polymers [37].

Because PCE is generally plastic and shows liquid-like behavior at room temperature, PCE has been composited with polymer matrix to produce a modified composite GPE which is called PCPE. Choi et al. developed a thin, deformable, and safety-reinforced plastic crystal polymer electrolyte (called N-PCPE in their work) by combining a PCPE with a porous polyethylene terephthalate (PET) nonwoven [35]. As shown in Figure 3.9a, PCPE was fabricated through the *in situ* UV-crosslinking of ETPTA monomer directly inside the PET nonwoven skeleton in the co-presence of PCE. The time evolution of

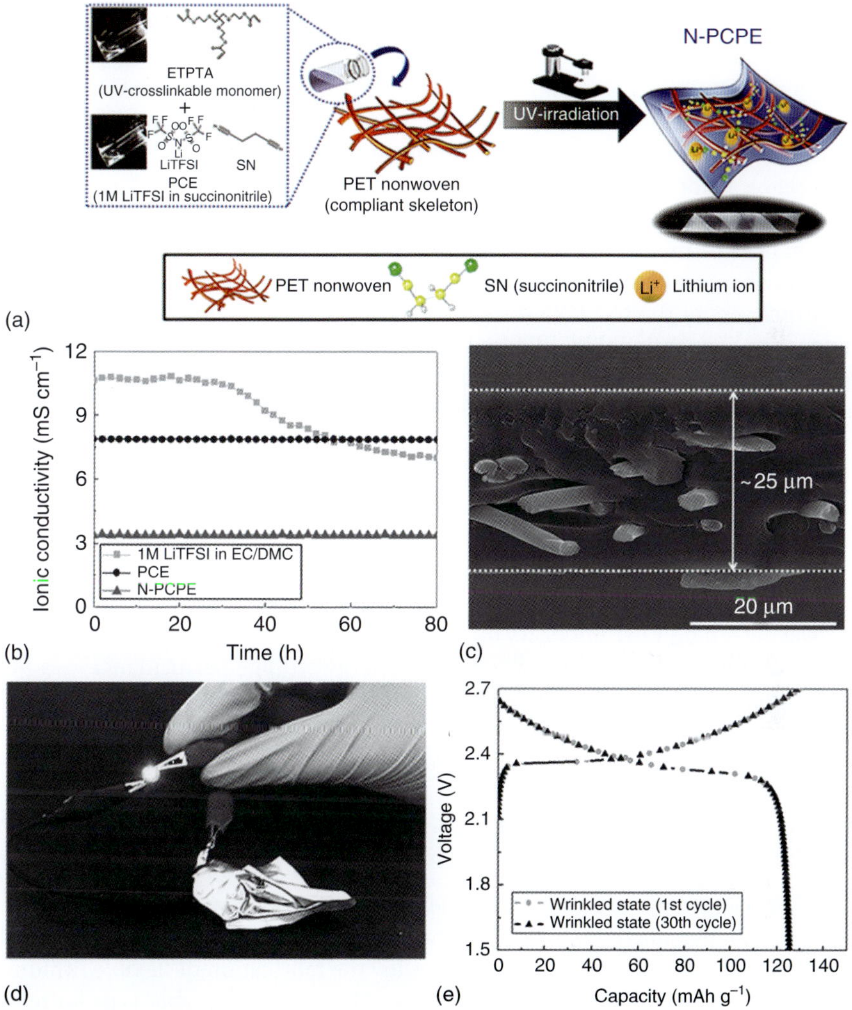

Figure 3.9 (a) Schematic illustration of the fabrication process and chemical structure of N-PCPE, along with a photograph showing its mechanical flexibility. (b) Time evolution of ionic conductivity for PCE, N-PCPE, and conventional carbonate-based liquid electrolyte at 80 °C. (c) SEM image of N-PCPE (thickness ≈ 25 μm) with the plastic crystal polymer electrolyte matrix successfully impregnated into PET nonwoven skeleton. (d) Photograph of an LED lightened by an N-PCPE-based flexible lithium ion battery under severely wrinkled state. (e) Charge/discharge profiles at current density of 0.2 C of the wrinkled N-PCPE-based flexible lithium ion battery before and after cycling. *Source:* Choi et al. 2014 [35]. Copyright 2013. Reprinted with permission from John Wiley & Sons.

ionic conductivities for different electrolytes in Figure 3.9b confirms that the polymer electrolyte exhibited higher thermal stability compared with liquid electrolyte. The produced N-PCPE possesses three-dimensional reticulated matrix structures and the porous PET nonwoven is well-impregnated with the plastic crystal polymer electrolyte matrix (as shown in Figure 3.9c). Owing to this structural uniqueness with PET nonwoven playing a role of a compliant skeleton to enhance the mechanical strength of N-PCPE, the N-PCPE shows highly improved deformability. As shown in Figure 3.9d, a flexible lithium ion battery based on N-PCPE can be extremely wrinkled and can simultaneously power an LED without any short circuit occurring between electrodes. Moreover, the advantageous features of the PCE, such as high ionic conductivity and thermal stability, can be well maintained during the deformations (as shown in Figure 3.9e). Undoubtedly, both the high ionic conductance and the excellent deformability of N-PCPE play functional roles in improving the energy storage performances of the flexible lithium ion batteries particularly at high rates and during mechanical deformation.

The thickness of the N-PCPE discussed earlier is about 25 μm, which is much thinner than conventional PCPE film (thicker than 100 μm). However, this thickness still cannot meet the demands of light weight and thin film flexible lithium ion batteries used for nano- or microelectromechanical systems at the nano- or microscale. To solve this problem, lithium phosphorus oxynitride (LiPON) electrolyte, which is physically deposited as amorphous thin film electrolyte, has been investigated [38]. In Lee's work, LiPON with thickness of about 2 μm has been prepared to play roles of both solid electrolyte and separator, which enables the realization of diverse flexible lithium ion batteries based on various electrode materials. The produced flexible lithium ion battery shows high robustness by powering a blue LED under its bending state. During repeated bending processes with different bending radiuses, the energy storage performance can be well maintained, which demonstrated the excellent mechanical stability of the flexible lithium ion batteries. Of course, it should be noted that the solid-state LiPON electrolyte also suffers from some disadvantages, such as low ionic conductivity, limited reaction interface with electrodes, and low power density and these accordingly limit its practical application at present [39].

3.2.1.4 Special Structural Design-Based Flexible Lithium–Ion Battery

From a special perspective, cable-like lithium ion batteries also possess excellent flexibility and thus provide a novel concept for the fabrication of flexible lithium ion batteries [40, 41]. Figure 3.10a shows one of the simplest cable-type battery designs, consisting of an inner electrode (anode), surrounding a separator and an outer electrode (cathode) [40]. Shin and coworkers modified the battery with a hollow spiral structure [41]. As shown in Figure 3.10b, the lithium battery is composed of a hollow spiral, spring-like anode (nickel/tin-coated copper wire), a separator membrane (modified PET), cathode composite ($LiCoO_2$)-coated spiral Al wire and packaging insulator with coaxial structure. The hollow-spiral structure highly improved the energy storage performance of the battery (Figure 3.10c). This unique hollow and spring-like structure enables mechanical flexibility and

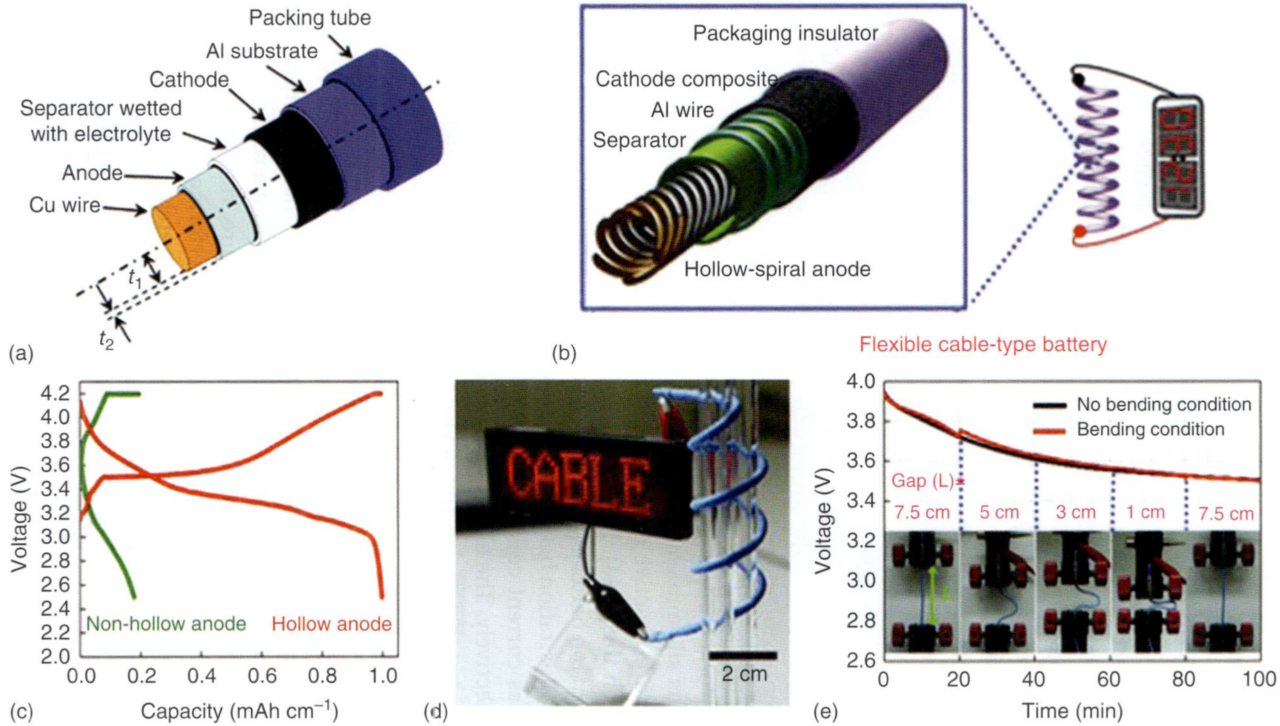

Figure 3.10 (a) Schematic illustration of a general cable battery. *Source:* Lee et al. 2013 [40]. Copyright 2013. Reprinted with permission from Royal Society of Chemistry. (b) Schematic illustration of the cable battery with hollow-helix anode having multiple helix structure. (c) First charge and discharge profiles of cable batteries. (d) Photographs of a highly flexible cable battery under bent state. (e) The cable battery exhibited stable operation even when bent. *Source:* Kwon et al. 2012 [41]. Copyright 2012. Reprinted with permission from John Wiley & Sons.

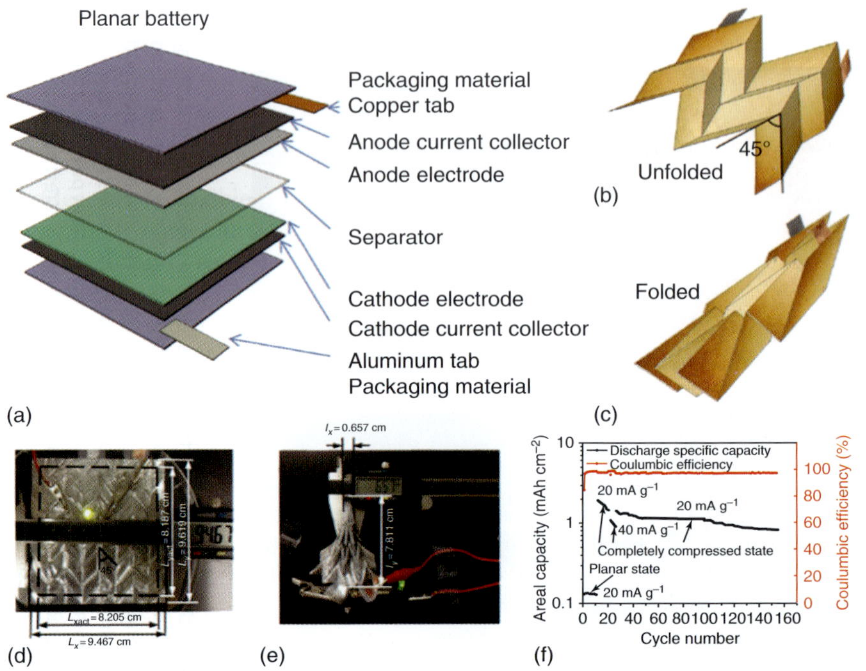

Figure 3.12 (a) Exploded view of the multilayer structure of conventional lithium ion batteries (LIBs) in the planar state. (b) Illustration of 45° Miura folding in the unfolded and (c) folded states. (d) Photograph of an unfolded origami battery powering a green LED. (e) Photograph of a completely compressed origami battery powering a green LED. (f) Capacity retention (left axis, black) and coulombic efficiency (right axis, red) as a function of cycle number under planar and completely compressed state at 20 and 40 mA g^{-1}, respectively. *Source:* Song et al. 2014 [45]. Copyright 2014. Reprinted with permission from Nature Publishing Group.

In Jiang's work, the origami battery is carried out by slurry coating $LiCoO_2$- and $Li_4Ti_5O_{12}$-electrode active materials onto CNT papers current collectors and packaging the battery with aluminized polyethylene. Then the fabricated planar battery is folded in a Miura pattern that divides the battery into many units into an identical parallelogram through predefined creases (as shown in Figure 3.12b,c). Figure 3.12d,e indicates that the Miura pattern makes the battery compressible and collapsible. Because the deformability of the origami battery was derived from the prescribed creases, the origami pattern did not sustain large strain, which endows the battery with linear and areal deformability, large twistability, and bendability. Therefore, the resulting origami battery exhibited a stable performance during the planar and completely compressed configurations that are indicated by Figure 3.12f. Moreover, the LEDs powered by this origami battery under planar and completely compressed state also confirm this. This strategy enables the deformation of rigid origami-based devices at the system level, which highly extends the stretchable lithium ion battery electrode material sources.

3.3 Fiber-Shaped Flexible Lithium Ion Batteries

To keep up with the quick development of wearable devices, it is highly required to develop wearable power sources including wearable supercapacitors, lithium ion batteries, and so on. Fiber electrode plays a critical role in fabricating wearable energy storage devices. To realize fiber electrodes, CNT fibers, graphene fibers are mainly explored. Here, we mainly concentrate on the fiber-shaped lithium ion batteries employing the CNT fibers, which are incorporated with active materials, as cathodes or anodes.

3.3.1 Bendable Fiber-Shaped Lithium Ion Battery

Possessing both high electrical conductivity and excellent mechanical properties, CNT fibers play an important role in fabricating bendable fiber-shaped lithium ion batteries. The CNT fiber has been composited with various active materials such as metal oxide and silicon to prepare composite fiber electrode for bendable fiber-shaped lithium ion battery. Owing to high specific capacities, the resulting composite fibers demonstrate high electrochemical performances. At the same time, benefiting from the high conductivity of CNT fibers, no binders are required in these fiber electrodes and these binder-free composite fiber electrodes exhibit excellent rate capability and power density.

In 2012, a novel fiber-shaped lithium ion battery was realized using the composite fiber of an aligned CNT fiber and MnO_2 as the cathode [46]. In this work, the aligned CNT fiber is spun from spinnable CNT array that is prepared through the chemical vapour deposition (CVD) method. The diameter of the fiber is controlled to range from 20 to 30 μm. The MnO_2 nanoparticles are composited with CNT fibers through electrochemical deposition. Adjusting the cycle number of electrochemical deposition, the content of MnO_2 is controlled between 0.5 and 8.6 wt%. The schematic illustration of the proof-of-concept fiber-shaped lithium ion battery is displayed in Figure 3.13a. It should be noted that no extra metal current collectors or binders are used in these fiber-shaped batteries, which endows them with better portability and flexibility. The structures of a bare CNT fiber and CNT/MnO_2 composite fiber-shaped electrode are characterized through SEM as shown in Figure 3.13b–d. Though the flower-like MnO_2 nanoparticles are uniformly grown onto the surface of the CNT fiber, this hybrid structure is very stable and thus no obvious damage in structure integrity nor electrical conductivity decay are traced during the bending process, which ensures a decent flexibility of the composite fiber. The energy storage properties of the novel fiber-shaped battery are detected through a charge–discharge process as shown in Figure 3.13e with discharge platform of 1.5 V and specific capacity of 109.62 $mAh\,cm^{-3}$ or 218.32 $mAh\,g^{-1}$ based on the whole CNT/MnO_2 electrode at 5×10^{-4} mA. On increasing the current, the specific capacity endured obvious decrease to 8.45 $mAh\,cm^{-3}$ at 5×10^{-3} mA (Figure 3.13f). In this work, due to the excellent mechanical and electronic

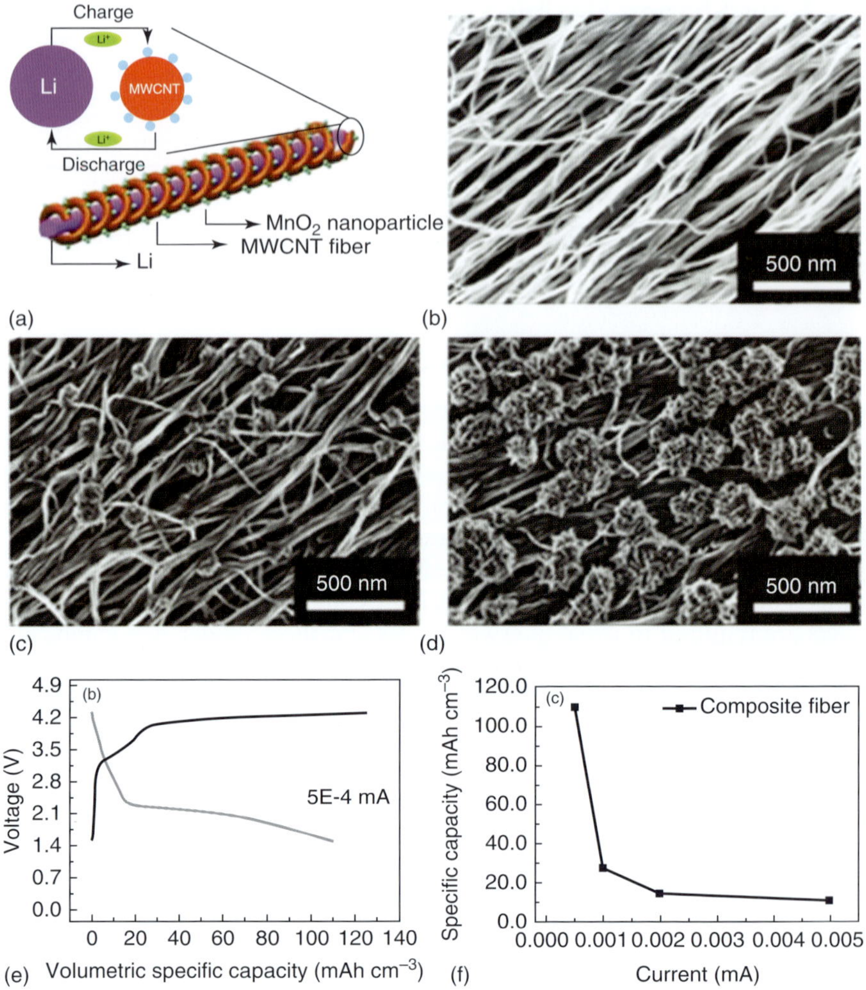

Figure 3.13 (a) Schematic illustration to a fiber-shaped lithium ion battery with lithium wire as anode and CNT/MnO_2 composite fiber as cathode. (b) SEM image of aligned CNT fiber. (c, d) SEM images of CNT/MnO_2 composite fibers with MnO_2 weight contents of 0.5% and 4.1%, respectively. (e) Voltage profiles of the fiber-shaped lithium ion batteries with 4.1 wt% of the MnO_2 in CNT/MnO_2 composite fiber during charge and discharge process at a current of 5×10^{-4} mA. (f) Dependence of specific capacitance for a battery wire derived from the multi-walled carbon nanotube (MWCNT)/MnO_2 composite fiber with MnO_2 weight percentage of 4.1. *Source:* Ren et al. 2013 [46]. Copyright 2013. Reprinted with permission from John Wiley & Sons.

properties of the aligned CNT/MnO_2 composite fibers, the fiber-shaped lithium ion battery displayed a good electrochemical performance. The flexible one-dimensional structure also enables promising applications pertaining to wearable devices. However, lithium wire is used as a negative electrode, which limited the practical application of such a fiber-shaped battery.

The flexibility and wearability of fiber-shaped devices is acclaimed as their characteristic property distinct from their planar counterparts. However, it is seriously weakened with the presence of lithium metal in lithium ion batteries. Therefore, it is highly needed for preparing an eligible fiber electrode to displace the lithium wire as anode in a fiber-shaped lithium ion battery. To replace lithium wire negative electrode, silicon possessing the theoretical capacity of 4200 $mAh\,g^{-1}$ and lithiation potential within 0.2–0.4 V is investigated for compositing with CNT fiber possessing outstanding flexibility [7, 47]. Silicon is incorporated into the CNT fiber through the electron beam evaporation technology. In this method, silicon is first vaporized and then deposited onto an aligned two-dimensional CNT sheet that is drawn from the spinnable CNT array and paved on a substrate. The produced CNT/Si composite sheet is rolled into a CNT/Si composite fiber as illustrated in Figure 3.14a. The amount of silicon that is deposited is controlled by the sputtering time. The structure, characterized through SEM images, shows uniform structure and increased diameter of the CNT/Si composite fiber as 60 μm as compared with that of bare CNT fiber as 30 μm (Figure 3.14b,c). The CNTs in the composite fiber remained orientated as expected. Benefiting from the uniform and stable structure, the composite fiber exhibited a decent flexibility that enabled the fiber to endure the repeated deformation without loss of significant properties. In addition, the space in the hybrid fiber can accommodate the volume expansion of the silicon and alleviate its negative effect on the longevity of the battery, as discussed next. Here, lithium wire is used to examine the electrochemical performance of CNT/Si composite fibers. The resulting fiber battery is charge–discharged between 0.02 and 1.20 V (Figure 3.14d). The CNT/Si hybrid fiber delivered a voltage plateau around 0.4 V and significantly higher capacity as compared with that delivered by the bare CNT fiber. However, the capacity retained by 32% after 100 cycles of draining and recharging, which was attributed to the volume change of silicon. The CNT/Si hybrid fiber exhibited flexibility that enables itself to sustain repeated bending for 100 cycles without sacrificing its electrochemical performance. In fact, the capacity retained by 94% after enduring deformations, which suggests that it survived more than 80% after 20 cycles (Figure 3.14e).

Hu and coworkers also designed and synthesized a composite of silicon and CNT fiber electrode [48]. As shown in Figure 3.15a, CNT fibers are spun from CNT array and Si is coated onto twofold spun CNT fibers. The resulting composite fibers are highly flexible and can be weaved into wearable textiles as shown in Figure 3.15b,c. Using this composite fiber electrode as anode, half coin-cell lithium ion batteries were fabricated and a capacity of 2200 $mAh\,g^{-1}$ was achieved which is five times higher than that of commercial graphite electrodes (Figure 3.15d). Benefiting from the high electrical conductivity of CNT fibers, the composite fiber anode can work without copper current collectors. Moreover, the CNT fiber possesses excellent strength and flexibility and the resulting CNT/Si composite anode fiber is as strong and flexible as widely used fibers in the textile industry.

To fabricate full fiber-shaped lithium ion batteries, Weng et al. assembled fiber-shaped full lithium ion battery through compositing CNT fibers with Si and $LiMn_2O_4$ (LMO) as anode and cathode, respectively (as shown in

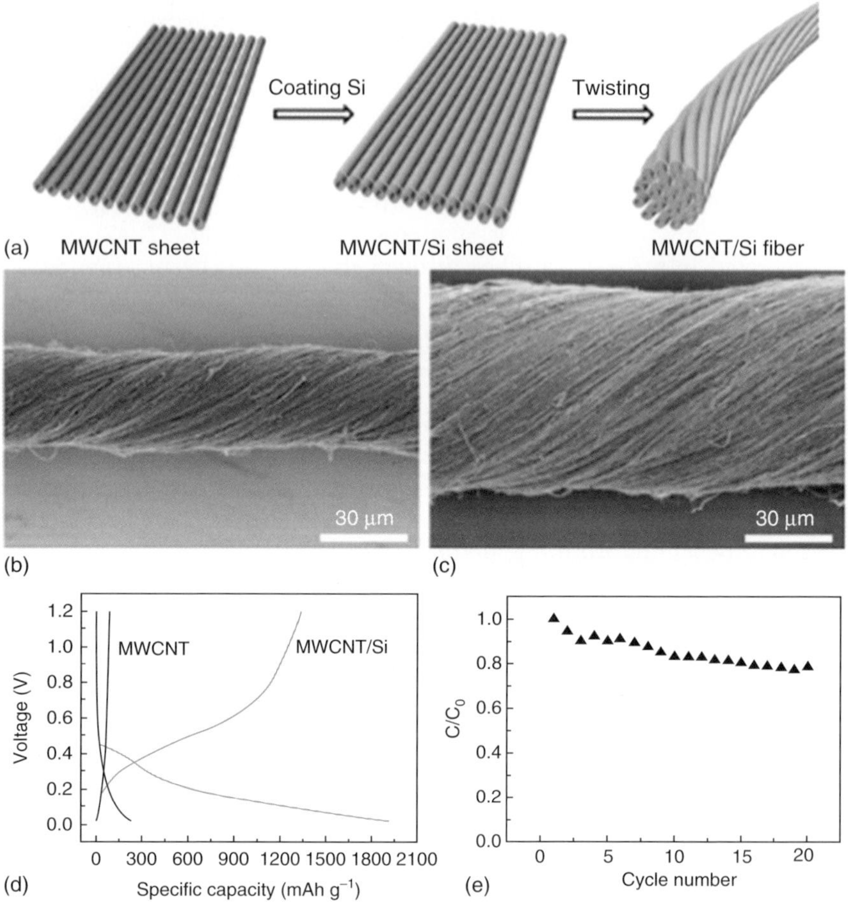

Figure 3.14 (a) Schematic illustration to the preparation of the aligned CNT/Si composite fiber. (b, c) SEM images of a bare aligned CNT fiber and an aligned CNT/Si composite fiber with Si weight percentage of 38.1%, respectively. (d) Typical charge–discharge curves of a half lithium-ion battery based on the bare aligned CNT and aligned CNT/Si composite fibers with a Si weight percentage of 38.1% at 1 A g^{-1}. (e) Dependence of specific capacity on cycle number for the composite fiber after bending for 100 cycles at 2 A g^{-1}. C_0 and C correspond to the specific capacities at the first and following cycle, respectively. *Source:* Lin et al. 2014 [47]. Copyright 2013. Reprinted with permission from John Wiley & Sons.

Figure 3.16a) [49]. As demonstrated, Si nanoparticles are deposited onto aligned CNT sheets that are then sandwiched by two layers of aligned CNT sheets and the resulting CNT-Si/CNT multilayer structure is rolled into fibers. In this work, the volume expansion of Si nanoparticles during the charge–discharge process is effectively restricted because of the sandwich structure before rolling into fibers. As shown in Figure 3.16b, a fiber-shaped full battery is fabricated by sequentially winding CNT-$LiMn_2O_4$ and CNT-Si/CNT composite fibers onto a commercial cotton fiber. In this fiber-shaped full battery, gel electrolyte is used

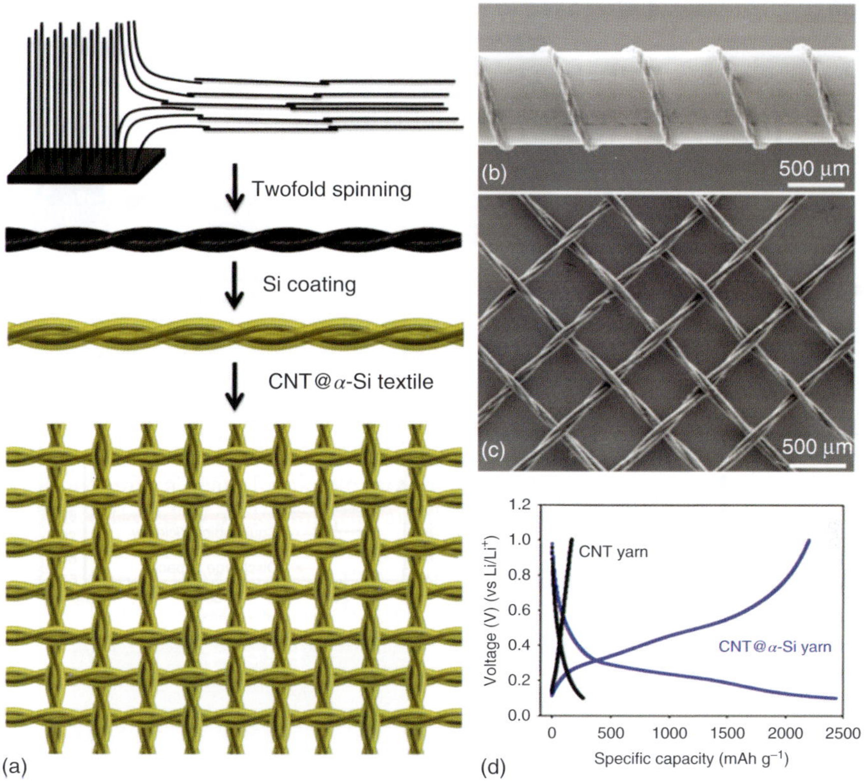

Figure 3.15 (a) Schematic illustrations of the fabrication process of weavable CNT/Si composite fiber and textile. (b) SEM image of CNT/Si composite fibers wrapped round a copper rod, showing excellent flexibility. (c) SEM image of a textile woven from CNT/Si fibers. (d) Voltage profiles of CNT and CNT/Si composite fiber with Li metal as the counter electrode at 1st cycle with an operation voltage range of 0.1–1.0 V and a charge rate of C/5. *Source:* Sun et al. 2013 [48]. Copyright 2013. Reprinted with permission from Elsevier Ltd.

which also plays the role of separator to separate two electrodes and prevent short circuiting. The produced full battery was galvanostatically cycled at a voltage range of 2.0–4.3 V at a current density of 1 C (148 mA g^{-1}) relative to the cathode. The voltage profiles are as shown in Figure 3.16c, which demonstrated an average voltage of 3.4 V derived from the potential difference between CNT-LMO (4.0 V vs Li/Li^{+}) and CNT-Si/CNT (0.6 V vs Li/Li^{+}) composite electrodes. The high stability of CNT-LMO and CNT-Si/CNT produced excellent cyclic stability of the fiber-shaped full battery. As shown in Figure 3.16d, the specific discharge capacity of the first cycle achieved 106.5 mAh g^{-1} and 87% of the capacity could be maintained after 100 cycles. The energy density is further calculated to be 27.7 mWh g^{-1}. Based on these excellent energy storage properties, the fiber-shaped full battery can light up an LED as shown in Figure 3.16e.The fiber-shaped full battery can be simultaneously woven into a

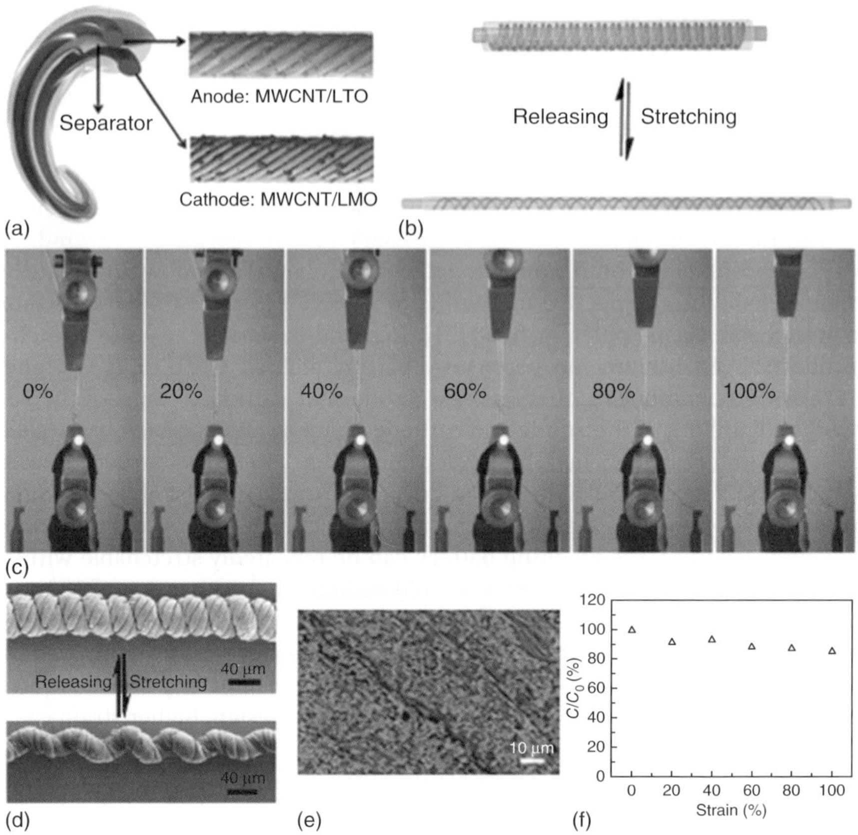

Figure 3.17 (a) Schematic illustration of the flexible wire-shaped lithium-ion battery. The aligned CNT/LTO and CNT/LMO composite fibers are paired as the anode and cathode, respectively. (b) Schematic illustration for the stretching and releasing process of a lithium ion battery wrapped on a heat-shrinkable tube. (c) Photographs of a battery powering a light-emitting diode at increasing strains. *Source:* Ren et al. 2014 [51]. Copyright 2014. Reprinted with permission from John Wiley & Sons. (d) SEM image of a spring-like fiber under unstretched and stretched states. (e) SEM image of an over-twisted CNT/LTO composite fiber at high magnification. (f) Evolution of specific capacitance with strain. *Source:* Zhang et al. 2014 [53]. Copyright 2014. Reprinted with permission from John Wiley & Sons.

3.4 Perspective

Flexible lithium ion battery has been widely investigated with the rapid development of portable devices. In this chapter, we describe representative flexible lithiumion batteries. Endowing the battery with flexibility is not easy, even making them to be wearable. It is recognized that flexible electrode plays a vital role in constructing flexible batteries. The requirements for flexibility, electrical conductivity, and electrochemical compatibility make carbon materials and flexible substrates the preferred choice for use in flexible batteries. Anode and/or cathode active materials are used to improve the electrochemical energy storage properties. These strategies have realized flexible lithium ion batteries and have greatly improved their properties.

However, further investigation for the fabrication of flexible lithium ion batteries that can fully meet the requirements of application is still badly needed. A foreseeable challenge is, in fact, addressing the inferior flexibility and stability during deformation. Most reported flexible lithium ion batteries suffer property loss during repeated bending or other deformation processes. The other problem is poor energy storage properties compared with that of traditional lithium ion batteries. All of these problems are calling for new materials and new designing for structures.

References

1 Whittingham, M.S. (2004). *Chem. Rev.* 104: 4271.
2 Abu-Lebdeh, Y. (2013). Introduction. In: *Nanotechnology for Lithium–Ion Batteries* (ed. Y. Abu-Lebdeh and I. Davidson), 2–11. New York: Springer-Verlag.
3 Tarascon, J.M. and Armand, M. (2001). *Nature* 414: 359.
4 Cheng, X., Zhang, R., Zhao, C. et al. (2016). *Adv. Sci.* 3: 1500213.
5 Cabana, J., Monconduit, L., Larcher, D., and Palacín, M (2010). *Adv. Mater.* 22: E170.
6 Yoshio, M. and Noguchi, H. (2009). A review of positive electrode materials for lithium–ion batteries. In: *Lithium–Ion Batteries: Science and Technologies* (ed. M. Yoshio, R.J. Brodd and A. Kozawa), 11. New York: Springer-Verlag.
7 Li, H., Wang, Z., Chen, L., and Huang, X. (2009). *Adv. Mater.* 21: 4593.
8 Winter, M., Besenhard, J.O., Spahr, M.E., and Novák, P. (1998). *Adv. Mater.* 10: 725.
9 Babu, G., Kalaiselvi, N., and Bhuvaneswari, D. (2014). *J. Electron. Mater.* 43: 1062.
10 Singh, G., Thomas, R., Kumar, A., and Katıyar, R. (2012). *J. Electrochem. Soc.* 159: A410.
11 Li, Z.H., Wang, L.Q., Li, K.Y., and Xue, D.F. (2013). *J. Alloys Compd.* 580: 592.
12 Pan, F., Wang, W.-L., Li, H. et al. (2011). *Electrochim. Acta* 56: 6940.
13 Jin, Y., Yang, C.P., Rui, X.H. et al. (2011). *J. Power Sources* 196: 5623.
14 Zhao, Y., Peng, L.L., Liu, B.R., and Yu, G.H. (2014). *Nano Lett.* 14: 2849.
15 Dey, A. (1971). *J. Electrochem. Soc.* 118: 1547.
16 Etacheri, V., Marom, R., Elazari, R. et al. (2011). *Energy Environ. Sci.* 4: 3243.
17 Szczech, J.R. and Jin, S. (2011). *Energy Environ. Sci.* 4: 56.
18 Zhang, W.-J. (2011). *J. Power Sources* 196: 13.
19 Xu, K. (2004). *Chem. Rev.* 104: 4303.
20 Hu, L., Choi, J.W., Yang, Y. et al. (2009). *Proc. Natl. Acad. Sci. U.S.A.* 106: 21490.
21 Izadi-Najafabadi, A., Yasuda, S., Kobashi, K. et al. (2010). *Adv. Mater.* 22: E235.
22 Pushparaj, V.L., Shaijumon, M.M., Kumar, A. et al. (2007). *Proc. Natl. Acad. Sci. U.S.A.* 104: 13574.

23 Hu, L., Wu, H., La Mantia, F. et al. (2010). *ACS Nano* 4: 5843.
24 Liu, B., Zhang, J., Wang, X. et al. (2012). *Nano Lett.* 12: 3005.
25 Hou, X., Liu, B., Wang, X. et al. (2013). *Nanoscale* 5: 7831.
26 Li, W., Wang, X., Liu, B. et al. (2013). *Nanoscale* 5: 10291.
27 Liu, B., Wang, X., Chen, H. et al. (2013). *Sci. Rep.* 3: 1622.
28 Chen, Z., Ren, W., Gao, L. et al. (2011). *Nat. Mater.* 10: 424.
29 Xu, Y., Lin, Z., Huang, X. et al. (2013). *Adv. Mater.* 25: 5779.
30 Li, N., Chen, Z., Ren, W. et al. (2012). *Proc. Natl. Acad. Sci. U.S.A.* 109: 17360.
31 Wang, X., Liu, B., Hou, X. et al. (2014). *Nano Res.* 7: 1073.
32 Masquelier, C. (2011). *Nat. Mater.* 10: 649.
33 Ihlefeld, J.F., Clem, P.G., Doyle, B.L. et al. (2011). *Adv. Mater.* 23: 5663.
34 Kil, E.-H., Choi, K.-H., Ha, H.-J. et al. (2013). *Adv. Mater.* 25: 1395.
35 Choi, K.-H., Cho, S.-J., Kim, S.-H. et al. (2014). *Adv. Funct. Mater.* 24: 44.
36 MacFarlane, D.R., Huang, J., and Forsyth, M. (1999). *Nature* 402: 792.
37 Alarco, P.-J., Abu-Lebdeh, Y., Abouimrane, A., and Armand, M. (2004). *Nat. Mater.* 3: 476.
38 Koo, M., Park, K.-I., Lee, S.H. et al. (2012). *Nano Lett.* 12: 4810.
39 Liu, Y., Zhang, W., Zhu, Y. et al. (2013). *Nano Lett.* 13: 293.
40 Lee, S.-Y., Choi, K.-H., Choi, W.-S. et al. (2013). *Energy Environ. Sci.* 6: 2414.
41 Kwon, Y.H., Woo, S.-W., Jung, H.-R. et al. (2012). *Adv. Mater.* 24: 5192.
42 Li, X., Gu, T., and Wei, B. (2012). *Nano Lett.* 12: 6366.
43 Hu, L., Pasta, M., Mantia, F.L. et al. (2010). *Nano Lett.* 10: 708.
44 Xu, S., Zhang, Y., Cho, J. et al. (2013). *Nat. Commun.* 4: 1543.
45 Song, Z., Ma, T., Tang, R. et al. (2014). *Nat. Commun.* 5: 3140.
46 Ren, J., Li, L., Chen, C. et al. (2013). *Adv. Mater.* 25: 1155.
47 Lin, H., Weng, W., Ren, J. et al. (2014). *Adv. Mater.* 26: 1217.
48 Sun, C.-F., Zhu, H., Baker Iii, E.B. et al. (2013). *Nano Energy* 2: 987.
49 Weng, W., Sun, Q., Zhang, Y. et al. (2014). *Nano Lett.* 14: 3432.
50 Liu, B., Wang, X., Liu, B. et al. (2013). *Nano Res.* 6: 525.
51 Ren, J., Zhang, Y., Bai, W. et al. (2014). *Angew. Chem. Int. Ed.* 53: 7864.
52 Zhang, Y., Bai, W., Ren, J. et al. (2014). *J. Mater. Chem. A.* 2: 11054.
53 Zhang, Y., Bai, W., Cheng, X. et al. (2014). *Angew. Chem. Int. Ed.* 53: 14564.

4

Flexible Sodium Ion Batteries: From Materials to Devices

Shengyang Dong, Ping Nie, and Xiaogang Zhang

Nanjing University of Aeronautics and Astronautics, College of Material Science and Engineering, 29 Jiangjun Avenue, Nanjing 211106, China

4.1 Introduction to Flexible Sodium Ion Batteries (SIBs)

Recently, with the increasing development of wearable and flexible electronic devices, including conformable skin sensors, implantable medical devices, and wearable sensors, technologies involving flexible energy storage have attracted increasing attention due to the potential enormous market [1, 2]. These flexible and wearable applications usually integrate with the requirements of the human body to provide smart functions. Moreover, the wearable electronic devices can also remarkably boost their development in the Internet of Things fields such that people, devices, and data are connected and integrated to enrich our daily life [3]. Toward this, the power supply sources require light weight, flexible construction, and conformal design.

It is well known that lithium ion batteries (LIBs) have achieved great success in our daily life and have dominated the market in applications from mobile electronics to electric vehicles (EVs), hybrid electric vehicles (HEVs), and other electric public utilities, because LIBs have a high operating voltage and long cycling life as well as a large energy density, and are less likely to have undesirable influence caused by a memory effect [4, 5]. However, the rapid consumption of lithium resources driven by the tremendous demand in EVs and large-scale energy storage may eventually render lithium resources unaffordable since lithium is rare and its distribution is geographically restricted. Thus, pursuing sustainable flexible power technology for flexible and wearable electronic applications becomes increasingly important. In contrast to lithium, sodium is the sixth most abundant element in the Earth's crust ($\approx$2.6%), and virtually unlimited sodium resources are available in sea water [6]. Additionally, sodium has physical and chemical properties similar to those of lithium. Therefore, sodium ion batteries (SIBs) have attracted great attention as a promising alternative to LIBs [7, 8].

Similarly to LIBs, SIBs, including the flexible SIBs, also have a "rocking-chair" working mechanism. In a typical flexible SIB configuration: flexible anode, flexible cathode, and electrolyte are the three major components, as shown in Figure 4.1. The SIBs function by converting chemical energy into electrical

Flexible Energy Conversion and Storage Devices, First Edition. Edited by Chunyi Zhi and Liming Dai.

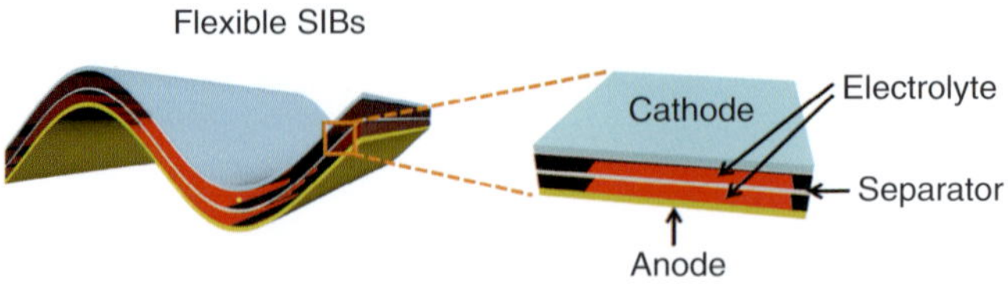

Figure 4.1 Schematic of flexible sodium ion batteries.

energy *via* Faradaic reactions, accompanied by insertion/extraction of sodium ions into/from the crystalline framework of electro-active materials [9, 10]. On charging, sodium ions de-intercalate from the cathodes, pass across the electrolyte, and intercalate into the anodes. The above process is reversed during the discharge process. Sodium ions diffuse through bulk diffusion while electrons move in the opposite direction and pass around the external circuit.

In general, the specific capacity, cycle life, rate capability, and mechanical flexibility are the most important performance indicators for flexible SIBs, which are strongly dependent on the properties of the electro-active materials and the design of flexible electrodes [11]. As far as the flexible electrodes are concerned, the morphology, crystal structure, and composition can influence the rate properties and transfer processes of ions/electrons and can be manipulated to alter the overall electrochemical properties of the flexible device [12]. Therefore, the key challenges in achieving flexible SIBs lie in the selection and design of bendable electrodes with good mechanical flexibility and the fabrication of flexible electro-active materials with a high capacity and excellent conductivity. The ductile and flectional electrodes combine with the nanostructured electro-active materials rendering them as promising candidates for flexible SIBs [13]. Additionally, the electrochemical reactions of batteries are commonly determined by ion and electron transfer through the bulk of the electrodes. By employing nanostructured materials, the fast charge/discharge properties can be highly enhanced by shortening the diffusion pathway of ion and electron [14].

4.2 The Key Scientific Issues of Flexible SIBs

In order to construct a reliable flexible device, flexible SIBs must possess highly reversible electrochemical behavior and mechanical properties, before, during, and after the process of deformation [15]. SIBs usually consist of several main components including cathode, anode, separator, current collector, electrolyte, and package materials, all of which must be capable of being bent, stretched, even compressed so as to be compatible for flexible SIBs. Therefore, the crucial aspects toward the design of flexible SIBs include the development of advanced electro-active materials, deformable substrates, and mechanically durable electrodes and circuits. Besides these, novel processing technologies and system integration are also very important [3].

4.2.1 Design of Advanced Active-Materials

Generally, developing novel elastic materials and structural designs are main routes to achieving flexible devices. However, developing intrinsically flexible materials is extremely difficult, because it needs to meet many different technical demands [15]. Compared to conventional inorganic materials, organic materials have better stretchability, and possibly lower cost, which enable devices to have some flexibility [16]. Although many works have focused on the design of organic material-based flexible devices, their electrochemical performance is still incomparable to that of inorganic material-based devices. Inorganic materials, such as metal oxides, hard carbon, and activated carbon, usually suffer from extremely low elastic limit (usually <0.1%) [15]. They are intrinsically brittle, and are not capable of accommodating a harsh mechanical deformation. Thus, conventional inorganic materials are the major challenge to developing flexible SIBs.

4.2.2 Design of Flexible Substrates and Electrodes

The conventional electrode is usually prepared by using the slurry-coating technology. The electro-active materials, such as metal oxides, hard carbon, and other inorganic materials are made into slurry with conductive additives and polymer binders, and further coated on Al or Cu foil current collectors that are rigid. Besides, these electro-active materials are intrinsically brittle with an extremely low elastic limit [17]. As a result, the electro-active materials may easily peel off from the Al or Cu foils during repeated mechanical deformation. Currently, flexible electrodes are typically prepared by incorporating electro-active materials onto flexible or/and conductive substrates [18–20]. For example, the unique physical, chemistry, mechanical, and electrical properties of graphene make it a promising candidate for incorporation in flexible substrates to produce flexible SIBs [21, 22].

Currently, flexible devices can be classified into three categories according to their loading type – bendable, stretchable, and compressible devices [15], as shown in Figure 4.2. Among them, bendable SIBs are the most investigated. The key issue for bendable SIBs is to develop reliable flexible electrodes. Suo investigated the strain in a film coated on a substrate in detail. The strain in the top surface ($\mathcal{E}_{\text{top}}$) can be obtained by the equation [23, 24]:

$$\mathcal{E}_{\text{top}} = \left(\frac{d_{\text{f}} + d_{\text{s}}}{2R}\right)\frac{(1 + 2\eta + \chi\eta^2)}{(1 + \eta)(1 + \chi\eta)} \tag{4.1}$$

$$\eta = \frac{d_{\text{f}}}{d_{\text{s}}} \tag{4.2}$$

$$\chi = \frac{Y_{\text{f}}}{Y_{\text{s}}} \tag{4.3}$$

where d_{f} and d_{s}, Y_{f} and Y_{s} are the thickness and Young's moduli of the film (electro-active materials) and substrate, respectively, and R is the bending radius.

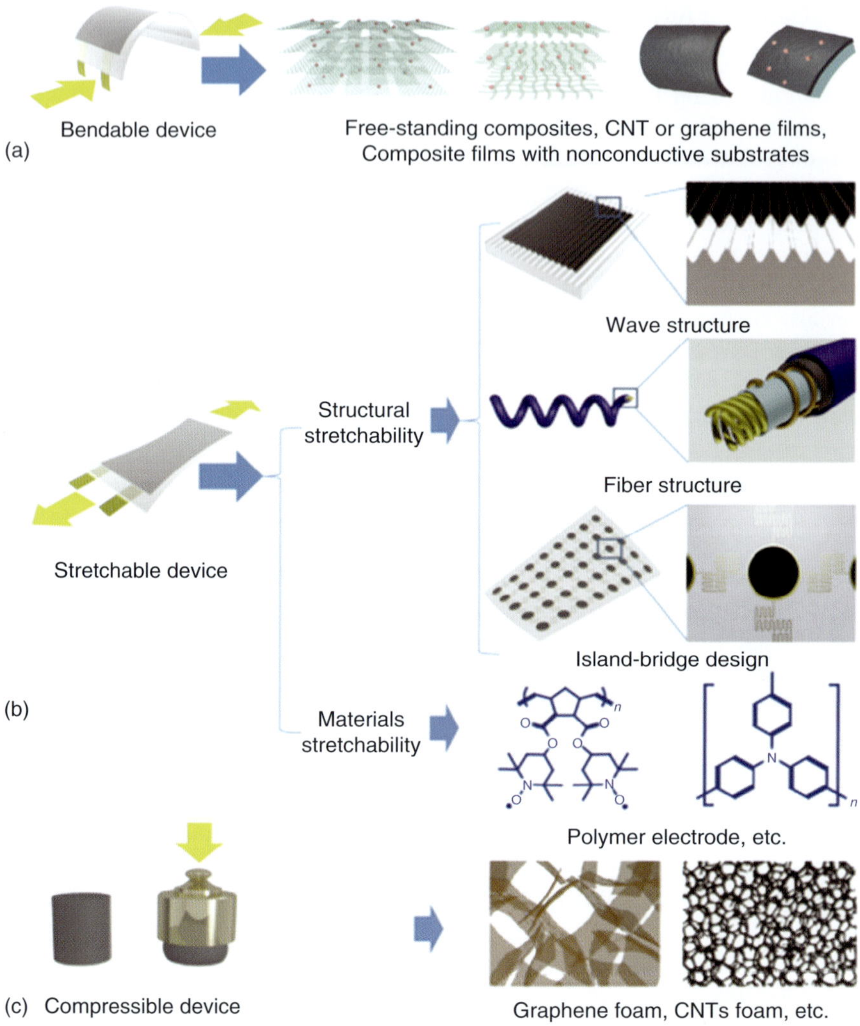

Figure 4.2 Classification of flexible electrodes. *Source:* Wen et al. 2016 [15]. Copyright 2016. Reproduced with permission from John Wiley and Sons.

Specially, when the flexible film and substrate have the same Young's modulus, the $\mathcal{E}_{top}$ can be simplified as follows:

$$\mathcal{E}_{top} = \frac{(d_f + d_s)}{2R} \tag{4.4}$$

From the theoretical model, Equations (4.1) and (4.4), we can summarize that depositing nanomaterials with a small thickness onto flexible substrates will be a feasible strategy to developing bendable SIBs.

The stretchable device is another common flexible configuration [25]. However, developing stretchable SIBs is more difficult than developing bendable devices. Designing intrinsically stretchable materials for flexible SIBs is extremely challenging. Another feasible strategy to fabricate stretchable SIBs is using stretchable electrode structures. Three novel structures have been developed and demonstrated in flexible LIBs and supercapacitors: wavy layout, coiled fiber structure, and island-bridge design [26–28]. Although the stretchability can be obtained by wavy layout and helically coiled design, such structures can only be unidirectionally stretched. The island-bridge structure circumvents this limitation, but the assembly processes are complicated and the stability obtained is unsatisfactory. In fact, coiled fiber, wavy layout, and island-bridge structures possess a little compressibility to some extent. However, so far, there has been no relevant works on stretchable SIBs. Cui and coworkers [29] reported a novel method to design stretchable electrodes, by using sugar as a template to generate the highly elastic polydimethylsiloane (PDMS) scaffolds with 3D porous sponge-like structure. They used the $Li_4Ti_5O_{12}$ anode and $LiFePO_4$ cathode and demonstrated the practicability of the stretchable PDMS sponges in LIBs. Actually, fabrication of stretchable electrodes is a versatile strategy [29]. Zhang and coworkers developed a simple strategy to prepare nitrogen-doped carbon foam (NCF) with compressible 3D electrode architectures by direct carbonization of commercially available melamine foam [30]. They used it as a compressible substrate for supercapacitor applications.

4.2.3 Developing Novel Processing Technologies

Conventional energy storage devices usually have simple shapes, such as coins, cylinders, prism, and pouch types [31, 32]. Stress accumulation caused by the geometrical shape during deformation is another challenge for achieving flexible SIBs. To avoid internal stress between components, the processing technology of conventional energy storage devices should be changed [15]. Therefore, novel processing technologies should be developed to allow flexible SIB operation under repeated mechanical deformation. In previous reports, PDMS is the most widely used flexible polymer for fabricating elastic electrodes and packaging devices due to its unique properties, such as wonderful flexibility, stretchability, and chemical stability [29, 33]. In very recent years, printing technologies, especially 3D printing technology, are drawing much attention owing to their large-scale and high-throughput production capabilities [34, 35]. That may be a promising choice to developing flexible SIBs in the near future.

4.3 Design of Advanced Materials for Flexible SIBs

4.3.1 Inorganic Anode Materials for Flexible SIBs

For inorganic anode materials, we herein mainly focus on the recent processes involving carbon materials, alloy anodes, metal oxides/sulfides based on conversion reaction and intercalation-based titanium-based compounds as flexible electrodes for flexible SIB applications.

Unlike LIBs, graphite has poor electrochemical performance for sodium storage, which might be related to the larger radius of Na^+ and also because of thermodynamic issues [36, 37]. Fortunately, disordered carbon, such as hard carbon, soft carbon, and graphene, is particularly attractive due to its disordered structure and larger interlayer distance that could accommodate sodium. Yu and coworkers reported flexible and lightweight porous carbon nanofibers (P-CNFs) with 3D interconnected structures using simple electro-spun technology and subsequent pyrolysis process [38]. The P-CNFs can be directly used as anodes for SIBs, and exhibit a high reversible capacity of 266 $mAh\,g^{-1}$ after 100 cycles at a current of 50 $mA\,g^{-1}$. Singh and coworker first investigated the electrochemical behavior of reduced graphene oxide (rGO) paper electrode on lithium/sodium storage under different annealing temperatures [39]. They found that the lithium storage capacity increased with increasing thermal treatment temperature in Ar gas from 300 to 900 °C. However, the sodium storage capacity was highest at 500 °C [39], which is due to the decreased interlayer spacing and increasing order of rGO paper during increasing temperatures. Further, they found that both the fracture strength and corresponding strain to failure decreased for rGO paper prepared at increasing treatment temperatures [39]. In order to improve the sodium storage performance of graphene, Shao and coworkers prepared sulfur-doped flexible graphene films (SFGs) by using thiourea as the reducing agent and sulfurizing agent [40]. The SFGs showed a high reversible capacity of 377 $mAh\,g^{-1}$ at 100 $mA\,g^{-1}$, which might be related to S doping as it induces larger interspacing layers and creates additional redox sites for reaction with sodium [40].

2D layered transition metal dichalcogenides are promising anodes for energy storage applications [41]. In particular, molybdenum disulfide (MoS_2) has received extraordinary attention in recent years as a new anode for SIBs due to its moderate potential, high specific capacity and highly flexible layered structure [42]. The electrochemical reaction of MoS_2 is based on intercalation and conversion, where Na^+ firstly intercalates into the S–S layered space to form Na_xMoS_2, then the Na_xMoS_2 converts into the Mo and Na_2S phase. David et al. [43] reported a self-standing flexible electrode composed of acid-exfoliated few-layer MoS_2 and rGO flakes for SIBs, which is synthesized through vacuum filtration of homogeneous dispersions of acid-treated MoS_2 flakes and GO aqueous solution followed by thermal reduction (Figure 4.3a–e). Furthermore, mechanical tests involving static uniaxial tension reveal good mechanical strength of about 2–3 MPa and high failure strain (approximately 2%) of this rGO/MoS_2 composite paper (Figure 4.3f,g). When evaluated as an electrode for SIB, the film shows highly reversible capacity with a stable charge capacity of approximately 230 $mAh\,g^{-1}$ [43]. For a cost-effective large-scale production method for MoS_2 exfoliation/dispersion, Hu and coworkers [44] reported an efficient method for MoS_2 exfoliation by using a green dispersant, 2,2,6,6-tetramethylpiperidine-1-oxyl (TEMPO)-oxidized nanofibrillated cellulose in an aqueous solution. Multifunctional MoS_2 films were fabricated by filtering with carbon nanotubes (CNTs) and were employed as flexible electrodes for Li/SIBs [44]. Gogotsi, Wang, and coworkers designed a freestanding anode material consisting of molybdenum disulfide nanosheets aligned vertically on carbon paper derived from commercial paper towels [45]. The MoS_2 nanosheets were firstly deposited on the flexible substrate, a carbon paper, under hydrothermal process at 180 °C for 24 h, then calcined at 800 °C in Ar atmosphere. The unique structure favors fast electron transfer and electrolyte diffusion, enabling

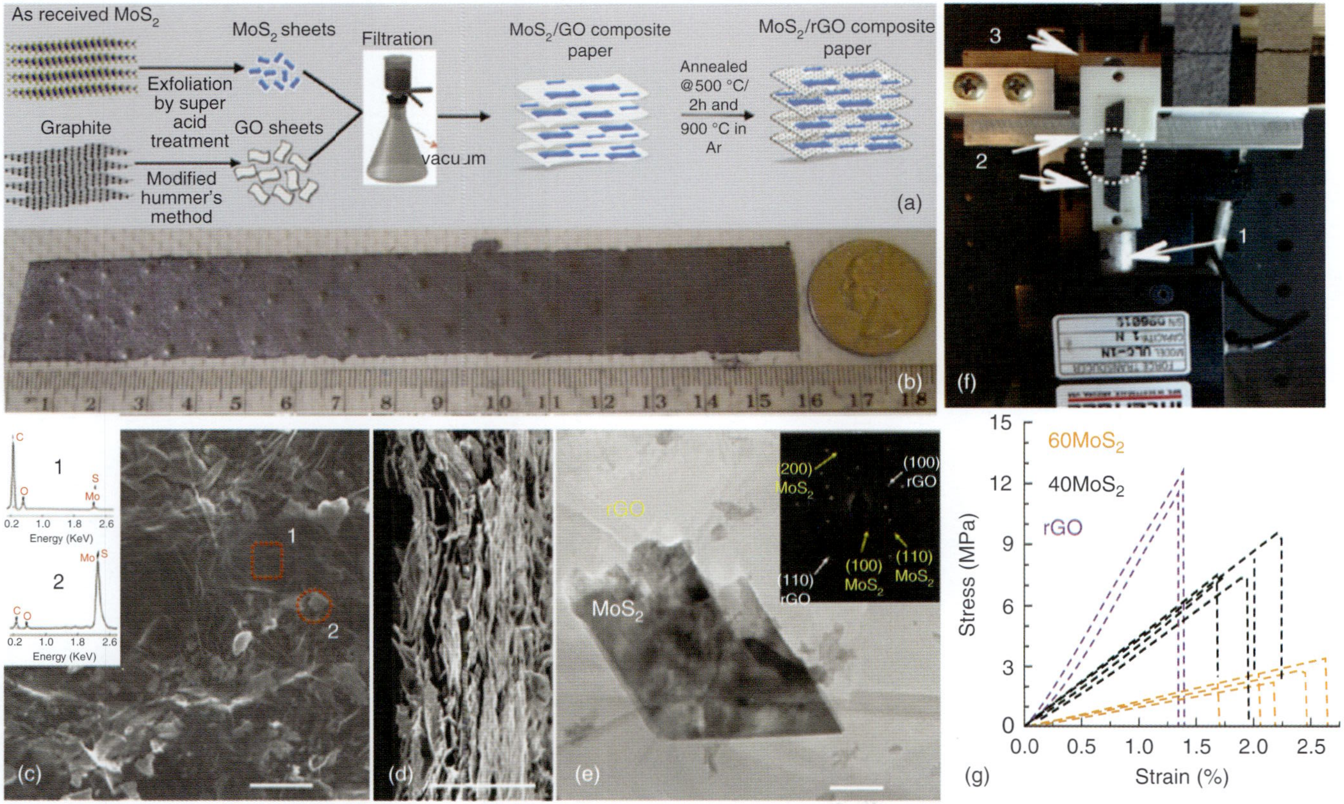

Figure 4.3 (a) Schematic representation showing synthesis of rGO/MoS_2 composite paper. (b) Digital picture showing large area composite paper prepared through vacuum filtration. (c) SEM top-view image of 60MoS_2 paper; inset shows the EDX spectra of spots in the SEM image indicating the material to be rGO (square) and MoS_2 (circle). The scale is 10 μm. (d) Corresponding SEM cross-sectional images show the morphology of the paper. Average thickness of this paper was observed to be ~20 μm. (e) TEM image and SAED pattern of 60 MoS_2. (f) Tensile test setup (1, load cell, fixed; 2, clamps, top clamp not shown; 3, computer-controlled movable translation stage) with sample after fracture from loading (inset shows zoomed-in view of two such specimens). (g) Engineering stress strain plot for rGO, 40MoS_2, and 60MoS_2 free-standing papers. *Source:* David et al. 2014 [43]. Copyright 2014. Reproduced with permission from American Chemical Society.

high initial Coulombic efficiency (79.5%), reversible capacity, good cycling performance, and excellent rate capability, delivering a capacity of 205 mAh g^{-1} at 1000 mA g^{-1}. The reversible phase transition from 2H-MoS_2 to 1T-MoS_2 is observed by using in situ Raman spectroscopy during the electrochemical cycling [45]. Self-standing electrode of MoS_2 nanosheets assembled on graphene foam, which were grown on nickel foam by chemical vapor deposition (CVD) method was achieved by a single microwave-assisted hydrothermal method. With a high conductive network as well as the interconnected channel properties of the 3D graphene foam, which enable fast electron transport, efficient penetration of electrolyte and short sodium ion diffusion paths, the resulting composites exhibit superior rate performance and high specific capacity of 290 mA h g^{-1} after 50 cycles at a rate of 0.1 A g^{-1} in SIBs [46]. As a close analogue to MoS_2, $MoSe_2$ has smaller band gap and larger layer space, thus showing better electric conductivity and Coulombic efficiency for SIB application. Zhang et al. reported $MoSe_2$ nanosheets grown on carbon cloth by a solvothermal method as the flexible electrode for SIBs for the first time [47]. When used as a self-supporting anode directly, the composite delivers high capacity of 452.6 mA h g^{-1} at 0.2 A g^{-1}, superior rate ability at 5 A g^{-1}, and keeps 85.5% capacity retention after 100 cycles [47].

Layered molybdenum trioxide (MoO_3) has been studied as an attractive electrode in SIBs due to its high specific capacity, low cost, and resource abundance [48]. However, poor cycling stability has limited its application originating from its low electronic/ionic conductivity and irreversible phase transition. Yao and coworkers investigated partially reduced MoO_{3-x} grown on flexible carbon cloth substrate, where the oxygen vacancy significantly improved the cycling stability and the carbon cloth increased the electronic conductivity of MoO_3 material, as shown in Figure 4.4a. The capacity retention is 92% after 2000 cycles at 1 A g^{-1} [49]. Compared to a conventional electrode, the above flexible electrode exhibits lighter weight, favorable mechanical strength, and higher electric conductivity without carbon additive, binder, and metal current collector, demonstrating promising perspective and great potential for high energy and power SIB application (Figure 4.4b,c). Similarly, flexible carbon-coated FeS on carbon cloth substrate was prepared by a facial hydrothermal method combined with a carbonization treatment [50]. When used as free-standing electrode for sodium storage, it could deliver a reversible capacity of 365 mAh g^{-1} after 100 cycles at 0.15 C [50].

Tin and Tin-based composites attracted remarkable attention due to their high specific capacity, low cost, and environmental friendliness [51]. However, this kind material usually suffers from deteriorating capacity and cycling stabilities owing to large volume expansion, leading to pulverization of active materials, loss of electrical contact with current collector, and unstable solid electrolyte interphase (SEI) layer [52]. Wang and coworkers [53] reported SnS_2@graphene nanosheet arrays grown on carbon cloth substrate as binder-free anodes for SIBs. The composite is achieved by using a facile two-step solvothermal route. The integrated structure facilitates efficient electron transfer and fast kinetics, and improved rate and cycling capability was observed for this composite [53]. Very recently, tin nanodots (1–2 nm) encapsulated in porous nitrogen-doped carbon

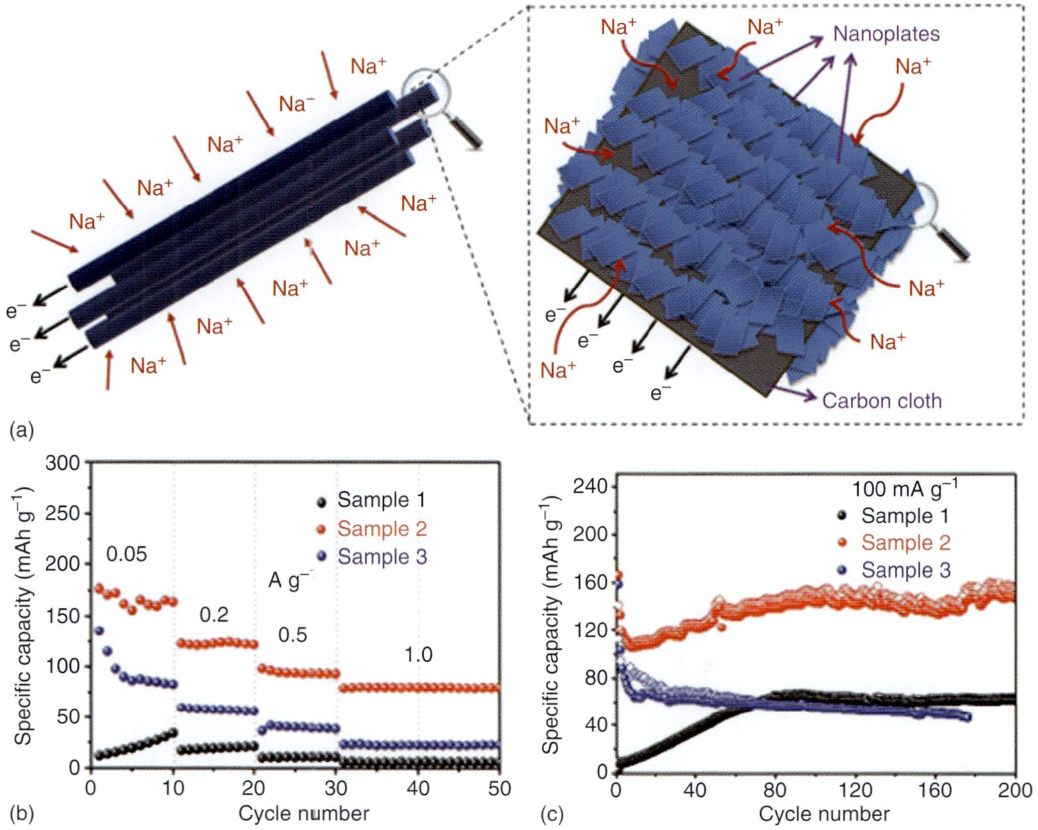

Figure 4.4 (a) Schematic diagram of partially reduced MoO_{3-x} deposited on flexible carbon cloth to enable facile ion diffusion and electron conduction. (b, c) The rate and cycling performance. *Source:* Li et al. 2016 [49]. Copyright 2016. Reproduced with permission from Royal Society of Chemistry.

nanofibers were synthesized through feasible electro-spinning techniques and thermal treatment processes using polyacrylonitrile, poly(methyl methacrylate), and $SnCl_2$ as precursors, as shown in Figure 4.5 [54]. The size and content of Sn particles can be controlled by changing their synthesis condition. The as-obtained nanofibers with flexible membrane could be directly used as binder- and current collector-free anode for SIBs. As expected, the hybrid film exhibits

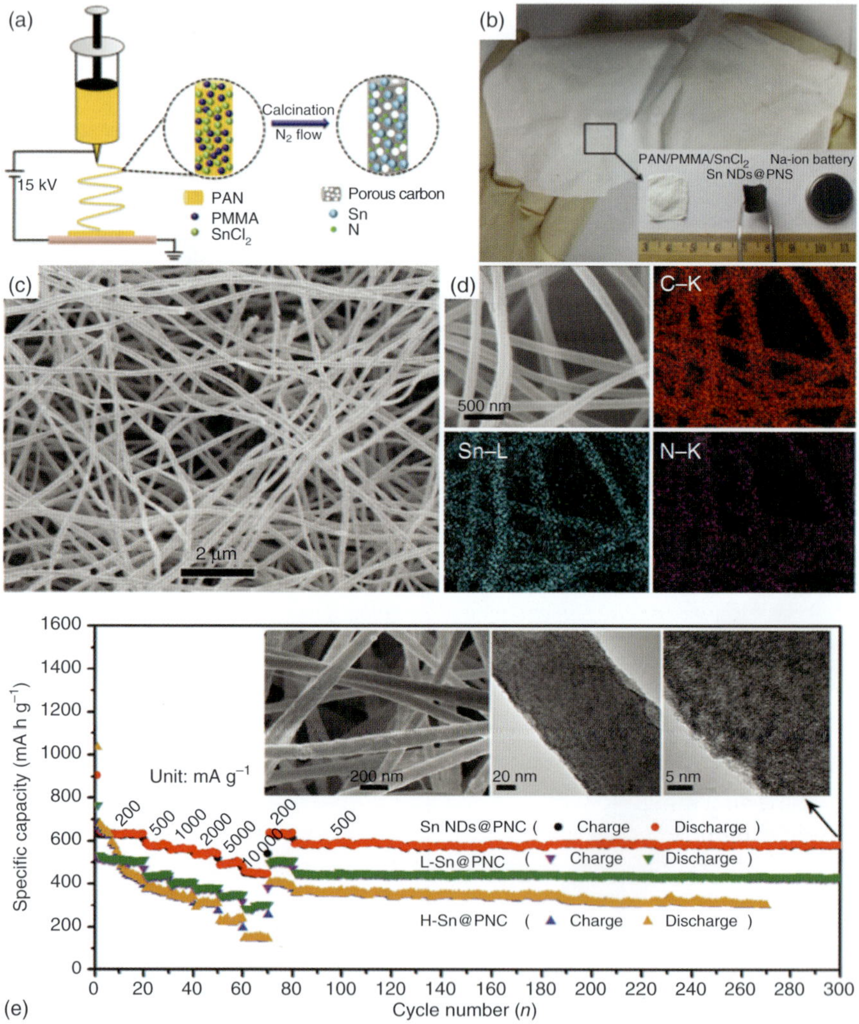

Figure 4.5 (a) Schematic illustration of the preparation process for Sn NDs@PNC nanofibers. (b) Digital photos of the as-spun PAN/PMMA/$SnCl_2$ fiber membrane and the calcinated self-supported Sn NDs@PNC electrode for Na ion batteries. (c) SEM, (d) EDS mapping. (e) Rate capability and cycling performance of Sn NDs@PNC, lower Sn content (L-Sn@PNC), and higher Sn content (H-Sn@PNC) electrodes, inset: SEM, TEM, and HRTEM images of Sn NDs@PNC after 300 cycles. *Source:* Liu et al. 2015 [54]. Copyright 2015. Reproduced with permission from John Wiley and Sons.

outstanding rate capability and ultralong cycling performance, and a high specific capacity of 450 mAh g^{-1} can be delivered at a current density of 10 A g^{-1}. When cycling at 2 A g^{-1} for 1300 cycles, only 10% capacity loss is observed at such a high rate, which is the highest performance reported for tin electrode in SIBs. They further evaluate its practicability in a sodium ion full cell. When coupled with $NaVPO_4F/C$ cathode, the full cell delivers an initial capacity of 540 mAh g^{-1} with a capacity retention of 85.2% over 100 cycles [54]. Among the reported alloy-based materials, Sb seems to the most promising anode candidates for SIBs due to its large capacity of 660 mAh g^{-1} and high electrical conductivity. However, the large volume change during cycling is difficult for the application of Sb-based materials in SIBs. Wang and coworkers also fabricated antimony/carbon fibers electrode by a simple electro-spinning method, where ~30 nm Sb nanoparticles were well dispersed inside the carbon fibers [55]. The binder-free film exhibits high specific capacity and excellent cycling stability with 92% capacity retention after 100 cycles and 75% over 300 cycles [55].

Phosphorus appears to be the most attractive anode candidate for SIBs because of its ultrahigh theoretical capacity of 2596 mAh g^{-1}, attractive potential plateau, and low cost [56]. The main disadvantage is its low conductivity and rapid capacity degradation caused by large volume expansion (>490%) during cycling. Zhang et al. synthesized amorphous P@N-doped graphene film via the well-known "phase-transformation" approach with superior capacity retention of 85% over 350 cycles and rate capability; a high specific capacity of 809 mAh g^{-1} can be obtained at 1.5 A g^{-1}, as shown in Figure 4.6 [57]. For the metal oxide anode, there is an increasing interest in developing flexible electrodes for SIBs, such as hollow γ-Fe_2O_3 nanoparticles/CNTs composite film [58], γ-Fe_2O_3film grown directly on a copper sheet [59], and CuO quantum dots embedded in carbon nanofibers [60].

Due to their low toxicity, low cost, wide abundance, and acceptable voltage, titanium-based materials including TiO_2 and Na–Ti–O systems are attractive for sodium ion storage. TiO_2 has achieved great success as an anode material for LIBs. Compared to lithium ion storage, the operating potential for TiO_2 must be lowered to more negative values for sodium ion insertion. Mitlin and coworkers [61] successfully employed anatase TiO_2 nanocrystals as anode materials for SIBs for the first time. The mesoporous TiO_2 nanocrystals exhibited a stable reversible sodium ion storage capacity of 150 mAh g^{-1} at the cycling rate of 50 mA g^{-1}. The redox potential is about 0.60 (cathodic) and 0.95 V (anodic) [61]. Tirado and coworkers [62] reported 1D self-supported TiO_2 nanotube arrays through oxidation of a titanium substrate at high voltage of 100 V. The TiO_2 nanotube obtained by this method is amorphous at room temperature but it becomes crystallized to anatase or rutile by annealing. The sodium storage mechanism of faradic and pseudocapacitive reaction was studied for anatase TiO_2 nanotube by cyclic voltammetry (CV) technology. According to their investigations, the electrochemical reaction is dominated by pseudocapacitance (surface redox reaction) in the region over 0.5 V and by irreversible faradic reaction below 0.3 V. The pseudocapacitive contribution was also confirmed by ^{23}Na MAS NMR spectra. In order to enhance the electrochemical performance of TiO_2, Cao and coworkers [63] employed rGO as a flexible and conductive

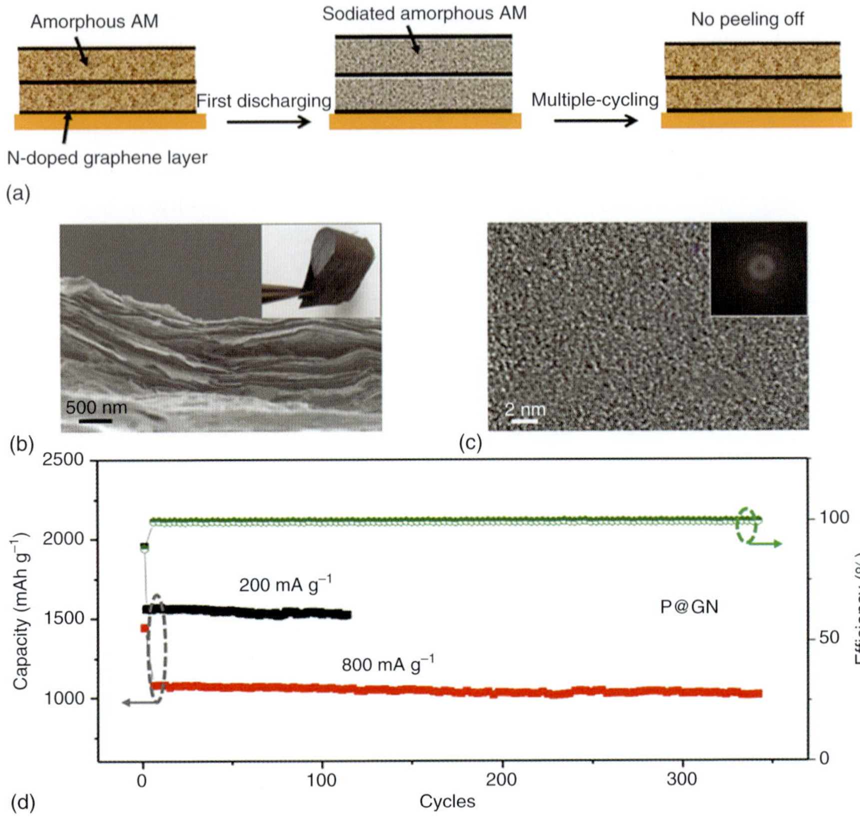

Figure 4.6 (a) Illustrative scheme of the designed novel "butter-bread"-like anode structure consisting of amorphous P layer@N-doped graphene frameworks. (b) SEM image of the cross-section of a P@GN paper, the inset shows its paper-like appearance. (c) HRTEM image and the corresponding FFT pattern of the P@GN portion, confirming its amorphous structure. (d) Cyclic performance and Coulombic efficiency of P@GN at 200 and 800 mA g^{-1}. *Source:* Zhang et al. 2016 [57]. Copyright 2016. Reproduced with permission from American Chemical Society.

supporter by a facile spray-drying method. The TiO_2@rGO electrode exhibited high reversible capacity of 170 mAh g^{-1} at 0.1 C (1 C = 330 mA g^{-1}) and high rate capability of 55 mAh g^{-1} at 10 C. After 300 cycles at 0.2 C, the capacity is about 87%.

Recently, a few layered (or tunnel structures) titanium-based oxides, Na–Ti–O compounds, were investigated for SIB anodes utilizing the topotactic intercalation mechanism [8, 64]. Most of these compounds have the general structural features of $Na_2Ti_3O_7$ and $Na_2Ti_6O_{13}$ in that they consist of TiO_6 octahedra connected by edges and corners to form either layered or tunnel structures [64]. Palacín and coworkers [65] first reported that two sodium ions could be intercalated into the layered $Na_2Ti_3O_7$ with a theoretical capacity of 200 mAh g^{-1} at an average insertion potential of 0.3 V. However, Jiao and coworkers revealed that the theoretical capacity of $Na_2Ti_3O_7$ was 311 mAh g^{-1} by first-principles geometry analysis [66]. They presumed that the resultant difference between the practical capacity and simulation values might be due to the charge–discharge current density being too high to allow the complete intercalation reaction of sodium [66]. Recently, Li et al. developed a series of self-supported $Na_2Ti_3O_7$ nanotube arrays on titanium foil by hydrothermal reaction and subsequent hydrogenation or sulfidation treatment. These self-supported materials can be used directly as the working electrode without any conductive agent and binder. Although the electrochemical performance of the self-supported electrodes is good, the flexibility of the electrodes is defective. In order to fabricate a flexible electrode, Wang and coworkers [67] employed carbon cloth as a flexible and conductive substrate and then designed an ultralong $Na_2Ti_3O_7$ nanowires/carbon cloth (L-NTO NWs@CC) electrode by a one-pot hydrothermal strategy, as shown in Figure 4.7. The interwoven nanowires and carbon networks enable

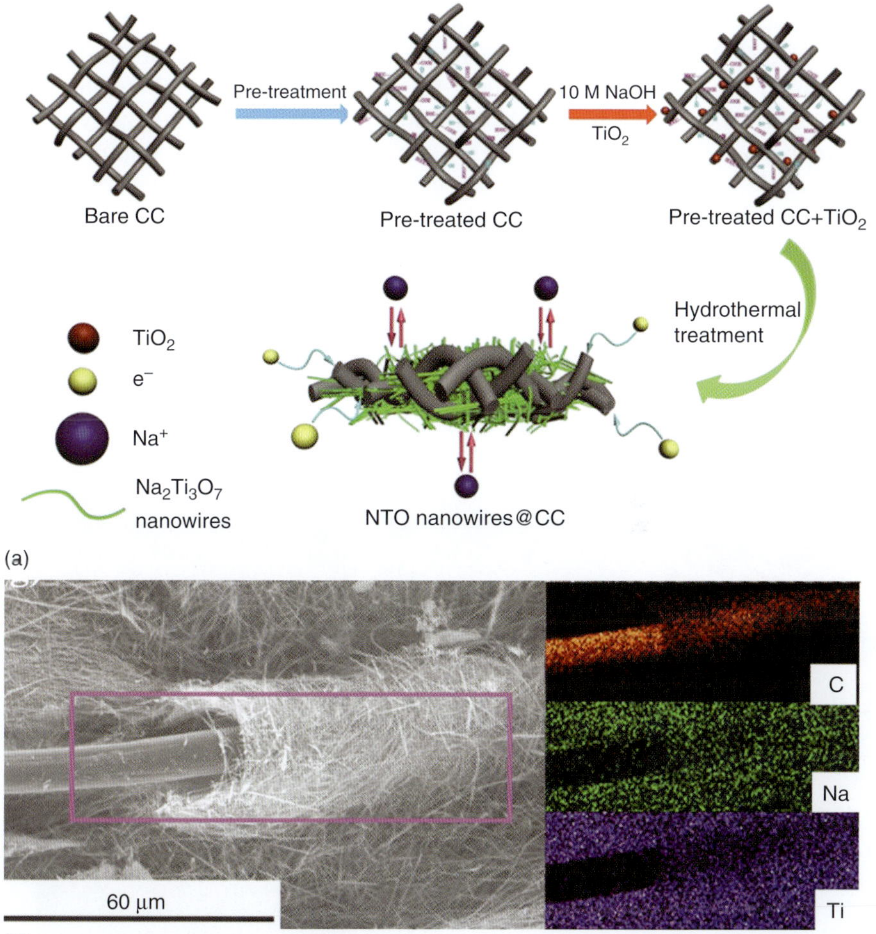

Figure 4.7 (a) Schematic illustration demonstrating the fabrication procedure of the L-NTO NWs@CC composites. (b) SEM image and corresponding EDX mapping of C, Na, and Ti. *Source:* Li et al. 2016 [67]. Copyright 2016. Reproduced with permission from Royal Society of Chemistry.

sufficient electro-active material/electrolyte contact and fast charge transportation. As a self-supported flexible electrode for SIB, this L-NTO NWs@CC anode exhibits a high specific capacity of 170 $mAh\,g^{-1}$ and outstanding cycling stability (96% of capacity retention after 200 cycles at a current density of 2 C). Even at 3 C (1 C = 177 $mA\,g^{-1}$), the discharge capacity still remains as 100.6 $mAh\,g^{-1}$ after 300 cycles.

$Na_2Ti_6O_{13}$ is another most investigated titanium-based material with a tunnel structure. Two sodium ions are located in each tunnel but there are only three cubic sites in all, suggesting that only one sodium can insert into $Na_2Ti_6O_{13}$, corresponding to approximately 50 $mAh\,g^{-1}$. Brousse and coworkers [68] first reported the electrochemical properties of $Na_2Ti_6O_{13}$ in sodium storage and found that only half a mole of sodium could effect reversible insertion/extraction. Though the capacity of $Na_2Ti_6O_{13}$ is quite low, it exhibits impressive cycling performance of over 5000 cycles and high rate capability (over 30 C) [69]. Recent works have shown that a combination of the features of the layered and tunnel-like structures is expected to enhance the structural stability and rate performance of the $Na_2Ti_3O_7$ and $Na_2Ti_6O_{13}$. Wang et al. [70] developed self-supported sodium titanate (NTO) nanotube arrays on the titanium substrate, as shown in Figure 4.8, which exhibit long cycle life (5000 cycles) with a reversible capacity of 55 $mAh\,g^{-1}$ [70].

4.3.2 Inorganic Cathode Materials for Flexible SIBs

Since the resurgence of research on SIBs in 2010, various positive electrode materials have been prepared and tested for SIB applications. The recent progress in positive materials for SIBs mainly focused on layered transition metal oxide

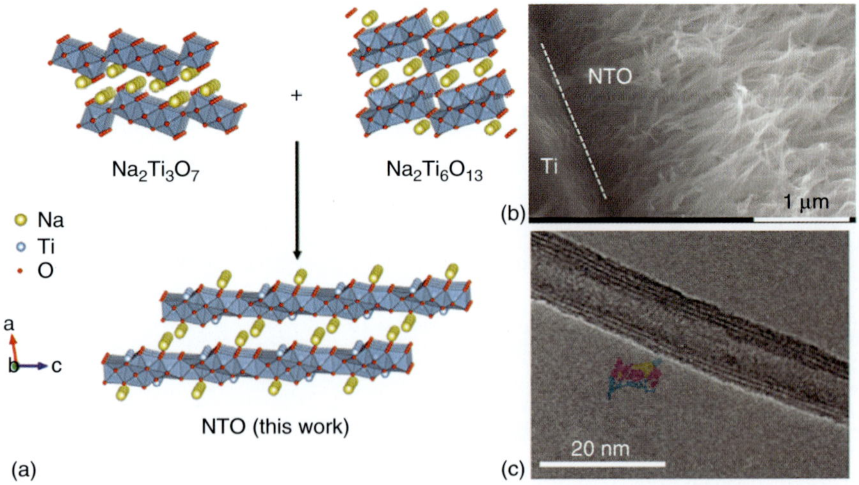

Figure 4.8 (a) The schematic structures of some sodium titanates. SEM (b) and TEM (c) images of sodium titanate nanotube arrays. *Source:* Wang et al. 2015 [70]. Copyright 2015. Reproduced with permission from Elsevier.

materials (P2-type, O3-type layered metal oxide), polyanionic compound (phosphates, fluorophosphates, fluorosulfates, sulfates, pyrophosphates), Prussian blue analogues, and the promising organic cathodes [6]. Nonetheless, studies on flexible SIB cathode are still in their infancy compared with LIB technology. In this section, we briefly present the most recent progress and discovery of promising cathode materials and their sodium storage mechanisms for flexible SIBs as well as the efforts to boost their electrochemical properties.

Flexible and binder-free electrode consisting of $Na_{2/3}[Ni_{1/3}Mn_{2/3}]O_2$ cathode and rGO has been fabricated by a simple vacuum filtration process. The π—π bonds of rGO impart strong intercalation between rGO sheets and $Na_{2/3}[Ni_{1/3}Mn_{2/3}]O_2$ particles, which significantly enhance electrical conductivity of $Na_{2/3}[Ni_{1/3}Mn_{2/3}]O_2$ particles [71]. The electrode delivered a specific capacity of 86 mAh g^{-1} at 0.1 C, good excellent stability of 200 cycles and keeps a capacity retention of 68.4% at 10 C current density [71]. Polyanionic compounds are another kind of potential cathode material for SIBs because of their structural robustness and stability. $Na_{2+2x}Fe_{2-x}(SO_4)_3$@P-CNF hybrid films were synthesized by the combined use of electro-spinning and electro-spraying techniques and employed as self-supported and flexible electrodes for SIBs (Figure 4.9) [72]. By combining the merits of the flexible structure with decreased inert components, nanoparticle morphology, highly conductivity, and excellent stability, the free-standing hybrid film exhibits superior sodium intercalation kinetics and ultralong life of 500 cycles with the capacity retention of >95% at 40 C [72]. Zhang et al. reported the hierarchical mixed polyanion material $Na_7V_4(P_2O_7)_4(PO_4)$/C nanorod–graphene composite as flexible electrode for lithium and sodium storage (Figure 4.10a) [73]. The composite consists of carbon coated $Na_7V_4(P_2O_7)_4(PO_4)$/C nanorod wrapped by a rGO sheet, which approaches high initial reversible specific capacities with reasonable discharge potentials over 3.8 V, high rate capabilities, and good cycling properties with capacity retentions of 95% and 83% after 200 cycles at a 1 C rate in sodium and lithium battery systems, respectively. The enhanced electrochemical properties can be contributed to the bicontinuous electron and ion pathways of the open network structure, improved conductivity, and large surface area (Figure 4.9b) [73]. Owing to the large radius of sodium (0.97 vs 0.68 Å for lithium) and heavy atomic mass, the sodium intercalation/de-intercalation in conventional polyanionic compounds often leads to severe structure degradation and sluggish diffusion kinetics. Vanadium oxides with a layered structure have been recently investigated as promising candidates for SIBs because of their high theoretical specific capacity, large interlayer space, low cost, and abundant sources. Mai's group studied the flexible vanadium oxide hydrate $H_2V_3O_8$ nanowire membrane synthesized under a hydrothermal and suction filtration process method as binder-free cathode for SIB application, which shows high specific capacity and long cycling performance for about 280 cycles [74]. They also investigated the sodium storage mechanisms by using CV, ex situ X-ray diffraction (XRD), Fourier transform infrared spectroscopy (FTIR), and X-ray photoelectron spectroscopy (XPS) measurements. Similarly, free-standing carbon quantum dot coated VO_2 interwoven nanowires as flexible cathode for lithium and SIBs were presented by Tong's group for the first time [75], where the VO_2 nanowires were firstly

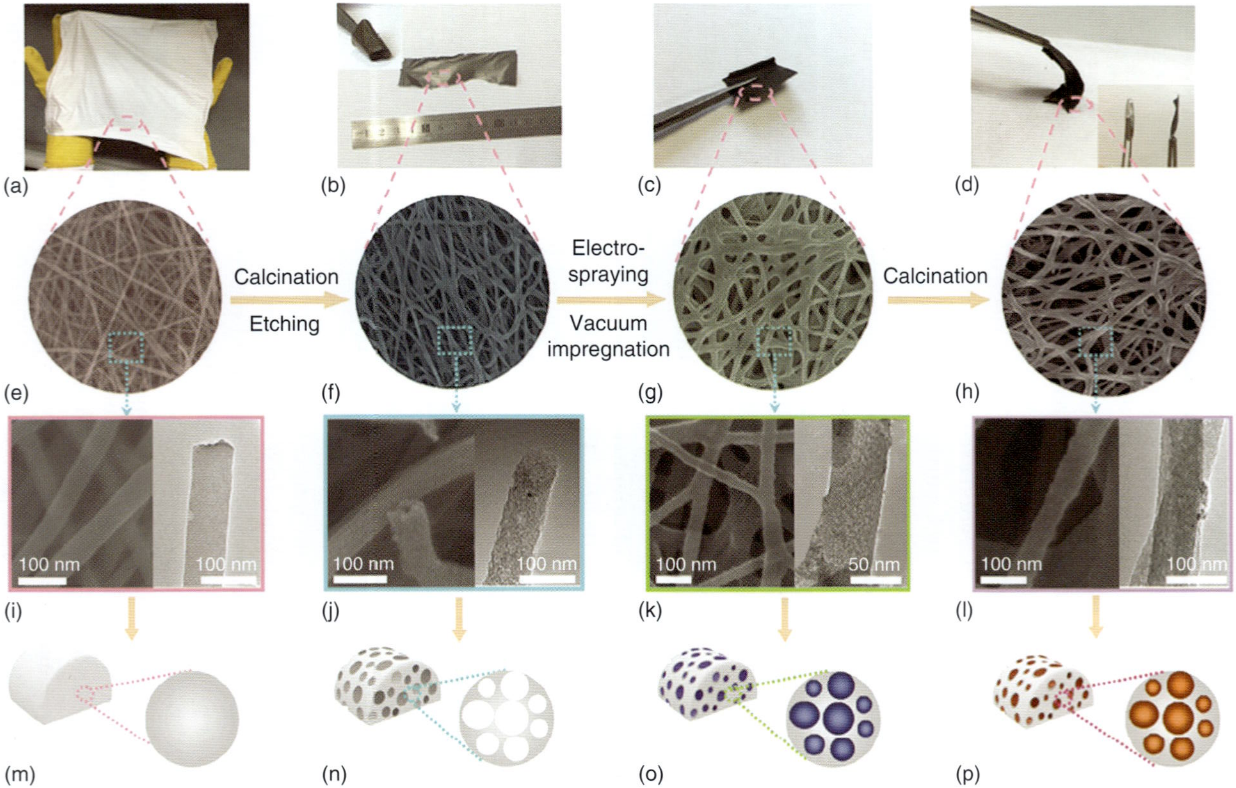

Figure 4.9 Fabrication of the alluaudite $Na_{2+2x}Fe_{2-x}(SO_4)_3$@porous carbon nanofiber (PCNF) hybrid film. (a–d) Digital photos, (e–h) SEM images, (i–l) TEM images and (m–p) schematic illustrations of the (e–h) porous carbon nanofiber. *Source:* Yu et al. 2016 [72]. Copyright 2016. Reproduced with permission from Royal Society of Chemistry.

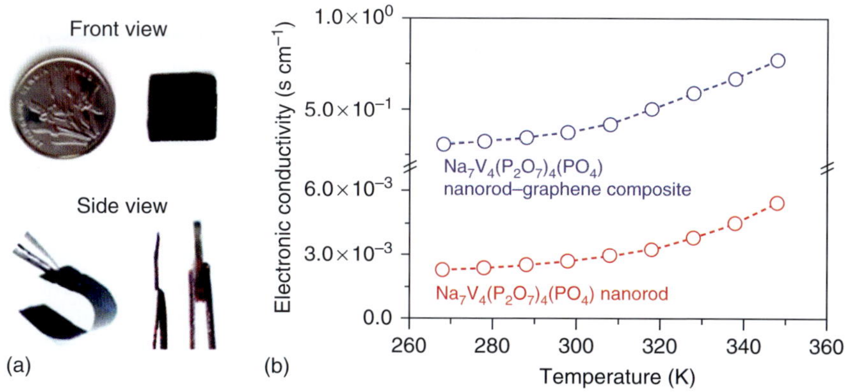

Figure 4.10 Front view and side view of the flexible and binder-free $Na_7V_4(P_2O_7)_4(PO_4)$/C nanorod–graphene thin film. (b) Temperature dependence of the electronic conductivity of the $Na_7V_4(P_2O_7)_4(PO_4)$/C nanorod–graphene film and the $Na_7V_4(P_2O_7)_4(PO_4)$/C nanorod pellet. *Source:* Zhang et al. 2014 [73]. Copyright 2014. Reproduced with permission from Royal Society of Chemistry.

directly grown on a carbon cloth substrate and followed by surface engineering using carbon quantum dots. The carbon quantum dots could serve the function of protecting the nanowire surface and play an important role for charge transfer, improving electrical conductivity and exhibiting favorable kinetics during battery operation [75]. Free-standing bulky paper fabricated by suction filtration of V_2O_5 nanobelts obtained by ambient dissolution–recrystallization process and multi-walled carbon nanotubes (MWCNTs) also show good electrochemical performance for LIBs and SIBs [76].

Prussian blue and its analogues perform as excellent alkali-ion rechargeable battery electrodes owing to their inherently open framework structure with large interstitial sites, low cost, and easy accessibility [77, 78]. Importantly, Prussian blue analogues have been demonstrated as potential hosts for reversible di–/tri-valent ion insertion, such as Mg^{2+}, Ca^{2+}, Pb^{2+}, Zn^{2+}, Al^{3+}, and Y^{3+} in aqueous/nonaqueous electrolytes. However, their application is hampered by insufficient cycle life and low energy efficiency in nonaqueous electrolytes for SIBs, associated with the low electrical conductivity and structural imperfection. Zhang and coworkers reported a significant advance in the design and preparation of Prussian blue analogue $FeFe(CN)_6{\cdot}xH_2O$ nanoparticles on a flexible carbon fiber paper substrate as a binder-free cathode in both organic and aqueous electrolytes by using a simple solution precipitation method [79]. Its sodium storage mechanism and improved electrochemical performance were examined by ex situ ^{57}Fe Mossbauer spectroscopy and CV technology. With the merits of flexible structure and high conductivity, the resulting $FeFe(CN)_6{\cdot}xH_2O$ nanocomposite electrode exhibits a reversible specific capacity of 82 mAh g^{-1} at 0.2 C, excellent cycling performance for 1000 cycles with 81.2% capacity retention, and good high rate response up to 10 C.

In addition, sodium–air batteries have emerged as potential alternatives to lithium–air batteries owing to resource abundance, cost competition, safety, and environmental friendliness. Kim, Ko, and coworkers synthesized a binder-free and robust electrode by directly growing hierarchical urchin-shaped α-MnO_2 on rGO-coated carbon microfiber (MGC) and used it as an air electrode for aqueous sodium–air battery [80]. Scanning electron microscopy (SEM) and transmission electron microscopy (TEM) images showed urchin-shaped MnO_2 microspheres composed of radially oriented MnO_2 nanowires with diameters of 20–25 nm uniformly covered on the carbon fibers. Owing to the urchin-shaped structure of the MnO_2, high surface area and the superior electrocatalytic activity of the MGC, the assembled aqueous Na–air battery exhibited excellent recharge ability up to 20 cycles with a low over potential gap of 0.7 V and high round-trip efficiency (81%), which is comparable to that of the battery using the noble metal Pt.

4.3.3 Organic Materials for Flexible SIBs

Organic electrode materials are attracting increasing attention as alternatives to conventional inorganic electrode materials due to their sustainability, universal availability, better compatibility with flexible substrates, and simpler packaging, which enable flexible SIBs to have better flexibility [81, 82]. Different from inorganic materials, which is naturally brittle, organic materials possess flexible and soft structures. The volume change of organic materials could be negligible after intercalation of sodium [83]. Although some innovative works have been developed in the design of organic material-based SIBs, their performance is still far behind that of inorganic material-based devices, especially in terms of cycling performance. It is very important to highlight that the physical and chemical properties of organic materials can be adjusted by chemical syntheses with different functional groups. The working voltage can be regulated through controlling the lowest unoccupied molecular orbital (LUMO) energies [84]. For example, adding electron-donating (such as –ONa, $-CH_3$, $-NH_2$, etc.) or electron-withdrawing (such as –F, –Cl, –Br, $-NO_2$, $-OCH_3$, etc.) groups, the LUMO energies can be rationally enhanced or lowered, thus leading to reduced or enhanced working voltage. Increasing the active groups can enhance the theoretical specific capacity. The dissolution in electrolyte can be inhibited by formatting salt or/and elevating the intermolecular forces. The conductivity can be enhanced by extending the conjugated structures in the core unit.

Quinone-based organics is a promising cathode material for flexible SIBs. However, the most quinone-based cathode usually suffers from low working voltage (under 2.5 V vs Na/Na^+). Park, Kang, and coworkers reported a quinone-derived organic cathode, tetrachloro-1,4-benzoquinone ($C_6Cl_4O_2$), by introducing electronegative elements with high energy density of 580 Wh kg^{-1} and the highest working voltages among organic materials (2.72 V vs Na/Na^+) [85]. Density functional theory (DFT) calculations demonstrated that the introduction of electronegative elements (here it is Cl) into the quinone structure lowered the LUMO energies of the $C_6Cl_4O_2$, which in turn increased the sodium storage voltage [85]. To inhibit the dissolution of tetrachloro-1,4-benzoquinone, they incorporated $C_6Cl_4O_2$ molecules into mesoporous carbon (CMK). The $C_6Cl_4O_2$/CMK delivered higher capacity and longer cycling stability.

Most of the organic electrode materials were tested in a very slow charge/discharge rate (e.g. 10 $mA\,g^{-1}$) and were observed to suffer from significant capacity degradation under high rate [86, 87], which is related to the low conductivity of organics. To improve the rate performance, Lei et al. reasonably designed sodium 4,4′-stilbene-dicarboxylate (SSDC) for anode material by extension of the π-conjugated structure [88]. As shown in Figure 4.11, the extension of the π-conjugated system, SSDC, delivered a high reversible capacity of 220 $mAh\,g^{-1}$ at a current density of 50 $mA\,g^{-1}$. Even at ultrahigh current density of 10 $A\,g^{-1}$, a reversible capacity of 72 $mAh\,g^{-1}$can be obtained. The capacity retention is over 70% after 400 cycles at 1 $A\,g^{-1}$. The wonderful electrochemical performance of SSDC, probably due to the extension of π-conjugated structure can: (i) enhance the ion/electron transport and (ii) increase the stability of the charged and discharged states [88]. Recently, Lei and coworkers systematically investigated the important factors for high performance organic electrode materials by using disodium rhodizonate salt (DSR) as a model [84]. They found that the solubility, morphology, electrochemical window and binder are very important for high performance SIBs, especially, with high reversible capacity and long-term cycling stability under fast-charge/discharge reaction [84].

4.3.4 Other Major Components for Flexible SIBs (Electrolyte, Separators, etc.)

Electrolytes and separators are also very important for flexible SIBs, especially in terms of safety, because flexible devices usually integrate with the human body. In conventional liquid organic electrolytes, sodium dendrite issue is an intractable problem, which will cause disastrous security problems. In this regard, solid electrolytes may be an ideal solution to these crucial issues.

Although solid-state ceramic electrolytes have been developed for many years and show good safety in LIBs, they are inapplicable for flexible devices due to their rigid nature and low ionic conductivity (<0.1 $mS\,cm^{-1}$) [89]. Gel polymer electrolytes (GPEs) may be a better choice for flexible SIBs, which usually consist of conventional liquid electrolytes and polymers with good mechanical strength and high ionic conductivity [90–92]. The polymers usually act as separators, while the liquid electrolytes act as charge carriers. Wu and coworkers [93] reported a sodium ion conducting gel polymer electrolyte (S-GPE) based on poly(vinylidene difluoride-co-hexafluoropropylene), P(VDF-HFP) porous membrane. Then the P(VDF-HFP) membrane was soaked in conventional organic electrolyte to obtain the S-GPE. The GPE possesses high ionic conductivity of 0.60 $mS\,cm^{-1}$ at ambient temperature, good mechanical strength and thermostability, and high safety.

In order to further increase the ionic conductivity, developing a polymer/oxide composite membrane separator is an alternative strategy. Archer and coworkers [94] reported a sandwich-type composite membrane by laminating a nanoporous sheet of γ-Al_2O_3 with P(VDF-HFP). Although the membrane had an acceptable ionic conductivity and a good mechanical modulus, it lacked flexibility [94]. Lately, Goodenough and coworkers [95] developed a polymer P(VDF-HFP)-polyvinylpyrrolidone (PVP)/Sb_2O_3 composite membrane separator for SIBs with ultraflexible, dendrite-blocking and low-cost characteristics

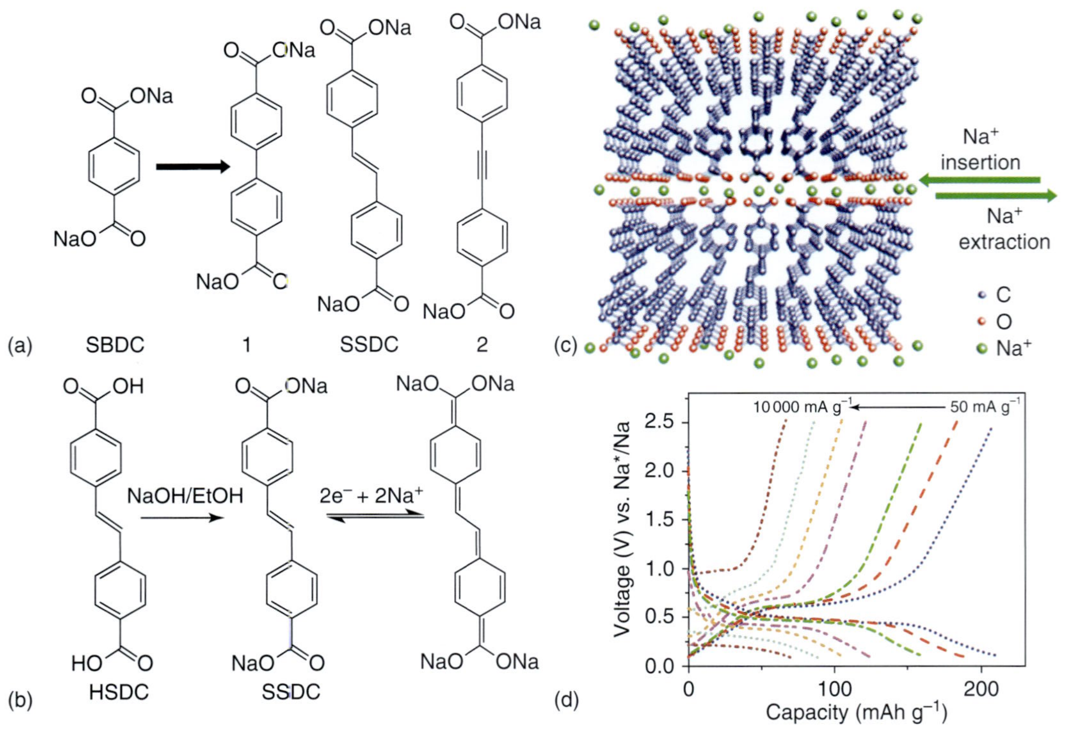

Figure 4.11 (a) Schematic chemical structure of starting molecule sodium benzene-dicarboxylate (SBDC) and three possible linkages to extend its π-conjugation, forming 1, SSDC, and 2, respectively. (b) Schematic diagram for the synthesis and reversible Na ion insertion/extraction mechanism of SSDC. (c) Schematic molecular packing of SSDC: the molecules are stacking layer-by-layer with strong π–π intermolecular interaction in plane and carboxylate group situated on the surface of the layers, forming a channel exactly at the active center for insertion/extraction of sodium ions between layers. (d) Rate capability of SSDC. *Source:* Wang et al. 2015 [88]. Copyright 2015. Reproduced with permission from American Chemical Society.

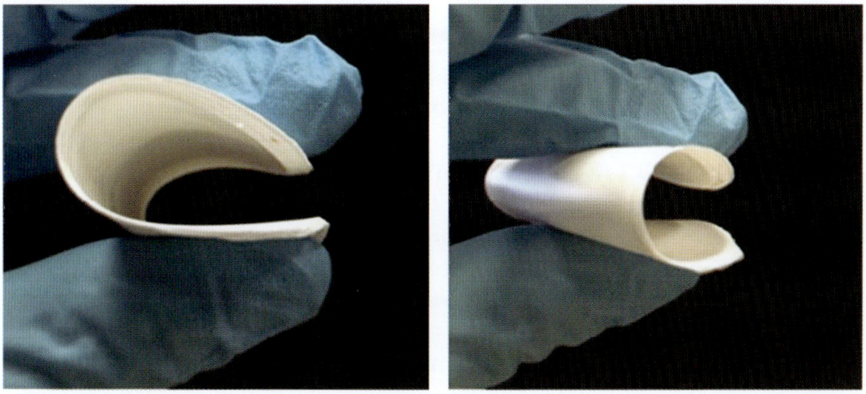

Figure 4.12 Digital picture of P(VDF-HFP)- (PVP)/Sb_2O_3 composite membrane. *Source:* Ansari et al. 2014 [95]. Copyright 2014. Reproduced with permission from The Electrochemical Society.

(Figure 4.12). The larger particles of 5 μm Sb_2O_3 provide a better flexibility than a nanoporous sheet of γ-Al_2O_3. The PVP forms pores that are large enough to incorporate an organic electrolyte and are small enough to retain the Sb_2O_3 particles [95]. The composite membrane separator has a wide electrolyte window, and wonderful mechanical and thermal stabilities when soaked with 1 M $NaClO_4$ ethylene carbonate/diethyl carbonate (EC/DEC) electrolyte. The sodium ion conductivity of this S-GPE is 6.4 $mS\,cm^{-1}$ at 25 °C. Even at 250 °C, the P(VDF-HFP)-(PVP)/Sb_2O_3 composite membrane is stable.

Although using oxides, such as Al_2O_3, SiO_2, TiO_2, etc., can improve the ionic conductivity of the polymer electrolytes [94, 96–98], this method also causes poor thermal and mechanical stability to some extent. To address this problem, using ionic liquids to replace the conventional metal oxides may be a good strategy due to their wide liquidus range, good electrochemical stability, high ionic conductivity, and excellent thermal stability. Besides, ionic liquids also can act as the supplier of free ion charge carriers. Singh et al. [98] developed a novel S-GPE membrane based on polymer, poly(ethyleneoxide) (PEO); ionic liquid, 1-butyl-3-methylimidazolium methylsulfate (BMIM-MS); and sodium salt, sodium methyl sulfate salt (NaMS), by solution cast technique. The as-prepared polymer electrolyte membranes are free-standing and flexible with good mechanical stability. This optimized polymer electrolyte shows high electrochemical potential window of 4–5 V with good thermal stability over 300 °C [98].

4.4 Design of Full Cell for Flexible SIBs

We have reviewed the development of the major components for flexible SIBs. Next, we will focus on the design of full cell for SIBs. There is an urgent need but it is also a critical challenge to design flexible SIBs full cell with high performance.

Cui and coworkers first prepared a flexible porous polyimide graphite film (PGF) by laser drilling pore technique [99]. The PGF possesses considerable mechanical strength, high planar electrical conductivity, and enhanced diffusion kinetics. They further designed a novel full cell based on PGF anode and $Na_3V_2(PO_4)_3$ cathode. The electrolyte is 1 M $NaPF_6$ in diglyme. This unique energy storage system makes the best of co-intercalation behavior of sodium ion and solvent into a porous PGF host. The SIB full cell delivers improved energy density of 130.7 Wh kg^{-1} based on the total electrode materials and prolonged cycling behavior of over 500 cycles. The porous PGF is flexible, but the full cell is also rigid. Yan and coworkers [100] reported a flexible full cell assembled with N-doped carbon sheets (NDCS)//$Na_3V_2(PO_4)_3$(NVP) with a voltage window of 0.7–4.2 V. The anode material, NDCS, with a high N content of about 9.89 at.% was prepared by using biomass byproduct okara as the sources of both C and N [100]. The cathode material NVP was synthesized by the solution-based carbo-thermal reduction method, which delivered 99.5 mAh g^{-1} with the plateau of 3.6 V. The full cell delivered a stable capacity of 103 mAh g^{-1}at the current density of 1.5 C (1 C = 375 mA g^{-1}) where the electrolyte was 1 M $NaClO_4$ in propylene carbonate (PC) and fluoroethylene carbonate (FEC, 5%). It is important to highlight that the NDCS//NVP was easy to power a wearable light-emitting diode (LED) screen, as can be seen in Figure 4.13a,b.

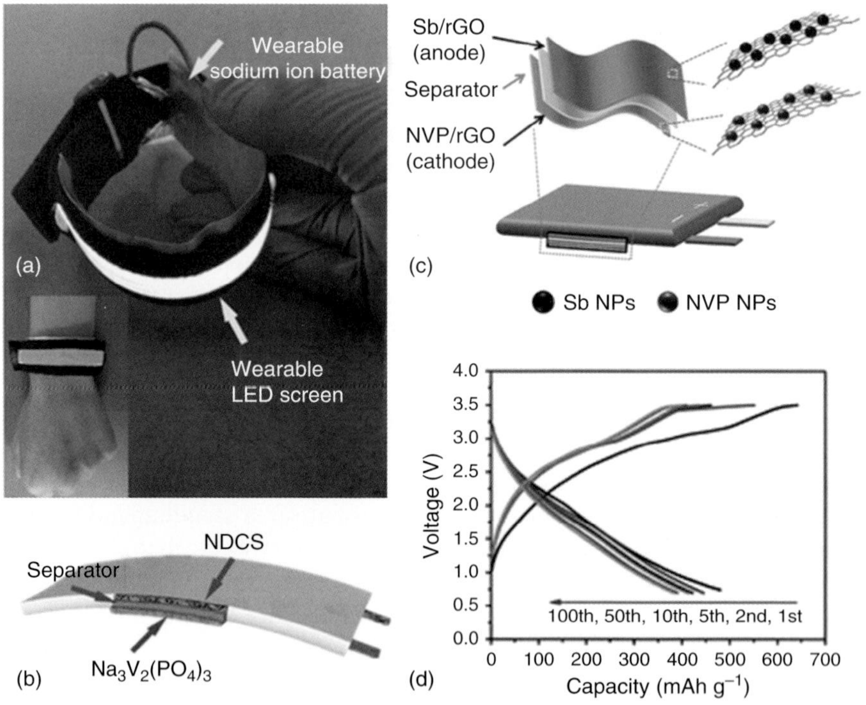

Figure 4.13 (a) The flexible SIB full cell (NDCS//NVP) was linked to a flexible/wearable LED screen. (b) The schematic plan of full cell assembled with NDCS//NVP. *Source:* Yang et al. 2016 [100]. Copyright 2016. Reproduced with permission from John Wiley and Sons. (c) Schematic illustration for the fabrication of the flexible Sb/rGO//NVP/rGO SIB full cell. (d) Galvanostatic charge/discharge curves at a current density of 100 mA g^{-1} over the voltage range of 0.7–3.5 V. *Source:* Zhang et al. 2015 [101]. Copyright 2015. Reproduced with permission from John Wiley and Sons.

Hu and coworkers [101] designed the flexible and binder-free anode and cathode of nanocomposites of Sb/rGO and NVP/rGO for flexible SIBs. The unique interconnected electrode network, excellent electrical conductivity, and homogenous loading of Sb and NVP nanoparticles enable short ion/electron transport path and volume flexibility, thus leading to superior reversible capacities and cycling stability. As shown in Figure 4.13c,d, the flexible SIB, which was fabricated by soft package technology, delivered a high capacity of 400 $mAh\,g^{-1}$ at 100 $mA\,g^{-1}$ after 100 cycles. Later, Chen and coworkers [102] assembled a flexible sodium ion full battery with $MnFe_2O_4$(MFO)@C//$Na_3V_2(PO_4)_2F_3$ (NVPF)/C using a similar Al-plastic soft package technology, as shown in Figure 4.14a,b. The MFO@C nanofibers, MFO nanodots encapsulated in porous N-doped carbon nanofibers, were prepared by an electro-spinning strategy, which can be directly employed as flexible free-standing anode with ultrahigh rate character of 305 $mAh\,g^{-1}$ even at 10 $A\,g^{-1}$ and ultralong cycling life over 4200 cycles. When combined with NVPF/C cathode at an optimized electro-active material mass ratio of anode to cathode at 1 : 5, the flexible SIB delivered a stable capacity of 400 $mAh\,g^{-1}$ based on MFO@C and a high energy density of 77.8 $Wh\,kg^{-1}$ based on the total mass of the flexible SIB (6.72 g in this work) [102].

Recently, Zhang and coworkers [11] developed a novel sodium ion energy storage device, called flexible sodium ion capacitor (SIC), based on flexible $Na_2Ti_3O_7$ nanosheets/carbon texibles (NTO/CT) anode and flexible reduced graphene oxide films (GFs) cathode. In this flexible device, sodium ions are intercalated into NTO/CT anodes during charging; meanwhile, anions (ClO_4^- in this work) are adsorbed on the surface of the GFs cathode [11]. The above process is reversed during the discharge process. This novel device integrated the advantages of both flexible SIBs and flexible supercapacitors. The flexible NTO/CTs//GFs configuration was sealed with PDMS with a voltage window of 1–3 V. Taking the fully packed flexible SIC into consideration, the volumetric energy density and power density can reach up to 1.3 $mWh\,cm^{-3}$ and 70 $mW\,cm^{-3}$, respectively. It must also be mentioned that the flexible device exhibited a stable electrochemical performance even under various mechanical deformation (Figure 4.14c,d).

Safety is also a key issue for flexible energy storage device in practical applications. However, all the above mentioned flexible sodium ion energy storage devices use the liquid electrolytes, where sodium salt dissolved into organic solvent. When flexible devices are assembled with liquid organic electrolyte, it may give rise to leakage or produce a short circuit in the resulting flexible energy storage devices under mechanical deformation. In this regard, developing solid electrolytes would be a perfect solution to these crucial issues. Therefore, Zhang, Yu, and coworkers. [103] designed a flexible quasi-solid-state sodium ion capacitor (QSS-SIC) based on urchin-like NTO anode and peanut shell derived carbon (PSC) cathode with a kind of sodium ion conducting gel polymer as the separator and electrolyte. Such QSS-SIC can exhibit high-energy and high-power characteristics and long cycling life because of the integrated storage mechanisms of batteries and supercapacitors. Based on the total mass of the electrode materials, the energy density and power density can reach to 111.2 $Wh\,kg^{-1}$ and 11 200 $W\,kg^{-1}$, respectively. The QSS-SIC also delivers a wonderful cycling stability with capacity retention of 86% after 3000 cycles.

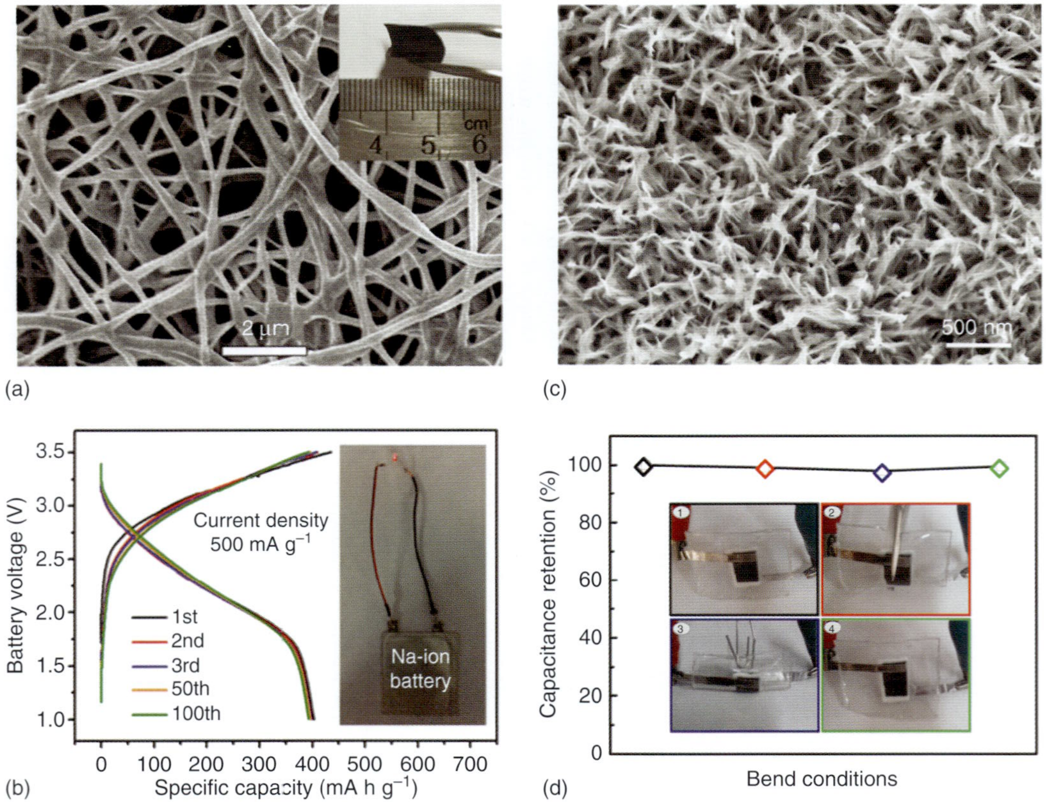

Figure 4.14 (a) SEM image and digital photo (inset) of the MFO@C electrode charged at 3.0 V after 300 cycles. (b) Charge/discharge curves of flexible SIB full cell with (MFO)@C//NVPF/C. *Source:* Liu et al. 2016 [102]. Copyright 2016. Reproduced with permission from American Chemical Society. (c) SEM images of NTO/CT composite. (d) Capacitance retention of flexible NIC measured under different bent conditions. *Source:* Dong et al. 2016 [11]. Copyright 2016. Reproduced with permission from John Wiley and Sons.

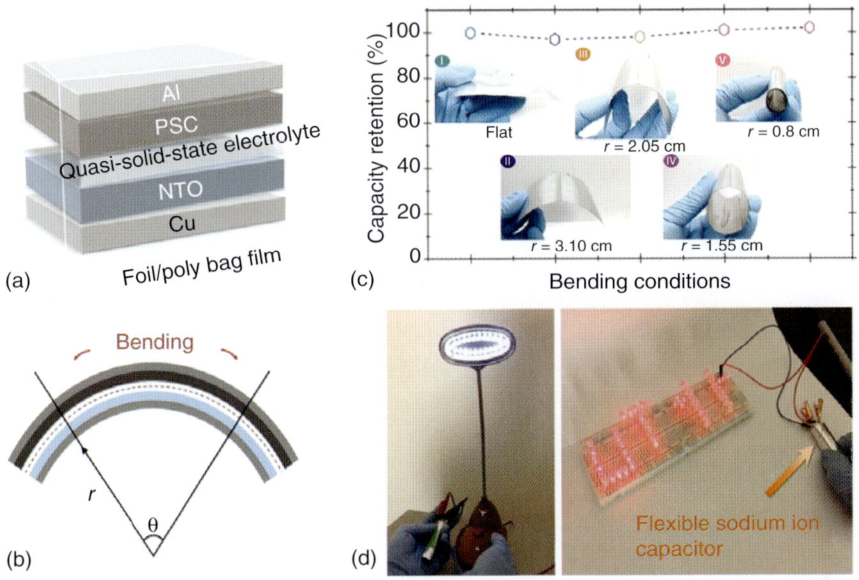

Figure 4.15 (a) Schematic illustration of the structure of the flexible quasi-solid-state SIC. (b) Diagram of the flexible QSS-SIC under various bending conditions. (c) Stability of capacitive performance of the QSS-SIC collected at different bending conditions. (d) A commercial desk lamp and a UT capacitor logo consisting of 50 red LEDs powered by the assembled QSS-SIC under rolled state. *Source:* Li et al. 2016 [103]. Copyright 2016. Reproduced with permission from American Chemical Society.

Furthermore, a flexible QSS-SIC was fabricated by using foil/poly bag film as the package material. It is robust enough to operate under various deformations without significant capacity loss (Figure 4.15) [103]. These findings provide a common strategy for the development of flexible quasi-solid-state energy storage devices. Kim and coworkers [104] reported a NASICON ($Na_3Zr_2Si_2PO_{12}$)-based hybrid solid electrolyte (HSE) for flexible solid-state SIBs for the first time. The HSE shows low interface resistance and high thermal and electrochemical stability. A pouch-type flexible solid-state sodium ion full cell was fabricated comprising $NaFePO_4$/HSE/hard carbon (HC) and using carbon paper as flexible current collector [104]. The $NaFePO_4$/HSE/HC full cell has an average voltage of 2.6 V and achieves a high reversible capacity of 120 mA h g^{-1}. After 200 cycles, it also delivers good capacity retention of 96%. When the solid-state full cell is bent several times, it can continue to power an LED lamp. This means that the flexible solid-state sodium ion full cell possesses wonderful mechanical deformability.

Recently, Yu and coworkers reported an all-stretchable-component SIB full cell based on graphene-modified PDMS sponge electrodes and an elastic gel membrane [105]. This full cell delivered high capacity, good rate capability, long cycling stability, and good mechanical deformation.

4.5 Summary and Outlook

Without doubt, flexible SIBs have become a kind of competitive flexible energy storage devices for flexible and wearable electronic applications in the near future. The key scientific issues and the progress that have been achieved according to the design of electro-active materials, exploration of electrode structure, and fabrication of full cell, and so on have been highlighted. As compared with conventional energy storage devices, in order to realize flexibility of SIBs, the main components should be flexibility and stretchability, as well as the anode, cathode, electrolyte, current collector, and packaging materials. Although some achievements have been made, current flexible SIBs are still far away from satisfying the requirements for flexible and wearable electronic devices. The development in the field of flexible SIB systems is still at the early stage of academic validation. Hence, exploring novel approaches for flexible SIBs remains a greatly academic and industrial challenge. In order to improve the energy density, power density, and reliability of flexible SIBs, several areas can be investigated as follows.

1) *Developing advanced electro-active materials with satisfactory electrochemical performance.* The most commonly used methods include design of nanostructure, conductive coating, and doping with hetero-atoms. Compared with brittle inorganic materials, organic electrode materials have better compatibility with flexible devices, possibly simpler packaging and lower cost. In addition, with rational design, organic materials are able to achieve competitive electrochemical performance.
2) *Design of innovative electrode structure with admirable mechanical and chemical stability.* At present, the major works just focus on design of nanomaterials grown on bendable substrates or as composites. The stretchability and compressibility are still unsatisfied. In order to develop the advanced electrode structure with wonderful bendability as well as stretchability and compressibility, new processes for fabricating flexible electrodes, such as lithography and 3D printing, are needed.
3) Flexible SIBs must sustain severe mechanical deformations that also need advanced flexible electrolytes. Besides, safety is another very important factor for flexible energy storage devices in practical applications. Compared with conventional organic liquid electrolyte, solid-state electrolytes possess better safety and compatibility. However, flexible solid-state electrolytes that have comparable ionic conductivity with liquid electrolyte have not yet been realized.
4) Currently, packaging materials, especially for stretchable packages, with flexible and stretchable features have not attracted sufficient attention. Therefore, modification of reliable packaging materials and optimization of device structures to protect the integrity of energy storage system under various environments should also be considered urgently.
5) Moreover, the development of flexible micro-SIBs with enhanced electrochemical characters also plays an interesting direction in miniaturized flexible energy storage applications, such as microelectronics, microdisplay, implantable medical devices, etc.
6) Modern portable flexible devices not only strongly rely on flexible energy storage devices, but also urgently need multifunctional devices. Integration of

flexible SIBs with other components, for example, with solar cells or sensors, will be very attracting.

Although there are enormous challenges, advanced flexible SIBs with unique properties will provide promising applications in flexible and wearable electronics field in the near future.

References

1 El-Kady, M.F., Strong, V., Dubin, S., and Kaner, R.B. (2012). *Science* 335: 1326.
2 Li, L., Wu, Z., Yuan, S., and Zhang, X.-B. (2014). *Energy Environ. Sci.* 7: 2101.
3 Bao, Z. and Chen, X. (2016). *Adv. Mater.* 28: 4177.
4 Sasaki, T., Ukyo, Y., and Novák, P. (2013). *Nat. Mater.* 12: 569.
5 Shen, L., Dong, S., Zhang, X., and Cao, G. (2017). *Black TiO_2 Nanomaterials for Energy Applications*, 249–273. World Scientific.
6 Kim, H., Kim, H., Ding, Z. et al. (2016). *Adv. Energy Mater.* 6: 1600943.
7 Luo, W., Shen, F., Bommier, C. et al. (2016). *Acc. Chem. Res.* 49: 231.
8 Guo, S., Yi, J., Sun, Y., and Zhou, H. (2016). *Energy Environ. Sci.* 9: 2978.
9 Pan, H., Hu, Y.-S., and Chen, L. (2013). *Energy Environ. Sci.* 6: 2338.
10 Slater, M.D., Kim, D., Lee, E., and Johnson, C.S. (2013). *Adv. Funct. Mater.* 23: 947.
11 Dong, S., Shen, L., Li, H. et al. (2016). *Adv. Funct. Mater.* 26: 3703.
12 Kim, S.W., Seo, D.H., Ma, X. et al. (2012). *Adv. Energy Mater.* 2: 710.
13 Zhang, Y., Zhao, Y., Ren, J. et al. (2015). *Adv. Mater.* 28: 4524.
14 Dong, S., Shen, L., Li, H. et al. (2015). *J. Mater. Chem. A* 3: 21277.
15 Wen, L., Li, F., and Cheng, H.M. (2016). *Adv. Mater.* 28: 4306.
16 Zhao, Q., Lu, Y., and Chen, J. (2016). *Adv. Energy Mater.* doi: 10.1002/aenm.201601792.
17 Hu, Y., Zhao, T., Zhu, P. et al. (2016). *J. Mater. Chem. C* 4: 5839.
18 Shen, L., Che, Q., Li, H., and Zhang, X. (2014). *Adv. Funct. Mater.* 24: 2630.
19 Xu, J., Wang, Q., Wang, X. et al. (2013). *ACS Nano* 7: 5453.
20 Xue, J., Zhao, Y., Cheng, H. et al. (2013). *Phys. Chem. Chem. Phys.* 15: 8042.
21 Shao, Y., El-Kady, M.F., Wang, L.J. et al. (2015). *Chem. Soc. Rev.* 44: 3639.
22 Kim, S.D., Rana, K., and Ahn, J.-H. (2016). *J. Mater. Chem. A* 4: 19197.
23 Suo, Z., Ma, E., Gleskova, H., and Wagner, S. (1999). *Appl. Phys. Lett.* 74: 1177.
24 Wong, W.S. and Salleo, A. (2009). *Flexible Electronics: Materials and Applications*. Springer Science & Business Media.
25 Shi, G., Zhao, Z., Pai, J.H. et al. (2016). *Adv. Funct. Mater.* 26: 7614.
26 Zhang, Y., Huang, Y., and Rogers, J.A. (2015). *Curr. Opin. Solid State Mater. Sci.* 19: 190.
27 Peng, H. (2015). *Fiber-Shaped Energy Harvesting and Storage Devices*. Springer.
28 Song, J. (2015). *Curr. Opin. Solid State Mater. Sci.* 19: 160.

29 Liu, W., Chen, Z., Zhou, G. et al. (2016). *Adv. Mater.* 28: 3578.
30 Shen, L., Wang, J., Xu, G. et al. (2015). *Adv. Energy Mater.* 5: 1400977.
31 Pang, G., Nie, P., Yuan, C. et al. (2014). *J. Mater. Chem. A* 2: 20659.
32 Cao, W., Luo, J., Yan, J. et al. (2017). *J. Electrochem. Soc.* 164: A93.
33 Li, N., Chen, Z., Ren, W. et al. (2012). *Proc. Natl. Acad. Sci. U.S.A.* 109: 17360.
34 Kokkinis, D., Schaffner, M., and Studart, A.R. (2015). *Nat. Commun.* 6: 8643.
35 MacDonald, E. and Wicker, R. (2016). *Science* 353: aaf2093.
36 Wen, Y., He, K., Zhu, Y. et al. (2014). *Nat. Commun.* 5: 4033.
37 Cao, Y., Xiao, L., Sushko, M.L. et al. (2012). *Nano Lett.* 12: 3783.
38 Li, W., Zeng, L., Yang, Z. et al. (2014). *Nanoscale* 6: 693.
39 David, L. and Singh, G. (2014). *J. Phys. Chem. C* 118: 28401.
40 Deng, X., Xie, K., Li, L. et al. (2016). *Carbon* 107: 67.
41 Hu, X., Zhang, W., Liu, X. et al. (2015). *Chem. Soc. Rev.* 44: 2376.
42 Liu, Y., He, X., Hanlon, D. et al. (2016). *ACS Nano* 10: 8821.
43 David, L., Bhandavat, R., and Singh, G. (2014). *ACS Nano* 8: 1759.
44 Li, Y., Zhu, H., Shen, F. et al. (2015). *Nano Energy* 13: 346.
45 Xie, X., Makaryan, T., Zhao, M. et al. (2015). *Adv. Energy Mater.* 6: 1502161.
46 Xiang, J., Dong, D., Wen, F. et al. (2016). *J. Alloy. Compd.* 660: 11.
47 Zhang, Y., Liu, Z., Zhao, H., and Du, Y. (2016). *RSC Adv.* 6: 1440.
48 Xia, W., Xu, F., Zhu, C. et al. (2016). *Nano Energy* 27: 447.
49 Li, Y., Wang, D., An, Q. et al. (2016). *J. Mater. Chem. A* 4: 5402.
50 Wei, X., Li, W., Shi, J.-A. et al. (2015). *ACS Appl. Mater. Interfaces* 7: 27804.
51 Luo, B., Qiu, T., Ye, D. et al. (2016). *Nano Energy* 22: 232.
52 Zhao, M., Zhao, Q., Qiu, J. et al. (2016). *RSC Adv.* 6: 95449.
53 Xu, W., Zhao, K., Zhang, L. et al. (2016). *J. Alloy. Compd.* 654: 357.
54 Liu, Y., Zhang, N., Jiao, L., and Chen, J. (2015). *Adv. Mater.* 27: 6702.
55 Zhu, Y., Han, X., Xu, Y. et al. (2013). *ACS Nano* 7: 6378.
56 Sun, J., Lee, H.-W., Pasta, M. et al. (2016). *Energy Storage Mater.* 4: 130.
57 Zhang, C., Wang, X., Liang, Q. et al. (2016). *Nano Lett.* 16: 2054.
58 Koo, B., Chattopadhyay, S., Shibata, T. et al. (2013). *Chem. Mater.* 25: 245.
59 Sun, B., Bao, S.J., Le Xie, J., and Li, C.M. (2014). *RSC Adv.* 4: 36815.
60 Wang, X., Liu, Y., Wang, Y., and Jiao, L. (2016). *Small* 12: 4865.
61 Xu, Y., Lotfabad, E.M., Wang, H. et al. (2013). *Chem. Commun.* 49: 8973.
62 González, J.R., Alcántara, R., Nacimiento, F. et al. (2014). *CrystEngComm* 16: 4602.
63 Zhu, X., Li, Q., Fang, Y. et al. (2016). *Part. Part. Syst. Char.* doi: 10.1002/ppsc.201500216.
64 Doeff, M.M., Cabana, J., and Shirpour, M. (2014). *J. Inorg. Organomet. Polym Mater.* 24: 5.
65 Senguttuvan, P., Rousse, G.L., Seznec, V. et al. (2011). *Chem. Mater.* 23: 4109.
66 Wang, W., Yu, C., Lin, Z. et al. (2013). *Nanoscale* 5: 594.
67 Li, Z., Shen, W., Wang, C. et al. (2016). *J. Mater. Chem. A* 4: 17111.
68 Trinh, N.D., Crosnier, O., Schougaard, S.B., and Brousse, T. (2011). *ECS Trans.* 35: 91.

69 Rudola, A., Saravanan, K., Devaraj, S. et al. (2013). *Chem. Commun.* 49: 7451.
70 Wang, X., Li, Y., Gao, Y. et al. (2015). *Nano Energy* 13: 687.
71 Yang, D., Liao, X.-Z., Shen, J. et al. (2014). *J. Mater. Chem. A* 2: 6723.
72 Yu, T., Lin, B., Li, Q. et al. (2016). *Phys. Chem. Chem. Phys.* 18: 26933.
73 Zhang, S., Deng, C., and Meng, Y. (2014). *J. Mater. Chem. A* 2: 20538.
74 Xu, X., An, Q., and Mai, L. (2016). *Phys. Chem. Chem. Phys.* 18: 12074.
75 Balogun, M.-S., Luo, Y., Lyu, F. et al. (2016). *ACS Appl. Mater. Interfaces* 8: 9733.
76 Rui, X., Tang, Y., Malyi, O.I. et al. (2016). *Nano Energy* 22: 583.
77 Wu, X., Luo, Y., Sun, M. et al. (2015). *Nano Energy* 13: 117.
78 Jiang, Y., Yu, S., Wang, B. et al. (2016). *Adv. Funct. Mater.* 26: 5315.
79 Nie, P., Shen, L., Pang, G. et al. (2015). *J. Mater. Chem. A* 3: 16590.
80 Khan, Z., Park, S., Hwang, S.M. et al. (2016). *NPG Asia Mater.* 8: e294.
81 Wang, Y., Ding, Y., Pan, L. et al. (2016). *Nano Lett.* 16: 3329.
82 Wu, S., Wang, W., Li, M. et al. (2016). *Nat. Commun.* 7: 13318.
83 Wang, C., Jiang, C., Xu, Y. et al. (2016). *Adv. Mater.* 28: 9182.
84 Wang, C., Fang, Y., Xu, Y. et al. (2016). *Adv. Funct. Mater.* 26: 1777.
85 Kim, H., Kwon, J.E., Lee, B. et al. (2015). *Chem. Mater.* 27: 7258.
86 Castillo-Martínez, E., Carretero-González, J., and Armand, M. (2014). *Angew. Chem. Int. Ed.* 53: 5341.
87 Luo, W., Allen, M., Raju, V., and Ji, X. (2014). *Adv. Energy Mater.* 4: 1400554.
88 Wang, C., Xu, Y., Fang, Y. et al. (2015). *J. Am. Chem. Soc.* 137: 3124.
89 Quartarone, E. and Mustarelli, P. (2011). *Chem. Soc. Rev.* 40: 2525.
90 Terabe, K., Hasegawa, T., Nakayama, T., and Aono, M. (2005). *Nature* 433: 47.
91 Zhou, L., Cao, Q., Jing, B. et al. (2014). *J. Power Sources* 263: 118.
92 Cao, C., Liu, W., Tan, L. et al. (2013). *Chem. Commun.* 49: 11740.
93 Yang, Y., Chang, Z., Li, M. et al. (2015). *Solid State Ionics* 269: 1.
94 Tu, Z., Kambe, Y., Lu, Y., and Archer, L.A. (2014). *Adv. Energy Mater.* 4: 1300654.
95 Ansari, Y., Guo, B., Cho, J.H. et al. (2014). *J. Electrochem. Soc.* 161: A1655.
96 Rajendran, S., Sivakumar, M., and Subadevi, R. (2004). *Mater. Lett.* 58: 641.
97 Appetecchi, G., Croce, F., and Scrosati, B. (1995). *Electrochim. Acta* 40: 991.
98 Singh, V.K., Chaurasia, S.K., and Singh, R.K. (2016). *RSC Adv.* 6: 40199.
99 Han, P., Han, X., Yao, J. et al. (2015). *Electrochem. Commun.* 61: 84.
100 Yang, T., Qian, T., Wang, M. et al. (2016). *Adv. Mater.* 28: 539.
101 Zhang, W., Liu, Y., Chen, C. et al. (2015). *Small* 11: 3822.
102 Liu, Y., Zhang, N., Yu, C. et al. (2016). *Nano Lett.* 16: 3321.
103 Li, H., Peng, L., Zhu, Y. et al. (2016). *Nano Lett.* 16: 5938.
104 Kim, J.-K., Lim, Y.J., Kim, H. et al. (2015). *Energy Environ. Sci.* 8: 3589.
105 Li, H., Ding, Y., Ha, H. et al. (2017). *Adv. Mater.* 29: 1700898.

5

1D and 2D Flexible Carbon Matrix Materials for Lithium–Sulfur Batteries

Tianyi Wang, Yushu Liu, Dawei Su, and Guoxiu Wang

University of Technology Sydney, School of Mathematical and Physical Sciences, Research Centre for Clean Energy Technology, 15 Broadway, Ultimo, NSW 2007, Australia

5.1 Introduction

Since the 1960s, the concept of lithium–sulfur (Li–S) batteries has excited people as a seemingly ideal way for energy storage and conversion. Li–S batteries can deliver high energy storage density, which is almost five times that of lithium–ion batteries. Compared with lithium–ion batteries, the mechanism of energy storage in Li–S batteries is entirely different. The operation of lithium–ion batteries is based on the intercalation of lithium ions into/out of cathode and anode hosts. Because the amount of lithium ions is limited in cathode materials, the theoretical capacity of lithium–ion batteries is limited to ~420 Wh kg^{-1} [1]. In contrast, energy conversion and storage of Li–S batteries is based on the conversion reaction of sulfur to form Li_2S on the cathode and lithium stripping and plating on the anode. The nontopotactic nature of these reactions gives the lithium anode and sulfur cathode a high theoretical specific capacity of 3860 and 1673 mAh g^{-1}, respectively [2–4]. In addition, the average working voltage of Li–S batteries is 2.15 V, which endows them with an energy density of ~2500 Wh kg^{-1} and ~2800 Wh l^{-1} [5]. Sulfur is an inexpensive material with a rich distribution throughout the world and therefore, is of low cost for commercial applications. However, the research on Li–S batteries was rather stagnant due to quick capacity decay, which impaired further development of Li–S batteries [6].

Recently, wearability and flexibility have become a new trend for electronic devices, which require to be powered by flexible power sources. In this chapter, we mainly discuss the recent development of flexible Li–S batteries, focusing on three aspects: electrode material (cathode and anode), electrolytes, and separators.

Flexible Energy Conversion and Storage Devices, First Edition. Edited by Chunyi Zhi and Liming Dai.

5.2 The Working Mechanism and Challenges of Li–S Batteries

The operation of Li–S batteries is based on the electrochemical reaction:

$$S_8 + 16Li \rightarrow 8Li_2S \quad (5.1)$$

During the discharge process, the lithium metal (anode material) is oxidized to lithium ions and electrons that travel through the electrolyte and outer circuit, respectively, to reduce sulfur (cathode material) [7, 8]. The discharge process usually shows two different discharge voltage plateaus [9, 10]. At the first high discharge plateau near 2.4 V, S_8 rings are opened and lithiated to intermediate or long-chain lithium polysulfides (PSs).

$$S_8 \rightarrow Li_2S_8 \rightarrow Li_2S_6 \rightarrow Li_2S_4 \quad (5.2)$$

Then, at the second (lower) discharge voltage plateau (about 2.1 V), the intermediate polysulfides are further converted to lithium sulfide.

$$Li_2S_4 \rightarrow Li_2S_2 \rightarrow Li_2S \quad (5.3)$$

The upper voltage discharge plateau contributes to about 25% of the theoretical capacity (~418 mAh g^{-1}) and the lower discharge plateau contributes to about ~75% of the theoretical capacity (about 1255 mAh g^{-1}). Because the long-chain polysulfides can be dissolved into the electrolyte easily while sulfur and lithium sulfides are insoluble, the whole reaction undergoes a solid → liquid → solid sequence. This complicated sequence increases the side reactions of Li–S batteries and has been hindering the practical applications of Li–S batteries [11–13].

Overall, there are four challenges facing the development of flexible lithium–sulfur batteries, which are as follows:

1) *The dissolution of active materials.* Since long-chain polysulfides are soluble in the electrolyte during the discharging process, a large amount of long-chain polysulfides may separate from the cathode [14]. This dissolution normally causes large amount of loss of active sulfur because these long-chain polysulfides are unable to return to the cathode host even during the charging process. This phenomenon causes quick capacity decay and limits the working life span of lithium–sulfur batteries.
2) *The "Shuttle effect".* Some of the long-chain polysulfides that are dissolved in the electrolyte, migrate through the separator to the anode side and react with the lithium anode. Because of the strong reducibility of elemental lithium, polysulfides can be reduced to Li_2S and Li_2S_2 quickly. Because of the insoluble characteristics of Li_2S_2 and Li_2S, they can passivate the anode surface. This effect is irreversible and causes severe damage to the structure of the anode.
3) *Large volume expansion.* During discharge process, the volume expansion can reach ~80%. The quick volume expansion of sulfur can break its host material and result in the collapse of the whole cathode. The volume expansion of sulfur cathode during the discharge process also decreases mechanical integrity and the stability of electrodes over long time cycling.

4) *Low electronic conductivity of sulfur*. Owing to the insulting nature of elemental sulfur, sulfur cathodes usually have a high internal resistance, leading to poor electrochemical performances [15].

5.3 Flexible Cathode Hosts for Lithium–Sulfur Batteries

To suppress the shuttling effect and to overcome problems arising from the insulating nature of sulfur, various carbon materials have been applied as cathode materials to improve the capacity and cycling performance of Li–S batteries [16]. The mass of sulfur in the cathode determines the total capacity of Li–S batteries. Therefore, sulfur content in the composite electrode is of vital importance to improve the energy density of the battery system [17]. Generally, achieving high sulfur loading is still a challenge that needs to be resolved in Li–S batteries. Meanwhile, during the discharging process, the volume expansion of sulfur can reach 80% [18]. This expansion can damage the ordered structure of the cathode materials and result in irreversible capacity loss particularly for flexible cathodes. To solve these problems, many cathode modification methods have been investigated.

In general, carbon nanotubes (CNTs) and graphene can be used as the matrix materials for flexible lithium–sulfur batteries. In 2015, Sun et al. reported an aligned and laminated sulfur-absorbed carbon/CNT hybrid flexible cathode for high performance Li–S batteries (Figure 5.1a,b). In this battery, mesoporous carbon materials were used as the sulfur host to hinder the diffusion of polysulfides. The capacity of the as-synthesized cathode material can reach 1226 $mAh\,g^{-1}$ with a capacity retention of 75% (100 cycles at 0.1 C). In addition, they also obtained a high capacity of around 900 $mAh\,g^{-1}$ at 20 $mg\,cm^{-2}$, which is the highest sulfur loading reported to date. Moreover, the aligned and laminated hybrid cathode endows the battery with high flexibility and its electrochemical performances are well maintained under bending and after being folded for 500 cycles. Furthermore, aligned CNTs provide better electronic conductivity than freestanding CNTs (Figure 5.1c) [19].

Novel structures, can improve conductivity and tolerate the sulfur volume change, which are desirable properties for impregnating sulfur and trapping the dissolved PSs during cycling. Porous carbon and porous graphene are three dimensional graphene and have high surface areas for loading sulfur. Considering the mass of aluminum foil typically used as a cathode substrate, the maximum sulfur loading is less than 30 wt% if the mass of Al substrate is taken into account [21]. However, in cases using freestanding porous carbon or porous graphene or three-dimensional graphene, the total sulfur loading can reach as high as 90 wt%. Wang et al. reported macroporous freestanding nanosulfur/graphene (S-rGO) materials, which can be directly employed as an electrode for Li–S batteries. The S-rGO flexible paper used in the batteries was fabricated through a facile freeze drying route followed by low-temperature heat treatment. The flexible S-rGO paper not only provides a conductive framework for electron transport, but also alleviates the volume change during cycling. The as-prepared S-rGO paper shows exceptional rate capability and cyclability. The

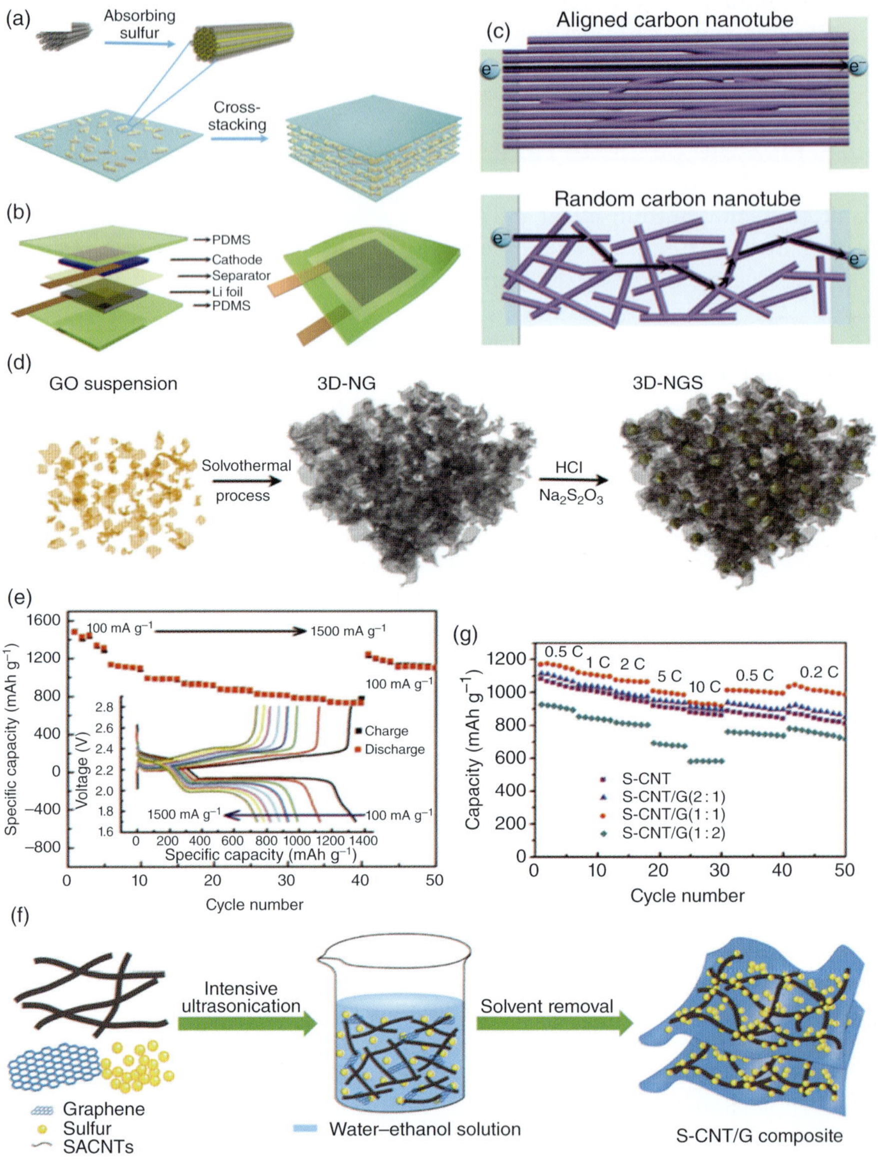

Figure 5.1 (a) The fabrication process of the laminated hybrid sulfur cathode and (b) schematic image of structure of the flexible Li–S battery. (c) Comparison to explain the differences in electronic conductivity between aligned CNTs and random CNTs. *Source:* (a–c) Sun et al. 2015 [19]. Copyright 2015, John Wiley & Sons. (d) The preparation method for the S-rGO paper. (e) The rate capability of 3D–NGS with current densities ranging from 100–1500 mAh g^{-1} and the first discharge–charge profiles at different current densities [12]. *Source:* Copyright 2014, Royal Society of Chemistry. (f) The preparation method for synthesizing S-CNT/G and (g) its electrochemical performance when used as flexible Li–S battery cathode. *Source:* (f, g) Sun et al. 2015 [20]. Copyright 2015, Royal Society of Chemistry.

specific discharge capacity is 800 mAh g^{-1} after 200 cycles at a current density of 300 mA g^{-1} and the capacity fading rate is only 0.035% per cycle. When the current density increased to 1500 mA g^{-1}, it still showed a good performance. This work ascribes the high performance of the S-rGO paper to the stable macroporous structure and strong interaction between sulfur nanoparticles and graphene [21].

Graphene with unique properties including large surface area, light weight, high electrical conductivity, and superior mechanical flexibility, is a promising material for fabrication of flexible Li–S batteries. It has been demonstrated as an effective sulfur immobilizer for Li–S batteries with high performance. Therefore, it is widely considered that graphene will play a great role in the development of flexible sulfur cathodes due to its triple functions in enhancing electronic conductivity, trapping soluble intermediates, and improving the mechanical properties. Despite some progresses, the mechanical properties of the as-fabricated flexible electrodes are still not satisfied and researchers are making efforts to improve it. For example, Sun et al. reported the employment of super-aligned CNT/graphene (CNT/G) hybrid materials as a 3D conducting framework for sulfur accommodation (Figure 5.1f) [20]. The CNT/G network acts as a skeleton to form a self-supporting cathode. Graphene with a 2D sheet structure extends in an additional dimension to provide improved restriction for sulfur/polysulfides. Moreover, the CNT/G hybrid framework enables better dispersion of sulfur and allows each sulfur particle to attach loosely to the conductive components, which greatly enhances the electrochemical conductivity and thereby approaches the full potential of the active materials. Compared with the sulfur doped carbon nanotube (S-CNT) composite, the S-CNT/G nanocomposite exhibited enhanced mechanical performance and electrochemical characteristics. The S-CNT/G nanocomposite with a superior structure achieved a high discharge capacity of 1048 mA h g^{-1} at 1 C with a capacity fading rate as low as 0.041% per cycle over 1000 charge–discharge cycles (Figure 5.1g) [20].

The electronic and chemical properties of graphene can be effectively modified by the incorporation of heteroatoms into graphene lattice. Therefore, continuous efforts have been made to fabricate doped graphene. Nitrogen-doped, boron-doped, and sulfur-doped graphene exhibit good electrochemical performances and are commonly employed in fuel cells and lithium ion batteries. The reason for the performance enhancement lies in the doped atoms, which have different electron negativities and can break the electro-neutrality of graphene to create charged active sites with increased conductivity to improve the discharging capacity of graphene in Li–S batteries [12]. Introducing 2D graphene sheets into a 3D macroscopic structure is also a unique approach for providing active sites and enough support space for stable cycling. Wang et al. found that hydrothermal and solvothermal treatment of graphene oxide (GO) is a highly efficient method for fabricating 3D–GO structures. A porous three-dimensional nitrogen-doped graphene (3D–NG) was prepared and used as an interconnected framework for sulfur in Li–S batteries. The 3D–NG-sulfur composite (3D–NGS) with a high sulfur content of 87.6 wt% was synthesized via a facile one-pot solution method and sulfur was well dispersed (Figure 5.1d). The 3D–NGS composite demonstrated excellent rate capability and cyclability. The discharge specific capacity is 792 mA h g^{-1} after 145 cycles at a current density of 600 mA g^{-1} and the capacity fading rate is 0.05% per cycle. Even at a high rate of 1500 mA g^{-1}, the composite still shows a good cycle performance with a capacity of 671 mAh g^{-1} after 200 cycles (Figure 5.1e). The outstanding electrochemical performance is ascribed to the flexible porous 3D structure and N-doping in graphene. The flexible 3D–NG can provide a conductive framework for electron transport and alleviate the volume expansion during cycling. N-doping can facilitate the diffusion of Li ions across the graphene sheets and restrain sulfur due to the strong chemical bonding between S and the nearby N atoms [12].

The key parameter for a good flexible electrode is high tensile strength. However, the poor flexibility and weak adhesion between active materials and metal foil substrates, leads to the pulverization of active materials and performance degradation during repeated bending. Thus, conventional battery electrodes fail to meet the requirements for flexibility. Carbon cloth, carbonized cellulose, and textiles have been commonly applied to construct flexible electrodes and exhibit high mechanical strength and flexibility. Particularly, CNTs and graphene films provide a flexible platform for Li–S batteries. However, until now, flexibility contradicts electrochemical performance, and compromises have to be made, such as relatively fast capacity decay and low sulfur loading, while achieving high flexibility. However, the energy density of the electrodes is lowered by a high mass of the flexible substrates with low active mass content. The energy density of flexible Li–S batteries can be effectively enhanced by employing freestanding electrodes. Vacuum filtration for the preparation of graphene or CNTs as the matrix is becoming common for fabricating flexible electrodes. On the other hand, electrospinning has also been extensively explored for the fabrication of freestanding porous carbon nanofiber films. However, these films exhibited a low breaking strength (usually lower than 2 MPa with 0.5–1.5% strain) and have failed to meet the demands of practical applications. For sulfur-based cathodes, porous carbon nanofibers have been employed to support sulfur [12]. However, the filled sulfur tends to bond rigidly in the micropores of carbon nanofibers, which degrades the flexibility and is detrimental to the structural stability of the flexible films. Song et al. have demonstrated a novel technique to fabricate flexible and freestanding sulfur cathodes with enhanced mechanical strength using a matrix constructed from graphitized nitrogen-doped mesoporous carbon nanofibers (NPCFs) [22]. The matrix was made by electrospinning followed by carbonization at 1000 °C (Figure 5.2a). Because of the unique micro/mesoporous structure and highly graphitic carbon, the NPCF film can accommodate more sulfur, and maintain substantially high mechanical strength and flexibility after high sulfur loading, compared with traditional microporous carbon nanofiber films (Figure 5.2b). As a freestanding and flexible cathode for Li–S batteries, the robust composite film showed excellent rate capability (540 mAh g^{-1} at 5 C) and cycling stability (76.5% retention after 500 cycles at 5 C) [25].

Chen et al. fabricated a robust and flexible cathode for Li–S batteries with high electrochemical performance and tensile strength (72.0 MPa)

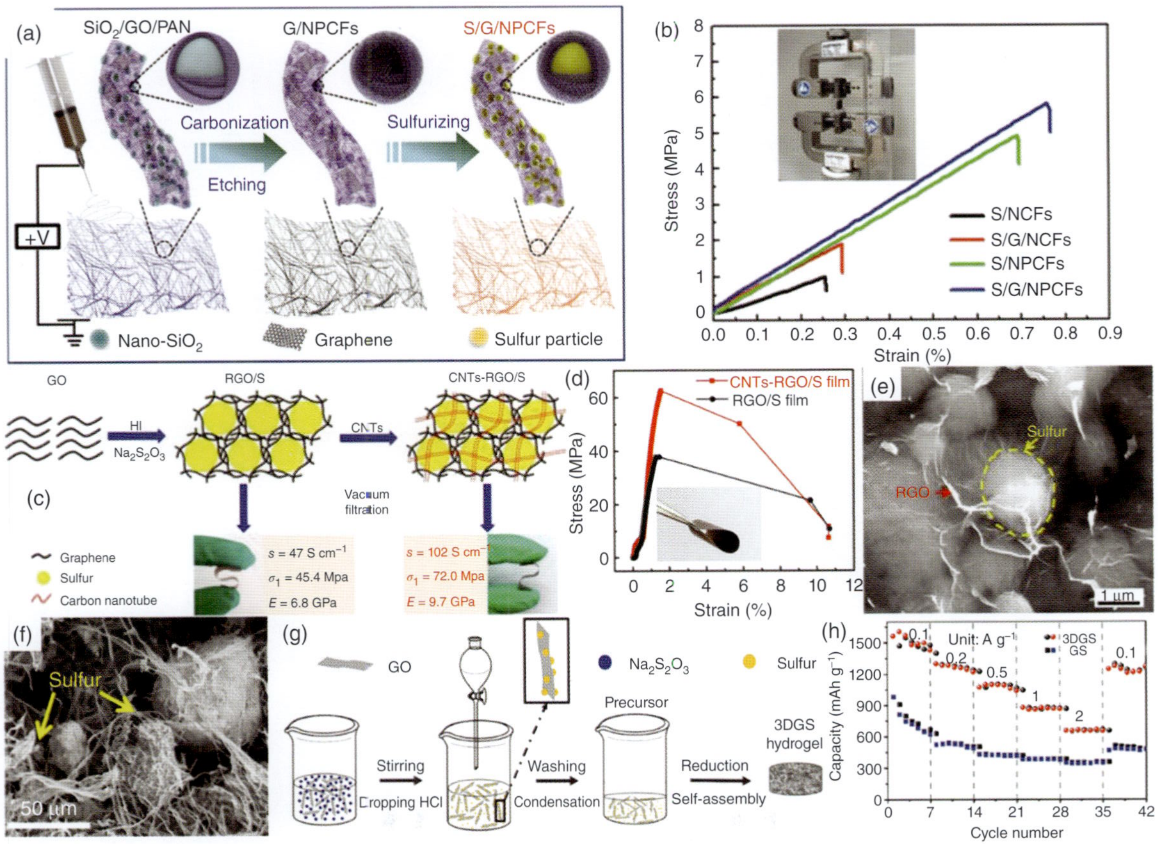

Figure 5.2 (a) The preparation process of NPCFs by an electrospinning method. (b) Typical tensile stress–strain curves of the as-prepared membrane and photograph of the tensile stress–strain testing instrument. *Source:* (a,b) Song et al. 2017 [22]. Copyright 2017, Royal Society of Chemistry. (c) The preparation process of the CNT–rGO/S composite. (d). The physical performance of CNT–rGO/S. (e) SEM image of sulfur particles and holder materials. *Source:* (c–e) Chen et al. [18]. Copyright 2015, American Chemical Society. (f) The sulfur particles that are protected by GTS. (g) The preparation process of the 3DGS hydrogel. *Source:* Guo et al. [23]. Royal Society of Chemistry. (h) The electrochemical performance of a Li–S battery using 3DGS hydrogel as cathode material at various current densities. *Source:* Lin et al. 2016 [24]. Copyright 2016, Royal Society of Chemistry.

based on interpenetrative composites of carbon nanotubes–graphene/sulfur (CNTs-rGO/S)(Figure 5.2c) [18]. The CNTs and graphene act as both the cathode matrix and separator to inhibit the migration of the dissolved polysulfide intermediates in the electrolyte. Consequently, the robust and flexible CNTs-rGO/S cathode with 53 wt% sulfur showed a reversible capacity of 772 $mAh\,g^{-1}$ at a current rate of 0.2 C and 613 $mAh\,g^{-1}$ at 1 C, respectively. It possesses a high electrode capacity of ~409 $mAh\,g^{-1}$. Meanwhile, this flexible cathode has good physical flexibility to tolerate shape changes (Figure 5.2d). Such an improved rate capability and cycle stability could be attributed to the unique graphene-wrapped sulfur particles and the enhanced electrical conductivity and mechanical strength from the incorporation of CNTs. Based on the intrinsic properties of graphene, CNTs are proposed to act as an alternative carbonaceous cathode scaffold for Li–S batteries. The inherent open topology of CNTs enables sulfur particles to be easily diffused as well as to be more efficient for in situ immobilization of sulfur and geometric confinement of polysulfides in the interior face of the tube (Figure 5.2e). Moreover, the highly elastic tubular walls of CNTs is supposed to facilitate radial expansion to accommodate volume variation of enclosed sulfur particles during the lithiation process [18]. Guo et al. used a modified, facile and efficient synthesis strategy to fabricate high-quality 1D graphene nanoscrolls/graphene nanosheet (GNS) in high yield from GO sheets through fast quenching combined with lyophilization [23]. The as-synthesized GNS have an ordered 1D CNT-like nanostructure with an excellent electronic conductivity and especially a large specific surface area (SSA) of 545 $m^2\,g^{-1}$, which is one of the highest value among the reported GNS materials. Owning to the large SSA, the as-synthesized GNS can be employed as an ideal carrier to incorporate sulfur for cathode assembly. A very high sulfur loading of up to 81 wt% is achievable in the resultant composite cathode (denoted as S@GNS). Three-dimensional (3D) graphene sponge has a continuous conductive structure and numerous pores, which is beneficial for sulfur utilization and anchoring (Figure 5.2f). However, strategies for the fabrication of 3D graphene sponges (3D GS) composited with sulfur nanoparticles are either energy consuming or involve toxic reagents [23]. Thus, a new 3DGS was fabricated via a reduction-induced self-assembly method, which is simple but facile and scalable (Figure 5.2g) [24]. The 3D GS enables fast Li^+ transport, superior electrolyte absorbability, and effective electrochemical redox reactions of sulfur. Consequently, this 3D GS delivered a stable capacity of 580 $mAh\,g^{-1}$ after 500 cycles at a high rate of 1.5 $A\,g^{-1}$, corresponding to a low capacity fading rate of 0.043% per cycle (Figure 5.2h) [24]. According to the previous studies, it has been effectively demonstrated that 3D construction strategies are effective for fabricating flexible high performance Li–S batteries. Three-dimensional current substrates possess a high SSA compared with the conventional two-dimensional current collectors, providing higher interfacial interaction.

An et al. reported an ultralight flexible carbon foam (MFC) with a 3D interconnected hollow network, which was prepared through a special heat treatment program from melamine-formaldehyde foam (Figure 5.3a) [26]. The MFC can be employed as a favorable host material to obtain both ultrahigh sulfur loading and satisfactory electrochemical performance in Li–S batteries. Without conductive

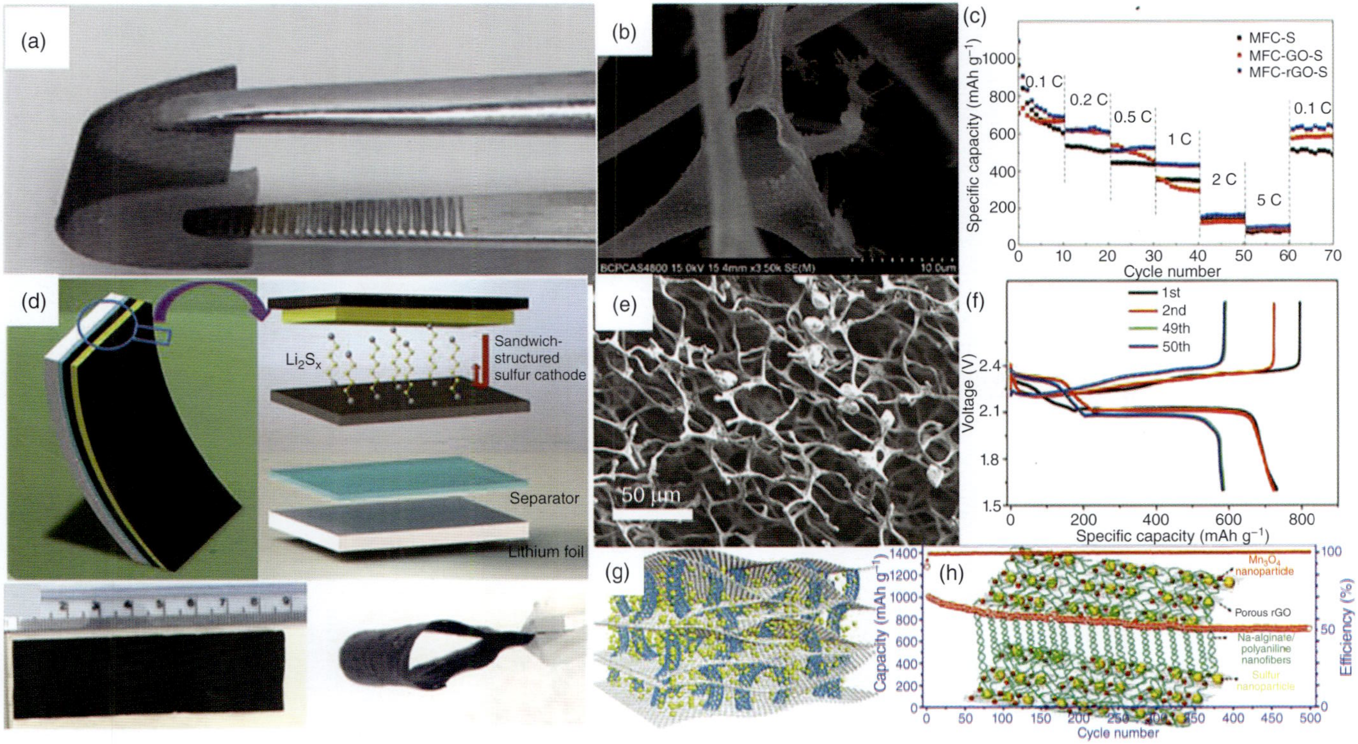

Figure 5.3 (a) The flexibility of material. (b) The SEM image of porous and inner-connected structure. (c) The discharging capacity at various current densities. *Source:* (a, c) An et al. 2016 [26]. Copyright 2016, Royal Society of Chemistry. (d) Li–S battery with sandwich-structured cathode that can be curved or bent. *Source:* Song et al. 2014 [27]. Copyright 2014, Royal Society of Chemistry. (e) SEM view of sulfur particles that are attached on carbon foam. (f) The charge and discharge curve of the battery. *Source:* (e, f) Milroy and Manthiram 2016 [28]. Copyright 2016, Royal Society of Chemistry. (g) The structure of 3DCNT/Graphene-Li_2S aerogel. (h) Electrochemical performance of scaffold material. *Source:* (g, h) Ghosh et al. 2016 [29]. Copyright 2016, American Chemical Society.

additive and binders, the sulfur content in the MFC-based cathode could easily reach 75 wt%. The MFC has a high content of N and O atoms, which can strongly absorb the polysulfides. The flexible and sparse skeleton can accommodate large volume changes during electrochemical reactions. GO or rGO was also used to enhance the MFC skeleton to further improve the performance of the electrode. The MFC-based electrodes also showed excellent cycling performances. The initial capacity of the MFC-S, MFC-GO-S, and MFC-rGO-S, are 1011, 1183, and 1135 mA h g^{-1} at 0.1 C, respectively (the capacities were calculated based on the mass of the whole electrodes) (Figure 5.3b). After the irreversible capacity fading in the first few cycles, the capacity retention is over 95% after 100 cycles, indicating that the carbon foam may be an excellent cathode matrix for Li–S batteries for high sulfur loading [26].

Song et al. constructed a novel sulfur cathode consisting of a layer of sulfur-active materials sandwiched by two flexible porous carbon membranes [27]. Compared with the previously reported sulfur cathodes, this novel sandwich-structured cathode design shows many advantages. Firstly, the internally connected carbon foam provides large surface area and many active sites for polysulfide binding. Secondly, the charge transfer resistance is significantly decreased compared to conventional sulfur cathodes not only due to reduced contact resistance between sulfur cathodes and current collectors but also due to intimate contact between sulfur and the two layers of highly conductive carbon. Thirdly, this carbon-supported sandwich-structured cathode is flexible, which can potentially extend Li–S batteries to other applications such as wearable devices, and implantable biomedical devices [27]. Milroy and Manthiram synthesized nitrogen-doped mesoporous carbon cathode materials, which can promote chemical adsorption between the sulfur and oxygen functional groups and effectively immobilize the sulfur species and minimize sulfur loss. They further noticed that the carbon foam calcined at 800 °C displayed the best stability, which was closely related to the foam structure [28].

Wang et al. have developed a simple drop coating method to prepare a Li_2S/rGO cathode paper for flexible Li–S batteries as a freestanding and binder-free cathode without the use of viscous slurries or metal substrates [30]. It showed both high rate capability and long cycling life. He et al. synthesized 3D CNT/graphene-Li_2S (3D CG-Li_2S) cathodes (81.4 wt% Li_2S loading) by solvothermal reaction and a subsequent liquid–infiltration–evaporation impregnation method [31]. The highly flexible and conductive 3D mesoporous interconnected network based on two-dimensional (2D) graphene nanosheets and one-dimensional (1D) CNTs provides highly efficient channels for electron transfer and ionic diffusion, leading to a low solubility of polysulfides during electrolytes in charges/discharges. Without polymeric binders or conductive additives, the freestanding 3D CG-Li_2S cathode exhibited excellent electrochemical performances including reversible discharge capacities of 1124 and 915 mAh g^{-1} at current densities of 4.5 and 3.4 mAh g^{-1}, 0.02% long-term capacity decay per cycle and a high rate capacity of 514 mAh g^{-1} at 4 C. He et al. also reported the other 3D material, 3D LG, (three-dimensional Li_2S/graphene hierarchical architecture with Li_2S nanoparticles deposited into a 3D graphene fiber (GF) network). They fabricated 3D LG via drop casting and heating. The electrode delivered a high discharge capacity of 895 mAh g^{-1} at 0.1 C and maintained 87.7% of the initial reversible capacity after 300 cycles (0.2 C) and a high rate capacity of 599 mAh g^{-1} up at 4 C. 3D LG sets a new record and has a great potential for high-performance Li–S batteries (Figure 5.3g) [31, 32].

The design of flexible cathode scaffolds that can maintain good interfacial contact between all the building blocks for high-performance Li–S batteries must meet the following requirements. (i) enhancement of electronic conductivity throughout the whole cathode scaffold; (ii) a low rate of polysulfides dissolution; (iii) high utilization of active materials; and (iv) high mechanical strength to resist large volume changes during repeated charge/discharge processes. Therefore, more and more novel materials have been investigated to satisfy the needs of cathodes with good cycle life for high-performance Li–S batteries. Huang et al. have reported a flexible 3D carbonized bacterial cellulose (CBC) aerogel [33], as the sulfur support. The 3D CBC has a highly nanofibrous structures with good mechanical stability and high electrical conductivity. The unique interconnection network of CBC provides satisfactory electrical conductivity as well as a solid framework, which can maintain the tension produced by volume changes of the active sulfur. This interconnected macroporous structure allows lithium ions to migrate rapidly. In addition, the ultralight CBC layers between the sulfur cathode and the separator membrane lessen the resistance of the entire electrode. More significantly, the CBC layer can effectively relieve the excess accumulation of sulfur on the cathode surface. Even at a high sulfur loading content, the CBC composite still has enough free space to accommodate the volume expansion of sulfur in the lithiation process. In addition, the CBC layers inserted between the sulfur cathode and the separator promotes active material utilization, cycle stability, and significantly improved coulombic efficiency. The CBC layer can also be used as an extra collector of sulfur to prevent the excessive concentration of insulating sulfur on the cathode surface. The good electrochemical properties can be attributed to the rational design of the battery configuration and the flexible three-dimensionally interconnected nanostructure of the carbon framework. Ghosh et al. investigated the interaction between the particle size with mechanical and electrochemical properties (Figure 5.3g) [29]. They reported a bottom-up approach, creating a unique independent type of flexible cathode matrix. It is composed of porous graphene oxide, nanosized sulfur, and Mn_3O_4 nanoparticles, which are mixed with polyaniline (PANI) with a sodium alginate (SA) matrix for three-dimensional mutual connection. A capacity of 1098 mAh g^{-1} is achieved after 200 cycles at a current rate of 2 A g^{-1} with 97.6% of the initial capacity maintained. Interestingly, at the higher current density of 5 A g^{-1}, the composite electrode still exhibited an initial capacity of 1015 mAh g^{-1} and retained 71% of the original capacity after 500 cycles. In situ Raman study confirms the polysulfide absorption capability of Mn_3O_4. The synergistic effect of each component resulted in excellent electrochemical performances including high rate capability, high specific capacity, and unprecedented cycling stability. It is believed that the simple and inexpensive electrode design concept can open new avenues for practical industrial applications of Li–S batteries with high energy density.

At present, flexible Li–S battery systems mainly use carbon-based materials to accommodate sulfur. Most research studies focus on modification of individual components of batteries, especially electrode materials. However, the fluidity of liquid electrolytes restricts the development of flexible Li–S batteries, because the bending and stretching processes make the cell more prone to leakage, resulting in corrupted safety, thus limiting its application in flexible electronics [23]. Therefore, we also summarize the development of electrolyte for flexible Li–S batteries.

5.4 Electrolyte Membranes for Flexible Li–S Batteries

Conventional Li–S batteries usually consist of liquid electrolytes, which show excellent integrated properties such as high Li ionic conductivity (about 10^{-2} S cm^{-1}) and form a satisfactory contact interface with solid electrodes. However, because of their fluidity, electrolytes may leak once cells are bent or stretched in flexible batteries. Their fluidity may also distort the size and shape of the battery, which cannot meet the demands of flexible and microminiaturized electronic devices. Therefore, the development of solid but flexible electrolytes is a focal point of research for flexible Li–S batteries. Remarkable improvements have been achieved through enormous efforts made by several groups [34, 35].

Compared with liquid electrolytes, using solid electrolytes have been recognized as one of the most promising approaches for high safety requirements because they can eliminate concerns about internal short-circuiting and leaking when facing large stresses in wearable devices [36]. Moreover, solid electrolytes can also isolate the positive and negative electrodes of a battery (working as a separator), effectively avoiding internal shorting and increasing the energy density of the battery. There are still some obstacles for solid electrolyte applications [37, 38]. Some solid electrolytes cannot achieve good contacts with solid electrodes, resulting in poor cycle life. Another major disadvantage of solid electrolytes is that they exhibit low Li ionic conductivities (about 10^{-6} S cm^{-1}) at room temperature [39]. In contrast, gel polymer electrolytes (GPEs) containing liquid traditional electrolytes in polymers, exhibit high ionic conductivities, and therefore are more suitable for flexible Li–S batteries. Both composite polymer electrolytes (CPEs) and GPEs show promising properties in flexible Li–S batteries [40].

GPEs are usually prepared by immobilizing liquid electrolytes in a polymer matrix. Liquid electrolytes provide a high ionic conductivity while the polymer backbone provides good mechanical strength and flexibility [40, 41]. At the same time, GPEs can integrate with electrode materials closely after the monomer polymerization process. Thus, the flexible Li–S batteries can withstand more severe deformation and their lifespan can be extended. Figure 5.4a–c shows the electrochemical performance of GPE-based batteries and their immobilization mechanisms for polysulfides during charging and discharging cycles.

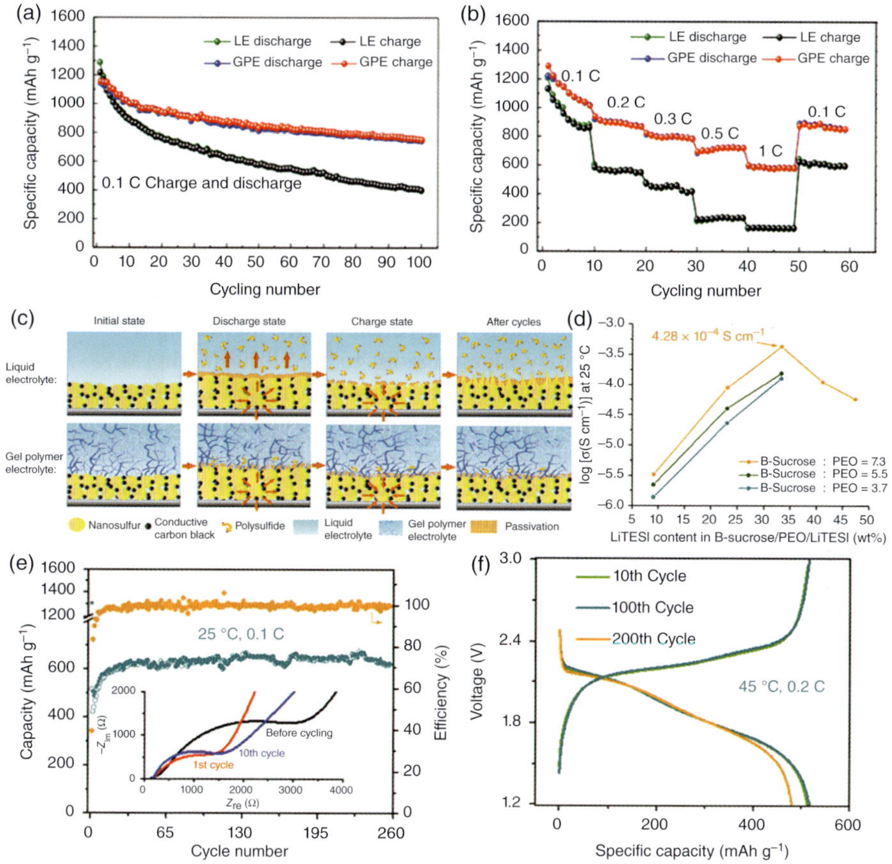

Figure 5.4 (a) Cycling performance of GPE-based batteries, and those of liquid-based ones at 0.1 C. (b) Rate performance of GPE-based batteries and the liquid-based ones from 0.1 to 1 C. (c) The polysulfide immobilization mechanism of GPE-based batteries and the liquid-based ones. Dendrites continually grow on the sulfur electrodes' surface in the liquid electrolyte based batteries while the GPE-based ones can maintain a smooth morphology. *Source:* (a–c) Liu et al. 2016 [42]. Copyright 2016, Elsevier B.V. (d) With the increasing weight ratio of sucrose in the sucrose-PEO electrolyte, the ionic conductivity increases obviously. (e) Cycling performance of Li–S batteries with sucrose-PEO electrolyte and the electrochemical impedance spectroscopy (EIS) spectrum of the battery. (f) Specific capacity of cells using the sucrose-PEO electrolyte. The value remained almost constant after 200 cycles. *Source:* (d–f) Liu et al. 2017 [2]. Copyright 2017, Electrochemical Society.

5.4.1 Solid Polymer Electrolytes for Flexible Li–S Batteries

Polyethylene oxide (PEO) is one of the most popular solid polymer electrolyte (SPE) for Li–S batteries [43, 44], in which the –C–O–C– bond units facilitate ion conduction. Much enhancement has been made to improve the performances of PEO [45]. However, the ion conduction is hindered at low temperatures [46], which is a major barrier for applications.

Al_2O_4 [57], and TiO_2 [55, 58] to obtain CPEs. Their physicochemical properties mainly depend on their modification process [59, 60]. Kurc and Jesionowski used TiO_2–SiO_2 submicron size powders as fillers in CPEs to create a more stable TiO_2–SiO_2 modified CPE. The desired ionic conduction could be attained with the combination of crystalline structured TiO_2 and amorphous SiO_2. SiO_2 increased the surface area while TiO_2 increased thermal stability and mechanical strength simultaneously [55]. The ionic conductivity of TiO_2–SiO_2 modified CPE was 9.8×10^{-4} S cm^{-1} at 298 K and also exhibited good cyclability and Coulombic efficiency. Figure 5.6c shows that the cell using this electrolyte lost only 0.02% of its capacity per cycle during cycling processes. SEM images (Figure 5.6d) show the porous structure of the electrolyte, which hosts the electrolyte solution and has a flexible cell geometry [55].

Although SPEs can eliminate the safety problems associated with liquid electrolytes, their ionic conductivity is low and flexibility may be limited due to their solid state nature. GPEs, with better mechanical strength and higher ionic conductivity, can withstand more severe deformation. CPEs can achieve both high ionic conductivity and excellent mechanical strength.

5.5 Separator for Flexible Li–S Batteries

Compared with conventional batteries, flexible batteries have more requirements on flexibility, elasticity, and chemical stability of separators, because the break of separators can cause severe damage to the entire battery system [61]. In a Li–S battery, the separator normally needs other functions and more characteristics:

1) The surface facing the cathode can limit the diffusion of polysulfide to improve the life span and reduce capacity decay [62, 63].
2) The separator surface can improve the conductivity and the cycling ability of active materials. This has been found helpful for improving columbic efficiency and working life span [64, 65].
3) The surface facing the anode material can prevent the growth of needle-like crystals (dendrites), which is recognized as a fatal problem for battery safety [66, 67].

In 2015, Zhou et al. invented a flexible sulfur–graphene–polypropylene (PP) separator integrated electrode system by combining a cathode electrode and separator together. As shown in Figure 5.7a, it is different from other freestanding metal or graphene collectors in that a graphene membrane was directly coated onto a commercial PP membrane. The separator not only divides the cathode and anode, but also supports graphene and sulfur. Its excellent flexibility and mechanical strength make this graphene@PP membrane dynamic and robust. The graphene in this membrane functions as an internal collector to support 70% sulfur loading. SEM image shows that the graphene coating is about 30 μm thick (Figure 5.7b). They claim that this material can replace Al foil, which always accounts for approximately 5–9% of a battery weight. Additionally, the presence of graphene can retain the dissolved polysulfides. The modified

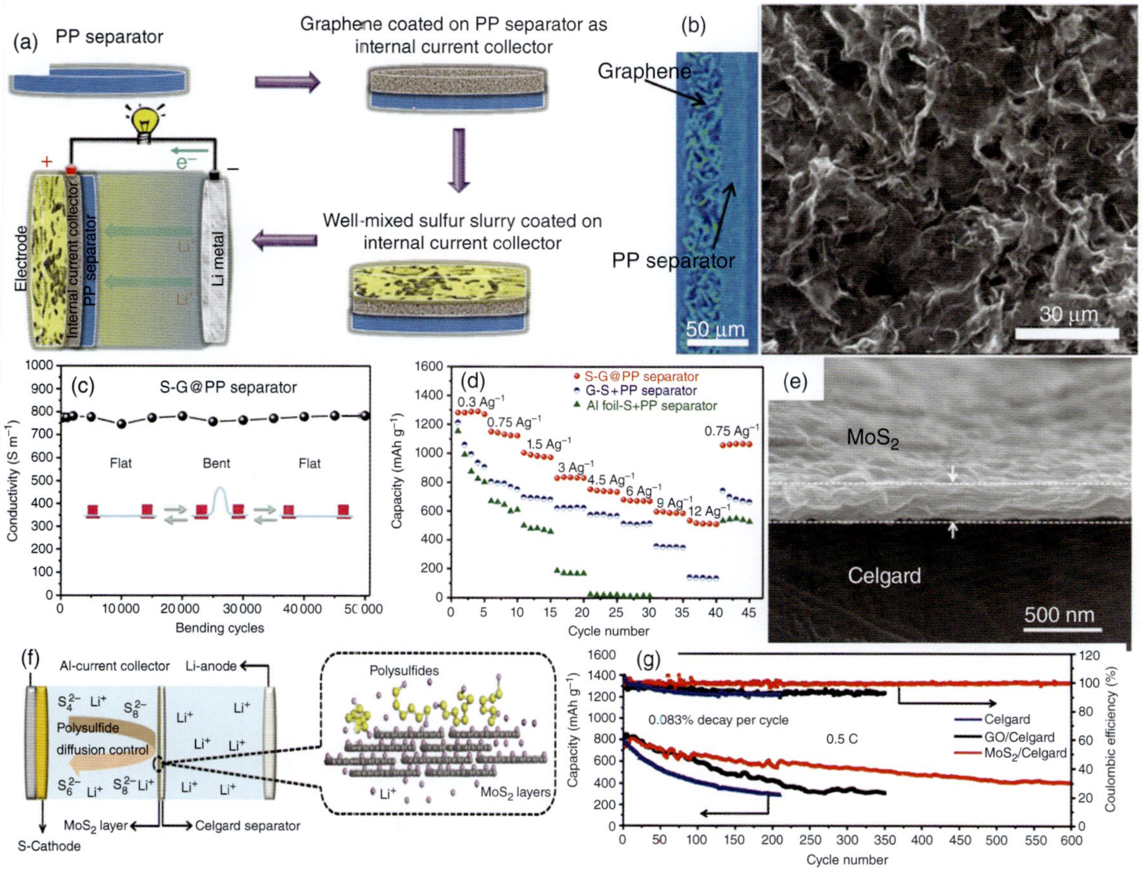

Figure 5.7 (a) Schematic of electrode configuration illustrating the process of combining sulfur and graphene@PP membrane to form an integrated electrode–separator structure. (b) Cross-section SEM image of sulfur electrode and corresponding EDS spectrum. (c) Stability of electrode–separator combination in terms of conductivity after 5000 bending cycles. (d) Rate performance of the S-G@PP separator, G-S + PP separator and Al foil-S + PP separator batteries at different current densities. *Source:* (a–d) Zhou et al. 2015 [68]. Copyright 2015, Wiley. (e) Cross section of MoS_2-Celgard layer. (f) Schematic Li–S battery configuration of MoS_2 Celgard™ separator. (g) The cycling performance for Celgard™, GO/Celgard™, and MoS_2/Celgard™ separators at 0.5 C. *Source:* (e–g) Ghazi et al. 2017 [25]. Copyright 2017, Wiley.

adhesion between sulfur and graphene membrane can significantly decrease polarization and the interfacial contact resistance. After being bent for many times, the flexible capacity and stable conductivity still remained (Figure 5.7c). To further confirm the excellent electrical contact between sulfur and the separator, they compared S-G@PP membrane, G-S + PP membrane and Al foil-S + PP membrane batteries cycled at various current densities (Figure 5.7d). The highest initial discharge capacity of the battery using the S-G@PP separator reached 1278 $mAh\,g^{-1}$ at 0.3 $A\,g^{-1}$ while the Li–S battery with a traditional separator battery showed a quick capacity decay [68].

Many people attempted to modify Celgard™ membranes to meet the requirements of flexibility and stronger diffusion suppression of polysulfides without losing their ion transport ability. For example, Ghazi et al. deposited a MoS_2 layer on the surface of a Celgard™ separator (Figure 5.7e). MoS_2 is a kind of A–B–A sandwich-like 2D material, which has both a superior chemical stability and a higher SSA. Because of the high density of lithium ions on the MoS_2 surface, this composite separator shows fast lithium diffusion and facile lithium transference. When used in a Li–S battery, the separator is proven to be highly efficient for restricting polysulfide shuttle and increasing the service life. When the sulfur loading was 65%, the battery capacity was 808 $mAh\,g^{-1}$ after 600 cycles. The coulombic efficiency remains above 99.5% during all 600 cycles (Figure 5.7f). As shown in Figure 5.7g, the battery with MoS_2 Celgard™ separator delivers an initial discharge capacity of 804 $mAh\,g^{-1}$, which is higher than the battery with an unmodified plain separator. They found that the diffusion coefficient for the MoS_2 Celgard™ separator at lower voltages is higher than both GO/Celgard™ and pristine Celgard™. At lower voltages, the high-order polysulfides convert to solid lithium sulfides. These polysulfides could easily precipitate at the cathode–separator interface as an insulating layer and then hinder lithium ion diffusion. They deduced that the MoS_2 nanosheets provide voids to accommodate various redox species and prevent the formation of such an insulating layer, leading to facile lithium ion diffusion [25].

Jiang et al. attempted to inhibit the shuttle effect by coating GO on a membrane. They found that compared with pristine Celgard™ membrane, less sulfur and polysulfide species were left on the GO-coated separator membrane. The graphene oxide modification also created a three dimensional porous structure, while maintaining the ideal flexibility of pristine Celgard™ membrane (Figure 5.8a). They declared that the graphene modification is a good way to construct a separator membrane with an ideal electrochemical performance [69].

Abbas et al. fabricated a bifunctional separator by spray coating poly(3,4-ethaylenedioxythiophene):poly(styrene sulfonate) (PEDOT:PSS) on the surface of the separator. They found that after the modification, the surface of the separator can transform from hydrophobic to hydrophilic, resulting in greater wettability, better electrolyte intake, and improved electrochemical performance. This structure could greatly increase the cycling stability of Li–S batteries with a capacity loss as low as 0.0364% per cycle after 1000 cycles at a current rate of 0.25 C. More importantly, this membrane has unique flexibility and can retain a stable conductivity even when bearing large structure bending (Figure 5.8b,c). The modified separator has abundant sulfonate functional groups (from

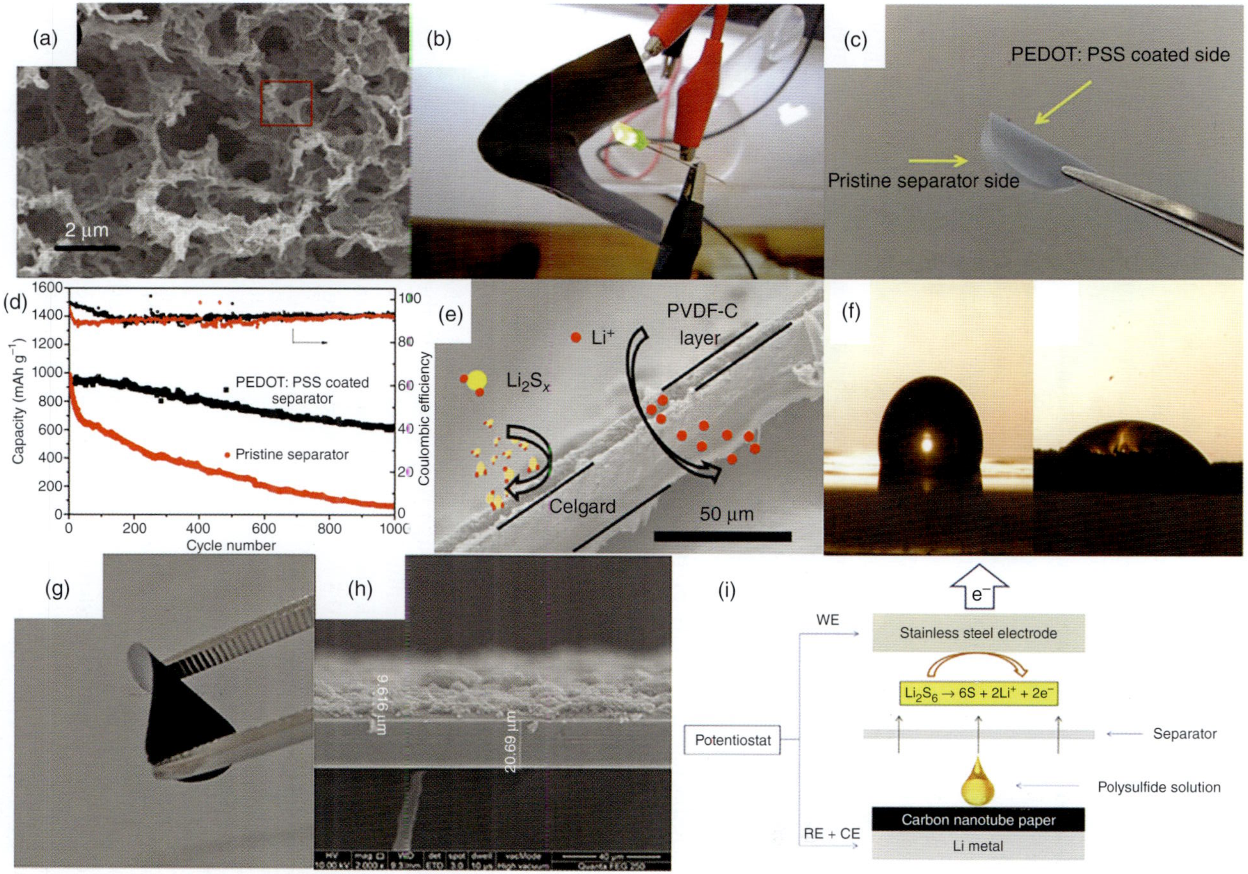

Figure 5.8 (a) The porous structure of a graphene-modified separator [69]. *Source:* Copyright 2017, Elsevier B.V. (b) The flexibility of a Li–S battery using the separator and metal Li as anode material. (c) The flexibil ty of the separator. (d) Schematic diagram of a modified Celgard™ separator restricting the polysulfide migration while maintaining good lithium ion transportation. (d) Comparison of hydrophilic separator (before and after modification). *Source:* (b–d) Abbas et al. 2016 [70]. Copyright 2015, Royal Society of Chemistry. (g, h) The structure of the modified membrane [2]. (i) The schematic diagram of the experimental battery. *Source:* Cui and Fu 2015 [71]. Copyright 2015, Elsevier B. V.

PSS). The stable specific capacity can be ascribed to PSSs' negative charges, which effectively suppresses the shuttle effect through mutual coulombic repulsion. Meanwhile, the PEDOT also supports chemical interactions and forms a chelated coordination structure. PSS can be considered as a dopant in PEDOT:PSS, which can facilitate lithium ion transport, and provide a "polar solvent-philic" environment [70]. Wei et al. demonstrated another way to modify Celgard™ separators with a polyvinylidene fluoride-C (PVDF-C) layer, whereby the polysulfides generated during discharge processes can be anchored on the cathode side (Figure 5.8d). Similar to Abba's discovery, the hydrophilic performance yields a large intensification (Figure 5.8e,f). Li–S batteries based on this modified separator achieved a high discharge capacity of 919 $mAh\,g^{-1}$ after 100 cycles at a rate of 0.5 C. Even at a current rate of 5 C, the batteries were able to maintain a discharge capacity approximately 400 $mAh\,g^{-1}$. The exceptional electrochemical performance can be attributed to the enhanced adsorption capability of the micropores. Meanwhile, the presence of C–C double bonds, and the conductivity of the carbon networks in the PVDF-C layer also compensate for the low conductivity of elemental sulfur. This economical and simple strategy to overcome polysulfide dissolution provides a commercially feasible carbon method for the development of flexible Li–S batteries [72].

Zhang et al. presented a separator with nitrogen-doped porous hollow carbon sphere coating, which showed enhanced electrochemical performance. The high initial discharge capacity of 1656 $mAh\,g^{-1}$ and a low fading rate of 0.11% per cycle within 500 cycles (1 C) were achieved. This is ascribed to the chemical and physical adsorption properties of porous and nitrogen-doped NHCs. The separator demonstrated an excellent flexibility and chemical stability as shown in Figure 5.8g,h [7].

Cui and Fu reported a "linear voltage" sweep method to quantitatively measure anodic (oxidization) currents of polysulfides crossing the separators. As shown in Figure 5.8i, when the liquid polysulfide solution was added to the CNT paper, the polysulfides migrated through the separator to the electrode and were oxidized to form elemental sulfur. The electrical charge in Coulombs produced by the oxidation of polysulfides is linearly related to the concentration of polysulfides within a certain range ($\leq$0.5 M). This method provides guidance for developing new separators for flexible Li–S batteries [71].

5.6 Summary

In this chapter, we discussed the development of flexible Li–S batteries, including cathode materials, electrolytes, and separators. Carbonaceous materials are widely used as cathode hosts for flexible Li–S batteries. In particular, graphene and CNTs were intensively applied as flexible cathodes for Li–S batteries. However, carbonaceous materials cannot obtain the same enhancement as lithium ion batteries because of their nonpolar characteristics. Heteroatom doping is a good approach, which can not only strengthen the affinity between sulfur and carbon materials but also provide many active sites for anchoring polysulfides. Applicable electrolytes and separators are needed for flexible Li–S batteries.

In general, SPEs or GPEs are required to avoid leakage during bending for flexible Li–S batteries. The modification of separators can significantly enhance the performances of flexible Li–S batteries. With the successful fabrications of flexible sulfur cathodes, electrolyte membranes, and modified separators, high-performance flexible Li–S batteries can be fully developed for powering flexible and wearable electronic devices.

References

1 Vizintin, A., Patel, M., Genorio, B., and Dominko, R. (2014). Effective separation of lithium anode and sulfur cathode in lithium–sulfur batteries. *ChemElectroChem* 1: 1040–1045.

2 Liu, K., Lin, Y., Miller, J.D. et al. (2017). Study of sucrose based room temperature solid polymer electrolyte for lithium sulfur battery. *J. Electrochem. Soc.* 164: A447–A452.

3 Luo, J., Lee, R., Jin, J. et al. (2017). A dual-functional polymer coating on a lithium anode for suppressing dendrite growth and polysulfide shuttling in Li–S batteries. *Chem. Commun.* 53: 963–966.

4 Peled, E. and Yamin, H. (1981). Solid electrolyte interphase (SEI) electrodes: lithium/polysulfide-THF system. *J. Power Sources* 8: 101–115.

5 Cao, R., Xu, W., Lv, D. et al. (2015). Anodes for rechargeable lithium–sulfur batteries. *Adv. Energy Mater.* 5: 1402273–1402295.

6 See, K.A., Wu, H., Lau, K. et al. (2016). Effect of hydrofluoroether cosolvent addition on Li solvation in acetonitrile-based solvate electrolytes and its influence on S reduction in a Li–S battery. *ACS Appl. Mater. Interfaces* 8: 34360–34371.

7 Zhang, Z., Li, Q., Zhang, K. et al. (2015). Titanium-dioxide-grafted carbon paper with immobilized sulfur as a flexible free-standing cathode for superior lithium–sulfur batteries. *J. Power Sources* 290: 159–167.

8 Zhou, G., Wang, D., Li, F. et al. (2012). A flexible nanostructured sulphur–carbon nanotube cathode with high rate performance for Li–S batteries. *Energy Environ. Sci.* 5: 8901–8906.

9 Zeng, L., Pan, F., Li, W. et al. (2014). Free-standing porous carbon nanofibers-sulfur composite for flexible Li–S cathode. *Nanoscale* 6: 9579–9587.

10 Zeng, L., Zeng, W., Jiang, Y. et al. (2015). A flexible porous carbon nanofibers-selenium cathode with superior electrochemical performance for both Li–Se and Na–Se batteries. *Adv. Energy Mater.* 5: 1401377–1401386.

11 Thieme, S., Brueckner, J., Bauer, I. et al. (2013). High capacity micro-mesoporous carbon–sulfur nanocomposite cathodes with enhanced cycling stability prepared by a solvent-free procedure. *J. Mater. Chem. A* 1: 9225–9234.

12 Wang, C., Su, K., Wan, W. et al. (2014). High sulfur loading composite wrapped by 3D nitrogen-doped graphene as a cathode material for lithium–sulfur batteries. *J. Mater. Chem. A* 2: 5018–5023.

13 Wang, H., Zhang, W., Liu, H., and Guo, Z. (2016). A strategy for configuration of an integrated flexible sulfur cathode for high-performance lithium–sulfur batteries. *Angew. Chem. Int. Ed.* 55: 3992–3996.

14 Schuster, J., He, G., Mandlmeier, B. et al. (2012). Spherical ordered mesoporous carbon nanoparticles with high porosity for lithium–sulfur batteries. *Angew. Chem. Int. Ed.* 51: 3591–3595.

15 Meyer, W. (1998). Polymer electrolytes for lithium–ion batteries. *Adv. Mater.* 10: 439–448.

16 Lee, K., Black, R., Yim, T. et al. (2012). Surface-initiated growth of thin oxide coatings for Li–sulfur battery cathodes. *Adv. Energy. Mater.* 2: 1490–1496.

17 Li, D., Han, F., Wang, S. et al. (2013). High sulfur loading cathodes fabricated using peapodlike, large pore volume mesoporous carbon for lithium–sulfur battery. *ACS Appl. Mater. Interfaces* 5: 2208–2213.

18 Chen, Y., Lu, S., Wu, X., and Liu, J. (2015). Flexible carbon nanotube-graphene/sulfur composite film: free-standing cathode for high-performance lithium/sulfur batteries. *J. Phys. Chem. C* 119: 10288–10294.

19 Sun, Q., Fang, X., Weng, W. et al. (2015). An aligned and laminated nanostructured carbon hybrid cathode for high-performance lithium–sulfur batteries. *Angew. Chem. Int. Ed.* 54: 10539–10544.

20 Sun, L., Kong, W., Jiang, Y. et al. (2015). Super-aligned carbon nanotube/graphene hybrid materials as a framework for sulfur cathodes in high performance lithium sulfur batteries. *J. Mater. Chem. A* 3: 5305–5312.

21 Wang, C., Wang, X., Wang, Y. et al. (2015). Macroporous free-standing nano-sulfur/reduced graphene oxide paper as stable cathode for lithium–sulfur battery. *Nano Energy* 11: 678–686.

22 Song, X., Wang, S., Bao, Y. et al. (2017). A high strength, free-standing cathode constructed by regulating graphitization and the pore structure in nitrogen-doped carbon nanofibers for flexible lithium–sulfur batteries. *J. Mater. Chem. A* 5: 6832–6839.

23 Guo, Y., Zhao, G., Wu, N. et al. (2016). Efficient synthesis of graphene nanoscrolls for fabricating sulfur-loaded cathode and flexible hybrid interlayer toward high-performance Li–S batteries. *ACS Appl. Mater. Interfaces* 8: 34185–34193.

24 Lin, C., Niu, C.J., Xu, X. et al. (2016). A facile synthesis of three dimensional graphene sponge composited with sulfur nanoparticles for flexible Li–S cathodes. *Phys. Chem. Chem. Phys.* 18: 22146–22153.

25 Ghazi, Z.A., He, X., Khattak, A.M. et al. (2017). MoS_2/Celgard separator as efficient polysulfide barrier for long-life lithium–sulfur batteries. *Adv. Mater.* 29: 1606817–1606822.

26 An, Y.B., Zhu, Q.Z., Hu, L.F. et al. (2016). A hollow carbon foam with ultra-high sulfur loading for an integrated cathode of lithium–sulfur batteries. *J. Mater. Chem. A* 4: 15605–15611.

27 Song, J., Yu, Z., Xu, T. et al. (2014). Flexible freestanding sandwich-structured sulfur cathode with superior performance for lithium–sulfur batteries. *J. Mater. Chem. A* 2: 8623–8627.

28 Milroy, C. and Manthiram, A. (2016). An elastic, conductive, electroactive nanocomposite binder for flexible sulfur cathodes in lithium–sulfur batteries. *Adv. Mater.* 28: 9744–9751.

29 Ghosh, A., Manjunatha, R., Kumar, R., and Mitra, S. (2016). A facile bottom-up approach to construct hybrid flexible cathode scaffold for high-performance lithium–sulfur batteries. *ACS Appl. Mater. Interfaces* 8: 33775–33785.

30 Wang, C., Wang, X., Yang, Y. et al. (2015). Slurryless Li_2S/reduced graphene oxide cathode paper for high-performance lithium sulfur battery. *Nano Lett.* 15: 1796–1802.

31 He, J., Chen, Y., Lv, W. et al. (2016). Three-dimensional CNT/graphene–Li_2S aerogel as freestanding cathode for high-performance Li–S batteries. *ACS Energy Lett.* 1: 820–826.

32 He, J., Chen, Y., Lv, W. et al. (2016). Three-dimensional hierarchical reduced graphene oxide/tellurium nanowires: a high-performance freestanding cathode for Li–Te batteries. *ASC Nano* 10: 8837–8842.

33 Huang, Y., Zheng, M., Lin, Z. et al. (2015). Flexible cathodes and multifunctional interlayers based on carbonized bacterial cellulose for high-performance lithium–sulfur batteries. *J. Mater. Chem. A* 3: 10910–10918.

34 Armand, M. and Tarascon, J. (2008). Building better batteries. *Nature* 451: 652–657.

35 Chen, S., Xin, Y., Zhou, Y. et al. (2014). Self-supported $Li_4Ti_5O_{12}$ nanosheet arrays for lithium ion batteries with excellent rate capability and ultralong cycle life. *Energy Environ. Sci.* 7: 1924–1930.

36 Khurana, R., Schaefer, J., Archer, L., and Coates, G. (2014). Suppression of lithium dendrite growth using cross-linked polyethylene/poly(ethylene oxide) electrolytes: a new approach for practical lithium-metal polymer batteries. *J. Am. Chem. Soc.* 136: 7395–7402.

37 Ramesh, S. and Chai, M. (2007). Conductivity, dielectric behavior and FTIR studies of high molecular weight poly(vinylchloride)-lithium triflate polymer electrolytes. *Mater. Sci. Eng., B* 139: 240–245.

38 Ramesh, S., Winie, T., and Arof, A. (2007). Investigation of mechanical properties of polyvinyl chloride–polyethylene oxide (PVC–PEO) based polymer electrolytes for lithium polymer cells. *Eur. Polym. J.* 43: 1963–1968.

39 Tarascon, J. and Armand, M. (2001). Issues and challenges facing rechargeable lithium batteries. *Nature* 414: 359–367.

40 Zhang, Y., Zhao, Y., and Bakenov, Z. (2014). A novel lithium/sulfur battery based on sulfur/graphene nanosheet composite cathode and gel polymer electrolyte. *Nanoscale Res. Lett.* 9: 137–143.

41 Zhao, Y., Zhang, Y., Bakenov, Z., and Chen, P. (2013). Electrochemical performance of lithium gel polymer battery with nanostructured sulfur/carbon composite cathode. *Solid State Ionics* 234: 40–45.

42 Liu, M., Zhou, D., He, Y. et al. (2016). Novel gel polymer electrolyte for high-performance lithium–sulfur batteries. *Nano Energy* 22: 278–289.

43 Barchasz, C., Molton, F., Duboc, C. et al. (2012). Lithium/sulfur cell discharge mechanism: an original approach for intermediate species identification. *Anal. Chem.* 84: 3973–3980.

electronic conductivity and rate performance. This ternary composite cathode sustains 960.7 mAh g^{-1} reversible specific discharge capacity after 40 cycles at 0.1 C, and 523.2 mAh g^{-1} at 0.5 C.

Flexible Cathode Based on Super-aligned Carbon Nanotube Arrays As one special type of aligned CNTs, the super-aligned CNTs (SACNTs) have the advantages of uniform arrangement of CNTs, high aspect ratio, and clean surface. The SACNTs can be easily processed into film or yarn by direct drawing or spinning. After a facile ultrasonic treatment [44], SACNTs can serve as the support skeleton with a high aspect ratio, clean surface, and strong interaction among the tubes [45, 46]. The S–SACNT composite cathode exhibits excellent high-rate performance due to the high conductivity and the efficient electrolyte penetration into the ordered porous structure. The adsorption force of the SACNT frame can limit the migration of sulfur and its lithiated products, thereby inhibiting the shuttle effect. In order to enhance the loading and cycling performance of sulfur, Wang and coworkers introduced porous CNTs (PCNTs) into the SACNT via controlled oxidation in air at high temperature [47]. Compared with the pristine SACNTs, PCNTs of improved dispersibility and more mesopores to host sulfur favor its performance in sulfur cathodes (Figure 6.4a,b). While PCNTs maintain high flexibility, their high surface area and large pore volume make high sulfur loading amount and high sulfur utilization possible. When used as cathodes for Li–S batteries, a high sulfur loading of 60 wt% is achieved, with superb reversible capacities of 866 and 526 mAh g^{-1} based on the weights of sulfur and electrode, respectively. In addition, the same research group obtained a composite separator with a high porosity, superb electrical conductivity, and extraordinary flexibility using the SACNT/graphene hybrid material as the three-dimensional conductive framework for sulfur storage [49]. CNT network provides flexible freestanding skeleton, and graphene hosts sulfur and polysulfides. The CNT/G hybrid framework allows better dispersion of sulfur and allows each sulfur particle to adhere tightly to the conductive network. By tuning the CNT/G ratio in the electrode, the S-CNT/G nanocomposite exhibits enhanced mechanical properties and good electrochemical properties.

Flexible Cathode Based on Nitrogen-doped CNTs The heteroatom doping of CNTs may introduce tunable surface chemistry for CNTs. Nitrogen-doped CNTs are one of the most attractive doped CNTs [50]. As nitrogen affords additional electrons, the conductivity of N-CNTs might be improved in comparison with pristine CNTs. The N-CNTs also afford a tunable surface with high chemical adsorption ability toward polysulfides [51–55]. Bakenov and coworkers proposed a S/N-CNT composite electrode with a coaxial tubular structure [48]. The sulfur particles were dispersed in N-CNTs networks to form freestanding S/N-CNT composites (Figure 6.4c). N-CNTs provide the carbon skeleton, forming a stable interconnected network. In addition, the formation of the freestanding and flexible S/N-CNT composite film allows high sulfur content and contributes to superb cycle performance and rate performance. A reversible capacity of 1098 mAh g^{-1} was obtained in the second cycle at 0.2 C. The as-obtained cell retained about 73.5% of its initial reversible discharge capacity after 100 cycles; the coulombic efficiency was maintained above 93%.

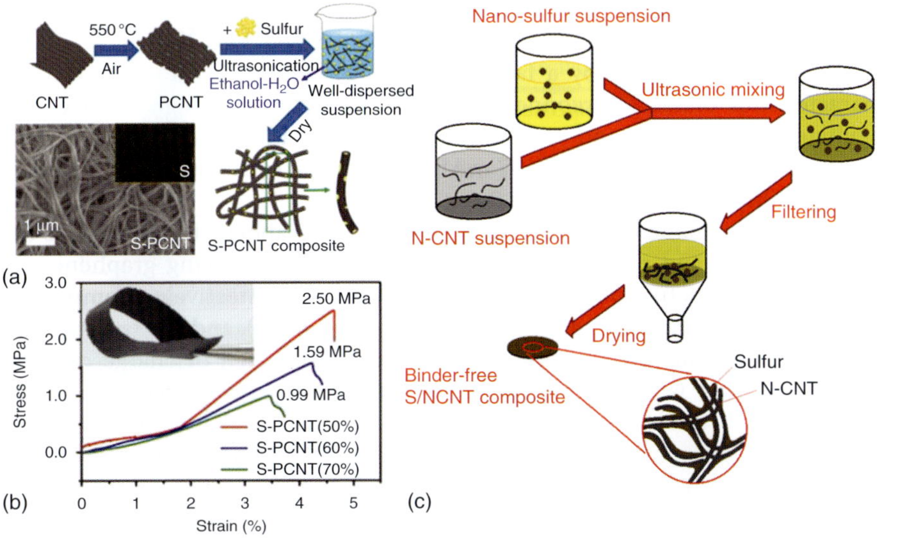

Figure 6.4 Flexible sulfur cathode with SACNTs and N-CNTs. (a) Synthesis and characterization of the S–PCNT composites; (b) mechanical properties of the S–PCNT composites. Inset shows the photographs of the S-PCNT composite. *Source:* Sun et al. 2016 [47]. Copyright 2016, American Chemical Society. (c) Schematic diagrams of the synthesis of S/N–CNT composite. *Source:* Zhao et al. 2015 [48]. Copyright 2015, Springer.

6.3.1.2 Flexible Cathode Based on Carbon Nanofibers

Carbon nanofibers (CNFs), as one type of high strength materials, can also be woven or integrated into an interconnected flexible matrix for electrochemical applications. Using CNFs as mechanical skeleton network affords even better long range mechanical strength than CNT electrodes.

Aurbach and coworkers reported a simple method for the preparation of freestanding carbon–sulfur cathodes by direct impregnation of microporous-activated carbon fibers with elemental melted sulfur [56]. The as-obtained monolithic carbon fiber cloth/sulfur composite allows a high loading of sulfur per cathode area, while maintaining good electrical conductivity for high utilization of sulfur active materials. Manthiram and coworkers applied a facile and LBL strategy for high-areal-capacity sulfur cathodes. The sulfur powders were directly splinted in between porous carbon nanofiber (PCNF) layers in the assembly process, which contributed to an areal sulfur loading amount of 11.4 mg cm^{-2} and a sulfur content as high as 56.3 wt%, respectively [57]. Manthiram's research group also proposed a robust ultratough flexible cathode based on CNTs, in which the active material was stacked between two buckypapers [58]. The paper exhibited a fibrous structure in which the CNTs were firmly adhered on the CNF skeleton. The flexible buckypaper sandwich can effectively block polysulfides and inhibit the loss of active materials. The hybrid carbon skeleton realizes rapid lithium ion transportation, and also activates and utilizes sulfur materials effectively. The buckypaper/sulfur/buckypaper cathode preserved a high capacity retention in the long-term cycle. Yu and coworkers fabricated freestanding S/(PCNF-CNT) composite electrode by penetrating the sulfur into the microporous carbon nanofiber (PCNF)-CNT composites [59].

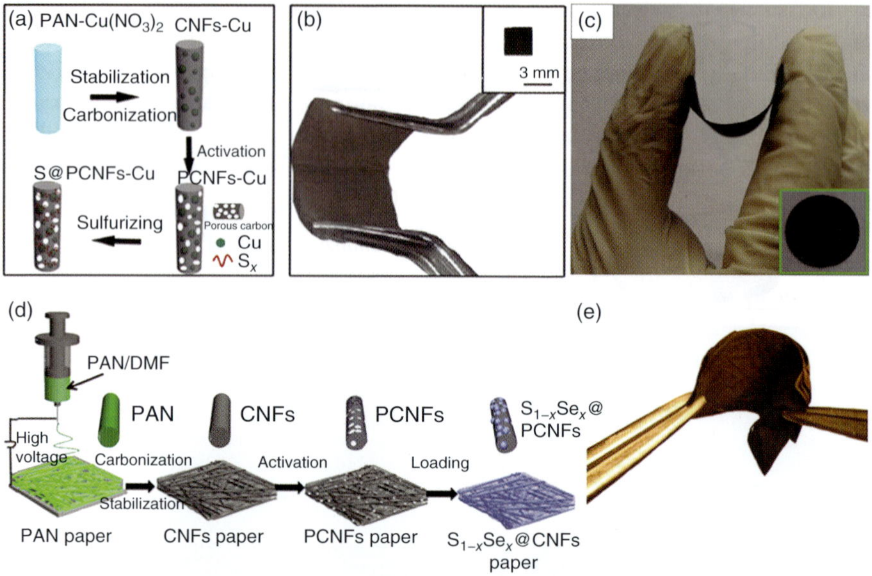

Figure 6.6 Flexible sulfur cathode with Cu, TiO_2, and Se-modified CNFs. (a) Schematic illustration of the synthesis for the S@PCNFs-Cu electrode; (b) photograph of the freestanding and flexible S@PCNFs-Cu electrode. *Source:* Zeng et al. 2015 [63]. Copyright 2015, Royal Society of Chemistry. (c) Digital photograph of CP@TiO_2-S membrane indicating its flexibility, with the inset showing a top view. *Source:* Zhang et al. 2015 [65]. Copyright 2015, Elsevier. (d) Schematic illustration of the synthesis for $S_{1-x}Se_x$@PCNFs electrode; (e) photograph of free-standing and flexible $S_{1-x}Se_x$@PCNFs electrode. *Source:* Zeng et al. 2016 [66]. Copyright 2016, Elsevier.

The combination of metal oxide and CNFs can effectively improve the sulfur utilization and cycling stability of the sulfur cathode. The metal oxide can be used as an adsorbent in the sulfur cathode to capture polysulfide to achieve high coulombic efficiency. Zhang and coworkers produced the flexible cathode using filter papers [65]. Titanium dioxide grafted carbon paper (CP@TiO_2) with sulfur can be obtained by simple pyrolysis of filter paper treated by tetrabutyl titanate. The resulting CP@TiO_2 film consists of interwoven carbon fibers and TiO_2 nanoparticles in the voids of carbon fibers. The as-obtained film retains the robust and flexible structure of the CP membrane and the sulfur particles are uniformly distributed in the CP@TiO_2 film (Figure 6.6c). Owing to the synergistic effect between the interpenetrated carbon fibers of CP film and TiO_2 nanoparticles, the CP@TiO_2 film can trap the polysulfides physically and chemically, thus rendering the electrode excellent electrochemical performance, displaying a reversible discharge capacity of 850 mAh g^{-1} after 200 cycles at 0.5 C and a specific capacity of 660 mAh g^{-1} at 5 C.

6.3.1.3 Flexible Cathode Based on Polymer Fibers

Flexible sulfur wires (SW) obtained by blending a polymer with active sulfur can be woven into a textile cell. Kumta and coworkers used a very simple electrospinning process to produce flexible sulfur wires with unique yarn morphology [67]. Polymers used in the spinning process act as binders during the spinning process.

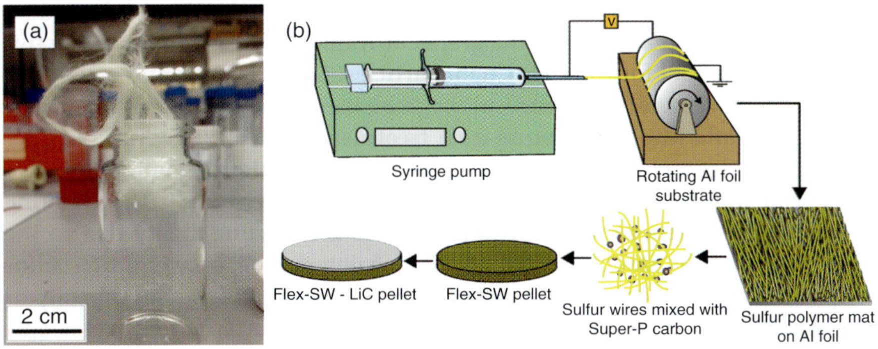

Figure 6.7 Standalone Flex-SW battery electrodes. (a) Image showing the yarn-like nature of the flexible sulfur wires (Flex-SWs); (b) procedure to prepare flexible sulfur wires (Flex-SWs), Flex-SWs-pellets, and Flex-SWs-pellets coated with lithium ion conductor (LIC)-Flex-SW-LIC pellets. *Source:* Hanumantha et al. 2016 [67]. Copyright 2016, Elsevier.

At the same time, the sulfur and polymer are tightly connected (Figure 6.7). The polymer-sulfur interface serves as a physical barrier to the liquid lithium electrolyte, thereby reducing the dissolution of polysulfides. Such a cathode exhibits stable capacity of ~650 $mAh\,g^{-1}$ and an impressive low fade rate of ~0.003% fade/cycle by overcoming the problem of initial capacity degradation through the introduction of a lithium ion conductive layer on the top of the flexible sheet.

The major roles of the one-dimensional material in flexible cathodes includes: (i) the interconnected networks provide a mechanically robust and flexible skeleton; (ii) the highly conductive networks can be served as rapid electron transportation; (iii) The porous structure is suitable for electrolyte immersion facilitating lithium ion transportation; (iv) facile accommodation of sulfur composites in the network; and (v) extra chemical interactions or physical barriers to limit the migration of polysulfides.

6.3.2 Flexible Cathode Based on Two-dimensional Materials

Graphene materials are with highly tunable surface properties and microstructures. Graphene materials (such as hierarchical porous graphene [68] and chemically activated nanoporous graphene [69]) are incorporated with sulfur for high efficiency of sulfur cathode materials in a routine slurry-coating process [70, 71]. Graphene-based materials were also efficiently applied as functional layers attaching with separators to mitigate the parasitic shuttle of polysulfides in lithium sulfur cells [72–75]. Herein, graphene papers can be used directly as flexible substrates to support active sulfur materials in a flexible electrode. The three-dimensional (3D) porous structure assembled from graphene sheets combines the advantages of graphene and 3D porous structures [76–81]. The cathodes based on graphene papers and graphene foams (GFs) are widely applied as flexible sulfur electrodes.

6.3.2.1 Flexible Cathode Based on Graphene Paper

Flexible cathode materials are obtained by direct combination of active material and graphene paper. Wen and coworkers fabricated a freestanding

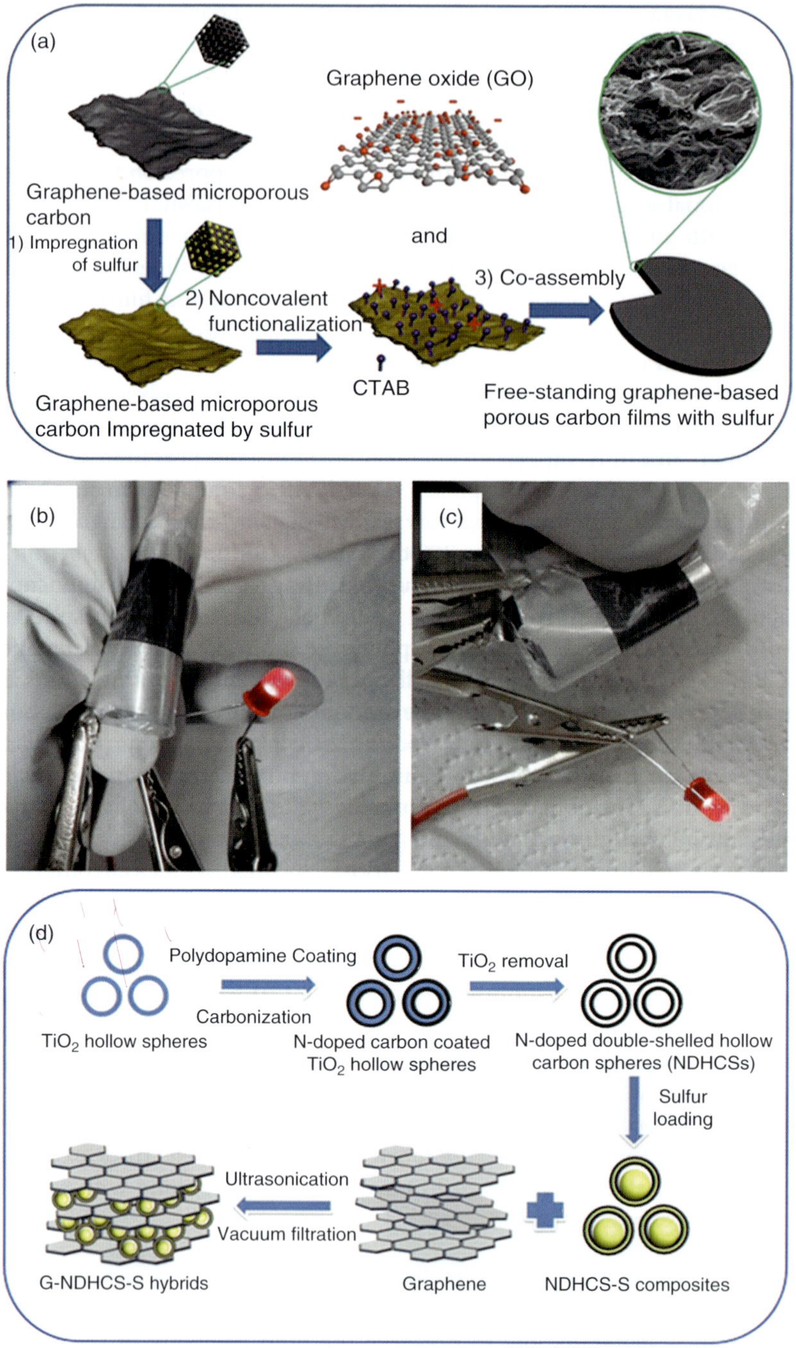

Figure 6.9 Freestanding graphene-based films with functionalized graphene sheets and hollow carbon spheres. (a) Fabrication of freestanding graphene-based porous carbon (GPC) films; (b,c) optical images of a red LED lightened by the Li–sulfur battery under the bending and folding states, respectively. *Source:* Wu et al. 2015 [87]. Copyright 2015, Royal Society of Chemistry. (d) Schematic illustration of the fabrication of the NDHCS–S composites and G–NDHCS–S hybrid paper. Source: Zhou et al. 2015 [88]. Copyright 2015, John Wiley & Sons.

CVD, followed by the coating of PDMS. By the incorporation of mixed sulfur slurry (including pure sulfur, carbon black and polyvinylidene fluoride), a flexible cathode material can be easily fabricated (Figure 6.10). The PDMS coating on the GF allows the interconnection network to be robust enough to ensure the flexibility of the cathode. The interconnected GF can provide a macroscopic effective electron path, as well as a large void space to accommodate a large number of

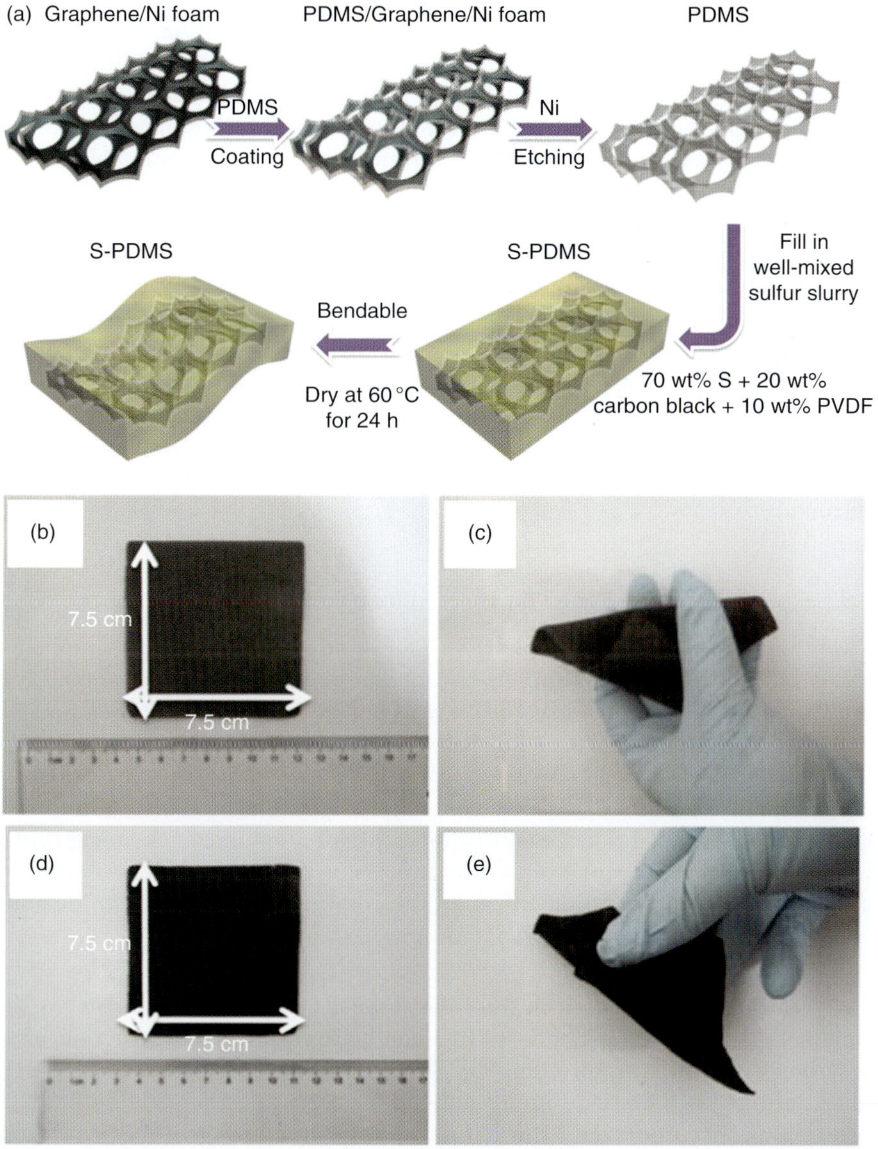

Figure 6.10 Flexible sulfur cathode with graphene foam. (a) Schematic of the procedure for fabricating PDMS/GF and S-PDMS/GF electrodes; (b,c) photographs of a PDMS/GF showing its good flexibility; (d,e) photographs of a flexible S-PDMS/GF electrode with 10.1 mg cm^{-2} sulfur loading. *Source:* [91], Copyright 2016, Elsevier Ltd.

active materials and sufficient electrolytes. The electrode with 10.1 $mg\,cm^{-2}$ sulfur loading could deliver an extremely high areal capacity of 13.4 $mAh\,cm^{-2}$. Meanwhile, the high sulfur-loaded electrodes retain a high rate performance with reversible capacities higher than 450 $mAh\,g^{-1}$ under a large current density of 6 $A\,g^{-1}$ and preserve stable cycling performance with ~0.07% capacity decay per cycle over 1000 cycles. Chen and coworkers also obtained the GF by CVD on nickel foam [92]. The highly crystallized Li_2S nanoparticles were then evenly deposited in the 3D GF network by the simple infiltration method. The material as an electrode without metal current collector and conductive additives and adhesives has good electrochemical properties. Zhou et al. proposed the use of a lightweight, porous nitrogen- and sulfur-codoped 3D graphene sponge derived through a hydrothermal assembly process as the freestanding electrode structure to host large amounts of dissolved lithium polysulfides [93]. The loading amount of active sulfur materials can reach 4.6 $mg\,cm^{-2}$ when using the graphene sponge as a flexible current collector.

By constructing different graphene frameworks, the following functions can be performed (i) build a robust and flexible electrode; (ii) provide a buffer space for the volume change during cycling; (iii) effectively promote the diffusion of electrons and ions to improve conductivity; (iv) provide a large number of pores to accommodate sulfur to increase the amount of loaded sulfur; (v) adsorb polysulfides to inhibit the shuttle effect; and (vi) obviates the requirement of conductive additives, adhesives, and current collector to increase the energy density of the electrode.

6.3.3 Flexible Cathode Based on Three-dimensional Materials

In addition to one-dimensional CNTs/ CNFs and two-dimensional graphene materials, 3D materials such as carbonaceous aerogels, carbon foam, and nickel foam also presents flexibility and can serve as matrix for flexible sulfur cathodes. Among these, carbon foam is mainly from the biomass and the polymer route. The advantages of biomass-derived carbon are very impressive in terms of environmental friendliness and low cost. The polymer foam is also of interest considering its cheap and facile fabrication process. The tunable chemical composition of the carbon foam is also attractive. Using commercial nickel foams as flexible substrates significantly enhances the conductivity of the electrode and greatly simplifies the electrode preparation process.

6.3.3.1 Flexible Cathode Based on Three-dimensional Carbon Foam Materials

Biomass such as bacterial cellulose (BC) can be used in the construction of a 3D carbon network structure [94, 95]. Sun and coworkers introduced the 3D carbon aerogel from the BC into lithium–sulfur battery as the cathode flexible skeleton [96]. 3D carbonized BC with highly interconnected nanofiber structure exhibits good electrical conductivity and mechanical stability (Figure 6.11a). The internal macroporous structure of the carbonized BC contributes to the uniform dispersion of sulfur and the increase in loading, and the carbonized BC composite contains sufficient free space to buffer the pressure caused by changes

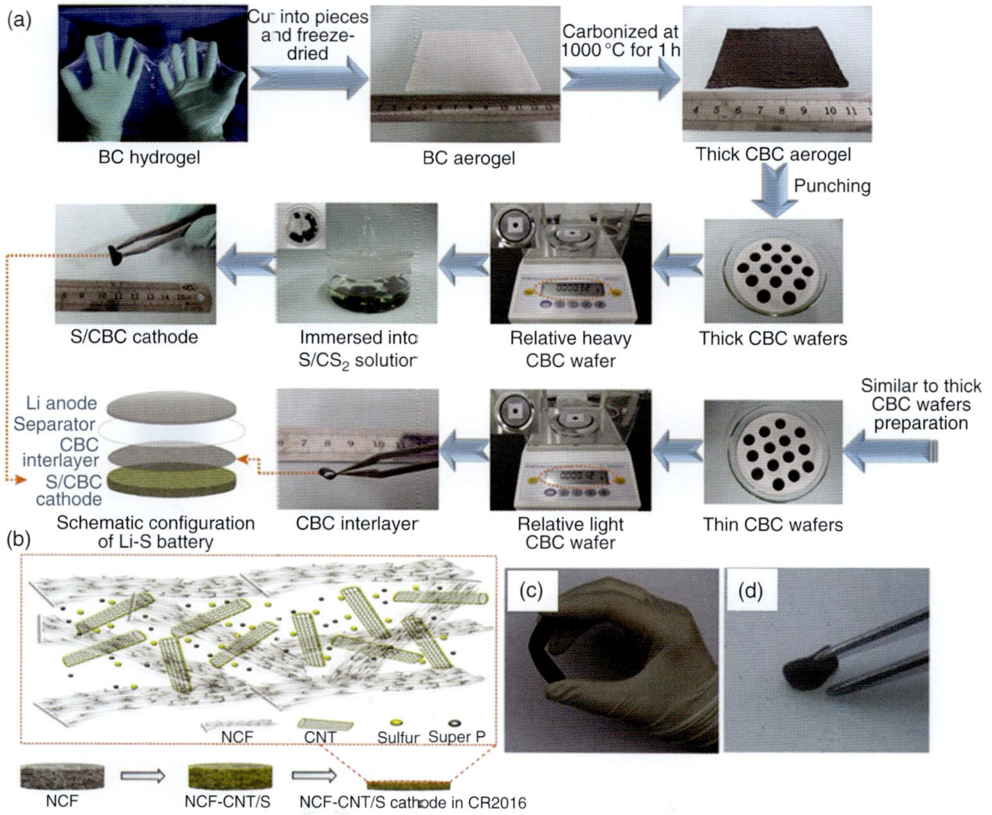

Figure 6.11 Flexible sulfur cathode with three-dimensional carbon foam. (a) The procedure for preparing the S/CBC cathode and CBC interlayer, and their specific location in the Li–S battery configuration. *Source:* Huang et al. 2015 [96]. Copyright 2015, Royal Society of Chemistry. (b) Structure illustration of the NCF-CNT/S cathode; (c) NCF bent by fingertips; (d) the NCF-CNT/S composite electrode, bent by tweezers. *Source:* Cao et al. 2016 [97]. Copyright 2016, Royal Society of Chemistry.

of electrode volume. The combined effects from such a configuration of carbonized BC in the cathode side result in high discharge capacity (1134 mAh g_{Sul}^{-1} at 200 mA g^{-1}), long-term cycle stability (700 mAh g_{Sul}^{-1} at 400 mA g^{-1} over 400 cycles), and good rate capability. Yushin and coworkers obtained a flexible porous conductive carbon layer by high-temperature carbonization of a biomass-derived porous cellulose sheet, and they fabricated the freestanding flexible matrix by impregnating the porous sheets with Li_2S-ethanol solution [98], wherein the Li_2S nanospheres with a diameter of about 50 nm were uniformly dispersed in most of the pores in the sheets. The composite exhibits high discharge capacity close to the theoretical value due to the pore structure and the high conductivity of the sheets. The good mechanical stability of the carbonized sheets assures the mechanical integrity of the prepared electrode.

The polymer foam with unique 3D structure can be directly used as the cathode material of lithium–sulfur battery after carbonization. Xu and coworkers obtained ultralight carbon foam with interconnected hollow network through the special heating of commercial melamine formaldehyde (MF) foam [99]. Owing to the strong adsorption capacity and the unique flexible skeleton, the integrated electrode exhibits excellent cycle stability. Similarly, Mi, Chen, and coworkers obtained porous carbon foams by carbonization of commercial melamine foams, and nitrogen-doped flexible carbon foam material by ultrasonic impregnation of sulfur slurry [100]. Yang and coworkers used polyurethane foam (PUF) as a carbon precursor to synthesize a new metal-free, nitrogen-doped flexible carbon foam (NCF) [97]. The derived 3D N-doped porous carbon network possesses a porous structure with high flexibility and high conductivity (Figure 6.11b–d). NCF current collector can also play the role of the interlayer to suppress the shuttle effect. The resulting batteries deliver a large reversible capacity of 1124 mAh g^{-1} at 0.5 C and retain a high specific capacity of 902.8 mAh g^{-1} after 100 cycles, with a coulombic efficiency of 98.6% throughout the cycles. The synthesis of carbon foams from polymer precursors is much easier than the process from nanomaterial building blocks, such as graphene and CNTs. The former can be prepared by simply carbonizing the polymer, while the latter always involves complex and tedious processes.

6.3.3.2 Flexible Cathode Based on Carbon/Binder Composites Materials

With suitable processing, the carbon particles and binders can also be integrated to present partial flexibility. Kaskel and coworkers proposed a solvent-free, highly versatile compression/heat treatment method for the rapid and reproducible production of freestanding carbon/sulfur composite cathode foils [101]. The mechanically stable positive electrode foils were obtained through combining sulfur, porous carbon material, and conductive additives of CNTs and binders (polytetrafluoroethylene) to agglomerate the particles. A stable capacity >740 mAh g^{-1}-S as well as high coulombic efficiency >96% is achieved over 160 cycles at a moderate rate of C/10. The obtained cathode material has the characteristics of adjustable sulfur loading, high in-plane electrical conductivity, high cycle stability, and high mechanical flexibility (Figure 6.12a–c).

Wang and coworkers proposed a sulfur cathode with porous carbon/sulfur/porous carbon membrane structure (Figure 6.12d) [102]. A high areal capacity

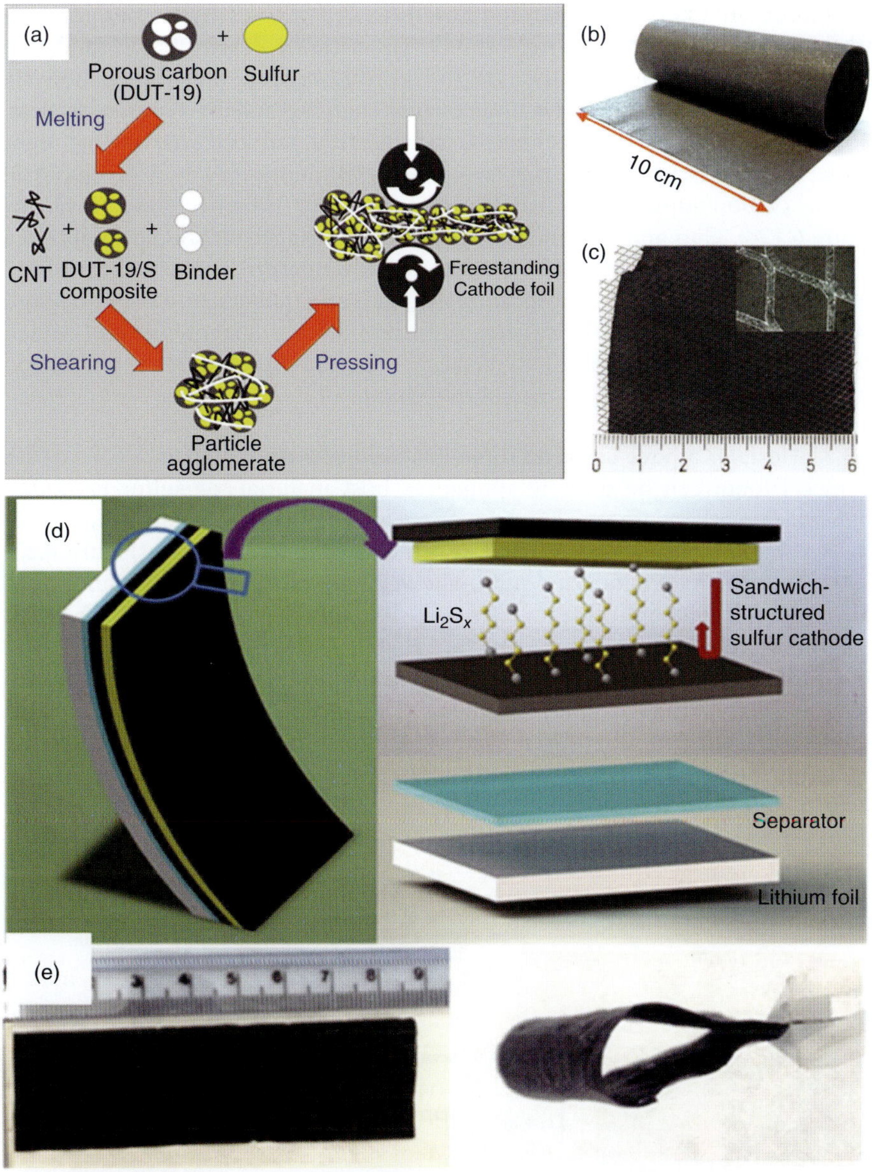

Figure 6.12 Flexible sulfur cathode with carbon/binder composites. (a) Schematic representation of the solvent-free processing approach; (b) photograph of the as-prepared freestanding cathode foil (~80 μm thick) with active material loading of c. 3.0–3.2 mg-S cm^{-2}; (c) electrode foil laminated onto the carbon-coated, expanded aluminum current collector for improved electrical conductivity and reduced contact resistance. *Source:* Thieme et al. 2013 [101]. Copyright 2013, Royal Society of Chemistry. (d) Schematic illustration of a Li–S cell with sandwich-structured cathode; (e) photo images of sandwich-structured electrodes at tiled (left) and bended state (right). *Source:* Song et al. 2014 [102]. Copyright 2014, Royal Society of Chemistry.

of ~4 $mAh\,cm^{-2}$ was demonstrated based on this new cathode configuration with high sulfur loading (~4 $mg\,S\,cm^{-2}$). The cathode design of the sandwich structure has many advantages: (i) the porous carbon film can effectively adsorb the soluble lithium polysulfide during charging and discharging process; (ii) The charge transfer resistance is significantly reduced due to the tight contact between sulfur and the two highly conductive carbon layers; And (iii) the carbon-supported sandwich structure also presents good flexibility (Figure 6.12e). Guo and coworkers used a simple scraper technique to obtain a carbon/sulfur/carbon sandwich structure on a polypropylene separator [103], wherein the carbon layer is composed of super P and binders, and the upper and bottom carbon layers of the electrode serve as a double current collector, which effectively improves the conductivity of the electrode, suppresses the diffusion of the polysulfide, and buffers the volume change of the electrode.

6.3.3.3 Flexible Cathode Based on Three-dimensional Metal Materials

3D nickel foam can be directly applied as host material for sulfur cathodes. Chen and coworkers prepared high-performance cathode material by electrodepositing highly active sulfur particles on the flexible nickel foam (Figure 6.13) [104]. The cathode of nanosized sulfur showed high rate performance and cyclic stability in electrochemical tests. The capacity still maintained 895 $mAh\,g^{-1}$ at 0.5 C after 300 cycles and 528 $mAh\,g^{-1}$ at 5 C after 1400 cycles. The one-step electrodeposition process shortened the preparation process for the electrode. Bai and coworkers fabricated sulfur/graphene/CNT composite by the hydrothermal co-assembly method [105]. When used as a cathode for lithium–sulfur batteries, the composite material can be directly pressed into nickel foam without binders

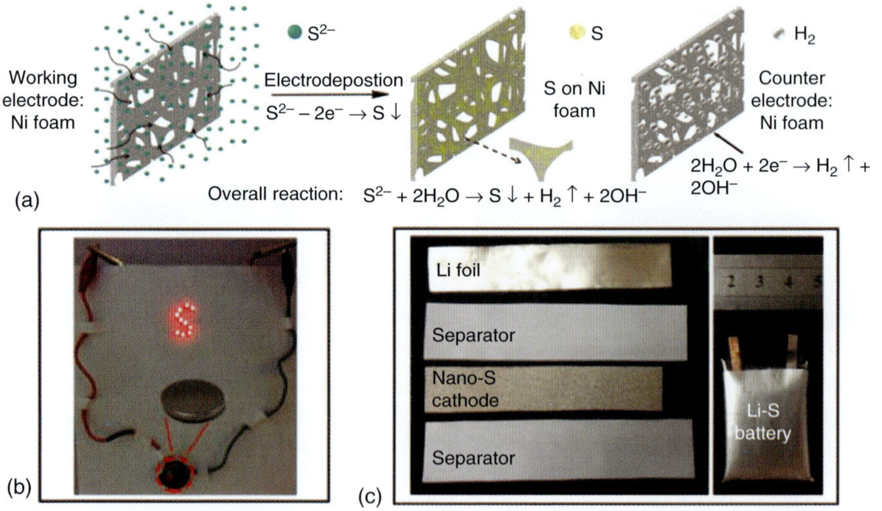

Figure 6.13 Flexible sulfur cathode with three-dimensional metal. (a) A schematic diagram of preparing sulfur nanodots on Ni foam with an electrodeposition method; (b) a facile application of Li–S cells: an "S"-shaped light that contains 10 LEDs lit by an assembled Li–S battery; (c) a flexible packaging Li–S battery with nano S cathode. *Source:* Zhao et al. 2015 [104]. Copyright 2015, American Chemical Society.

and conductive additives. The nickel foam support is highly efficient as a bulk conducting network. The combination of graphene/CNTs on Ni foam not only facilitates ion transport and electrolyte diffusion, but also provides a porous three-dimensional structure with open channels to effectively bind the soluble polysulfide. The resulting S/GN/CNT composite exhibited high and stable specific discharge capacities of 670 $mAh\,g^{-1}$ after 80 cycles at 0.2 C and good rate capability.

When the 3D carbon material and nickel foam are used as the host material for the positive electrode, the major advantages are that (i) the unique structure provides a robust framework for the composite material to provide the skeleton; (ii) it can store large amount of sulfur and provide strong adsorption to prevent the dissolution of polysulfides; (iii) it is more suitable than the traditional aluminum foil for the collector due to the good conductivity and light weight and (iv) its porous network helps the transfer and redistribution of lithium ions in the charge and discharge process. The production of host materials from biomass, polymer, or nickel foam is relatively easier and cheaper than flexible electrodes based on CNTs, CNFs, or graphene.

6.4 Summary and Prospect

This chapter summarizes the recent progress in the 3D flexible cathode of lithium–sulfur battery. Great improvements have been achieved in flexible sulfur cathodes with regard to their electrochemical and mechanical performances. Meanwhile, numerous challenges are still to be overcome before the practical application of flexible electrodes in lithium–sulfur batteries. The research on flexible cathode mainly focuses on the design of the electrode structure and the improvement of sulfur utilization. These flexible electrodes are mainly based on carbon materials such as CNFs, CNTs, and graphene. Some of the electrodes are based on inorganic and polymer materials.

Different electrode materials in flexible electrodes have their pros and cons. For example, CNT films exhibit good conductivity and flexibility, while the complex fabrication process and difficulty in continuous production hinder their applications. Polymer fibers and biomass material are of low cost and possess tunable surface chemistry to adsorb polysulfides, but the intrinsic low conductivity remain as obstacles. The assembly of different components into composite flexible cathodes may shed some light for the rational combination for better flexible electrodes.

Toward further improvements in battery performance and the practical applications of flexible lithium–sulfur batteries, the following major challenges should be further addressed:

1) The sulfur content in the flexible cathodes should be improved for high energy density of lithium–sulfur batteries. Although the flexible electrodes are free of current collector and binders, the sulfur content is generally low as a compromise to mechanical flexibility. The more efficient 3D conductive and mechanical network needs to be designed to host more sulfur.

2) The pore structure of the flexible cathode needs to be well designed for application. Generally, flexible electrodes are of high ratio of void spaces, which brings high specific capacity of the sulfur. In practical application, the voids would be immersed by electrolytes. The excessive pores would reduce the overall energy density of the whole cell. A balance between the cathode performance and the energy density of the full cell should be considered.
3) The evaluation in pouch cell and real folding test can give more information on the drawbacks of the current flexible cathodes. As most of the characterizations of flexible cathodes are based on static and coin cell level tests, the results cannot reflect the real behavior of electrodes in operational state. The tests at pouch cell scale in a dynamic bending/folding condition would provide more accurate findings.
4) In a flexible lithium–sulfur battery, other components such as flexible lithium metal anodes and flexible and safe electrolyte should also be developed to match the cathodes.

The concept of Li–S batteries was proposed more than 40 years ago. Great achievements have been made toward their application in recent years. The design and fabrication of flexible cathodes for lithium–sulfur batteries are also driven by the future demands in flexible devices. It is believed that with the combined effort in nanomaterials and electrochemistry, more strategies in building flexible lithium–sulfur batteries will be proposed. The insightful ideas are also versatile in other electrochemical systems based on multielectron conversion reactions.

References

1 Wang, X., Lu, X., Liu, B. et al. (2014). *Adv. Mater.* 26: 4763–4782.
2 Zhou, G., Li, F., and Cheng, H.-M. (2014). *Energy Environ. Sci.* 7: 1307–1338.
3 Lu, X., Yu, M., Wang, G. et al. (2014). *Energy Environ. Sci.* 7: 2160–2181.
4 Dunn, B., Kamath, H., and Tarascon, J.-M. (2011). *Science* 334: 928–935.
5 Simon, P. and Gogotsi, Y. (2008). *Nat. Mater.* 7: 845–854.
6 Winter, M. and Brodd, R.J. (2004). *Chem. Rev.* 104: 4245–4269.
7 Evers, S. and Nazar, L.F. (2013). *Acc. Chem. Res.* 46: 1135–1143.
8 Manthiram, A., Fu, Y., Chung, S.-H. et al. (2014). *Chem. Rev.* 114: 11751–11787.
9 Koo, M., Park, K.-I., Lee, S.H. et al. (2012). *Nano Lett.* 12: 4810–4816.
10 Kwon, Y.H., Woo, S.-W., Jung, H.-R. et al. (2012). *Adv. Mater.* 24: 5192–5197.
11 Ren, J., Zhang, Y., Bai, W. et al. (2014). *Angew. Chem. Int. Ed.* 53: 7864–7869.
12 Zhang, Y., Bai, W., Ren, J. et al. (2014). *J. Mater. Chem. A* 2: 11054–11059.
13 Manthiram, A., Chung, S.-H., and Zu, C. (2015). *Adv. Mater.* 27: 1980–2006.
14 Peng, H.J., Xu, W.T., Zhu, L. et al. (2016). *Adv. Funct. Mater.* 26: 6351–6358.
15 Hu, L., Wu, H., La Mantia, F. et al. (2010). *ACS Nano* 4: 5843–5848.
16 Choi, J.W., Hu, L., Cui, L. et al. (2010). *J. Power Sources* 195: 8311–8316.
17 Zhao, X., Hayner, C.M., Kung, M.C., and Kung, H.H. (2012). *Chem. Commun.* 48: 9909–9911.

18 Hu, L. and Cui, Y. (2012). *Energy Environ. Sci.* 5: 6423–6435.

19 Tobjork, D. and Osterbacka, R. (2011). *Adv. Mater.* 23: 1935–1961.

20 Nyholm, L., Nystrom, G., Mihranyan, A., and Stromme, M. (2011). *Adv. Mater.* 23: 3751–3769.

21 Li, N., Chen, Z., Ren, W. et al. (2012). *Proc. Natl. Acad. Sci. U.S.A.* 109: 17360–17365.

22 Liu, B., Zhang, J., Wang, X. et al. (2012). *Nano Lett.* 12: 3005–3011.

23 Lota, G., Fic, K., and Frackowiak, E. (2011). *Energy Environ. Sci.* 4: 1592–1605.

24 Park, S., Vosguerichian, M., and Bao, Z. (2013). *Nanoscale* 5: 1727–1752.

25 Zhang, Q., Huang, J.Q., Qian, W.Z. et al. (2013). *Small* 9: 1237–1265.

26 Landi, B.J., Ganter, M.J., Cress, C.D. et al. (2009). *Energy Environ. Sci.* 2: 638–654.

27 Zhang, Q., Huang, J.Q., Zhao, M.Q. et al. (2011). *ChemSusChem* 4: 864–889.

28 Chen, J., Minett, A.I., Liu, Y. et al. (2008). *Adv. Mater.* 20: 566–570.

29 Endo, M., Muramatsu, H., Hayashi, T. et al. (2005). *Nature* 433: 476–476.

30 Xu, G., Zhang, Q., Zhou, W. et al. (2008). *Appl. Phys. A* 92: 531–539.

31 Zhu, L., Zhu, W.C., Cheng, X.B. et al. (2014). *Carbon* 75: 161–168.

32 Cheng, X.B., Huang, J.Q., Zhang, Q. et al. (2014). *Nano Energy* 4: 65–72.

33 Huang, J.Q., Zhang, Q., Zhang, S.M. et al. (2013). *Carbon* 58: 99–106.

34 Zhou, G., Wang, D.-W., Li, F. et al. (2012). *Energy Environ. Sci.* 5: 8901–8906.

35 Yuan, Z., Peng, H.J., Huang, J.Q. et al. (2014). *Adv. Funct. Mater.* 24: 6105–6112.

36 Jin, K., Zhou, X., Zhang, L. et al. (2013). *J. Phys. Chem. C* 117: 21112–21119.

37 Yuan, G. and Xiang, J. (2013). *Ionics* 19: 1449–1453.

38 Huang, J.Q., Peng, H.J., Liu, X.Y. et al. (2014). *J. Mater. Chem. A* 2: 10869–10875.

39 Zhu, L., Peng, H.J., Liang, J. et al. (2015). *Nano Energy* 11: 746–755.

40 Shi, J.L., Peng, H.J., Zhu, L. et al. (2015). *Carbon* 92: 96–105.

41 Chen, Y., Lu, S., Wu, X., and Liu, J. (2015). *J. Phys. Chem. C* 119: 10288–10294.

42 Sun, Q., Fang, X., Weng, W. et al. (2015). *Angew. Chem. Int. Ed.* 54: 10539–10544.

43 Zhang, Y., Zhao, Y., The Nam Long, D. et al. (2013). *Solid State Ionics* 238: 30–35.

44 Sun, L., Li, M., Jiang, Y. et al. (2014). *Nano Lett.* 14: 4044–4049.

45 Sun, F.G., Wang, J.T., Chen, H.C. et al. (2013). *ACS Appl. Mater. Interfaces* 5: 5630–5638.

46 Luo, S., Wang, K., Wang, J.P. et al. (2012). *Adv. Mater.* 24: 2294–2298.

47 Sun, L., Wang, D., Luo, Y. et al. (2016). *ACS Nano* 10: 1300–1308.

48 Zhao, Y., Yin, F., Zhang, Y. et al. (2015). *Nanoscale Res. Lett.* 10: 450.

49 Sun, L., Kong, W., Jiang, Y. et al. (2015). *J. Mater. Chem. A* 3: 5305–5312.

50 Ayala, P., Arenal, R., Rümmeli, M. et al. (2010). *Carbon* 48: 575–586.

51 Sun, X.G., Wang, X.Q., Mayes, R.T., and Dai, S. (2012). *ChemSusChem* 5: 2079–2085.

52 Hou, T.Z., Peng, H.J., Huang, J.Q. et al. (2015). *2D Mater* 2: 014011.

53 Tang, C., Zhang, Q., Zhao, M.Q. et al. (2014). *Adv. Mater.* 26: 6100–6105.
54 Peng, H.J., Hou, T.Z., Zhang, Q. et al. (2014). *Adv. Mater. Interfaces* 1: 1400227.
55 Hou, T.Z., Chen, X., Peng, H.J. et al. (2016). *Small* 12: 3283–3291.
56 Elazari, R., Salitra, G., Garsuch, A. et al. (2011). *Adv. Mater.* 23: 5641–5644.
57 Qie, L. and Manthiram, A. (2015). *Adv. Mater.* 27: 1694–1700.
58 Chung, S.H., Chang, C.H., and Manthiram, A. (2016). *Small* 12: 939–950.
59 Zeng, L., Pan, F., Li, W. et al. (2014). *Nanoscale* 6: 9579–9587.
60 Huang, X., Sun, B., Li, K. et al. (2013). *J. Mater. Chem. A* 1: 13484–13489.
61 Yang, Z.Z., Wang, H.Y., Zhong, X.B. et al. (2014). *RSC Adv.* 4: 50964–50968.
62 Zhou, W., Guo, B., Gao, H., and Goodenough, J.B. (2016). *Adv. Energy Mater.* 6: 1502059.
63 Zeng, L., Jiang, Y., Xu, J. et al. (2015). *Nanoscale* 7: 10940–10949.
64 Zheng, S., Yi, F., Li, Z. et al. (2014). *Adv. Funct. Mater.* 24: 4156–4163.
65 Zhang, Z., Li, Q., Zhang, K. et al. (2015). *J. Power Sources* 290: 159–167.
66 Zeng, L., Yao, Y., Shi, J. et al. (2016). *Energy Storage Mater.* 5: 50–57.
67 Hanumantha, P.J., Gattu, B., Shanthi, P.M. et al. (2016). *Electrochim. Acta* 212: 286–293.
68 Huang, J.Q., Liu, X.F., Zhang, Q. et al. (2013). *Nano Energy* 2: 314–321.
69 Ding, B., Yuan, C., Shen, L. et al. (2013). *J. Mater. Chem. A* 1: 1096–1101.
70 Peng, H.J., Huang, J.Q., Zhao, M.Q. et al. (2014). *Adv. Funct. Mater.* 24: 2772–2781.
71 Zhao, M.Q., Zhang, Q., Huang, J.Q. et al. (2014). *Nat. Commun.* 5: 3410.
72 Huang, J.Q., Zhuang, T.Z., Zhang, Q. et al. (2015). *ACS Nano* 9: 3002–3011.
73 Peng, H.J., Wang, D.W., Huang, J.Q. et al. (2016). *Adv. Sci.* 3: 1500268.
74 Zhuang, T.Z., Huang, J.Q., Peng, H.J. et al. (2016). *Small* 12: 381–389.
75 Zhai, P.Y., Peng, H.J., Cheng, X.B. et al. (2017). *Energy Storage Mater.* 7: 56–63.
76 Evers, S. and Nazar, L.F. (2012). *Chem. Commun.* 48: 1233–1235.
77 Wang, B., Li, K., Su, D. et al. (2012). *Chem. Asian J.* 7: 1637–1643.
78 Lee, S.H., Kim, H.W., Hwang, J.O. et al. (2010). *Angew. Chem. Int. Ed.* 49: 10084–10088.
79 Chen, Z., Ren, W., Gao, L. et al. (2011). *Nat. Mater.* 10: 424–428.
80 Huang, X., Qian, K., Yang, J. et al. (2012). *Adv. Mater.* 24: 4419–4423.
81 Yin, S., Zhang, Y., Kong, J. et al. (2011). *ACS Nano* 5: 3831–3838.
82 Jin, J., Wen, Z., Ma, G. et al. (2013). *RSC Adv.* 3: 2558–2560.
83 Wang, C., Wang, X., Wang, Y. et al. (2015). *Nano Energy* 11: 678–686.
84 Zhou, G., Pei, S., Li, L. et al. (2014). *Adv. Mater.* 26: 625–631.
85 Wang, C., Wang, X., Yang, Y. et al. (2015). *Nano Lett.* 15: 1796–1802.
86 Ni, W., Cheng, J., Li, X. et al. (2016). *RSC Adv.* 6: 9320–9327.
87 Wu, C., Fu, L., Maier, J., and Yu, Y. (2015). *J. Mater. Chem. A* 3: 9438–9445.
88 Zhou, G., Zhao, Y., and Manthiram, A. (2015). *Adv. Energy Mater.* 5: 1402263.
89 Wu, F., Li, J., Su, Y. et al. (2016). *Nano Lett.* 16: 5488–5494.
90 Sun, H., Xu, Z., and Gao, C. (2013). *Adv. Mater.* 25: 2554–2560.
91 Zhou, G., Li, L., Ma, C. et al. (2015). *Nano Energy* 11: 356–365.
92 He, J., Chen, Y., Lv, W. et al. (2016). *J. Power Sources* 327: 474–480.

93 Zhou, G., Paek, E., Hwang, G.S., and Manthiram, A. (2015). *Nat. Commun.* 6: 7760.

94 Iguchi, M., Yamanaka, S., and Budhiono, A. (2000). *J. Mater. Sci.* 35: 261–270.

95 Huang, Y., Zhu, C., Yang, J. et al. (2014). *Cellulose* 21: 1–30.

96 Huang, Y., Zheng, M., Lin, Z. et al. (2015). *J. Mater. Chem. A* 3: 10910–10918.

97 Cao, Z., Zhang, J., Ding, Y. et al. (2016). *J. Mater. Chem. A* 4: 8636–8644.

98 Wu, F., Zhao, E., Gordon, D. et al. (2016). *Adv. Mater.* 28: 6365–6371.

99 An, Y., Zhu, Q., Hu, L. et al. (2016). *J. Mater. Chem. A* 4: 15605–15611.

100 Mi, L., Xiao, W., Cui, S. et al. (2016). *Dalton Trans.* 45: 3305–3309.

101 Thieme, S., Brueckner, J., Bauer, I. et al. (2013). *J. Mater. Chem. A* 1: 9225–9234.

102 Song, J., Yu, Z., Xu, T. et al. (2014). *J. Mater. Chem. A* 2: 8623–8627.

103 Wang, H., Zhang, W., Liu, H., and Guo, Z. (2016). *Angew. Chem. Int. Ed.* 55: 3992–3996.

104 Zhao, Q., Hu, X., Zhang, K. et al. (2015). *Nano Lett.* 15: 721–726.

105 Yuan, G., Wang, G., Wang, H., and Bai, J. (2015). *J. Nanopart. Res.* 17: 36.

7

Flexible Lithium–Air Batteries

Qing-Chao Liu, Zhi-Wen Chang, Kai Chen, and Xin-Bo Zhang

Chinese Academy of Sciences, Changchun Institute of Applied Chemistry, State Key Laboratory of Rare Earth Resource Utilization, 5625 Renmin Street, Changchun, Jilin 130022, China

7.1 Motivation for the Development of Flexible Lithium–Air Batteries

Worldwide demand for flexible electronics continues to grow rapidly, as these devices will open the door for the next-generation electronic environment based on lightweight, bendable, rugged, portable, rollable, and potentially foldable devices [1–7]. Up to now, many stretchable electronic devices have been extensively investigated in solar cells [8–15], flexible chemical batteries [16–20], and flexible supercapacitors [21–28] to meet both esthetic demands and the needs of specific environments. Great progress has been made in this field, however, the low energy density of these energy storage devices is the biggest obstacle for their long-range life and industrial applications. Therefore, the development of lightweight, high energy density, and flexible energy storage devices for comprehensive wearable electronics is highly significant. Fortunately, the recent development of the lithium–air battery with exceptionally high theoretical energy density of 3600 Wh kg^{-1} ($2Li^+ + O_2 + 2e^- = Li_2O_2$, 2.96 V vs Li/Li^+), will resolve the issue of energy density as they may be used as high energy storage devices with flexible features in the future [29–35]. A comparison of practical specific energy for several rechargeable batteries is presented in Figure 7.1 [36], from which it can be clearly seen that the specific energy of Li–air battery is several times greater than that of the common Li–ion battery. However, the development of lithium–air battery is still in the primary stages and numerous scientific and technological challenges need to be overcome, to say nothing of flexible ones. Currently, the most commonly studied ones are coin cell batteries [37–40] and Swagelok battery design [41–43] which are too heavy, thick, rigid, and bulky to match practical requirements of flexible ones. Therefore, there is an urgent need to firstly design new prototypes of flexible lithium–air batteries to effectively improve the practical energy density to promote their practical application in the flexible electronics field.

To achieve this goal, however, there are still significant challenges waiting to be solved, which mainly include two major factors. One is the essential

Flexible Energy Conversion and Storage Devices, First Edition. Edited by Chunyi Zhi and Liming Dai.

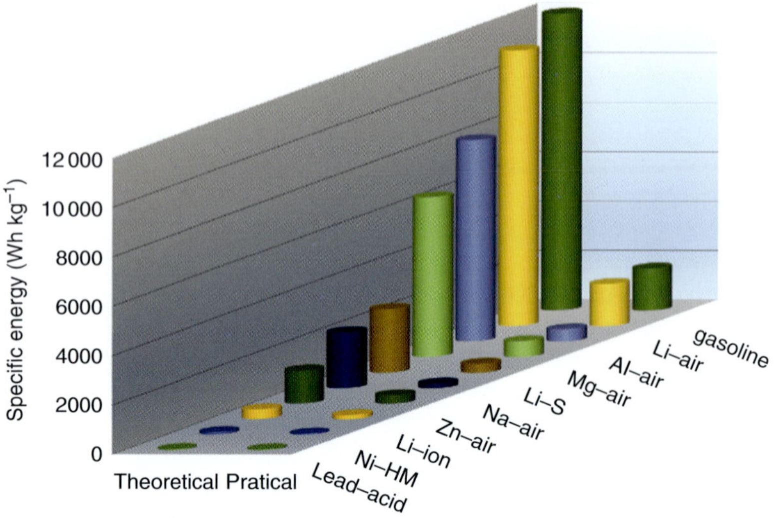

Figure 7.1 Comparison of gravimetric energy densities of various types of batteries and gasoline. The theoretical density is acquired from the thermodynamics of active materials.

electrochemical performances that includes capacity, cycle life, round-trip efficiency, rate performance, etc. [44–49]. The other is battery structure design consisting of the compatibility between battery body, packing materials, and assembly cost. This chapter serves as a primer for researchers interested in lithium–air battery and flexible electronics. We begin with a brief overview of flexible energy storage and conversion devices, followed by a deep-dive into the research status of flexible lithium–air batteries, which are basically divided into four parts: (i) planar battery, (ii) cable-type battery, (iii) woven-type battery pack, and (iv) battery array pack. Subsequent sections discuss the challenges and future work in the development of flexible lithium–air batteries and their potential application values. And finally, we end with a brief discussion on the development direction for flexible lithium–air batteries.

7.2 State of the Art for Flexible Lithium–Air Batteries

7.2.1 Overview of Flexible Energy Storage and Conversion Devices

Owing to their superior characteristics such as lightweight, bendability, portability, and potentially foldability, the flexible electronic devices became popular, following which some concepts and prototypes of flexible electronics have emerged, such as the Jawbone (https://jawbone.com/up), Philips Fluid flexible smartphone (http://www.designbuzz.com/philips-fluid-smartphone-with-flexible-oled-display/), Nokia Morph Concept (http://research.nokia.com/morph), iPhone ProCare14 (http://www.o-cellphones.com/tag/iphone-procare), and Samsung Youm flexible display (http://www.pcmag.com/article2/0,2414094,00.asp) (Figure 7.2) all of which allow users to bend the screen back

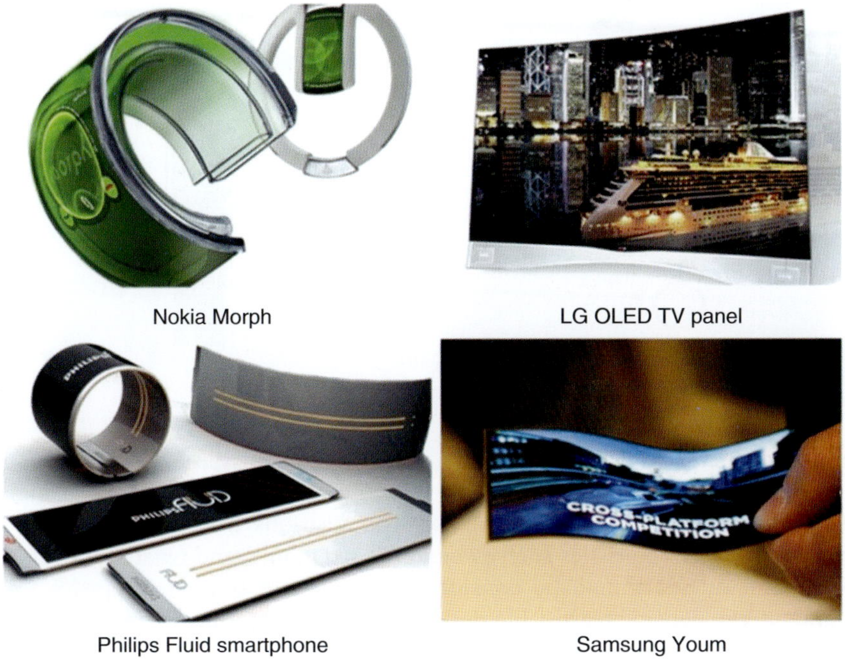

Figure 7.2 Nokia Morph Concept, LG OLED TV panel, Philips Fluid flexible smartphone, and Samsung Youm flexible display.

and forth with ease. It is believed that with the increasing application of portable flexible electronic products, more practical applications will be getting more attention and totally change our daily life.

The low energy density of the embedded energy storage devices, which are mainly based on lithium ion batteries, cannot be ignored although the currently used flexible electronic devices have been influencing our lives. Consequently, it is an urgent need to develop different kinds of high energy density storage devices to meet the present situation and the development of flexible lithium–air battery may resolve this issue.

7.2.2 Overview of Flexible Lithium–Air Batteries

The nonaqueous lithium–air battery, which is generally considered to have been initiated by Abraham and Jiang in 1996 [50], comprises a lithium conductive organic polymer electrolyte membrane, thin lithium metal foil anode, and a carbon composite air cathode. Currently, researches on lithium–air battery are mainly based on the coin cell and Swagelok battery design as displayed in Figure 7.3, but both types cannot accommodate the requirements of flexible electronic devices. As a consequence, the development of flexible lithium–air battery is imperative.

The development of a flexible lithium–air battery is similar to that of flexibile lithium ion battery, solar cells, and supercapacitors. Of note, the electrochemical performances of flexible lithium–air battery, such as capacity, cycle life,

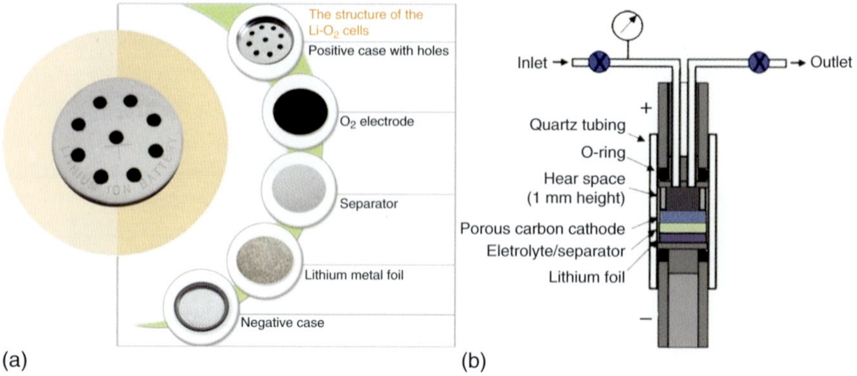

Figure 7.3 Typical structure of (a) coin cell and (b) Swagelok battery.

round-trip efficiency, rate performance, etc. need to be assured. Because these excellent performances are the premise for the development of flexible high energy storage devices. For its extensive applications in flexible electronic devices, other characteristics such as bending, stretching, and twisting or other deformation modes should be considered. Besides, in order to improve the integral energy density, reasonable design of battery structure and exploitation of lightweight assembling materials are absolutely necessary. And finally, safety is the most important factor that should be guaranteed, especially in various deformation conditions [51]. A typical laboratory flexible lithium–air battery type used by Zhang's group is shown in Figure 7.4. The structure of this assembly

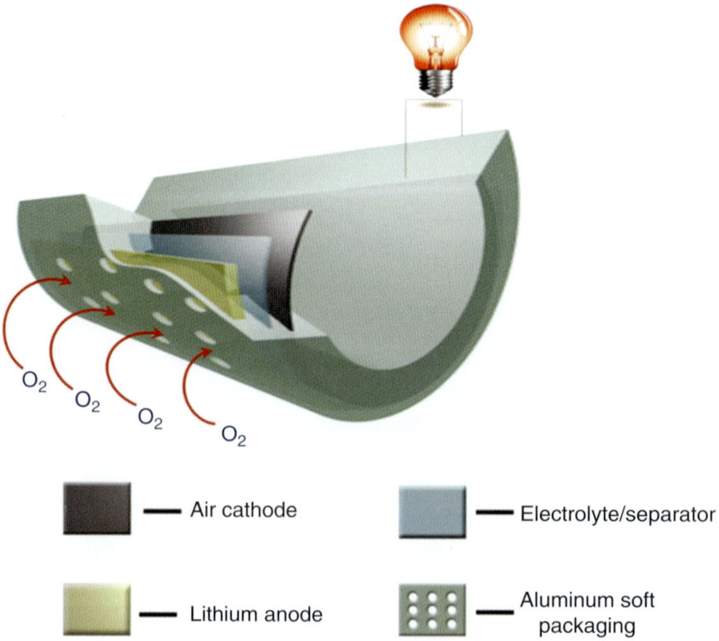

Figure 7.4 Schematic cell configurations for the flexible lithium–air battery.

structure still falls into "stack-type" category which is widely used in the lithium ion battery field and the only difference is packaging which possesses numerous holes for battery "breathing" and this type can be categorized as planar battery.

In this section, we will focus on introducing the similarities and differences among the cathode, anode, electrolyte, and current collector as well as the configuration between coin cell/Swagelok battery and flexible battery and provide some references for researchers though the reports in this field are rare.

7.2.2.1 Similarities Between Coin Cell/Swagelok Batteries with Flexible Battery

Similar to coin cell/Swagelok battery, the development of flexible lithium–air battery also involves aspects including cathode design, lithium anode protection, and electrolyte/separator adaptation. The above mentioned aspects could also refer to the coin cell/Swagelok battery.

Cathode Design It has been demonstrated that the electrochemical performance of lithium–air batteries is strongly dependent on air cathode (sometimes referred to as "positive electrode"), which contributes to the most round-trip efficiency of lithium–air battery [46, 52–54]. The development of cathode materials and the design of architectural structure influence batteries performance significantly.

Currently, most of the air cathode materials are based on the carbon substrate due to its excellent conductivity, definite catalytic properties, and relative stability [55–58]. In the design of a carbon-based cathode, a large surface area is necessary, which can provide more active sites for electrochemical reactions [46]. A further study reveals that the battery capacity is determined more by the volume of the pores residing in the mesopore range [59, 60]. As we all know, Li_2O_2 is formed only on the triple junctions where electrolyte, carbon, and oxygen coexist. Besides, too small or too large pore size will not benefit the electrochemical reaction. Too small a pore size will block reactants into the cathode interior and reduce the utilization of active materials while too large a pore size is usually flooded by electrolyte forming two-phase regions instead of three-phase regions [61, 62]. The schematic illustrations of the three cathodes with different pore sizes are shown in Figure 7.5. But beyond architecture design, developing efficient electrocatalysis to improve round-trip efficiency as well as noncarbon-based materials to improve cathode stability are of equal importance.

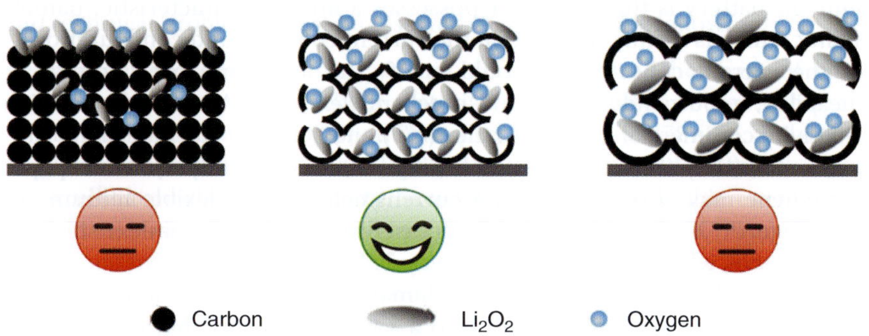

Figure 7.5 Schematic illustrations of the three cathodes morphology with discharge state.

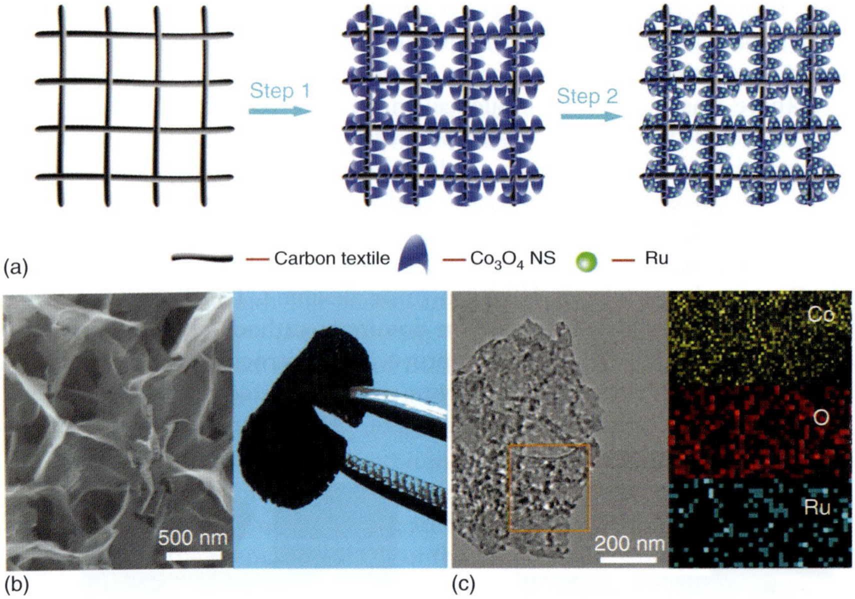

Figure 7.13 (a) Schematic representations for the design and preparation of the CT-Co_3O_4 NSs-Ru cathode. (b) SEM image and photograph of the CT-Co_3O_4 NSs-Ru cathode. (c) TEM image and its corresponding EDS mappings.

be observed that Ru NPs decorated Co_3O_4 NSs were uniformly grown on CT (left) and flexible property of the fabricated cathode was still reserved (right). Typical transmission electron microscopy (TEM) images and its corresponding energy-dispersive X-ray spectrometry (EDXS) characterization are exhibited in Figure 7.13c. It was found that Ru NPs with a size of about 2 nm were uniformly distributed on the surface of Co_3O_4 NSs.

A planar flexible lithium–air battery was assembled by using the fabricated CT-Co_3O_4 NSs-Ru cathode and the electrochemical performance was also detected as displayed in Figure 7.14. The assembled planar flexible lithium–air battery could endure more than 72 cycles with a discharge terminal voltage of 2.0 V, while the cell with the CT-Co_3O_4 could only run for 19 cycles (Figure 7.14a). Owing to the excellent electrocatalytic property of Ru, the assembled flexible lithium–air battery possesses superior rate capability (Figure 7.14b). To demonstrate its practical applications, flexible property was also investigated, the assembled battery was used to control a commercial red LED, which could work normally whether the battery was in planar or bending condition (Figure 7.14c,d).

Although CT as excellent flexible current collectors has been widely used [83, 88], the catalytic ability of CT is inferior. The addition of CT will increase the mass of cathode and reduce the energy density of the battery, and therefore the availability of air cathode needs to be improved. Inspired by the golden-toad spawning process as well as the unique structure of the eggs, Yin et al. reported a strategy to fabricate a flexible, self-standing, light-weight, hierarchical

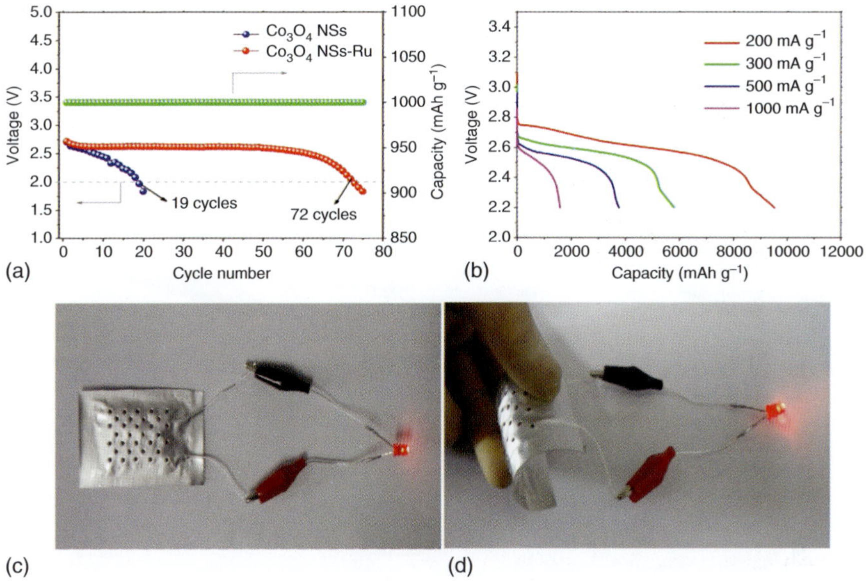

Figure 7.14 (a) Cycling performance and (b) rate capability of the planar flexible lithium–air battery. (c,d) Digital images of the red LED equipment turned on with the fabricated planar flexible lithium–air battery under planar and bent conditions.

macroporous active carbon fiber (MACF) cathode via an electrostatic spinning technique as shown in Figure 7.15a [90]. The fabricated MACF nonwoven textiles were self-standing with a 3D hierarchical structure and interlace with each other. Noteworthy, the as-synthesized MACF cathode also possesses excellent flexible properties (Figure 7.15b). It could be found that the obtained MACF cathode possesses abundant macroscopic carbon spheres with plentiful surface openings and interconnected pores (Figure 7.15c). To further promote the electrocatalytic ability, RuO_2 NPs were uniformly decorated on the surface of the MACF cathode (R-MACF) as displayed in Figure 7.15d. Compared with the TiO_2 NAs/CT cathode and CT-Co_3O_4 NSs-Ru cathode, this fabricated cathode does not needs current collector. As a benefit, the whole cathode acts as active material which inevitably improves the energy density of a flexible battery.

Owing to the flexible properties of the fabricated R-MACF cathode, another flexible planar flexible lithium–air battery was also assembled as shown in Figure 7.16. The assembled battery was used to control a commercial red LED, which could work normally whether the battery was in planar or bending conditions (Figure 7.16a–c). Besides, the electrochemical performance was also evaluated as displayed in Figure 7.16d and it was found that the cycling performance was still excellent even in bending conditions.

With the development of the flexible cathode, apart from the commonly used CT and nonwoven CT based cathode, some conceptual flexible cathodes, such as woven cotton, paper, etc. which are easily accessed, have been widely used as flexible current collectors [91, 92]. Inspired by the traditional Chinese calligraphy art, Liu et al. fabricated flexible free-standing paper-ink cathode (PI cathode) by

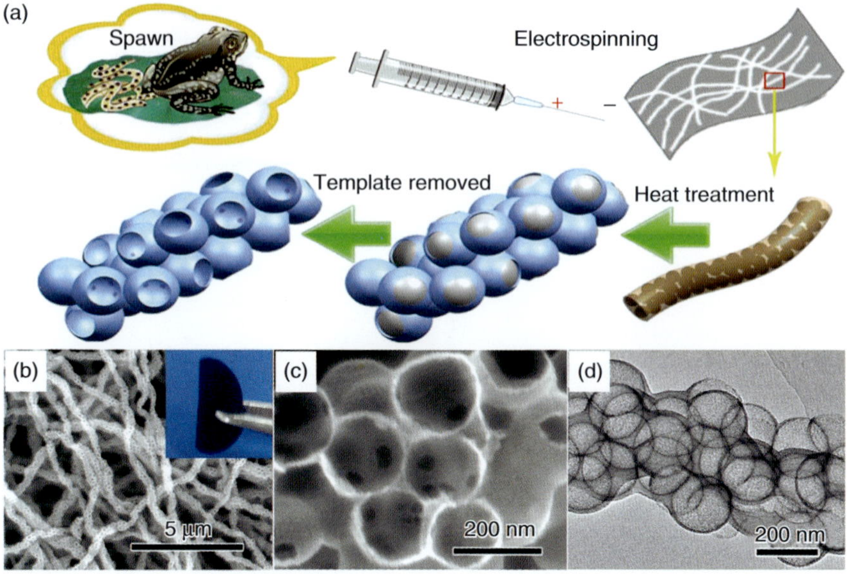

Figure 7.15 (a) Schematic illustration of the fabrication process of the macroporous hollow active carbon fiber cathode. (b) SEM image and photograph of the MACF cathode. (c) The enlargement of (b). (d) TEM image of the R-MACF cathode.

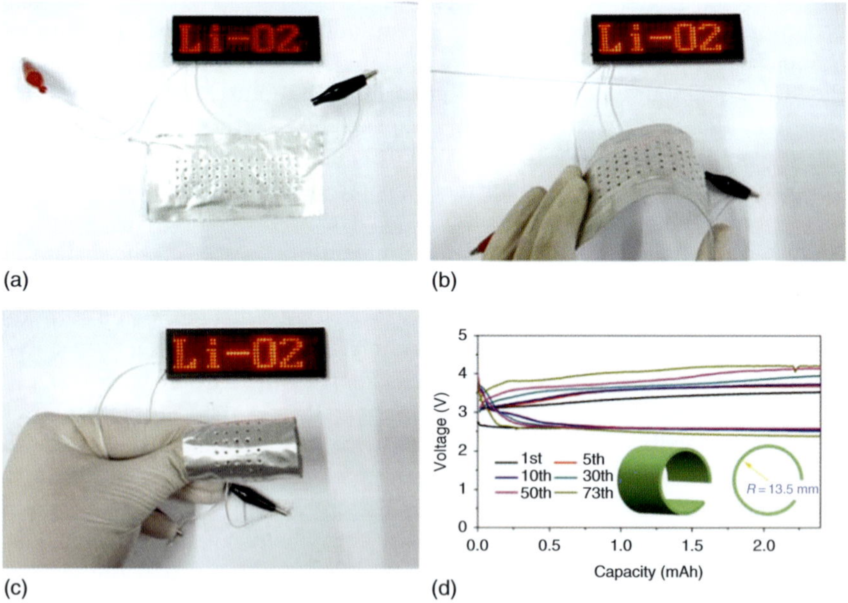

Figure 7.16 (a–c) Digital images of the red LED equipment turned on with the fabricated planar flexible lithium–air battery under planar and bent conditions. (d) Cycling performance of battery with R-MACF cathode under bend condition.

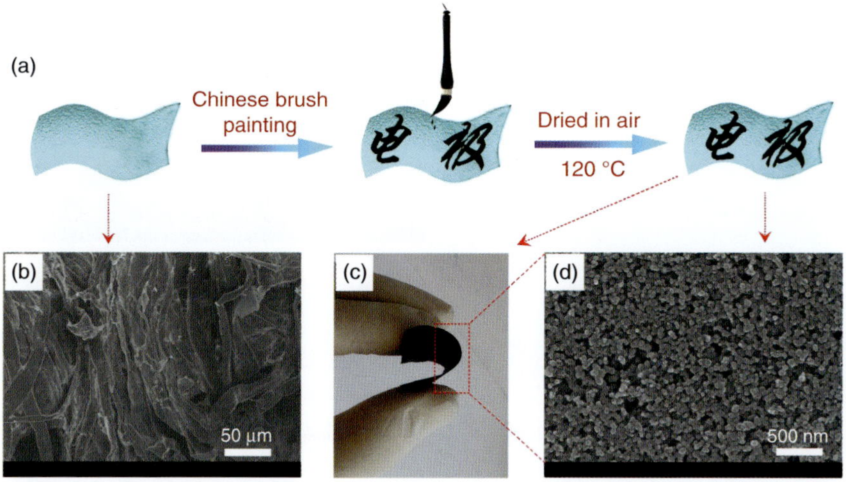

Figure 7.17 (a) Schematic illustration of the fabrication process of the PI cathode. (b) SEM image of a paper substrate. (c) Digital image of the obtained PI cathode. (d) SEM image of the obtained PI cathode.

drawing/painting on commonly used paper [93]. The schematic illustration of the fabrication process is displayed in Figure 7.17a. This cathode holds many advantages: (i) low cost of active material (ink) and cathode substrates (commonly used paper); (ii) the fabrication process is very facile; and (iii) the flexible feature of obtained cathode was reserved (Figure 7.17c). Compared with the cathode substrate (Figure 7.17b), carbon NPs (the main material of ink) were uniformly distributed on the paper sheet forming a freestanding architecture (Figure 7.17d).

Firstly, a common planar flexible lithium–air battery was assembled and the schematic is described in Figure 7.18a. The flexible battery can power commercial red LED equipment in both plane and bend conditions (Figure 7.18b). It is worth noting that the assembled flexible battery could run as high as circa 6500 $mAh\,g^{-1}$ even under conditions where pristine or bent cathode was used (Figure 7.18c). Furthermore, cycling life was also investigated. Bothbatteries with plane and bend cathodes could endure circa 50 cycles (Figure 7.18d).

All of the above fabricated planar flexible lithium–air batteries adopt the stack-type mode (Figure 7.9), similar to the coin cell or Swagelok design and the packing material of aluminum-plastic film is essential to provide pressure to the batteries and guarantee the normal operation of the battery. However, the addition of components will reduce the energy density of flexible batteries. In order to reflect the energy density advantage of lithium–air battery, a foldable lithium–air battery was assembled. The structure of the battery is displayed in Figure 7.19a. The assembled battery with four cathodes painted with ink and two anodes were wrapped into paper, two anodes were shared by four cathodes which could be also called lithium–air battery pack. This fabricated battery has its own advantages as depicted in Figure 7.18. All of the cathode substrate, separator as well as the outer package, is inexpensive flexible paper that reduces the cost compared to the above ones. The utilization of lithium metal anode

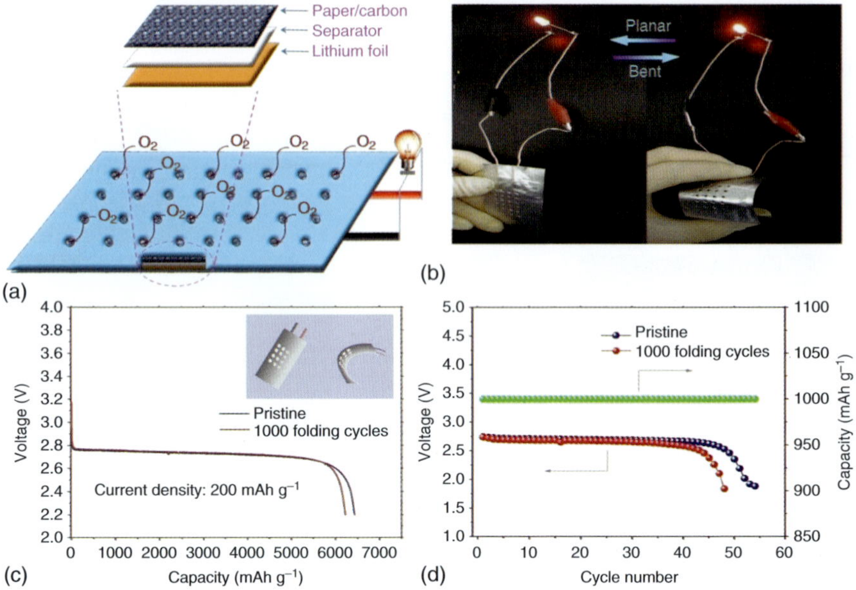

Figure 7.18 (a) Schematic cell configurations for the planar flexible lithium–air battery with PI cathode. (b) Digital images of the fabricated planar flexible lithium–air battery with plane and bent conditions turning on red LED equipment. (c) Discharge capacity of flexible lithium–air batteries with PI cathode (pristine and 1000 folding cycles). (d) Cycling performance of flexible lithium–air batteries with the two kinds of cathode (pristine and 1000 folding cycles).

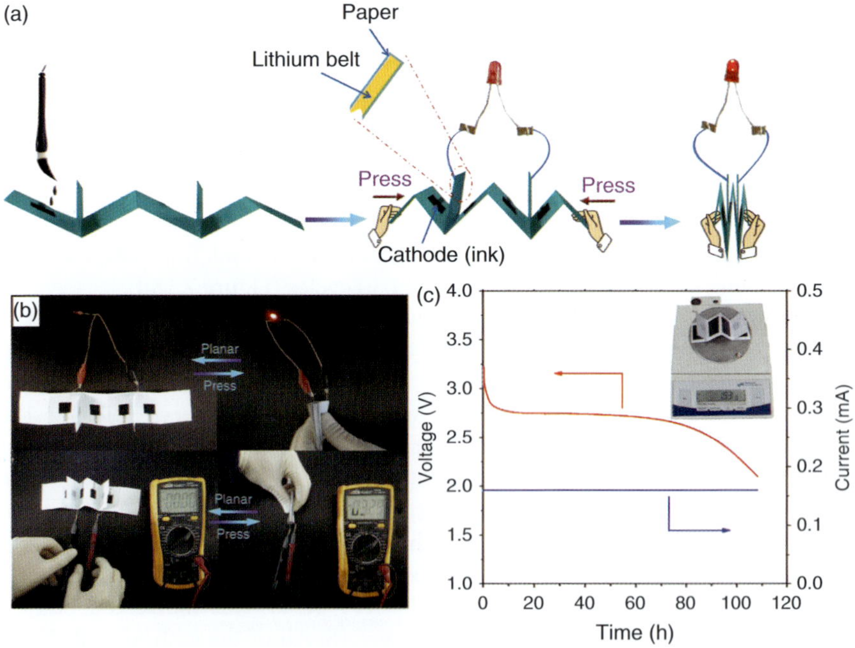

Figure 7.19 (a) Schematic of the fabrication and the working mechanism of a foldable lithium–air battery. (b) Photograph depicting the practical application of this battery. (c) Discharge voltage versus time and discharge current versus time, inset is the mass of the assembled foldable lithium–air battery.

was improved when one anode was shared by two cathodes. Owing to the special assembly structure, the energy density is also improved. Figure 7.19b demonstrated the feasibility of the practical application of this battery. The electrochemical performance was also evaluated as shown in Figure 7.19c. The fabricated fold battery (battery pack) could be operated normally about 20 h with the light weight being less than 2 g. This technique paves the way for cheap and easily manufactured flexible batteries.

7.2.3.2 Cable-type Battery

The commonly researched coin cell, Swagelok design, and planar flexible lithium–air battery are not suitable for extensive applications especially in the flexible electronics field. Recently, a cable-type design was found to possess advantages over the above ones [78–80]. Well-designed cable-type battery is omnidirectional, flexible, and twistable, which can be woven into many shapes and can be applied in many areas. The first cable-type air battery is the zinc–air battery [94] and inspired by this composition structure, lithium–air battery with cable-type was also fabricated. As the same with planar lithium–air battery, the assembled model of cable-type lithium–air battery is stack-type, too. Being different from other cable-type devices (lithium ion battery, supercapacitor, etc.), the lithium–air battery runs in an open system and the schematic battery configuration of lithium anode, electrolyte/separator, air cathode, air diffusion layer, and packing insulator is shown in Figure 7.20.

At present, the commonly used lithium anode is commercial lithium rod and the packing component are heat-shrinkable materials which mainly provide pressure on the battery. The electrolyte/separator, air cathode and air diffusion layer are the same with coin cell and Swagelok design. Liu et al. reported a strategy in fabricating a cable-type flexible lithium–air battery [95]. This battery consists of

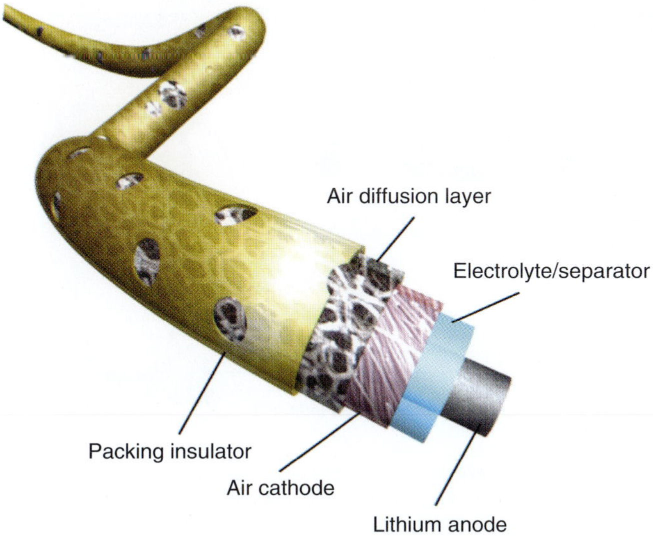

Figure 7.20 Schematic cell configurations for the cable-type air battery.

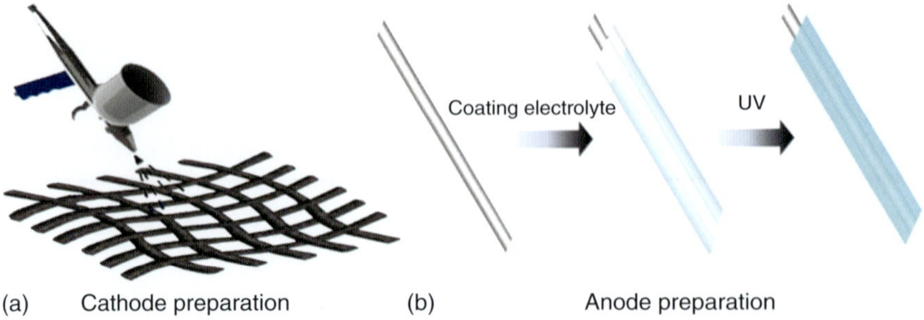

Figure 7.21 Schematic illustration of the fabrication process of the flexible cathode (a) and freestanding polymer electrolyte (b).

a flexible cathode by commercial Super P (SP) sprayed on CT, and a freestanding polymer electrolyte, and the fabricated process is displayed in Figure 7.21.

The flexibility of the fabricated cable-type lithium–air battery was investigated as shown in Figure 7.22. The working condition of this battery was not influenced when it was bended and twisted into any shape, such as arc, twist, and spiral (Figure 7.22a). Furthermore, the electrochemical performance was estimated, and the discharge capacity of fabricated cable-type battery that was bended and twisted into various shapes, was not changed significantly (Figure 7.22b) indicating the distinguished mechanical suppleness and electrochemical stability of the fabricated cable-type lithium–air battery. The results were further testified as shown in Figure 7.22c, which showed that the cable-type battery charging and discharging ran stably after being repeatedly bent and stretched. Even after

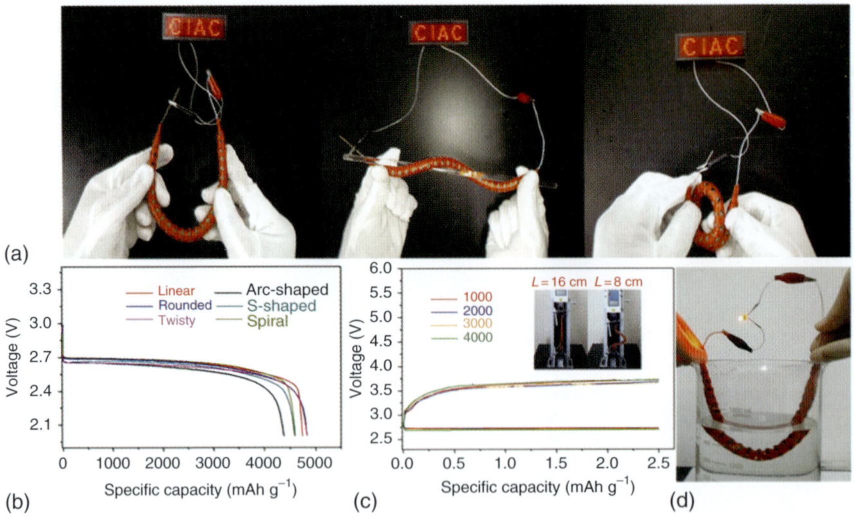

Figure 7.22 (a) Cable type lithium–air battery powering a commercial red LED display screen at various bended and twisted conditions. (b) First discharge curves of cable type lithium–air batteries with various conditions. (c) Charge–discharge curves of flexible lithium–air battery after being bended thousands of times. (d) Cable-type lithium–air battery powering a commercial red LED in water.

being bent for 4000 cycles, the overpotential of this fabricated battery increased negligibly.

It is well known that safety is the most important issue that should be noted because after all it is running in an open system and the anode is lithium metal. In this work, the anode lithium metal was protected via a layer of polymer electrolyte as displayed in Figure 7.21b. Unexpectedly, the obtained polymer electrolyte provides considerable ionic conductivity (2.843 $mS\,cm^{-1}$ at 30 °C) compared to organic liquid electrolyte (15.925 $mS\,cm^{-1}$ at 30 °C). Besides, the fabricated polymer electrolyte rendered hydrophobic property, which guarantees safety to some extent. Submerging test was carried out as shown in Figure 7.22d. It could be found that the cable-type lithium–air battery powered a commercial red light-emitting diode in water reposefully, indicating that the polymer electrolyte is an effective strategy to increase the safety of lithium–air battery.

Zhang et al. also reported a strategy in fabricating a cable-type flexible lithium–air battery [96], which consists of a lithium wire, solidified electrolyte layer, an aligned carbon nanotube (CNT) sheet and a packing tube, and the fabricated process is displayed in Figure 7.23. The cable-type flexible lithium–air battery exhibited a high flexibility which could be deformed into various shapes without damages in structure (Figure 7.24a). In addition, the dynamic stability of the battery was monitored, with the bending and releasing process of the battery, the discharge profile is constant with a voltage of about 2.7 V. To demonstrate its practical applications, the obtained cable type flexible lithium–air batteries were woven into a flexible powering textile and the discharge was stable as shown in Figure 7.24b. Besides, one fabricated cable-type flexible lithium–air battery

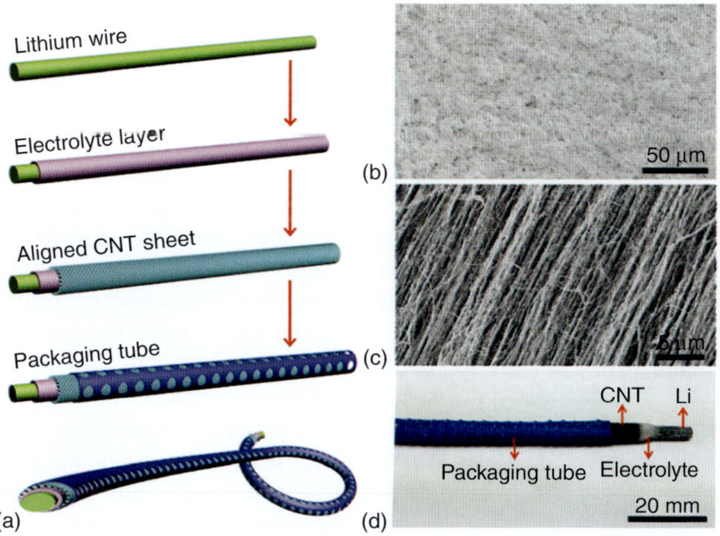

Figure 7.23 (a) Schematic illustration of the fabrication of the cable-type flexible lithium–air battery. (b) SEM image of fabricated gel electrolyte coated on the lithium wire. (c) SEM image of aligned CNT sheet wrapped as the outer layer. (d) Photograph of a cable-type flexible lithium–air battery.

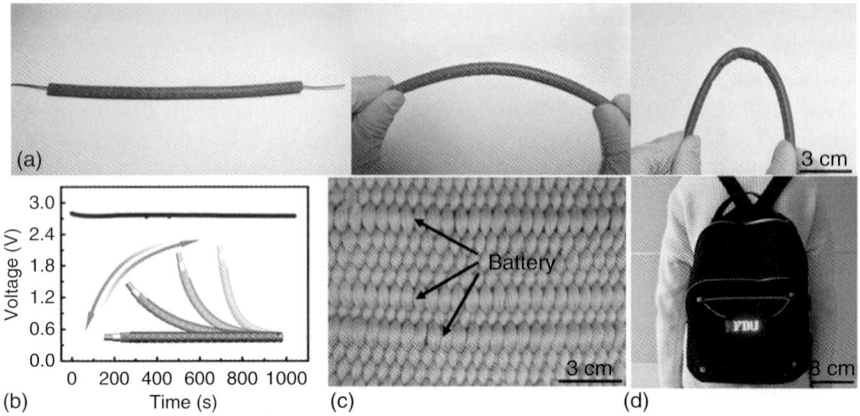

Figure 7.24 (a) The flexibility of the cable-type flexible lithium–air battery. (b) Discharge curve under a dynamic bending and releasing process. (c) Photograph of cable-type flexible lithium–air battery woven into flexible powering textile. (d) Photograph of flexibility of the cable-type flexible lithium–air battery integrated into a knapsack to power a commercial red LED display screen.

was integrated into a knapsack and successfully powered a commercial red LED display screen (Figure 7.24c,d).

7.2.3.3 Woven-type Battery Pack

The common type of researched batteries or supercapacitor are planar or cable-type and all fall into the "stack-type" category, which is the same as coin cell, Swagelok design, and columnar type. However, although the components used in these energy storage devices are not key components to contribute energy, they occupied a large portion of weight and this inevitably reduced the energy density. For the lithium–air battery, the influence of energy density caused by components is more severe since an additional air diffusion layer is needed in this system compared to lithium ion battery and supercapacitor.

A somewhat different ideology on the development of flexible lithium–air battery is an urgent need to improve its actual application of the energy density. Inspired by the structure of Chinese bamboo slips, Liu et al. reported a new strategy to fabricate a woven-type battery pack in an interlaced manner (Figure 7.25) [97]. In this system, all components are core components (lithium anode, polymer electrolyte, and air cathode), the air diffusion layer and packing material are avoided which occupy the main weight in a commonly assembled lithium–air battery. This fabricated device is not so much a battery but a battery pack since it is composed by multiple sets of batteries in parallel. In addition, this assembled manner possesses its own advantages compared with other reported flexible lithium–air batteries: (i) the tightly interwoven anodes and cathodes provide the necessary stress of lithium–air battery and avoid the need for fixation materials, thus the energy density was enhanced; (ii) gas could easily get access to the reaction sites via both sides of the weave, which improves the kinetics of the battery; and (iii) the fabricated lithium anode is protected by polymer electrolyte which avoids the sudden hazards posed by moisture or even water.

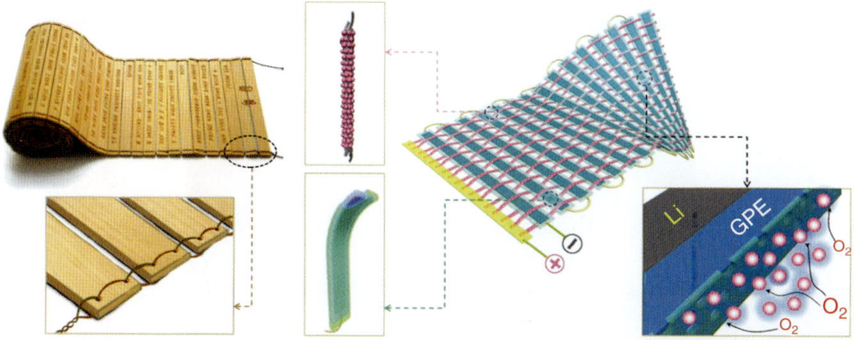

Figure 7.25 Inspired by the ancient bamboo slips, a flexible and wearable woven-type battery pack was fabricated.

To demonstrate the potential application of this woven-type battery pack, the as-obtained battery pack was used to power a commercial red LED display screen in various bending and twisting conditions as displayed in Figure 7.26a–f, indicating the mechanical suppleness of the fabricated woven-type battery pack. Besides, electrochemical stability of this battery pack was evaluated, too. About 5800 $mAh\,g^{-1}$ capacity was obtained from this fabricated battery pack and the mean voltage remained almost without any change (Figure 7.26g,h). Unprecedentedly, the fabricated woven-type battery pack could survive in water as showed in Figure 7.26i and this dramatically improves the safety of this battery pack.

Though several flexible lithium–air batteries were fabricated as discussed above, the energy density of the practical application is far from satisfactory because the proportion of the active material in this system is very low. The as-obtained woven-type battery pack could boost the energy density since the main components were air cathode and anode, the air diffusion layer and fixation materials were avoided. To further highlight the energy density advantage of woven-type battery pack, other flexible lithium–air batteries (planar battery and cable-type battery), and some coin-type batteries with different active materials were fabricated and their corresponding energy density values are listed in Figure 7.27. The energy density of the woven-type battery pack could achieve as high as 500 $Wh\,kg^{-1}$ which is far greater than other type batteries. However, there is a large gap to achieve its theoretical value and we still have a long way to go.

7.2.3.4 Battery Array Pack

With the development of flexible electronic devices, ultrathin and lightweight energy storage with high energy density is an urgent need in the fields of integrated electronic devices and implantable electronics. Liu et al. reported a "break up the whole into parts" strategy to build a novel battery array pack as showed in Figure 7.28 [98]. This battery array pack consisted of arrays of small-scale air cathode disks and lithium anode disks interconnected by carbon ropes and copper wires. The cathode and anode were separated by polymer

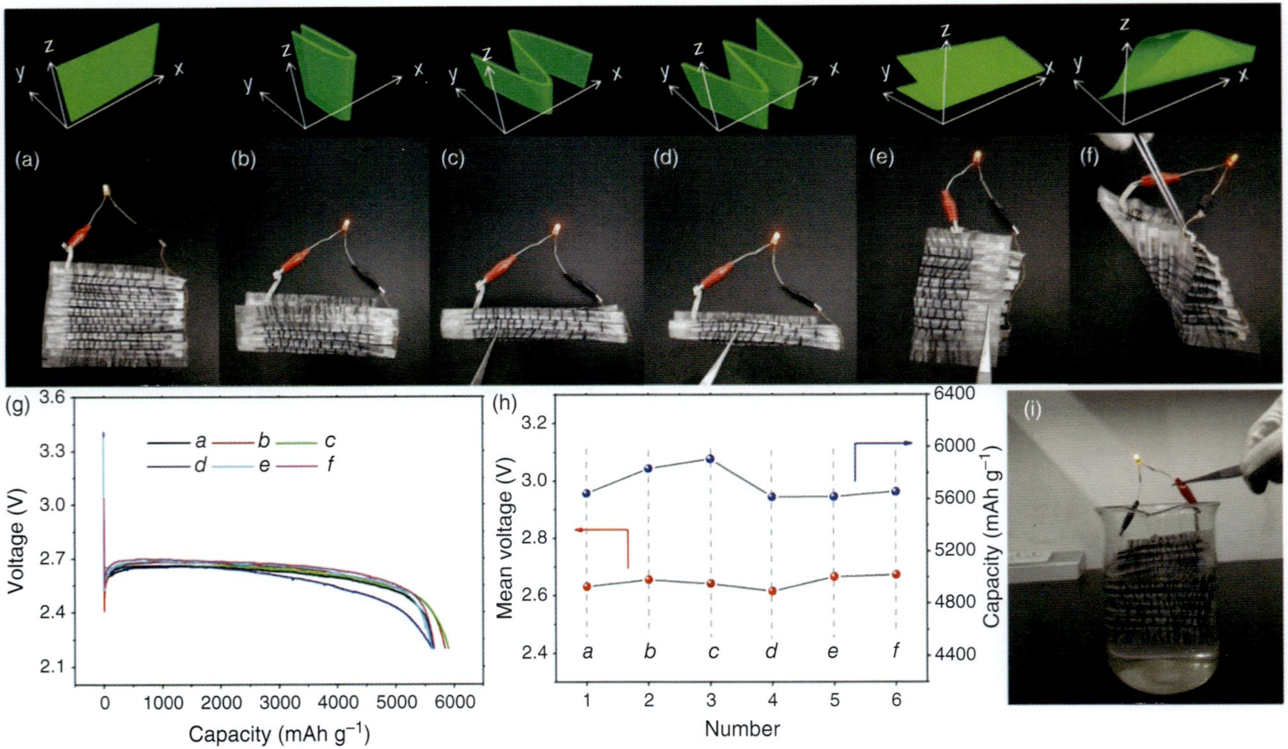

Figure 7.26 (a–f) Optic images of the fabricated woven-type battery pack powering a commercial red LED under various bending and twisting conditions; (g) first discharge curves of the woven-type battery pack under various bending and twisting conditions; (h) the contradistinction of mean discharge voltage and discharge capacity of the woven-type battery pack under various bending and twisting conditions; and (i) woven-type battery pack powering a commercial red LED immersed in water.

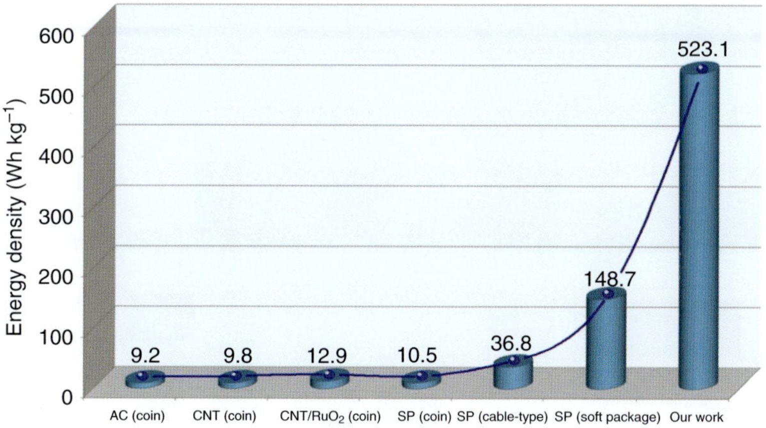

Figure 7.27 A rough comparison of energy densities with various types of lithium–air battery.

Figure 7.28 (a) Schematic illustration of the ultrathin and lightweight battery array pack. (b) Exploded view layout of the layers in the battery structure. (c) Photographs of the fabricated battery array pack integrated with a glove, powering a commercial LED. (d) Held by a branch of the asparagus fern. (e) Tested by a screw micrometer.

electrolyte. Significantly, compared to other strategies for fabricated flexible lithium–air battery, this obtained battery array pack holds many advantages: (i) all materials used are lightweight and thin and this endows this battery with high gravimetric/volumetric energy density and (ii) this battery can break up into many tiny units and remit the stress of battery which consequentially guarantees the integrity of the battery.

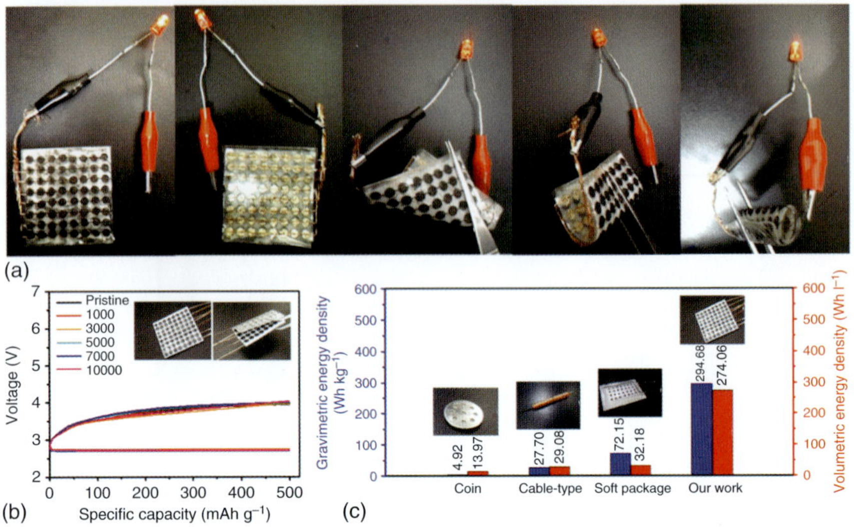

Figure 7.29 (a) A battery array pack powering a commercial red LED under various bending and twisting conditions. (b) Discharge–charge curves after being bended thousands of times; inset is the battery at planar and folded in half states. (c) The comparison of gravimetric energy density (left axis) and volumetric energy density (right axis) with four types of lithium–air batteries; insets are the photographs of the assembled lithium–air batteries.

To demonstrate the potential application of this battery array pack, the as-obtained battery was used to power commercial red LED display screens in various bending and twisting conditions as displayed in Figure 7.29a and the display screen could still run normally, indicating the mechanical suppleness of this battery array pack. To further confirm the influence caused by repetitive external force, the battery was folded and stretched several times and then its electrochemical performance was evaluated as displayed in Figure 7.29b, wherein it could be found that the overpotential was kept almost unchanged even after 10 000 cycles of folding/stretching.

For the development of flexible lithium–air battery, gravimetric energy density and volumetric energy density are two important indicators that must be considered. In this work, several other types of lithium–air battery also fabricated such as coin, cable-type, and soft package. From the result, the as-obtained battery array pack possesses the highest gravimetric energy density (294.68 Wh kg^{-1}) and volumetric energy density (274.06 Wh l^{-1}) as shown in Figure 7.29c.

7.3 Challenges and Future Work on Flexible Lithium–Air Batteries

Flexible lithium–air batteries exhibit much higher theoretical energy density than most other conventional flexible energy-storage systems such as Li-ion, lead-acid, and Ni-HM batteries because they are able to provide much higher specific energy. However, significant challenges still exist. The most important criteria of the flexible lithium–air batteries are the practical energy density (Wh kg^{-1} or Wh l^{-1}) and application. Before their practical application, the following essential challenges need to be overcome:

1) The electrochemical performances of the battery need further development and this is the most advantageous difference from other energy storage technologies. However, the electrochemical performances (capacity, rate capability, cycle life, efficiency, etc.) of the current lithium–air batteries are very poor, to say nothing of flexible ones. Therefore, before the development of flexible lithium–air batteries, electrochemical performances should be greatly improved involving developing and designing rational cathode, exploring accommodative electrolyte, and protecting lithium anode.
2) As it is operated in an open system, the anode of lithium–air battery will inevitably confront contaminations (H_2O, O_2, CO_2, etc.) derived from parasitic reaction or/and environment. These effects not only influence batteries electrochemical performances but also safety issues since lithium metal is too active. Besides, the battery must be able to endure various deformation modes. Otherwise, a short circuit may occur. Therefore, the lithium anode must be replaced to improve the safety of this battery and the lithium–silicon alloy with high capacity and low potential may be an appropriate candidate.
3) For the development of flexible and wearable lithium–air battery, the leakage of liquid electrolyte should be avoided especially when the battery is in deformation mode. And this will not only threaten the safety but also affect electrochemical performances. Therefore, the development of the polymer electrolyte with high ionic conductivity, accommodation for all battery components, and capability of enduring oxygen radicals attacks is critical.
4) The energy density (Wh kg^{-1} or Wh l^{-1}) of the current investigated lithium–air batteries is very low due to the mass proportion of the active material, which is inappreciable. The development of flexible lithium–air battery avoids the using of bulky and rigid stainless steel packing so that energy density is improved to some extent. However, it is far from a satisfactory value. Increasing active material proportion and reducing other components (anode, electrolyte/separator, and encapsulating materials) mass may be an appropriate way to improve their practical application energy density.
5) The limitations of the commonly researched flexible lithium–air battery still exists. In order to achieve the goal that batteries are widely used, some other new concepts of batteries also need to be exploited.

7.4 Concluding Remarks

Flexible lithium–air batteries are still at the infant stage, though some progress have been achieved in recent years and some battery models have been fabricated in the laboratory. Apart from the challenges stated above, insightful understanding on the mechanisms of oxygen electrochemistry during cycling is of paramount importance. In order to understand the mechanisms, both experimental and theoretical models are needed. The problems restraining the development of flexible lithium–air battery are not only scientific but also technical. This chapter briefly introduces the current development of the flexible lithium–air battery and the opportunities and challenges of future. The aim is to motivate more people to be interested in lithium–air battery research.

References

1 Rogers, J.A., Someya, T., and Huang, Y. (2010). Materials and mechanics for stretchable electronics. *Science* 327 (5973): 1603–1607.
2 De Volder, M.F.L., Tawfick, S.H., Baughman, R.H., and Hart, A.J. (2013). Carbon nanotubes: present and future commercial applications. *Science* 339 (6119): 535–539.
3 He, Y., Chen, W., Gao, C. et al. (2013). An overview of carbon materials for flexible electrochemical capacitors. *Nanoscale* 5 (19): 8799–8820.
4 Lee, S.-Y., Choi, K.-H., Choi, W.-S. et al. (2013). Progress in flexible energy storage and conversion systems, with a focus on cable-type lithium–ion batteries. *Energy Environ. Sci.* 6 (8): 2414–2423.
5 Li, L., Wu, Z., Yuan, S., and Zhang, X.-B. (2014). Advances and challenges for flexible energy storage and conversion devices and systems. *Energy Environ. Sci.* 7 (7): 2101–2122.
6 Lu, N. and Kim, D.-H. (2014). Flexible and stretchable electronics paving the way for soft robotics. *Soft Rob.* 1 (1): 53–62.
7 Zhou, G., Li, F., and Cheng, H.-M. (2014). Progress in flexible lithium batteries and future prospects. *Energy Environ. Sci.* 7 (4): 1307–1338.
8 Fu, Y., Lv, Z., Hou, S. et al. (2011). Conjunction of fiber solar cells with groovy micro-reflectors as highly efficient energy harvesters. *Energy Environ. Sci.* 4 (9): 3379–3383.
9 Lipomi, D.J. and Bao, Z. (2011). Stretchable, elastic materials and devices for solar energy conversion. *Energy Environ. Sci.* 4 (9): 3314–3328.
10 Zhang, S., Ji, C., Bian, Z. et al. (2011). Single-wire dye-sensitized solar cells wrapped by carbon nanotube film electrodes. *Nano Lett.* 11 (8): 3383–3387.
11 Kaltenbrunner, M., White, M.S., Głowacki, E.D. et al. (2012). Ultrathin and lightweight organic solar cells with high flexibility. *Nat. Commun.* 3: 770.
12 Lee, Y.-H., Kim, J.-S., Noh, J. et al. (2013). Wearable textile battery rechargeable by solar energy. *Nano Lett.* 13 (11): 5753–5761.
13 Li, H., Zhao, Q., Wang, W. et al. (2013). Novel planar-structure electrochemical devices for highly flexible semitransparent power generation/storage sources. *Nano Lett.* 13 (3): 1271–1277.
14 Liu, Z., Li, J., and Yan, F. (2013). Package-free flexible organic solar cells with graphene top electrodes. *Adv. Mater.* 25 (31): 4296–4301.
15 Yang, Z., Sun, H., Chen, T. et al. (2013). Photovoltaic wire derived from a graphene composite fiber achieving an 8.45% energy conversion efficiency. *Angew. Chem. Int. Ed.* 52 (29): 7545–7548.
16 Tominaka, S., Nishizeko, H., Mizuno, J., and Osaka, T. (2009). Bendable fuel cells: on-chip fuel cell on a flexible polymer substrate. *Energy Environ. Sci.* 2 (10): 1074–1077.

17 Hu, L., Wu, H., La Mantia, F. et al. (2010). Thin, flexible secondary Li-ion paper batteries. *ACS Nano* 4 (10): 5843–5848.
18 Cheng, Q., Song, Z., Ma, T. et al. (2013). Folding paper-based lithium-ion batteries for higher areal energy densities. *Nano Lett.* 13 (10): 4969–4974.
19 Kil, E.-H., Choi, K.-H., Ha, H.-J. et al. (2013). Imprintable, bendable, and shape-conformable polymer electrolytes for versatile-shaped lithium-ion batteries. *Adv. Mater.* 25 (10): 1395–1400.
20 Xu, S., Zhang, Y., Cho, J. et al. (2013). Stretchable batteries with self-similar serpentine interconnects and integrated wireless recharging systems. *Nat. Commun.* 4: 1543.
21 Gao, W., Singh, N., Song, L. et al. (2011). Direct laser writing of micro-supercapacitors on hydrated graphite oxide films. *Nat. Nanotechnol.* 6 (8): 496–500.
22 Wang, K., Zou, W., Quan, B. et al. (2011). An all-solid-state flexible micro-supercapacitor on a chip. *Adv. Energy Mater.* 1 (6): 1068–1072.
23 Xiao, X., Li, T., Yang, P. et al. (2012). Fiber-based all-solid-state flexible supercapacitors for self-powered systems. *ACS Nano* 6 (10): 9200–9206.
24 Fu, Y., Wu, H., Ye, S. et al. (2013). Integrated power fiber for energy conversion and storage. *Energy Environ. Sci.* 6 (3): 805–812.
25 Jost, K., Stenger, D., Perez, C.R. et al. (2013). Knitted and screen printed carbon-fiber supercapacitors for applications in wearable electronics. *Energy Environ. Sci.* 6 (9): 2698–2705.
26 Kim, D., Shin, G., Kang, Y.J. et al. (2013). Fabrication of a stretchable solid-state micro-supercapacitor array. *ACS Nano* 7 (9): 7975–7982.
27 Niu, Z., Zhang, L., Liu, L. et al. (2013). All-solid-state flexible ultrathin micro-supercapacitors based on graphene. *Adv. Mater.* 25 (29): 4035–4042.
28 Xu, Y., Lin, Z., Huang, X. et al. (2013). Flexible solid-state supercapacitors based on three-dimensional graphene hydrogel films. *ACS Nano* 7 (5): 4042–4049.
29 Girishkumar, G., McCloskey, B., Luntz, A.C. et al. (2010). Lithium–air battery: promise and challenges. *J. Phys. Chem. Lett.* 1 (14): 2193–2203.
30 Bruce, P.G., Freunberger, S.A., Hardwick, L.J., and Tarascon, J.-M. (2011). Li–O_2 and Li–S batteries with high energy storage. *Nat. Mater.* 11 (1): 19–29.
31 Scrosati, B., Hassoun, J., and Sun, Y.-K. (2011). Lithium–ion batteries. A look into the future. *Energy Environ. Sci.* 4 (9): 3287–3295.
32 Black, R., Adams, B., and Nazar, L.F. (2012). Non-aqueous and hybrid Li–O_2 batteries. *Adv. Energy Mater.* 2 (7): 801–815.
33 Cheng, F. and Chen, J. (2012). Metal–air batteries: from oxygen reduction electrochemistry to cathode catalysts. *Chem. Soc. Rev.* 41 (6): 2172–2192.
34 Hardwick, L.J. and Bruce, P.G. (2012). The pursuit of rechargeable non-aqueous lithium–oxygen battery cathodes. *Curr. Opin. Solid State Mater. Sci.* 16 (4): 178–185.
35 Hou, J., Yang, M., Ellis, M.W. et al. (2012). Lithium oxides precipitation in nonaqueous Li–air batteries. *Phys. Chem. Chem. Phys.* 14 (39): 13487–13501.
36 Lu, J., Li, L., Park, J.-B. et al. (2014). Aprotic and aqueous Li–O_2 batteries. *Chem. Rev.* 114 (11): 5611–5640.

37 Xu, J.-J., Wang, Z.-L., Xu, D. et al. (2013). Tailoring deposition and morphology of discharge products towards high-rate and long-life lithium–oxygen batteries. *Nat. Commun.* 4: 2438.

38 Xu, J.-J., Xu, D., Wang, Z.-L. et al. (2013). Synthesis of perovskite-based Porous $La_{0.75}Sr_{0.25}MnO_3$ nanotubes as a highly efficient electrocatalyst for rechargeable lithium–oxygen batteries. *Angew. Chem. Int. Ed.* 52 (14): 3887–3890.

39 Zhang, T. and Zhou, H. (2013). A reversible long-life lithium–air battery in ambient air. *Nat. Commun.* 4: 1817.

40 Xu, J.-J., Wang, Z.-L., Xu, D. et al. (2014). 3D ordered macroporous $LaFeO_3$ as efficient electrocatalyst for Li–O_2 batteries with enhanced rate capability and cyclic performance. *Energy Environ. Sci.* 7 (7): 2213–2219.

41 McCloskey, B.D., Scheffler, R., Speidel, A. et al. (2011). On the efficacy of electrocatalysis in nonaqueous Li–O_2 batteries. *J. Am. Chem. Soc.* 133 (45): 18038–18041.

42 Peng, Z., Freunberger, S.A., Chen, Y., and Bruce, P.G. (2012). A reversible and higher-rate Li–O_2 battery. *Science* 337 (6094): 563–566.

43 Chen, Y., Freunberger, S.A., Peng, Z. et al. (2013). Charging a Li–O_2 battery using a redox mediator. *Nat. Chem.* 5 (6): 489–494.

44 Shao, Y., Park, S., Xiao, J. et al. (2012). Electrocatalysts for nonaqueous lithium–air batteries: status, challenges, and perspective. *ACS Catal.* 2 (5): 844–857.

45 Lu, Y.-C., Gallant, B.M., Kwabi, D.G. et al. (2013). Lithium–oxygen batteries: bridging mechanistic understanding and battery performance. *Energy Environ. Sci.* 6 (3): 750–768.

46 Shao, Y., Ding, F., Xiao, J. et al. (2013). Making Li–air batteries rechargeable: material challenges. *Adv. Funct. Mater.* 23 (8): 987–1004.

47 Wang, J., Li, Y., and Sun, X. (2013). Challenges and opportunities of nanostructured materials for aprotic rechargeable lithium–air batteries. *Nano Energy* 2 (4): 443–467.

48 Luntz, A.C. and McCloskey, B.D. (2014). Nonaqueous Li–air batteries: a status report. *Chem. Rev.* 114 (23): 11721–11750.

49 Wang, Z.-L., Xu, D., Xu, J.-J., and Zhang, X.-B. (2014). Oxygen electrocatalysts in metal–air batteries: from aqueous to nonaqueous electrolytes. *Chem. Soc. Rev.* 43 (22): 7746–7786.

50 Abraham, K.M. and Jiang, Z. (1996). A polymer electrolyte-based rechargeable lithium/oxygen battery. *J. Electrochem. Soc.* 143 (1): 1–5.

51 Liu, S., Wang, Z., Yu, C. et al. (2013). A flexible TiO_2(B)-based battery electrode with superior power rate and ultralong cycle life. *Adv. Mater.* 25 (25): 3462–3467.

52 Xiao, J., Mei, D., Li, X. et al. (2011). Hierarchically porous graphene as a lithium–air battery electrode. *Nano Lett.* 11 (11): 5071–5078.

53 Wang, Z.-L., Xu, D., Xu, J.-J. et al. (2012). Graphene oxide gel-derived, free-standing, hierarchically porous carbon for high-capacity and high-rate rechargeable Li–O_2 batteries. *Adv. Funct. Mater.* 22 (17): 3699–3705.

54 Lim, H.-D., Park, K.-Y., Song, H. et al. (2013). Enhanced power and rechargeability of a Li–O_2 battery based on a hierarchical-fibril CNT electrode. *Adv. Mater.* 25 (9): 1348–1352.

55 Mirzaeian, M. and Hall, P.J. (2009). Preparation of controlled porosity carbon aerogels for energy storage in rechargeable lithium oxygen batteries. *Electrochim. Acta* 54 (28): 7444–7451.
56 Yang, X.-H., He, P., and Xia, Y.-Y. (2009). Preparation of mesocellular carbon foam and its application for lithium/oxygen battery. *Electrochem. Commun.* 11 (6): 1127–1130.
57 Kichambare, P., Kumar, J., Rodrigues, S., and Kumar, B. (2011). Electrochemical performance of highly mesoporous nitrogen doped carbon cathode in lithium–oxygen batteries. *J. Power Sources* 196 (6): 3310–3316.
58 Nakanishi, S., Mizuno, F., Nobuhara, K. et al. (2012). Influence of the carbon surface on cathode deposits in non-aqueous Li–O_2 batteries. *Carbon* 50 (13): 4794–4803.
59 Kuboki, T., Okuyama, T., Ohsaki, T., and Takami, N. (2005). Lithium–air batteries using hydrophobic room temperature ionic liquid electrolyte. *J. Power Sources* 146 (1–2): 766–769.
60 Mirzaeian, M. and Hall, P.J. (2010). Characterizing capacity loss of lithium oxygen batteries by impedance spectroscopy. *J. Power Sources* 195 (19): 6817–6824.
61 Song, M.-K., Park, S., Alamgir, F.M. et al. (2011). Nanostructured electrodes for lithium–ion and lithium-air batteries: the latest developments, challenges, and perspectives. *Mater. Sci. Eng., R* 72 (11): 203–252.
62 Sun, B., Chen, S., Liu, H., and Wang, G. (2015). Mesoporous carbon nanocube architecture for high-performance lithium–oxygen batteries. *Adv. Funct. Mater.* 25 (28): 4436–4444.
63 Li, F., Zhang, T., and Zhou, H. (2013). Challenges of non-aqueous Li–O_2 batteries: electrolytes, catalysts, and anodes. *Energy Environ. Sci.* 6 (4): 1125–1141.
64 Shui, J.-L., Okasinski, J.S., Kenesei, P. et al. (2013). Reversibility of anodic lithium in rechargeable lithium–oxygen batteries. *Nat. Commun.* 4: 2255.
65 Walker, W., Giordani, V., Uddin, J. et al. (2013). A rechargeable Li–O_2 battery using a lithium nitrate/*N*,*N*-dimethylacetamide electrolyte. *J. Am. Chem. Soc.* 135 (6): 2076–2079.
66 Lee, D.J., Lee, H., Song, J. et al. (2014). Composite protective layer for Li metal anode in high-performance lithium–oxygen batteries. *Electrochem. Commun.* 40: 45–48.
67 Kang, S.J., Mori, T., Suk, J. et al. (2014). Improved cycle efficiency of lithium metal electrodes in Li–O_2 batteries by a two-dimensionally ordered nanoporous separator. *J. Mater. Chem. A* 2 (26): 9970–9974.
68 Hassoun, J., Jung, H.-G., Lee, D.-J. et al. (2012). A metal-free, lithium–ion oxygen battery: a step forward to safety in lithium–air batteries. *Nano Lett.* 12 (11): 5775–5779.
69 Xu, K. (2004). Nonaqueous liquid electrolytes for lithium-based rechargeable batteries. *Chem. Rev.* 104 (10): 4303–4417.
70 Freunberger, S.A., Chen, Y., Drewett, N.E. et al. (2011). The lithium–oxygen battery with ether-based electrolytes. *Angew. Chem. Int. Ed.* 50 (37): 8609–8613.
71 Chen, Y., Freunberger, S.A., Peng, Z. et al. (2012). Li–O_2 battery with a dimethylformamide electrolyte. *J.Am. Chem. Soc.* 134 (18): 7952–7957.

72 Jung, H.-G., Hassoun, J., Park, J.-B. et al. (2012). An improved high-performance lithium–air battery. *Nat. Chem.* 4 (7): 579–585.
73 Kumar, J. and Kumar, B. (2009). Development of membranes and a study of their interfaces for rechargeable lithium–air battery. *J. Power Sources* 194 (2): 1113–1119.
74 Goodenough, J.B. and Kim, Y. (2010). Challenges for rechargeable Li batteries. *Chem. Mater.* 22 (3): 587–603.
75 Kumar, B., Kumar, J., Leese, R. et al. (2010). A solid-state, rechargeable, long cycle life lithium–air battery. *J. Electrochem. Soc.* 157 (1): A50–A54.
76 Allen, C.J., Hwang, J., Kautz, R. et al. (2012). Oxygen reduction reactions in ionic liquids and the formulation of a general ORR mechanism for Li–air batteries. *J. Phys. Chem. C* 116 (39): 20755–20764.
77 Liu, B., Zhang, J., Wang, X. et al. (2012). Hierarchical three-dimensional $ZnCo_2O_4$ nanowire arrays/carbon cloth anodes for a novel class of high-performance flexible lithium–ion batteries. *Nano Lett.* 12 (6): 3005–3011.
78 Fu, Y., Cai, X., Wu, H. et al. (2012). Fiber supercapacitors utilizing pen ink for flexible/wearable energy storage. *Adv. Mater.* 24 (42): 5713–5718.
79 Kwon, Y.H., Woo, S.-W., Jung, H.-R. et al. (2012). Cable-type flexible lithium ion battery based on hollow multi-helix electrodes. *Adv. Mater.* 24 (38): 5192–5197.
80 Viet Thong, L., Kim, H., Ghosh, A. et al. (2013). Coaxial fiber supercapacitor using all-carbon material electrodes. *ACS Nano* 7 (7): 5940–5947.
81 Li, R., Parvez, K., Hinkel, F. et al. (2013). Bioinspired wafer-scale production of highly stretchable carbon films for transparent conductive electrodes. *Angew. Chem. Int. Ed.* 52 (21): 5535–5538.
82 Koo, M., Park, K.I., Lee, S.H. et al. (2012). Bendable inorganic thin-film battery for fully flexible electronic systems. *Nano Lett.* 12 (9): 4810–4816.
83 Liu, Q.-C., Xu, J.-J., Xu, D., and Zhang, X.-B. (2015). Flexible lithium–oxygen battery based on a recoverable cathode. *Nat. Commun.* 6: 7892.
84 McCloskey, B.D., Speidel, A., Scheffler, R. et al. (2012). Twin problems of interfacial carbonate formation in nonaqueous Li–O_2 batteries. *J. Phys. Chem. Lett.* 3 (8): 997–1001.
85 Ottakam Thotiyl, M.M., Freunberger, S.A., Peng, Z., and Bruce, P.G. (2013). The carbon electrode in nonaqueous Li–O_2 cells. *J. Am. Chem. Soc.* 135 (1): 494–500.
86 Black, R., Oh, S.H., Lee, J.-H. et al. (2012). Screening for superoxide reactivity in Li–O_2 batteries: effect on Li_2O_2/LiOH crystallization. *J. Am. Chem. Soc.* 134 (6): 2902–2905.
87 Chang, Z.-W., Xu, J.-J., Liu, Q.-C. et al. (2015). Recent progress on stability enhancement for cathode in rechargeable non-aqueous lithium–oxygen battery. *Adv. Energy Mater.* 5 (21): 1500633.
88 Liu, Q.-C., Xu, J.-J., Chang, Z.-W. et al. (2016). Growth of Ru-modified Co_3O_4 nanosheets on carbon textiles toward flexible and efficient cathodes for flexible Li–O_2 batteries. *Part. Part. Syst. Char.* 33 (8): 500–505.
89 Liu, Q.-C., Xu, J.-J., Chang, Z.-W., and Zhang, X.-B. (2014). Direct electrodeposition of cobalt oxide nanosheets on carbon paper as free-standing cathode for Li–O_2 battery. *J. Mater. Chem. A* 2 (17): 6081–6085.

90 Yin, Y.-B., Xu, J.-J., Liu, Q.-C., and Zhang, X.-B. (2016). Macroporous interconnected hollow carbon nanofibers inspired by golden-toad eggs toward a binder-free, high-rate, and flexible electrode. *Adv. Mater.* 28 (34): 7494–7500.

91 Weng, Z., Su, Y., Wang, D.-W. et al. (2011). Graphene–cellulose paper flexible supercapacitors. *Adv. Energy Mater.* 1 (5): 917–922.

92 Hu, L. and Cui, Y. (2012). Energy and environmental nanotechnology in conductive paper and textiles. *Energy Environ. Sci.* 5 (4): 6423–6435.

93 Liu, Q.-C., Li, L., Xu, J.-J. et al. (2015). Flexible and foldable Li–O_2 battery based on paper-ink cathode. *Adv. Mater.* 27 (48): 8095–8101.

94 Park, J., Park, M., Nam, G. et al. (2015). All-solid-state cable-type flexible zinc–air battery. *Adv. Mater.* 27 (8): 1396–1401.

95 Liu, T., Liu, Q.-C., Xu, J.-J., and Zhang, X.-B. (2016). Cable-type water-survivable flexible Li–O_2 battery. *Small* 12 (23): 3101–3105.

96 Zhang, Y., Wang, L., Guo, Z. et al. (2016). High-performance lithium–air battery with a coaxial-fiber architecture. *Angew. Chem. Int. Ed.* 55 (14): 4487–4491.

97 Liu, Q.-C., Liu, T., Liu, D.-P. et al. (2016). A flexible and wearable lithium–oxygen battery with record energy density achieved by the interlaced architecture inspired by bamboo slips. *Adv. Mater.* 28 (38): 8413–8418.

98 Liu, T., Xu, J.-J., Liu, Q.-C. et al. (2016). Ultrathin, lightweight, and wearable Li–O_2 battery with high robustness and gravimetric/volumetric energy density. *Small* 13 (6): 1602952.

8

Nanodielectric Elastomers for Flexible Generators

Li-Juan Yin and Zhi-Min Dang

Department of Electrical Engineering, State Key Laboratory of Power System, Tsinghua University, Shuangqing Road, Haidian District, Beijing 100084, China

8.1 Introduction

Electro-active polymers (EAPs), a kind of soft materials that can respond to electrical fields by changing their shape, have great promise to mimic muscle-like behavior. EAPs can convert electrical energy to mechanical energy, acting as actuators, or vice versa acting as generators, namely flexible generators [1, 2]. Because of their advantages of light weight, high energy density, high conversion efficiency, and great strain, EAPs have attracted increasing attention from scientists, researchers, and enterprises, with more and more achievements emerging at the same time [3–5]. Traditionally, EAPs are divided into two main groups: ionic EAPs (activated by an electrically induced transport of ions and/or molecules) and electronic EAPs (activated by electrostatic forces). The greatest advantage of ionic EAPs is that they commonly require low driving voltages (as low as 1–2 V), while electronic EAPs require high driving voltages about 500 V to 10 kV, close to their breakdown strength. However, ionic EAPs require an electrolyte, which means that it is necessary to encapsulate the aqueous environment and maintain wetness. The advantages that electronic EAPs have over ionic EAPs are their higher energy density, large actuation forces, rapid response times, and long lifetimes [4–9]. Table 8.1 presents a detailed classification.

This chapter focuses on dielectric elastomers (DEs). Dielectric elastomer devices are soft or flexible capacitors, composed of a thin elastomer membrane sandwiched between two compliant electrodes. Initial studies focused mainly on dielectric elastomer actuators (DEAs), identifying the electro-mechanical principles and modifying material properties to improve device performance [10]. Nowadays, more and more scientists have shown great interest in dielectric elastomer generators (DEGs). In order to improve the performance of DEGs, materials and configurations are two main key factors. Therefore, in the following discussion, we will introduce the recent progresses made with regard to DEs, electro-mechanical coupling modes, and generators. However, we intend to first make a brief introduction of the electro-mechanical principles.

Flexible Energy Conversion and Storage Devices, First Edition. Edited by Chunyi Zhi and Liming Dai.

Table 8.1 EAPs classification.

EAPs class	Materials
Ionic EAPs	Polymer gels
	Ionic polymer–metal composites
	Conjugated polymers
	Carbon nanotubes
Electronic EAPs	Piezoelectric polymers
	Electrostrictive polymers
	Dielectric elastomers
	Liquid crystal elastomers
	Carbon nanotube aerogels

Source: Carpi et al. 2011 [83]. IOP.

8.2 Electro-Mechanical Principles

8.2.1 Electro-Mechanical Conversion

As the typical equivalent circuit model of DEGs is the capacitor of an elastomer sheet sandwiched by two flexible electrodes, the basic transduction principle of DEs is based on the electrical energy stored in capacitors, W, given by

$$W = \frac{1}{2}CU^2 \tag{8.1}$$

$$C = \varepsilon_0 \varepsilon_r \frac{S}{h} \tag{8.2}$$

where C is the capacitance of the equivalent model, U is the voltage of the capacitor, ε_0 is the permittivity of vacuum, ε_r is the relative permittivity of materials and s and h are the area and thickness of the sheet, respectively.

There are three kinds of energy harvesting modes for DEGs, namely constant charges, constant field, and constant voltage [11–13]. Their electro-mechanical conversion procedures are shown in Figure 8.1.

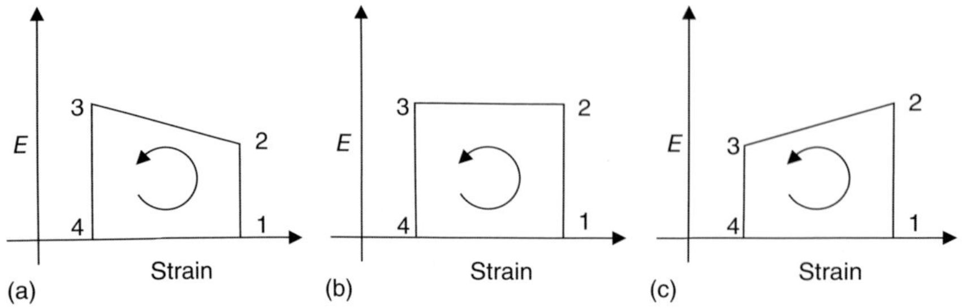

Figure 8.1 Schema of electric field and strain in different conversion cycles: (a) constant charge, (b) constant field, and (c) constant voltage.

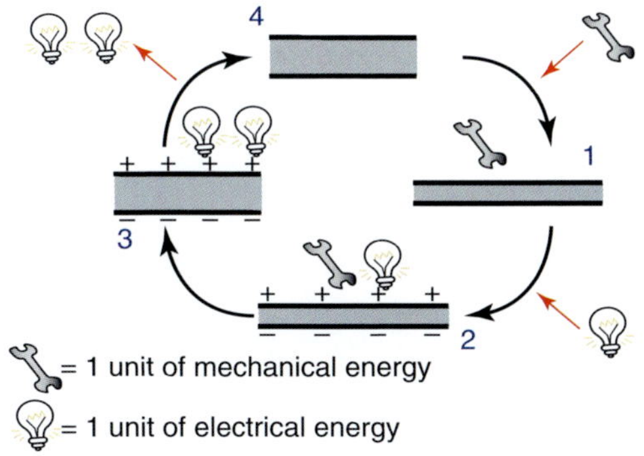

Figure 8.2 Schematic of a DEG energy cycle. *Source:* Romasanta et al. 2015 [10]. Elsevier.

Among them, the constant field cycle is most complex to control and difficult to achieve, but holds the maximal energy gain [14]. However, the constant charge cycle is easiest to comprehend and control, having therefore wide applications. In the following text, we will demonstrate the working principles of DEs in generator mode as shown in Figure 8.2 and deduce the electro-mechanical conversion equations of DEGs in constant charge cycle.

In generator mode, an input of mechanical energy (generally, strain) plus a small electrical charge results in an overall energy gain when the material relaxes. As for constant charge cycle, there are four key steps: firstly, an input of mechanical source stretches the DE membrane. Secondly, the film is charged. Thirdly, the DE film relaxes to a state of lower stretch, thus reducing its area while increasing its thickness because of the constancy of volume. The charge on the film is maintained constant at this step. That is why it is called constant charge cycle. Finally, the electrical energy, which has increased at the last step, is removed from the DE film. This is a complete loop.

8.2.2 Equations of DE Generators

Before deducing the electro-mechanical coupling equations, some assumptions need to be set: DE membranes are compressible, isotropic, and linear. In addition, loss of charge, piezoelectric effect, and failure of DEs can be neglected.
After stretch,

$$C_1 = \varepsilon_0 \varepsilon_r \frac{S_1}{d_1} \tag{8.3}$$

$$W_1 = \frac{1}{2} C_1 U_1^2 \tag{8.4}$$

After relaxing,

$$C_2 = \varepsilon_0 \varepsilon_r \frac{S_2}{d_2} \tag{8.5}$$

$$W_2 = \frac{1}{2} C_2 U_2^2 \tag{8.6}$$

$$\Delta W = W_2 - W_1 \tag{8.7}$$

where ΔW is the electrical energy gain.

Considering that the volume is unchangeable, when the DE membrane relaxes, the area decreases and the thickness increases. Therefore, the capacitance of the membrane is smaller than before. Besides,

$$Q = CU = C_1U_1 = C_2U_2 \tag{8.8}$$

therefore, $U_2 > U_1$ because $C_2 < C_1$. Plugging Equations (8.4), (8.6), (8.8) into Equation (8.7), we get the following:

$$\Delta W = \frac{1}{2}C_2U_2^2 - \frac{1}{2}C_1U_1^2 = \frac{1}{2}Q(U_2 - U_1) = \frac{1}{2}Q\Delta U \tag{8.9}$$

where ΔU is the voltage increment.

From another perspective, DE film is compressible, and permittivity of DEs is almost invariable. Therefore,

$$V = S_1d_1 = S_2d_2, \quad \frac{C_1}{C_2} = \frac{S_1d_2}{S_2d_1} = \frac{d_2^2}{d_1^2} = \frac{S_1^2}{S_2^2},$$

then,

$$\Delta W = \frac{1}{2}C_1U_1^2\left(\frac{d_2^2}{d_1^2} - 1\right) = W_1\left(\frac{d_2^2}{d_1^2} - 1\right) = W_1(\lambda^2 - 1) \tag{8.10}$$

where $\lambda = \frac{S_1}{S_2}$ is the strain ratio of the area, meaning the degree of membrane deformation.

From Equations (8.9) and (8.10), it is apparent that the electrical energy gain is proportional to charge quantity, voltage increment, electrical energy after charged, and square of strain ratio. These easy equations point out how to obtain large electrical energy gain, like increasing strain ratio.

Although the equations above all are based on the simplified model of plate capacitors, they still provide a foundation for other complex applications. Further refinements of this simplified model have taken nonlinear elastic materials [15, 16] or nonlinear electro-mechanical field theory [17] into consideration. These refinements are mostly used in modeling oriented studies and the initial model in material oriented studies [10].

In order to describe the mechanical properties, the Maxwell stress $p = \varepsilon_0\varepsilon_r E^2$ has been introduced. For low strains (e.g. <20%), the thickness strain $s_z = -\frac{p}{Y} = -\frac{\varepsilon_r\varepsilon_0 E^2}{Y}$, where Y is the elastic modulus, while for greater strains, Y generally depends on the strain itself [18]. Moreover, the mechanical property of DEs can be described accurately by the Yeoh model, the Mooney–Rivlin model, and the Ogden model [19].

8.3 Increasing the Performance of Dielectric Elastomers from the Materials Perspective

It can be seen from section 8.2.2 that the main parameters affecting the energy conversion are both experimental factors, such as the charging voltage, thickness of the film, and the degree of deformation, and material factors, such as permittivity and elastic modulus of DEs. The maximal electric energies generated per cycle have theoretically been predicted to be 1300 and 1700 $J\,kg^{-1}$ for natural rubber and acrylic elastomers [20], respectively, which correspond to their maximum strain possible. However, the highest energy density of VHB 4905, a commercial acrylic membrane, is 780 $J\,kg^{-1}$, having a conversion efficiency of 30%, so far [21]. Therefore, there is a lot of room for improvement. Previous studies reveal that there are two main methods with which we can improve the performance of DEs in generator mode, like raising electrical energy gain and energy density; from a materials perspective this would mean that one is increasing ε_r, and the other is decreasing Y. Therefore, we discuss them individually.

8.3.1 Increasing the Relative Permittivity of DEs

The need for high permittivity materials exists commonly among different applications, not only in generators, but also in actuators, and so on. As for DEs, the increase of the membrane permittivity can increase its capacitance directly, thus reducing the required electric field intensity. Therefore, numerous studies about it have sprung up. This strategy has been investigated through three approaches, namely elastomer composites, elastomer blends, and chemical modification of the elastomer backbone. We select some recent typical results to interpret these different ways.

8.3.1.1 Elastomer Composites

Employing elastomer composites is a traditional strategy to enhance the permittivity of given elastomers. The key point of this strategy is to combine a filler of high permittivity with the given elastomer, when a new system arises. Before experiments, the prediction of the permittivity of a new system is essential and significant. Accurate prediction is very difficult, although providing a scope has already been achieved [22]. Commonly, dielectric fillers can be considered as spherical particles while dielectric elastomer is an isolating matrix. The upper limit of permittivity after mixing can be simply and violently obtained by direct mixing rules, as shown in the following equation:

$$\varepsilon_c = \Phi_1\varepsilon_1 + \Phi_2\varepsilon_2 \tag{8.11}$$

where subscripts c, 1, and 2 refer to composite, matrix, and filler, respectively, and Φ refers to the volume fraction.

The lower limit can be determined by Equation (8.12) as follows:

$$\varepsilon_c = \frac{\varepsilon_1\varepsilon_2}{\Phi_2\varepsilon_1 + \Phi_1\varepsilon_2} \tag{8.12}$$

It can be seen from Equations (8.11) and (8.12) that the permittivity of composites may arise with the increase of volume fraction of the fillers. Nevertheless, the addition of fillers is likely to increase both the dielectric loss and the tensile stress value/elastic modulus, which will decrease the performance of DEs in generator mode. Thus, pursuit of improving permittivity without taking these factors into consideration may have an opposite effect. Satisfying these three characteristics (high dielectric permittivity, low dielectric losses, and elastic modulus) is a challenge and arriving at a compromise among them will be a better approach [10].

Fillers can be divided into organic fillers and inorganic fillers or conductive fillers and dielectric fillers. In this article, we choose the last classification method.

Conductive Filler/Elastomer Composites Generally speaking, the challenge with using conductive fillers is preventing percolation and short-circuiting of DEs. Conductive filler/elastomer composites exhibit a nonlinear increase of permittivity and conductivity as the concentration exceeds the percolation threshold (P_c). According to the percolation theory [23], at low concentration, the fillers are separated from each other and the electric properties of composites are still determined by the elastomer matrix. But, with the increase of the concentration, fillers start to form a three-dimensional network through the elastomer, resulting in a great increase of conductivity. Therefore, in order to obtain the increment in permittivity while maintaining dielectric property, the concentration has to be lower than the percolation threshold, which is called sub-percolation strategy.

The search for high permittivity elastomer composites has been conducted with different conductive fillers, ranging from metal [24, 25] to carbon-based microparticles [26, 27], and even to organic conductive polymers. For example, Park et al. [28] were pioneers to report that adding 0.5 wt% of commercial multiwalled carbon nanotubes (MWCNTs) increased the relative permittivity of a poly(dimethylsiloxane) (PDMS) matrix from $\varepsilon' = 3$ to $\varepsilon' = 5$, not only in low frequency of 50 Hz but also in high frequency of 100 kHz. Besides, the alignment of MWCNTs in polysulfone nanofibers by means of electrospinning produces a relative permittivity of up to $\varepsilon' = 58$, with 25 vol.% of MWCNTs [29]. Generally, fillers permit loadings of 1–5 wt% for carbon black, up to 0.5 wt% for MWCNTs, and up to 4 wt% for expanded graphite, depending on the methods of dispersion [5]. Conductive fillers may give rise to substantial loss, resulting in substantial challenges in achieving long-term stability. To overcome these challenges, Goswami et al. [30] introduced a more advanced mixing scheme to permit in situ entrapment of MWCNTs in silicone while maintaining controlled dispersion. The content of MWCNTs can be increased to up to 1 wt% while still keeping a dielectric matrix. However, at 1 wt% of MWCNTs, leakage of current becomes considerable at elevated temperatures, and breakdown strength is reduced to 20 V μm^{-1}.

A recent research has shown possible restacking of the graphene layers during sample preparation by self-assembling graphene oxide (GO) on carbon nanospheres (CNS) [31]. These hybrid nanoparticles, where GO forms the shell and the CNS the core, formed a dielectric network within an carboxylated nitrile rubber (XNBR) matrix, with relative permittivity increasing from $\varepsilon' = 25$ (raw XNBR matrix) to $\varepsilon' = 400$ (0.75 vol.% of nanohybrids) at 10^3 Hz, with dielectric losses still being low ($\varepsilon'' < 0.65$ at 10^3 Hz). In the above-mentioned studies, the contents of conductive fillers are almost lower than the percolation threshold. With the increase of permittivity, the breakdown strength decreases and dielectric loss increases, which is unfavorable to the usage of DEs. Nevertheless, some authors have observed increases of breakdown strength and decreases of dielectric loss [32, 33], which was ascribed to dispersion or agglomeration methods during composites preparation. Further research and efforts are necessary for fully understanding the dielectric behavior of nanocomposites and the role of the filler–polymer interface [10].

Dielectric Filler/Elastomer Composites It is a popular perception that mixing dielectric fillers of high permittivity can increase the dielectric permittivity of a given elastomer while maintaining its dielectric characteristics. The types of common fillers include organically modified montmorillonite [34], titanium dioxide TiO_2 (rutile: $\varepsilon' = 114$, anatase: $\varepsilon' = 31$) [35], $BaTiO_3$ ($\varepsilon' = 6000$) [36], boron nitride (BN) [37], and so on, while PDMS, polyurethane (PU), styreneic block copolymers (SEBS), ethylene/vinyl acetate (EVA) for elastomer matrix. A newer work has reported that using calcium copper titanate, $CaCu_3Ti_4O_{12}$ (CCTO) ($\varepsilon' > 3000$), whose frequency and temperature-independent giant dielectric constant of 10^4 (DC to 10^6 Hz), as a filler in PDMS of volume fractions up to 5.1 vol.%, can increase their permittivity while maintaining dielectric loss value and elastic modulus, showing a remarkable combination of dielectric and mechanical properties [38]. In general, silicones with $BaTiO_3$ provide similar dielectric permittivities ($\varepsilon' = 5 - 8$) compared to the TiO_2 at lower amounts [5]. However, the highest permittivity of $\varepsilon' = 300$ at 10 wt% of $BaTiO_3$ with a dielectric loss of only $\tan\delta = 0.22$, has been reported by Zhang et al. [39]. On the other hand, this system might have trouble in exhibiting long-term stability because of adding 35% plasticizer. A large number of lead zirconate-titanate (PZT) PDMS composites have been investigated for use in piezolectronics, showing relatively high permittivity, for instance, $\varepsilon' = 28$ with 40 vol.% of PZT [40] and $\varepsilon' = 160$ with 32.5 vol.% of PZT [41]. Other lead magnesium niobate lead titanate (PMN-PT) and PZT PDMS systems have been recently investigated, providing an exciting promise in wave harvesting applications using DEs due to their excellent overall properties [42].

In addition to the studies mentioned above, some researchers have proposed a novel filler with core–shell structure [43–46], which shows great advantages in increasing permittivity of polymers. Zhang and coworkers [47] introduced SiO_2-PDDA-GO (Figure 8.3) into high-temperature vulcanization silicone rubber (SR) using solution blending, and prepared SiO_2-PDDA-GO/SR dielectric elastomer composites. Experimental results show that this method can realize GO large surface coating on surface of SiO_2 to prevent GO from self-agglomerating. GO core–shell hybrid particles/SR composites were obtained without in situ thermal reduction because poly(diallyldimethylammonium chloride) (PDDA) can reduce GO, thereby simplifying the experimental scheme as well as affording environmental protection. The relative permittivity of SiO_2-PDDA-GO/SR dielectric composites increases to 21.53 with 60 wt% of SiO_2-PDDA-GO at 100 Hz, which is 11.6 times than that of SR, and dielectric losses remain at a low level (Figure 8.4), showing satisfactory properties combining the two values.

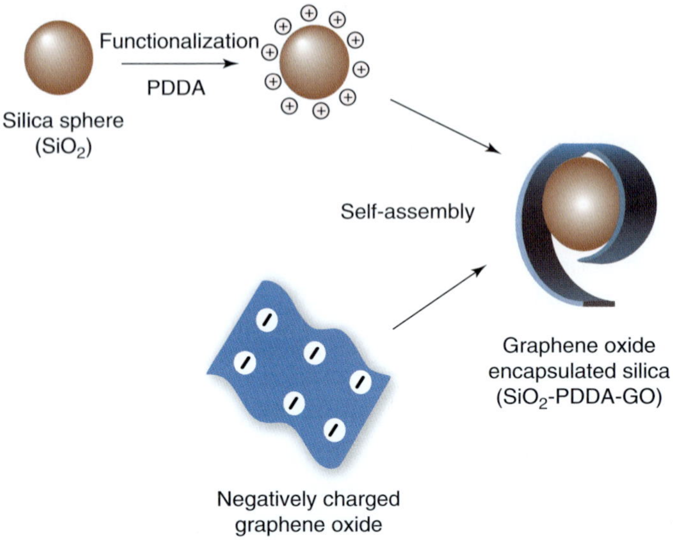

Figure 8.3 Fabrication process of SiO_2-PDDA-GO core–shell hybrid particles. *Source:* Wang et al. 2016 [47]. Taylor & Francis.

8.3.1.2 Elastomer Blends

Elastomer blends are similar to the composite approach, using two polymers, miscible or immiscible, rather than fillers and matrix, to obtain novel materials with satisfactory properties [5]. Blending a commercial PDMS (TC-5005 A/B-C, BJB Enterprises Inc., USA) with little contents (1–6 wt%) of a highly polarizable conjugated polymer, poly(3-hexylthiophene) (PHT), gave rise to attractive materials with a relatively large increased permittivity but a small increased dielectric loss and an unexpectedly decreased Young's modulus [48]. The high morphological integration between matrix and PHT results in a partial hindrance of the cross-linking behavior of the curing agent, reducing therefore the cross-linking degree of PDMS matrix. That is the reason why such reduction in elastic modulus, for example, from 100 to 20 kPa at 1 wt%, was observed. The dielectric loss increased with the increase of the PHT concentration, resulting in a progressive decrease in electro-mechanical performance. When the content of PHT was 1 wt%, the properties as a whole were noted to be perfect. As for blends of PDMS/PU, simply mixing these two elastomers at the volume rate of 60/40 gave rise to an amazing phenomenon where the permittivity of blends exceeded the permittivity of PDMS and PU at low frequencies, due to a strong interfacial polarization (Maxwell–Wagner–Sillars relaxation) [49]. Elastomer blends can increase the permittivity of a given elastomer matrix, and dielectric losses may increase at the same time. Madsen et al. have proposed adding chloropropyl-functional silicone oil (LMS-152 from Gelest Inc.) to the commercial silicone elastomer LR3043/50, showing an increase in permittivity $\varepsilon' = 4.4$ while there was a decrease in dielectric loss at 30 phr of LMS-152 [50]. Furthermore, LMS-152 has been proved to decrease the dielectric loss of a TiO_2-containing silicone elastomer. Thus, this study provided a positive

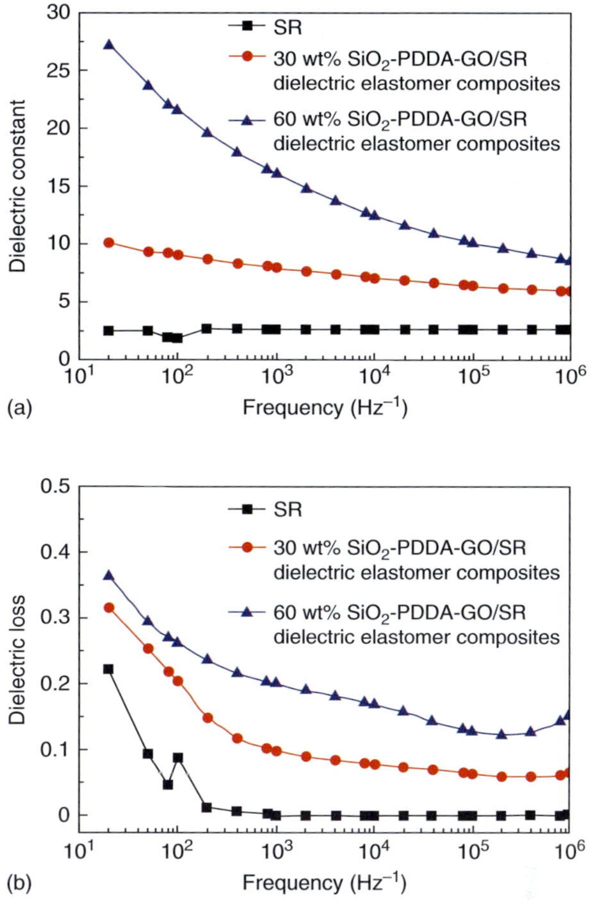

Figure 8.4 Variation curves of dielectric constant and dielectric loss for SR and SiO_2-PDDA-GO/SR dielectric elastomer. *Source:* Wang et al. 2016 [47]. Taylor & Francis.

step forward for improving the performance of DEs, because its experimental results supported the possibility of increasing permittivity while not increasing dielectric losses.

8.3.1.3 Chemical Modification

When it comes to improving the dielectric property of elastomers, the direct approach is the synthesis of new molecular structures or chemical modification of existing elastomers. The former is so difficult that attempts at it are very little, let alone success, while the latter is much more popular, and the approach is to chemically graft organic dipoles to an elastomer backbone. Molecules/chemical groups with high dipole moments increase the polarizability of the elastomer and thereby increase dielectric permittivity and energy density. This approach potentially leads to a more controlled network structure, because it does not rely on the efficient and perfect mixing of particles and blends, which is its advantage over elastomer composites and blends [5]. For example, Kussmaul et al. [51] used a push–pull dipole (*N*-allyl-*N*-methyl-*p*-nitroaniline), which possessed a vinyl functionalization, as high dielectric grafted molecule in a PDMS matrix, enabling

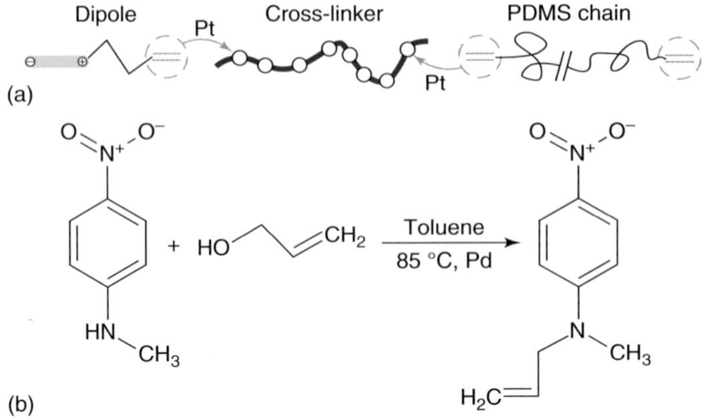

Figure 8.5 (a) Schematic representation of the direct grafting of a dipole to the cross-linker of a hydrosilylation reaction of a vinyl-terminated PDMS. (b) Push–pull dipole (*N*-allyl-*N*-methyl-*p*-nitroaniline) with vinyl functionalization. *Source:* Kussmaul et al. 2011 [51]. John Wiley & Sons.

the grafting of the dipoles and formation of the silicone in one step (Figure 8.5). The grafting of the push–pull dipole, in concentrations below 13.4 wt% to a silicone cross-linker, increased the relative permittivity from 3.0 to 5.9 at 1 kHz, and decreased the Young's modulus from 1900 to 550 kPa, obtaining a win-win result. They exploited this strategy further in their later work by changing the silicone matrix employed [52]. The grafting of *N*-allyl-*N*-methyl-*p*-nitroaniline dipoles to two commercial silicone networks in concentrations ranging from 0 to 10.7 wt% increased the relative permittivity from $\varepsilon' = 2.84$ to $\varepsilon' = 6.15$ for Sylgard 184 from Dow Corning and from $\varepsilon' = 3.17$ to $\varepsilon' = 5.57$ for Elastosil RT 625 from Wacker while decreasing the elastic modulus from 2490 to 850 kPa for Sylgard 184 and from 303 to 142 kPa for Elastosil RT 625. These authors have also modified other silicone elastomers [53]. However, these systems exhibit a decrease in breakdown strength, limiting their applications. Recent studies [54, 55] have evidenced the feasibility of increasing permittivity without decreasing breakdown strength.

Recently, Madsen et al. developed a very promising but much simpler silicone elastomer system based on chloropropyl-functional siloxane copolymers [50, 56]. These elastomers amazingly possessed a low elastic modulus (0.4–1 MPa, 75–90% lower than the utilized silicone reference), a high dielectric permittivity ($\varepsilon' = 4.7$ at 1 MHz), low dielectric losses as pure PDMS elastomers did, and a high breakdown strength (94.4 V μm^{-1}). Furthermore, the gel fractions and viscous losses maintained in the same order of magnitude as a PDMS reference elastomer, and therefore, the softness of the system did not compromise material stability [5].

In summary, with the concentrations of functional groups increasing, the permittivity of systems will increase while followed by an increase in Young's modulus and decrease in breakdown strength. However, recent researches have provided the possibility to overcome this challenge.

As for elastomer composites and blends, and chemically modified elastomers, it is a matter of great necessity to investigate long-term stability of materials, due to its close relationship to lifetime and stability over time of DE applications. Nevertheless, modified systems will most likely be more stable and reliable over time and less prone to aging effects than the others.

8.3.2 Decreasing Young's Modulus

In general, the mechanical properties of DEs can be modified easily with the use of plasticizers that can lower the mechanical stiffness of DEs. The pioneers to investigate the effect of concentrations and types of different hardeners (81-R, 81-F, and 81-VF from Suter-Kunststoffe) on the dielectric, mechanical, and electro-mechanical properties of a commercial silicone elastomer (DC3481 from Dow Corning) were Löwe et al. [57]. Zhang and coworkers have later investigated nontoxic addictives, and reported that the inclusion of 30 wt% of epoxidized soybean oil in hydrogenated acrylonitrile-butadiene rubber (HNBR) decreased the dielectric permittivity of matrix slightly from $\varepsilon' = 12.5$ to $\varepsilon' = 11.5$, but decreased Young's modulus considerably from 1.6 to 0.75 MPa [58]. The main disadvantage of using plasticizers is their volatile or migrating nature that would affect the durability of generators or the device application.

Another approach to reduce the mechanical stiffness of DEs is through the chemical modification of their molecular weight, stoichiometry, and cross-linking degree [5], similar to the chemical modification for increasing the dielectric permittivity. A recent study has reported a two-step controlled addition reaction to synthetize silicone elastomers with low Young's modulus [59]. The Young's modulus values were in the order of 20–30 kPa without adversely affecting the viscous dissipation through the development of a bimodal network.

There are numerous studies that can prove that the stiffness of DEs can be effectively reduced. But they tend to exhibit strong viscoelastic effects that might lead to hysteresis and relaxation, thereby decreasing breakdown strength values and reliability of DEGs.

8.3.3 Complex Network Structure

Finally, in this section we introduce systems with complex network structures-interpenetrating polymer networks (IPNs) and bimodal networks. The two subsections above are related to increasing dielectric permittivity and decreasing Young's modulus, respectively. But, nowadays, the aim of developing these structures is mainly at locking pre-strain. As there is relatively lack of research about these two approaches, we will just offer a brief conceptual introduction.

An IPN consists of at least two polymers that are cross-linked in situ, to develop interlaced networks. IPNs have different and often superior properties compared with the respective homo-polymers, which is ascribed to stable bulk and surface morphologies created by the interlaced networks. There are two main routes to prepare IPNs: sequentially and simultaneously. For the former, the IPN is formed in two successive steps, where initially a homo-network of polymer (I) is formed through cross-linking. This network is then swollen with monomer or polymer (II), cross-linker, and an initiator, and the IPN is formed either through in situ polymerization or cross-linking. The latter consists of forming two independent networks simultaneously. This is done by either cross-linking two sets of polymers or through polymerization and cross-linking two monomers through noninterfering orthogonal reactions [60, 61]. Brochu et al. prepared an all-silicone-based IPN using a soft room temperature vulcanizing (RTV) (Elastosil LR3003/70) silicone, which was used to preserve pre-strain in the host elastomer, fabrication process as shown in Figure 8.6 [62]. The second network high temperature vulcanizing (HTV) acts as a support for the pre-strained RTV, thereby eliminating external pre-strain structures. Another interesting feature of IPNs are self-healing properties, just as muscles, which allow broken materials

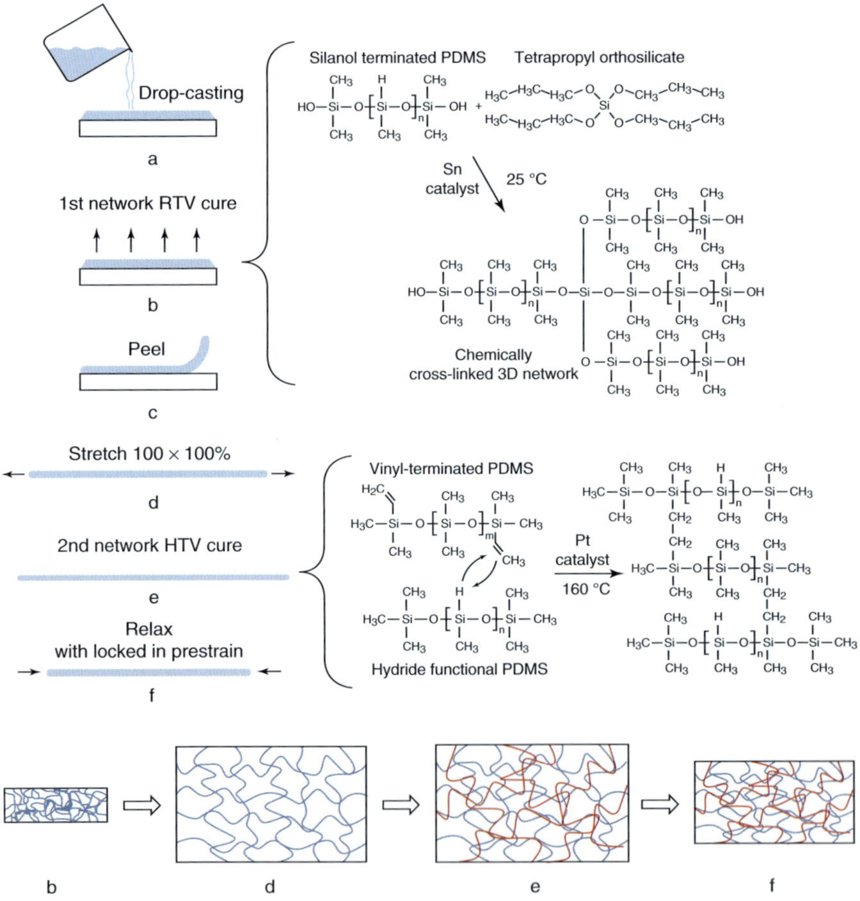

Figure 8.6 Schematic illustration of the formation of an all-silicone IPN with locked-in pre-strain; (a) solution is drop cast onto treated glass, (b) solvent is allowed to evaporate and the RTV silicone cures at room temperature, (c) film is peeled off the glass substrate, (d) film is stretched biaxially by 100% × 100%, (e) HTV silicone is cured at 180 °C for 30 min, and (f) film is relaxed and some pre-strain is preserved. *Source:* Brochu et al. 2013 [62]. IOP.

that have suffered dielectric breakdown to be repaired after moderate heating for a few hours [63, 64].

Bimodal networks, which are composed of a mixture of very short and relatively long chemically identical polymer chains, can act as a tool to enhance the mechanical properties of DEs, like tear and ultimate strength [5]. Bejenariu et al. have reported that compared with homogenous bimodal networks, heterogeneous bimodal networks had more advantages as DEs, like a lower Young's modulus without compromising other properties such as network stability and viscous loss, which is difficult to achieve by traditional methods [59, 65].

Most studies on IPNs as DEs concentrate on locking in pre-strain, and other properties like self-healing and the field of IPNs needs to be investigated.

In conclusion, approaches to improve dielectric, mechanical, and electromechanical properties of elastomers from the perspective of materials have been introduced. The first subsection is aimed at increasing dielectric permittivity, while the second one for Young's modulus and the third one for pre-strain. Currently, compared with the other two aspects, studies about locking in pre-strain are less and novel. For more academic researches, you can read reviews [5, 10], and articles of Madsen et al., Carpi et al., Zhang Liqun et al., Dang Zhimin et al., Pelrine et al., O'Brien et al., and so on.

8.4 Circuits and Electro-Mechanical Coupling Methods

The efficiency of DEs is given by the ratio of generated work to input energy, which is an important parameter to evaluate the performance of DEs. In general, the overall efficiency (η_t) depends on not only the electrical and mechanical efficiencies but also the electric driver circuit (η_d), given by [66].

$$\eta_t = \eta_e \times \eta_m \times \eta_d \tag{8.13}$$

where η_e and η_m are the electrical and mechanical efficiency of DEs, respectively, and relate directly to electrical and mechanical loss factors, respectively. η_d refers to the efficiency of the electric driver circuit, given by the electromechanical coupling efficiency (K^2), which is the electrical energy converted into mechanical work per cycle relative to the electrical energy applied per cycle. For constant dielectric permittivity and constant volume, K^2 is given as follows:

$$K^2 = -2s_z - s_z^2 \tag{8.14}$$

where s_z is the thickness strain of DEs. The theoretical value of η_t is 76.9%, which is unable to achieve in practice. Equations above are deduced in actuator mode, but also applied to generators. Up to now, the highest energy density of VHB 4905 is 780 J kg^{-1}, having a conversion efficiency of 30% [21], far from the theoretical value of 1700 J kg^{-1}. Therefore, many further studies are needed to improve the efficiency of DEs. In the following text, we will introduce the efforts researchers have made on aspects such as loading methods, and pre-strain, most of which are aimed at increasing mechanical efficiency and electro-mechanical coupling efficiency.

Suo et al. have done significant work on thermodynamics equations and energy calculations of DEs [17, 19, 20] and loading methods [67, 68]. They reported significant improvements in energy density (560 J kg^{-1} with a power density of 280 W kg^{-1} and an efficiency of 27%) using equi-biaxial stretching, a mechanical loading configuration. However, viscous losses within the acrylic elastomer limits the achievement of higher energy conversion efficiency of DEGs. As mentioned above, constant field control can realize maximum energy gain. Later, Shian et al. [21] have reported the maximum energy density and conversion efficiency of the DEs (780 J kg^{-1} and 30% using equi-biaxial loading for acrylic elastomer-3 M VHB 4900, respectively) so far. However, the power density is 170 W kg^{-1}, lower than the previous reported value of 280 W kg^{-1}, but it is easy to improve the power density. The electro-mechanical harvesting scheme is shown in Figure 8.7.

In addition to enhancing breakdown field [69], pre-train also can suppress the pull-in instability [70]. Taking acrylic elastomer into account, the performance of elastomers improves with increasing degree of pre-stretching. Nevertheless, as for silicones, like PDMS [71], there is an optimum pre-stretch ratio for

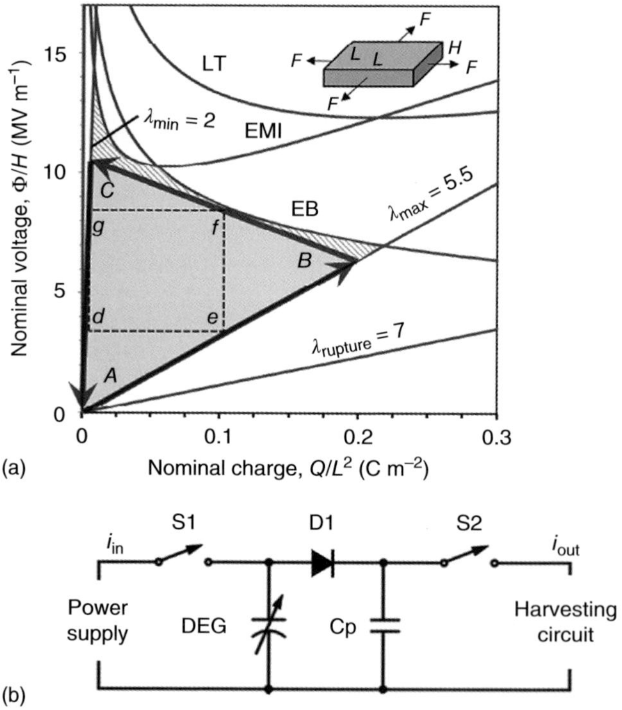

Figure 8.7 (a) The proposed electro-mechanical harvesting scheme is shown by the triangle A–B–C–A on the voltage-charge work-conjugate plane, and the rectangle d–e–f–g–d is the constant-voltage energy harvesting cycle for comparison. Also, the loci of the possible failure modes is shown by electrical breakdown (EB), electro-mechanical instability (EMI), loss of tension (LT), and rupture stretch($\lambda_{rupture}$). (b) Circuit diagram used to control the electro-mechanical cycle, where harvesting circuits block is to collect the electrical energy. *Source:* Shian et al. 2014 [21]. John Wiley & Sons.

electro-mechanical performance. The main shortcoming of the pre-strain technique is that a rigid frame must be used to maintain the tension in the DE film, thereby increasing the total weight and space of the DEGs, and eventually decreasing the effective energy density. As introduced above, IPNs can lock in pre-strain, eliminating the external frame.

In generator mode, the DE membrane must be charged for polarization by external circuit after stretched per cycle. However, due to charges being transferred to a load or being depleted by system losses, the auxiliary energy source needs regular replacement, thereby leading to complex configurations and nonautomation. How to utilize the electrical energy converted from mechanical energy for next polarization? Relevant studies on this are extremely lacking. McKay et al. [72–75] proposed a self-priming DEG circuit, that is able to replenish these charge losses from generated energy. Therefore, the external energy source no longer requires periodic replacement and the circuit no longer requires to be manually switched. The circuit diagram is shown in Figure 8.8. Shiju et al. [76] designed a self-filling circuit that can transfer portions

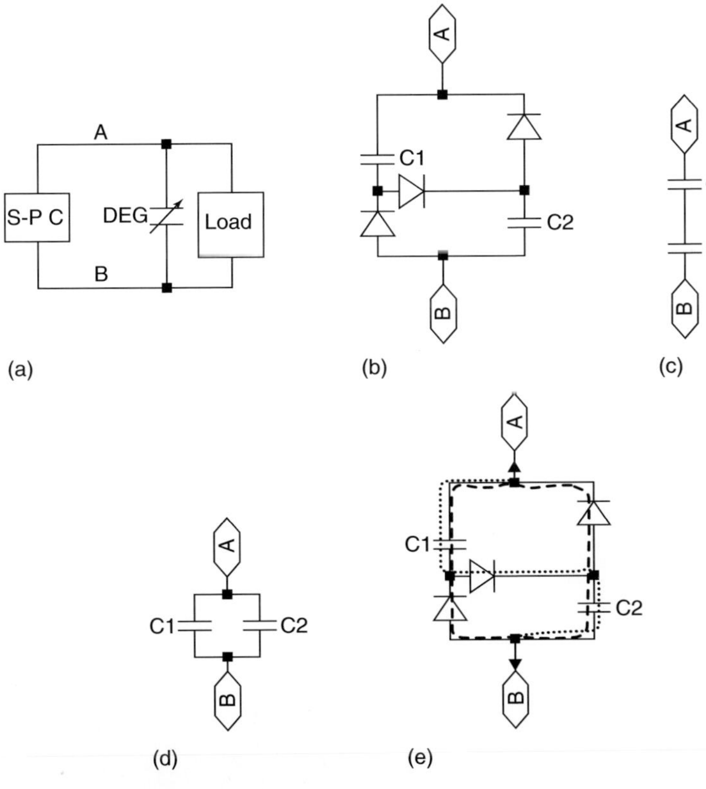

Figure 8.8 The self-priming circuit is connected in parallel with the DEG (a). A simple self-priming circuit is illustrated (b) along with its equivalent high voltage (c) and high charge (d) forms. With reference to (e), the circuit is in a high charge form when current flows along the dashed path and high voltage form when it follows the dotted path. *Source:* McKay et al. 2010 [72]. IOP.

of electrical energy generated back to the membrane as polarization charge, supplementing the charge loss.

In this section, we introduce the approaches to improve DEGs from the aspects of loading methods, pre-strain, eliminating external pre-strain frames, and circuits, most of which are intended to increase the mechanical and electromechanical coupling efficiencies.

8.5 Examples of Dielectric Elastomer Generators

In the past few years, DEGs have not only been tested in laboratories, but have also been tested in the actual environment. Indeed, DEs have been demonstrated to harvest energy ranging from human walking, ocean waves, flowing water, blowing wind, and pushing buttons [77]. For example, Suo et al. [78] have developed an experiment to monitor energy flows in DEGs and measure energy generation and efficiency of DEGs. A dielectric membrane of VHB 4910 in inflatable structure, coated with carbon grease as compliant electrodes showed an energy generated density of 102 $\mathrm{mJ\,g^{-1}}$ per cycle with electrical mechanical energy conversion efficiency of 7.5%. Mistral et al. [79] designed a generator located on knee to harvest 0.1 mJ per cycle at 1 Hz from human motion, enough to supply a low consumption system. Energy harvesting from an ocean wave buoy was reported to produce 35 mJ per cycle in a laboratory test by Kornbluh and coworkers [80]. And as for the same configuration, its energy production exceeded 25 J with energy density above 0.1 $\mathrm{J\,g^{-1}}$ in laboratory. However, when tested at sea, the yield would cut by half, with efficiency of energy harvesting being 78% (shown as Figure 8.9) [12]. Nevertheless, the highest energy density was achieved by a shoe generator (Figure 8.10) [12], as high as 0.3 $\mathrm{J\,g^{-1}}$. and the efficiency of shoe generators can reach 33%.

Some examples of DEGs have been introduced briefly above, of which, so far, the highest energy density is 300 $\mathrm{J\,kg^{-1}}$ attained by shoe generators. These studies prove that DEs as generators can convert mechanical energy to electrical

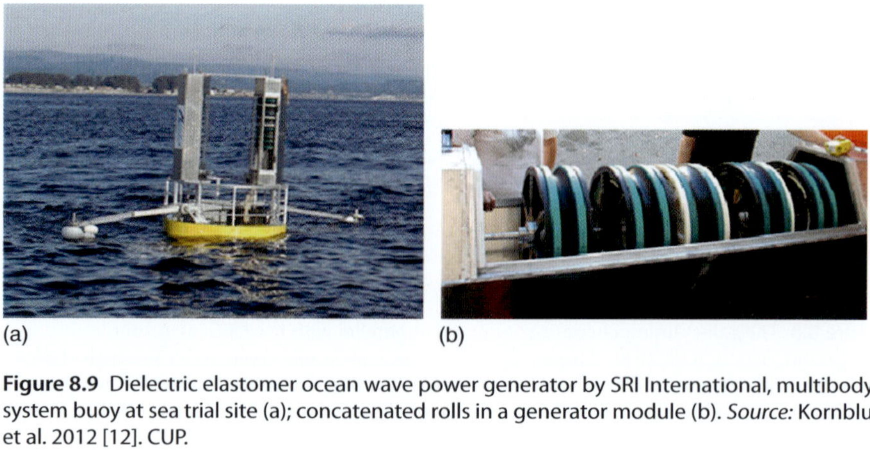

(a) (b)

Figure 8.9 Dielectric elastomer ocean wave power generator by SRI International, multibody system buoy at sea trial site (a); concatenated rolls in a generator module (b). *Source:* Kornbluh et al. 2012 [12]. CUP.

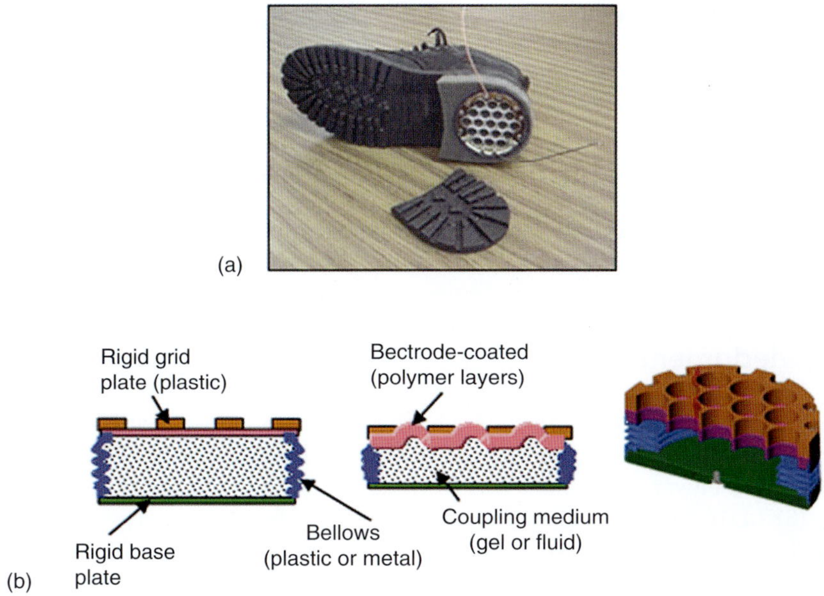

Figure 8.10 Shoe generators of dielectric elastomers from motion: photo of the device installed in boot (a) and a cross-section of the device (b). *Source:* Kornbluh et al. 2012 [12]. CUP.

energy in great scale, for example, waves and wind, or even in small scale, for example,motion.

8.6 Conclusion and Outlook

In this review, we make a relatively thorough but brief introduction of dielectric elastomer as generators from materials, electro-mechanical coupling methods, and circuits to generators. However, limited by its length, failure modes of DEs and compliant electrodes are not covered. Although studies about DEs are tremendous and emerge in an endless stream, some points deserve attention for the future. Although there are a considerable number of papers on improving materials, researchers mainly focus on one or two parameters, while other properties usually go bad as a result. And therefore, new materials may not have a competitive performance as a whole. For example, Yang et al. [81] reported a relative dielectric permittivity of $\varepsilon' = 27.5$ at 60 vol.% of $BaTiO_3$ in PMN, whereas a lower loading of 20 vol.% with $\varepsilon' = 5.2$ was optimal. McKay et al. [82] proposed a criterion for DEGs, given by,

$$F_{om}(\text{DEG}) = \frac{\varepsilon' \varepsilon_0 E_b^2}{2\varphi} \tag{8.15}$$

where E_b is the breakdown field and φ is the strain energy function of the elastomer. However, other properties such as viscous loss and long-term instability are not taken into account, which do have great effect on dielectric elastomer devices. Consequently, a new criterion has to be proposed to evaluate DEs, and it is the same with other aspects. Secondly, IPNs have great advantage over other modification methods, it may attract considerable interest of researchers. Thirdly, currently, energy harvesting of DEs has made huge progress, and Koh et al. [20] have pointed out that energy of conversion can be increased above 1.0 J g^{-1}, which is greater than the current level. Therefore, further research needs to be carried out. At last, to the possibility to eliminate external configurations for pre-strain and polarization has to be investigated further.

In summary, DEGs have provided a promising future for renewable and environmental source harvesting, and will bring interesting changes to the world.

Acknowledgments

This work was financially supported by National Nature Science Foundation of China (51425201, 51377010, and 51622701), State Key Laboratory of Electrical Insulation and Power Equipment (EIPE14204), State Key Laboratory of Power System (SKLD16Z02), the Talent Foundation of Tsinghua University for finical support (543320002), and Postgraduate Education Reform Project of Tsinghua University (53812000216).

References

1 Bar-Cohen, Y. (2009). EAP actuators for biomimetic technologies with humanlike robots as one of the ultimate challenges. *Adv. Sci. Technol.* 61: 1–7.
2 Pelrine, R., Kornbluh, R., Eckerle, J. et al. (2001). Dielectric elastomers: generator mode fundamentals and applications. Smart Structures and Materials 2001: Electroactive Polymer Actuators and Devices.
3 Carpi, F., Bauer, S., and De Rossi, D. (2010). Stretching dielectric elastomer performance. *Science* 330: 1759.
4 Brochu, P. and Pei, Q. (2010). Advances in dielectric elastomers for actuators and artificial muscles. *Macromol. Rapid Commun.* 31: 10.
5 Madsen, F.B., Daugaard, A.E., and Hvilsted, S. (2016). The current state of silicone-based dielectric elastomer transducers. *Macromol. Rapid Commun.* 37: 378–413.
6 Jager, E.W.H., Smela, E., and Inganas, O. (2000). Microfabricating conjugated polymer actuators. *Science* 290: 1540.
7 Jager, E.W.H., Inganas, O., and Lundstrom, I. (2000). Microrobots for micrometer-size objects in aqueous media: potential tools for single-cell manipulation. *Science* 288: 2335.
8 Bar-Cohen, Y. (2005). *Artificial Muscles Using Electroactive Polymers (EAP): Capabilities, Challenges and Potential.* Pasedena, CA: Jet Propulsion Laboratory, National Aeronautics and Space Administration.
9 Madden, J.D.W., Vandesteeg, N.A., and Anquetil, P.A. (2004). Artificial muscle technology: physical principles and naval prospects. *IEEE J. Ocean. Eng.* 29: 706.

10 Romasanta, L.J., Lopez-Manchado, M.A., and Verdejo, R. (2015). Increasing the performance of dielectric elastomer actuators: a review from the materials perspective. *Prog. Polym. Sci.* 51: 188–211.
11 Maas, J. and Graf, C. (2012). Dielectric elastomers for hydro power harvesting. *Smart Mater. Struct.* 21: 064006/1–12.
12 Kornbluh, R.D., Pelrine, R., Prahlad, H. et al. (2012). Dielectric elastomers: stretching the capabilities of energy harvesting. *MRS Bull.* 37: 246–253.
13 Graf, C. and Maas, J. (2010). Energy harvesting cycles based on electro active polymers. Proceedings of the SPIE, Vol. 7642, 764217.
14 Carpi, F., De Rossi, D., Kornbluh, R. et al. (2008). *Dielectric Elastomers as Electromechanical Transducers - Fundamentals, Materials, Devices, Models and Applications of an Emerging Electroactive Polymer Technology*. Elsevier. ISBN: 978-0-08-047488-5.
15 Kofod, G. (2001). Dielectric elastomer actuators. PhD thesis. Risø National Laboratory. The Technical University of Denmark, 130 pp.
16 Wissler, M. and Mazza, E. (2005). Modeling of a pre-strained circular actuator made of dielectric elastomers. *Sens. Actuators, A* 120: 184–192.
17 Suo, Z. (2010). Theory of dielectric elastomers. *Acta Mech. Solida Sin.* 23: 549–578.
18 Pelrine, R., Kornbluh, R., Pei, Q., and Joseph, J. (2000). High-speed electrically actuated elastomers with strain greater than 100%. *Science* 287: 5454.
19 Suo, Z.G., Zhao, X.H., and Greene, W.H. (2008). A nonlinear field theory of deformable dielectrics. *J. Mech. Phys. Solids* 56: 476–486.
20 Koh, S.J.A., Keplinger, C., Li, T. et al. (2011). Dielectric elastomer generators: how much energy can be converted? *IEEE/ASME Trans. Mechatron.* 16 (1): 33–41.
21 Shian, S., Huang, J., Zhu, S., and Clarke, D.R. (2014). Optimizing the electrical energy conversion cycle of dielectric elastomer generators. *Adv. Mater.* 26: 6617–6621.
22 Gallon, G., Carpi, F., De Rossi, D. et al. (2007). Dielectric constant enhancement in a silicone elastomer filled with lead magnesium niobate-lead titanate. *Mater. Sci. Eng., C* 27: 110.
23 Stauffer, D. and Aharony, A. (2003). *Introduction to Percolation Theory*, 2e, 179. London, UK: Taylor & Francis.
24 Yang, T.I., Brown, R.N.C., Kempel, L.C., and Kofinas, P. (2011). Controlled synthesis of core–shell iron–silica nanoparticles and their magneto-dielectric properties in polymer composites. *Nanotechnology* 22: 105601/1–8.
25 Kofod, G., Risse, S., Stoyanov, H. et al. (2011). Broad-spectrum enhancement of polymer composite dielectric constant at ultralow volume fractions of silica-supported copper nanoparticles. *ACS Nano* 5: 1623–1629.
26 Sahoo, B.P., Naskar, K., Choudhary, R.N.P. et al. (2012). Dielectric relaxation behavior of conducting carbon black reinforced ethylene acrylic elastomer vulcanizates. *J. Appl. Polym. Sci.* 124: 678–688.
27 Ismail, M.N. and Khalaf, A.I. (2011). Styrene-butadiene rubber/graphite powder composites: rheometrical, physicomechanical, and morphological properties. *J. Appl. Polym. Sci.* 120: 298–304.

28 Park, I.-S., Kim, K.J., Nam, J.-D. et al. (2007). Mechanical, dielectric, and magnetic properties of the silicone elastomer with multiwalled carbon nanotubes as a nanofiller. *Polym. Eng. Sci.* 47: 1396–1405.

29 Liu, H., Shen, Y., Song, Y. et al. (2011). Carbon nanotube array/polymer core/shell structured composites with high dielectric permittivity, low dielectric loss, and large energy density. *Adv. Mater.* 23: 5104–5108.

30 Goswami, K., Daugaard, A.E., and Skov, L. (2015). Dielectric properties of ultraviolet cured poly(dimethyl siloxane) sub-percolative composites containing percolative amounts of multi-walled carbon nanotubes. *RSC Adv.* 5: 12792.

31 Tian, M., Ma, Q., Li, X. et al. (2014). High performance dielectric composites by latex compounding of graphene oxide-encapsulated carbon nanosphere hybrids with XNBR. *J. Mater. Chem. A* 2: 11144–11154.

32 Nelson, J.K. (2010). Background, principles and promise of nanodielectrics. In: *Dielectric Polymer Nanocomposites* (ed. J.K. Nelson), 1–30. Berlin: Springer-Verlag.

33 Green, C. and Vaughan, A. (2008). Nanodielectrics how much do we really understand? *IEEE Electr. Insul. Mag.* 24: 6–16.

34 Razzaghi-Kashani, M., Gharavi, N., and Javadi, S. (2008). The effect of organo-clay on the dielectric properties of silicone rubber. *Smart Mater. Struct.* 17: 065035/1–9.

35 Maiolino, P., Galantini, F., Mastrogiovanni, F. et al. (2015). Soft dielectrics for capacitive sensing in robot skins: performance of different elastomer types. *Sens. Actuators, A* 226: 37–47.

36 Tomer, V. and Randall, C.A. (2008). High field dielectric properties of anisotropic polymer-ceramic composites. *J. Appl. Phys.* 104: 074106/1–7.

37 Kemaloglu, S., Ozkoc, G., and Aytac, A. (2010). Properties of thermally conductive micro and nano size boron nitride reinforced silicon rubber composites. *Thermochim. Acta* 499: 40–47.

38 Romasanta, L.J., Leret, P., Casaban, L. et al. (2012). Towards materials with enhanced electro-mechanical response: $CaCu_3Ti_4O_{12}$-polydimethylsiloxane composites. *J. Mater. Chem.* 22: 24705–24712.

39 Zhang, Z., Liu, L., Fan, J. et al. (2008). New silicone dielectric elastomers with a high dielectric constant. Proceedings of the SPIE, Vol. 6926, 692610–18.

40 Babu, I. and de With, G. (2014). Highly flexible piezoelectric 0-3 PZT-PDMS composites with high filler content. *Compos. Sci. Technol.* 91: 91.

41 Sharma, S.K., Gaur, H., Kulkarni, M. et al. (2013). PZT-PDMS composite for active damping of vibrations. *Compos. Sci. Technol.* 77: 42–51.

42 Bortot, E., Springhetti, R., and Gei, M. (2014). Enhanced soft dielectric composite generators: the role of ceramic fillers. *J. Eur. Ceram. Soc.* 34: 2623–2632.

43 Yang, D., Tian, M., Wang, W. et al. (2013). Controllable dielectric and electrical performance of polymer composites with novel core/shell-structured conductive particles through biomimetic method. *Electrochim. Acta* 87: 9–17.

44 Quinsaat, J.E.Q., Alexandru, M., and Nueesch, F.A. (2015). Highly stretchable dielectric elastomer composites containing high volume fractions of silver nanoparticles. *J. Mater. Chem. A* 3: 14675.

45 Quinsaat, J.E.Q., Nueesch, F.A., and Hofmann, H. (2013). Dielectric properties of silver nanoparticles coated with silica shells of different thicknesses. *RSC Adv.* 3: 6964.

46 Vudayagiri, S., Zakaria, S., Yu, L. et al. (2014). High breakdown-strength composites from liquid silicone rubbers. *Smart Mater. Struct.* 23: 105017.

47 Wang, M., Ning, N., Zhang, J. et al. (2016). Preparation and property of graphene oxide core-shell hybrid particles/silicone rubber dielectric elastomer composites. *Acta Mater. Compos. Sin.* 33 (6): 1192–1197. (in Chinese).

48 Carpi, F., Gallone, G., Galantini, F., and De Rossi, D. (2008). Silicone-poly(hexylthiophene) blends as elastomers with enhanced electromechanical transduction properties. *Adv. Funct. Mater.* 18: 235–241.

49 Gallone, G., Galantini, F., and Carpi, F. (2010). Perspectives for new dielectric elastomers with improved electromechanical actuation performance: composites versus blends. *Polym. Int.* 59: 400–406.

50 Madsen, F.B., Yu, L., Hvilsted, S., and Skov, A.L. (2015). Super soft silicone elastomers with high dielectric permittivity. Proceedings of the SPIE, Vol. 9430, 94301D-14.

51 Kussmaul, B., Risse, S., Kofod, G. et al. (2011). Enhancement of dielectric permittivity and electromechanical response in silicone elastomers: molecular grafting of organic dipoles to the macromolecular network. *Adv. Funct. Mater.* 21: 4589–4594.

52 Risse, S., Kussmaul, B., Kruger, H., and Kofod, G. (2012). A versatile method for enhancement of electromechanical sensitivity of silicone elastomers. *RSC Adv.* 2: 9029–9035.

53 Risse, S., Kussmaul, B., Krüger, H., and Kofod, G. (2012). Synergistic improvement of actuation properties with compatibilized high permittivity filler. *Adv. Funct. Mater.* 22: 3958–3962.

54 Madsen, F.B., Yu, L., Daugaard, A.E. et al. (2014). Silicone elastomers with high dielectric permittivity and high dielectric breakdown strength based on dipolar copolymers. *Polymer* 55: 6212–6219.

55 Zhang, L., Wang, D., Hu, P. et al. (2015). Highly improved electro-actuation of dielectric elastomers by molecular grafting of azobenzenes to silicon rubber. *J. Mater. Chem. C* 3: 4883–4889.

56 Madsen, F.B., Yu, L., Daugaard, A.E. et al. (2015). A new soft dielectric silicone elastomer matrix with high mechanical integrity and low losses. *RSC Adv.* 5: 10254–10259.

57 Löwe, C., Zhang, X., and Kovacs, G. (2005). Dielectric elastomers in actuator technology. *Adv. Eng. Mater.* 7: 361–367.

58 Yang, D., Zhang, L., Ning, N. et al. (2013). Large increase in actuated strain of HNBR dielectric elastomer by controlling molecular interaction and dielectric filler network. *RSC Adv.* 3: 21896–21904.

59 Bejenariu, A.G., Yu, L., and Skov, A.L. (2012). Low moduli elastomers with low viscous dissipation. *Soft Matter* 8: 3917.

60 Kornbluh, R.D., Flamm, D.S., Prahlad, H. et al. (2003). Shape control of large lightweight mirrors with dielectric elastomer actuation. Proceedings of the SPIE, Vol. 5051, pp. 143–158.

61 Jordi, C., Michel, S., Kovacs, G., and Ermanni, P. (2010). Scaling of planar dielectric elastomer actuators in an agonist-antagonist configuration. *Sens. Actuators, A* 161: 182–190.
62 Brochu, P., Stoyanov, H., Niu, X., and Pei, Q. (2013). All-silicone prestrain-locked interpenetrating polymer network elastomers: free-standing silicone artificial muscles with improved performance and robustness. *Smart Mater. Struct.* 22: 055022.
63 Yu, L., Madsen, F.B., Hvilsted, S., and Skov, A.L. (2015). High energy density interpenetrating networks from ionic networks and silicone. Proceedings of the SPIE, Vol. 9430, 94300T-11.
64 Yu, L., Madsen, F.B., Hvilsted, S., and Skov, A.L. (2015). Dielectric elastomers, with very high dielectric permittivity, based on silicone and ionic interpenetrating networks. *RSC Adv.* 5: 49739.
65 Bejenariu, A.G., Boll, M., Lotz, M.R. et al. (2011). New elastomeric silicone based networks applicable as electroactive systems. Proceedings of the SPIE, Vol. 7976, 79762V-18.
66 Shankar, R., Ghosh, T.K., and Spontak, R.J. (2007). Dielectric elastomers as next-generation polymeric actuators. *Soft Matter* 3: 1116–1129.
67 Huang, J., Shian, S., Suo, Z., and Clarke, D.R. (2013). Dielectric elastomer generator with equi-biaxial mechanical loading for energy harvesting. Proceedings of the SPIE, Vol. 8687, 86870Q-1.
68 Huang, J., Shian, S., Suo, Z., and Clarke, D.R. (2013). Maximizing the energy density of dielectric elastomer generators using equi-biaxial loading. *Adv. Funct. Mater.* 23: 5056–5061.
69 Kofod, G., Sommer-Larsen, P., Kornbluh, R., and Pelrine, R. (2003). Actuation response of polyacrylate dielectric elastomers. *J. Intell. Mater. Syst. Struct.* 14: 787–793.
70 Kofod, G. (2008). The static actuation of dielectric elastomer actuators: how does prestretch improve actuation? *J. Phys. D: Appl. Phys.* 41: 215405/1–11.
71 Akbari, S., Rosset, S., and Shea, H.R. (2013). Improved electromechanical behavior in castable dielectric elastomer actuators. *Appl. Phys. Lett.* 102: 071906/1–5.
72 McKay, T., O, Brien, B., Calius, E., and Anderson, I. (2010). Self-priming dielectric elastomer generators. *Smart Mater. Struct.* 19: 055025.
73 McKay, T., O, Brien, B., Calius, E., and Anderson, I. (2010). An integrated, self-priming dielectric elastomer generator. *Appl. Phys. Lett.* 97: 062911.
74 McKay, T., O, Brien, B., Calius, E., and Anderson, I. (2011). Soft generators using dielectric elastomers. *Appl. Phys. Lett.* 98: 142903.
75 McKay, T., O'Brien, B., Calius, E., and Anderson, I. (2011). Realizing the potential of dielectric elastomer generators. Proceedings of the SPIE, Vol. 7976, 79760B.
76 Shiju, E., Liu, A., Cao, J., and Ge, C. (2015). Research on energy harvesting circuit based on self-filling DEG. Proceedings of 2015 IEEE International Conference on Mechanical and Automation, 2–5 August 2015, Beijing, China.
77 Kornbluh, R.D., Pelrine, R., Prahlad, H. et al. (2011). From boots to buoys: promises and challenges of dielectric elastomer energy harvesting. Proceedings of the SPIE, Vol. 7976.

78 Kaltseis, R., Keplinger, C., Baumgartner, R. et al. (2011). Method for measuring energy generation and efficiency of dielectric elastomer generators. *Appl. Phys. Lett.* 99: 162904.

79 Mistral, C.-J., Basrourb, S., and Chaillout, J.-J. (2008). Dielectric polymer: scavenging energy from human motion. Proceedings of the SPIE, Vol. 6927, pp. 692716–692710.

80 Chiba, S., Waki, M., Kornbluh, R., and Pelrine, R. (2008). Innovative power generators for energy harvesting using electroactive polymer artificial muscles. Proceedings of the SPIE, Vol. 6927.

81 Yang, D., Zhang, L., Liu, H. et al. (2012). Lead magnesium niobate filled silicone dielectric elastomer with large actuated strain. *J. Appl. Polym. Sci.* 125: 2196–2201.

82 McKay, T.G., Calius, E., and Anderson, I.A. (2009). A dielectric elastomer actuator thin membrane rotary motor. Proceedings of the SPIE, Vol. 7287, 72870P-10.

83 Carpi, F., Kornbluh, R., Sommer-Larsen, P., and Alici, G. (2011). Electroactive polymer actuators as artificial muscles: are they ready for bioinspired applications? *Bioinspir. Biomim.* 6: 045006. (10 pp.)

9

Flexible Dye-Sensitized Solar Cells

Byung-Man Kim, Hyun-Gyu Han, Deok-Ho Roh, Junhyeok Park, Kwang Min Kim, Un-Young Kim, and Tae-Hyuk Kwon

Ulsan National Institute of Science and Technology (UNIST), Department of Chemistry, 50, UNIST-gil, Ulsan 44919, Republic of Korea

9.1 Introduction

Along with global advances in industry and technology, people are becoming increasingly aware of environmental issues. It is no longer a matter of choice to reduce the consumption of fossil fuels, whose combustion accelerates global warming and aggravates pollution [1]. At present, one of the most pressing challenges is to find ways to replace fossil fuels gradually and necessarily by sustainable resources. The total solar energy reaching the Earth is approximately 440 EJ h^{-1}, which is more than the global energy consumption per year [2]. For this reason, photovoltaics (PV), which deals with the immediate conversion of sunlight into electricity, has been a subject of active research and development. The global annual photovoltaic cell production has rapidly increased since 2000 (Figure 9.1).

Over 80% of commercial PV cells for outdoor deployment are crystalline silicon-based cells, which are classified as first-generation PV cells. Although such PV cells are widely employed and are based on proven technology, they have several limitations, which have motivated further studies on appropriate alternatives. These limitations include the following:

1) high cost associated with electrode thickness, use of rare materials, and purity grade;
2) fluctuating performance with incident light intensity and angle;
3) complicated fabrication process and heavy weight;
4) power conversion efficiency (PCE) close to the theoretical Shockley–Queisser limit [3].

Dye-sensitized solar cells (DSCs) based on a novel concept of using a semiconducting layer with a wide-bandgap nanostructure and Ru-complex sensitizers were reported in 1991 by O'Regan and Grätzel [4]. This development ushered in a new era of PV research. Subsequently, the PCE has been improved significantly, reaching 13% under a particular solar condition (AM1.5) [5].

Flexible Energy Conversion and Storage Devices, First Edition. Edited by Chunyi Zhi and Liming Dai.

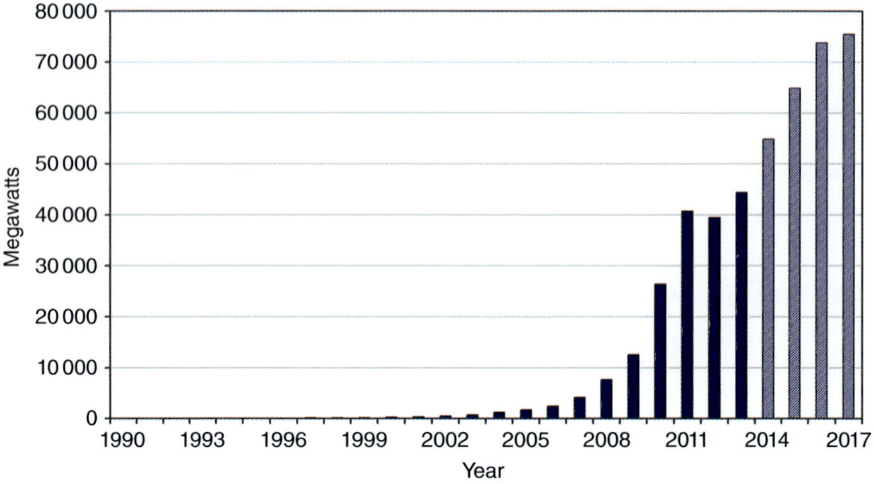

Figure 9.1 World annual solar photovoltaics cell production from 1990 to 2013, with projection to 2017 (Earth Policy Institute – www.earth-policy.org).

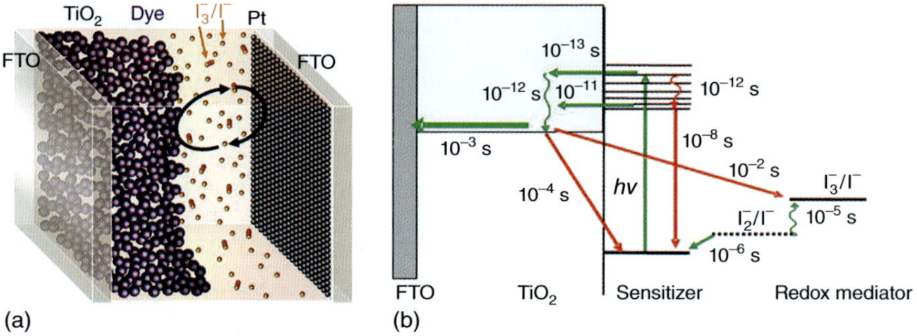

Figure 9.2 (a) Schematic of a typical dye-sensitized solar cell (DSSC) and (b) overview of processes and typical time constants under working conditions (one sun (AM 1.5, 1000 W m^{-2})) in a Ru-based DSC with I^-/I_3^- redox couples. *Source:* Roy-Mayhew and Aksay 2014 [8a] and Hagfeldt et al. 2010 [8b]. American Chemical Society.

DSCs are characterized by transparency, multiple colors, imperviousness to lighting condition, low cost, and simple fabrication. Moreover, they are flexible and lightweight, depending on the substrate used [6]. Owing to such unique features, DSCs are suitable for building-integrated photovoltaics (BIPV), automotive-integrated photovoltaics (AIPV), and portable and indoor power generators [7]. Figure 9.2 shows the structure and working principle of general liquid-type DSCs containing I^-/I_3^- redox couples [8].

Ground-state electrons in the sensitizers are excited from the highest occupied molecular orbital (HOMO) to the lowest unoccupied molecular orbital (LUMO) by photon energy and quickly injected into the conduction band of TiO_2 (10^{-11}–10^{-13} s). They keep migrating through the TiO_2 nanoparticle network until they reach the collecting contact (10^{-2}–10^{-4} s). The oxidized sensitizers are regenerated by electrons acquired from the reduced species, i.e. I^- in the electrolyte (10^{-6}–10^{-10} s). The oxidized redox species, i.e. I_3^- in the electrolyte, are reduced again at the Pt-based counter-electrode (Equation (9.1)), completing the circuit. The maximum open-circuit voltage (V_{OC}) can be determined from the difference between the Fermi level of TiO_2 and the redox potential of I^-/I_3^- couples. More precisely, an intermediate one-electron reaction (Equation (9.2)) rather than a two-electron reaction is responsible for dye regeneration [9].

$$3I^- \leftrightarrow I_3^- + 2e^- \quad (9.1)$$

$$2I^- \rightarrow I_2^{-\bullet} + e^- \quad (9.2)$$

During the operation, a number of charge losses occur through several routes, mainly via charge recombination between the photoinjected electrons and the oxidized dye molecules (10^{-4} s) or oxidized redox species, i.e. I_3^- in the electrolyte (10^{-2} s) [10].

General DSCs employ a rigid and heavy glass substrate, which limits the application range of the device and is not compatible with roll-to-roll production. Flexible devices have various advantages in terms of commercialization, such as lightness, impact resistance, and easy installation on curved surfaces. To meet the growing consumer demand, the most urgent challenge is the development of flexible dye-sensitized solar cells (FDSCs) for practical use. Toward this end, a suitable substrate is a prerequisite; conductive plastics or metallic foils are suitable alternatives to a glass substrate, each having their own advantages and disadvantages. The greatest advantage of metallic substrates is that they can withstand high temperatures, which means that most techniques that have been developed for conventional DSCs are easily applicable so that both flexibility and good performance can be achieved. However, if a metallic substrate is used at the anode side, sufficient light cannot pass through the photoactive film owing to the opacity of the substrate, leading to considerable photon energy loss. By contrast, an indium tin oxide (ITO) substrate on polyethylene terephthalate (PET) or polyethylene naphthalate (PEN), which are the most widely used plastic substrates, facilitates direct irradiation from the anode side. However, despite having various advantages for flexible devices, these materials present technical challenges. Considering their weak thermal stability, new electrode fabrication techniques are required to be developed at low temperatures (<150 °C). Moreover, a suitable electrolyte system and sensitizers should be developed. The desirable properties of the sensitizer are as follows:

1) high molar extinction coefficient and broad absorption range;
2) appropriate HOMO and LUMO levels for sufficient driving force of the charge transfer to the semiconducting metal oxide and from the electrolyte;
3) satisfactory stability over 10^8 cycles (equivalent to ~20 years of operation) [11].

For achieving completely bendable devices, the photoactive film should be thin to prevent film detachment. Therefore, the development of new sensitizers that perform well in thin photoactive films (<2 μm) would facilitate the production of highly efficient FDSCs. Electrolytes or hole transport materials for charge regeneration of the oxidized dye also have considerable scope for improvement. The major considerations for the electrolyte system in FDSCs are compatibility with sensitizers and long-term stability. Most of the previous studies on electrolytes for FDSCs have attempted to address the stability issue by optimizing solvent-free systems including ionic-liquid and quasi-solid types. However, relatively few studies have investigated the solid-state type.

The main objective of this chapter is to review the recent progress in FDSCs with regard to each component and thus present future research directions with the aim of overcoming the limitations of conventional DSCs. The remainder of this chapter is organized as follows. Section 9.2 describes electrode materials that act as the foundation of the device, including the substrate, photo-electrode, and counter-electrode. Sections 9.3 and 9.4 review sensitizers and electrolytes, respectively, for FDSCs from the viewpoint of future design goals. Finally, Section 9.5 states the overall conclusions and briefly explores the scope for future work.

9.2 Materials and Fabrication of Electrodes for FDSCs

In FDSCs, the substrate that acts as the foundation of each electrode should be conductive, transparent, bendable, and nonreactive with organic solvents and additives in the electrolyte. In general, conductive plastic substrates are commonly used owing to their light weight, low cost, and high transparency. However, they have some limitations. First, their maximum process temperature is around 150 °C, which limits the methods for coating the photoactive film and catalytic layer on the substrate. Second, plastic substrates are prone to water and oxygen penetration. Thin metal foils can overcome these limitations, but they suffer from opacity and corrosion. Thus, inexpensive, stable, and conductive carbon-based materials, such as graphene and carbon nanotubes (CNTs), have been researched as alternatives [12]. The following subsections describe the substrates and materials that have been used in electrodes for FDSCs.

9.2.1 Photo-electrode

9.2.1.1 Flexible Substrate for Photo-electrode

Flexible substrates for the photo-electrode should have important properties such as transparency and conductivity, because dyed TiO_2 film, which harvests incident light and generates photoelectron flow, is directly attached on the substrate. From this aspect, many types of substrates have been reported. Among them, ITO-coated polymer substrates, namely ITO/PEN and ITO/PET, are the most widely used substrates owing to their high transmittance and conductivity. Han et al. demonstrated ITO/PEN-based FDSCs with 4.3% PCE via ultrasonic spray deposition (USD) followed by isostatic pressing [13]. Zhang et al. demonstrated ITO/PET-based FDSCs with 3.3% PCE via hydrolysis of a titanium (Ti) precursor followed by UV-ozone treatment [14]. However, as stated above, polymer substrates cannot withstand high-temperature processes. Accordingly, complex processes have been employed for coating TiO_2 film without high thermal energy, but the results are not satisfactory. In addition, polymer substrates are prone to oxygen and moisture penetration.

For this reason, various metal substrates have been researched owing to their high thermal stability, high availability, and low sheet resistance. Their greatest advantage is that the processes for preparing TiO_2 film on the metal substrate include a wide range of methods regardless of the working temperature. In particular, Ti foil and stainless steel substrates are popular because of their low corrosiveness. Kang et al. reported high PCE (8.6%) of FDSCs using an ITO/SiO_x-sputtered stainless steel electrode coated with 10-μm TiO_2 film prepared by hydrolyzing a Titanium (IV) isopropoxide precursor [15]. As with a stainless steel electrode, a Ti foil electrode shows high performance and stability owing to its naturally formed surface-passivating oxide. Tsai et al. employed a H_2O_2-treated Ti foil coated with 17-μm TiO_2 film as a photo-electrode and compared its performance with that of an untreated counterpart. The H_2O_2-treated Ti foil enhanced electrical contact between the screen-printed TiO_2 nanoparticles owing to TiO_2 nanosheet formation, resulting in 7.1% PCE [16]. Similarly, Linnemann et al. carried out HNO_3 pretreatment of Ti foil, regardless of subsequent H_2O_2 treatment. Despite an obvious increase in surface area with H_2O_2 treatment, the best performance (6.1%) was achieved by simple HNO_3 treatment [17]. Although metal substrates show good performance and stability, they suffer from the drawback of opacity.

For this reason, they require a transparent substrate on the opposite side as well as back-illumination for light harvesting. To overcome this limitation, transparent metal meshes composed of a number of metal wires have been studied as flexible substrates. Fan et al. fabricated FDSCs (1.5% PCE) using a novel photo-electrode prepared by spraying TiO_2 colloid solution on both sides of a stainless steel mesh-based substrate under illumination by an infrared lamp, followed by sintering at 500 °C [18]. Recently, Bella et al. reported floating and transparent FDSCs composed of a 10-μm TiO_2-nanotube-array-grown Ti mesh photo-electrode, silicon cage, polymer electrolyte, and platinized Ti mesh counter-electrode. Such bifunctional FDSCs provide 5.4% PCE when front-illuminated and 5.2% PCE when back-illuminated [19]. Thus far, flexible substrate-based photo-electrodes have not performed as well as transparent conducting oxide (TCO) glass-based ones owing to their poor film quality. However, further improvement of processing techniques and substrate conditions are expected to enhance the device performance. Various studies including those introduced above are summarized in Table 9.1.

9.2.1.2 Nanostructured-photoactive Film

TiO_2 nanoparticles are widely used to prepare a photo-electrode for FDSCs owing to their good electrical and mechanical properties [22]. Typical TiO_2 pastes consist of TiO_2 nanoparticles, various additives, and organic solvents, which should undergo a high-temperature process near 500 °C to secure the interconnection between the TiO_2 nanoparticles, adhesion of TiO_2 film, and the porous structure [23]. However, plastic conducting electrodes such as ITO/PEN and ITO/PET

Table 9.1 PCE of FDSCs with different substrates for photo-electrode.

Substrate	Photoactive film preparation	Post-treatment	η (%)	References
ITO-PEN	Ultrasonic spray deposition	Cold isostatic pressing	4.3	[13]
	Doctor blading		4.0	[20]
	Electrophoretic deposition	Static mechanical compression	4.4	[21]
Stainless steel	Doctor blading	470 °C sintering	4.2	[15a]
	Doctor blading	600 °C sintering	8.0	[15b]
Ti foil	Screen-printing	500 °C sintering	7.1	[16]
	Screen-printing	500 °C sintering	6.1	[17]
Metal mesh	Air spray deposition	500 °C sintering	1.5	[18]
	Anodization growth	450 °C sintering	5.4	[19]

cannot withstand such environment [24]. In this regard, Huang et al. fabricated ITO/PEN-based FDSCs with submicrometer-sized TiO_2 building blocks, which were pre-sintered at 650 °C before being constructed on the substrates. Since sufficient thermal energy resulted in good connection between the TiO_2 nanoparticles, FDSCs based on TiO_2 building blocks provided enhanced PCE compared to that of FDSCs with normal 18 nm TiO_2 nanoparticles from 4.9% to 6.6%, which was further increased to 7.5% after $TiCl_4$ post-treatment of TiO_2 building blocks (Figure 9.3) [25].

Thus far, numerous studies have focused on the preparation of TiO_2 pastes with appropriate properties to produce efficient films on the plastic substrate. When preparing the paste, there are several important considerations: (i) interconnection of TiO_2 nanoparticles for smooth charge migration, (ii) adhesion between the TiO_2 film and the substrate for mechanical stability, and (iii) viscosity control for film deposition. The most commonly used pastes for low temperature process are mixtures of TiO_2 nanoparticles, additives, and solvent without organic binders [26]. Various compositions with additives were actively investigated, including Ti

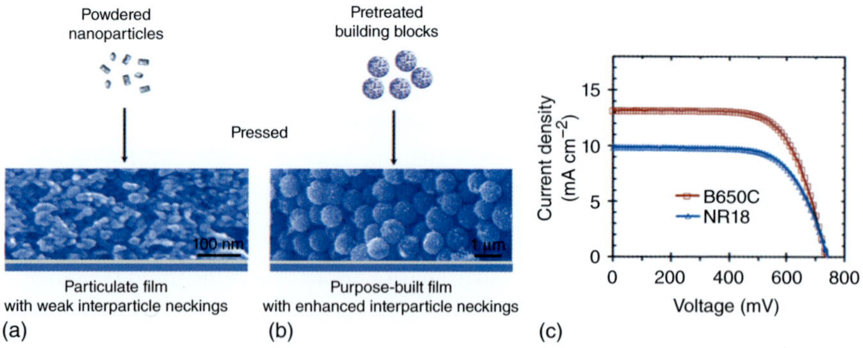

Figure 9.3 Schematic images of TiO_2 film construction from (a) only TiO_2 nanoparticles (NR18) and from (b) suggested building blocks (B650C), and (c) their J–V curves. *Source:* Huang et al. 2012 [25]. AIP.

precursors, polymers, and composites such as physical or chemical glue, to realize appropriate properties with low-temperature or room-temperature sintering [14, 27]. For example, only commercial P25 TiO_2 paste showed severe cracks in the TiO_2 film, but significant improvement was achieved by adding (Sb,In)SnO_2 to enhance the interconnection between the TiO_2 nanoparticles [27c]. The device performance was improved from 4.5% to 6.6% (Figure 9.4).

In addition, Weerasinghe et al. showed the rheological behavior of TiO_2 pastes by controlling the water percentage (Figure 9.5), and their films on ITO/PEN have different mechanical properties depending on the ratio of water, hydrochloric acid, and ammonia, as shown in Table 9.2, with PCEs varying from 4.2% to 4.9% [28]. The authors attributed enhanced device performances to increased

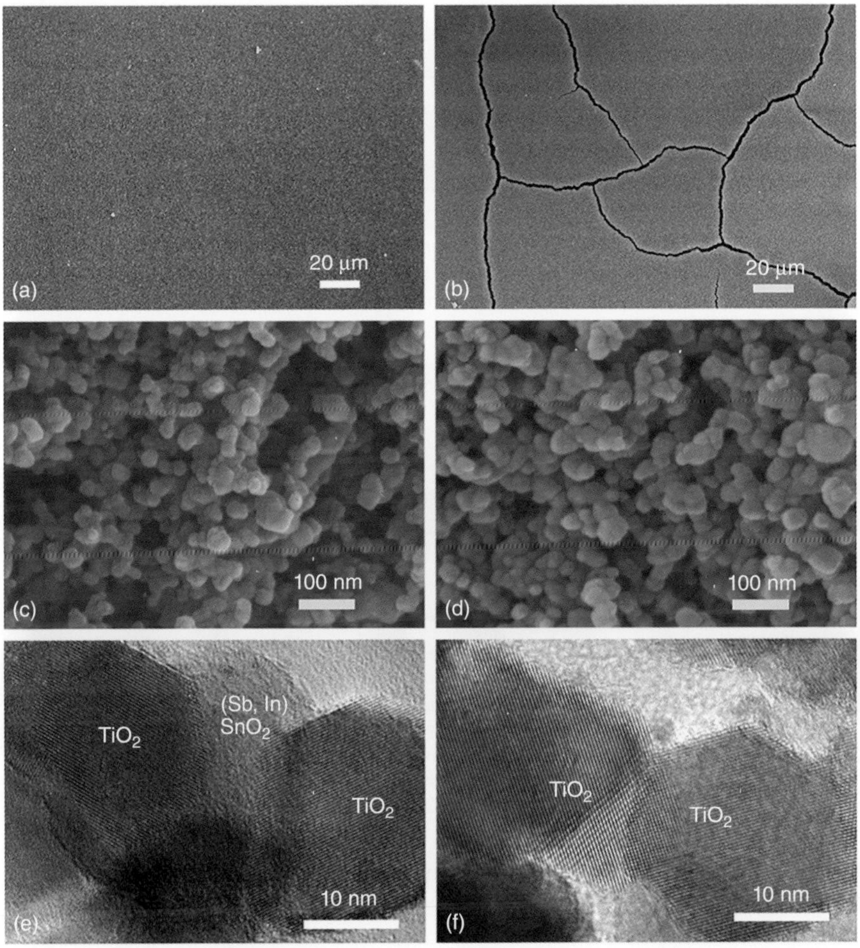

Figure 9.4 Images of (Sb,In)SnO_2-P25 ((a,c) SEM, (e) TEM) and only P25 (b,d – SEM, f – TEM) films. Photographs of bending property of SnO_2-P25 and only P25 on the plastic substrate. The effect of (Sb,In)SnO_2 additives on the device performance: (i) *I–V* curves and (j) Incident photon to current efficiency (IPCE) spectra. *Source:* Fu et al. 2013 [27c]. Elsevier.

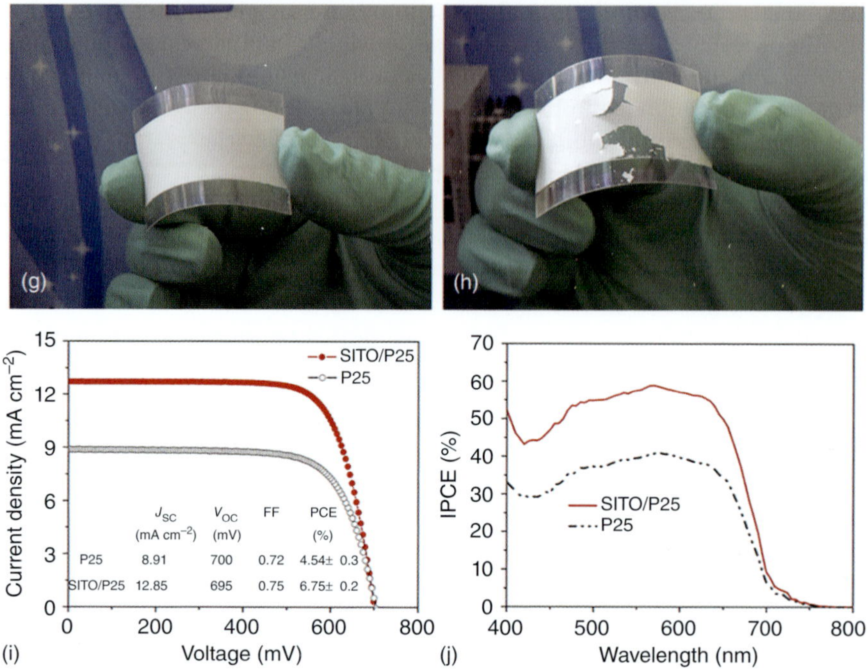

Figure 9.4 (*Continued*)

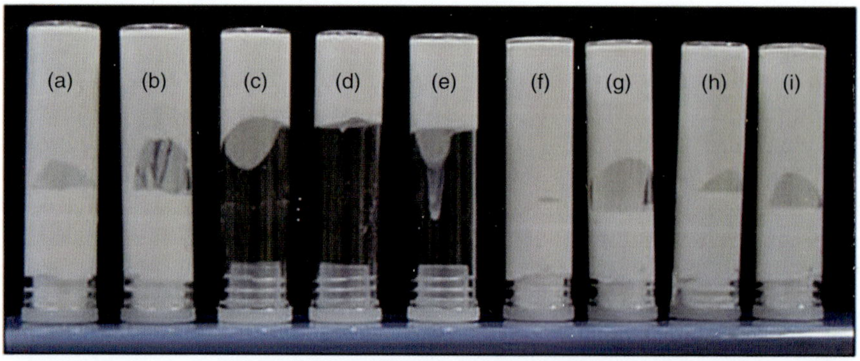

Figure 9.5 Image of the rheological behavior of ethanol-based TiO_2 paste containing various water percentages: a = 0%, b = 10%, c = 10%, d = 20%, e = 30%, f = 40%, g = 60%, h = 80%, and i = 100%. *Source:* Weerasinghe et al. 2010 [28]. Royal Society of Chemistry.

inter-particle connectivity, resulting from the flocculation of TiO_2 particles in paste due to decrease in zeta potential.

Moreover, to compensate the insufficient sintering in the low-temperature process, other treatments, such as compression of TiO_2 film, UV-assisted curing, and microwave irradiation, have been proposed [13, 29]. Among them, the pressing method is widely used for thermal-annealing-free processes, as they are simple and efficient. Figure 9.6 shows the huge difference in connectivity between TiO_2

Table 9.2 Averaged normal and steady state lateral forces of films prepared by different slurries [28].

Ethanol-based TiO_2 slurries for photoactive film	Maximum normal force (μN)	SD (σ)	Steady-state normal force (μN)	SD (σ)
Nontreated	472	95	333	62
5% Water	738	54	394	104
10% Eater	935	58	560	46
20% Water	1065	65	709	97
30% Water	1479	111	1064	77
0.025 M HCl	578	44	470	38
0.05 M HCl	1060	87	614	66
0.1 M HCl	2793	105	952	81
0.025 M ammonia	467	71	348	62
0.05 M ammonia	480	62	375	56
0.1 M ammonia	487	76	367	54

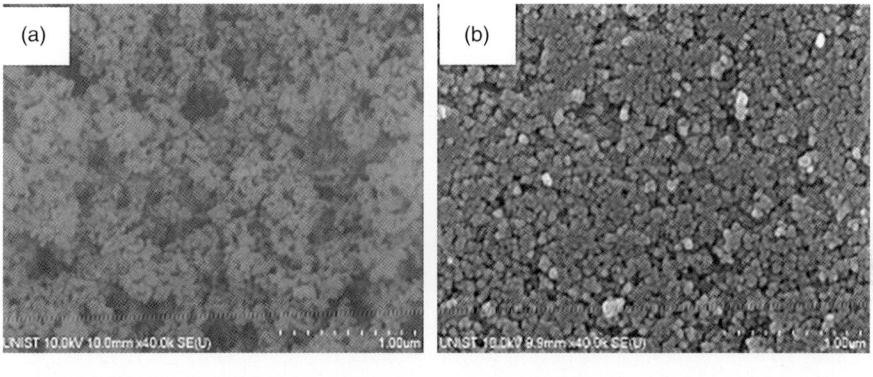

Figure 9.6 SEM image of TiO_2 layer by ultrasonic spray deposition (a) before and (b) after pressing. *Source:* Han et al. 2015 [13]. Nature Publishing Group.

nanoparticles before and after isostatic pressing. Yamaguchi et al. reported the gradual improvement of photocurrent density (J_{SC}) of ITO/PEN-based FDSCs depending on the applied pressure (Figure 9.7), suggesting the importance of interconnection between TiO_2 nanoparticles [29a].

Binder-free TiO_2 colloids normally show low viscosity, which is not suitable for large-area fabrication or mass production with roll-to-roll processes. To address this issue, several alternative deposition techniques, such as electrophoretic deposition (EPD) [30], electro-spray deposition (ESD) [31], USD [13], and chemical vapor deposition (CVD) [32], have been developed. In EPD, the flexible substrate is immersed in an electrically charged TiO_2 dispersion and the counter-electrode is connected. A voltage is applied to form the TiO_2 layer, and its thickness can be easily controlled according to the time of application. ESD also

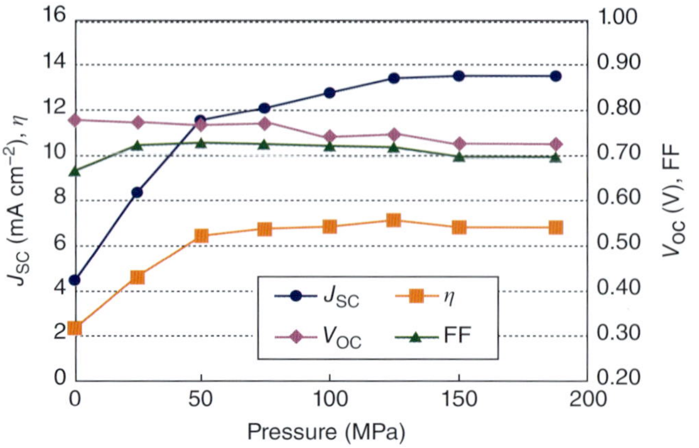

Figure 9.7 Relationship between the pressure applied to the TiO_2 film and the performance of the fabricated FDSC. *Source:* Yamaguchi et al. 2010 [29a]. Elsevier.

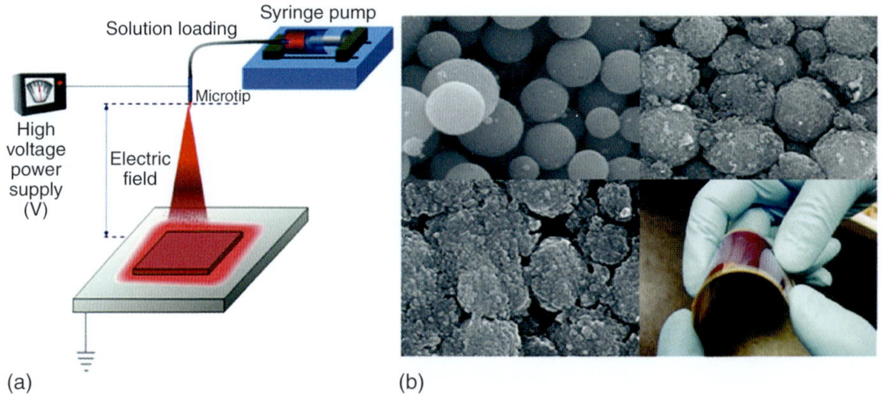

Figure 9.8 Electro-spray deposition (ESD) illustration (a) and SEM images of TiO_2 beads deposited by ESD and completed flexible photo-electrode (b). *Source:* Lee et al. 2012 [31]. American Chemical Society.

uses the charged solution. During ESD, the TiO_2 colloidal droplets are charged for their atomization in the presence of the electric field between the substrate and the solution. The high electric field facilitates the rapid solvent evaporation and thus the formation of TiO_2 beads (Figure 9.8). Wang et al. developed double-sided and TCO-free FDSCs that was consisted of a ZnO nanowire-coated metal-wire/Pt-wire/PET film. With CVD method, they could obtain well-aligned and high-density ZnO-nanowire arrays, fully covering the surface of metal-wire [32b].

These voltage-applied methods and CVD processes are relatively free for paste control, provide highly uniform film, and easily control the thickness of large-area. However, these processes require specific conditions, such as strict solution condition and extremely high applied voltage. Especially, CVD process requires high vacuum condition, meaning the adverse price competitiveness.

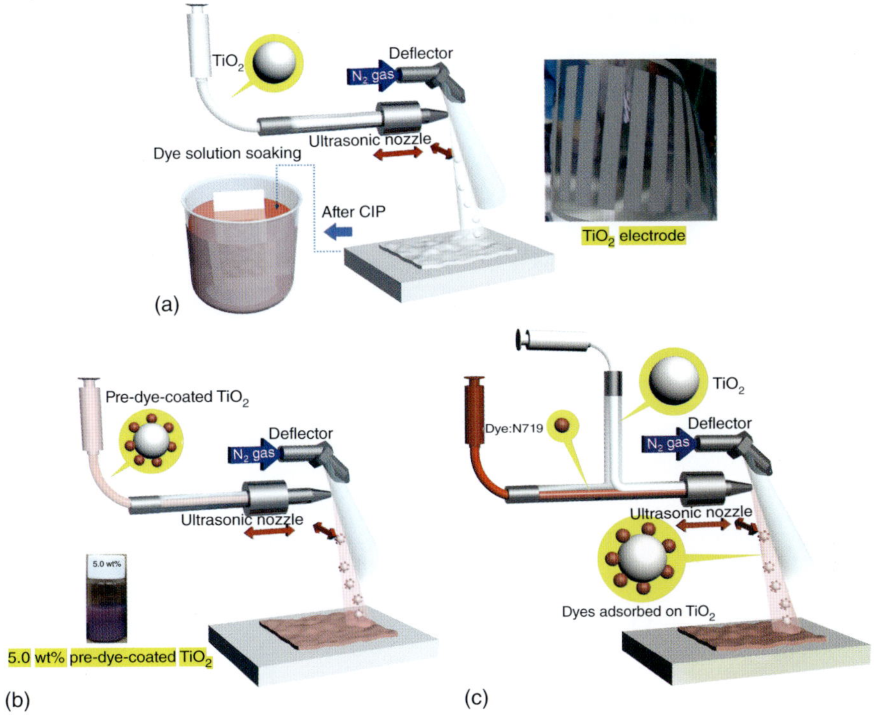

Figure 9.9 The fabrication of FDSCs using (a) a combination of dye soaking and ultrasonic spray technology, (b) pre-dye-coating method, and (c) co-deposition method. *Source:* Han et al. 2015 [13]. Nature Publishing Group.

Recently, USD has been emerging as another direct deposition method of TiO_2 colloids. The most noticeable advantages of USD are the formation of uniform droplets by ultrasonic wave during spraying. In addition, its multi-nozzle system allows us to skip the time-consuming soaking process, which facilitates the fabrication of high quality photo-electrode within a short time (Figure 9.9). After USD, the post-treatment of the photoactive film, including pressing, can be carried out to improve the device performance. These alternative approaches are successfully developed as new fabrication techniques for FDSCs, but they have not been fully optimized thus far. By further improving the processes and developing new techniques in combination with TiO_2 paste technologies, the fabrication of efficient large-area devices could be realized in the near future. Table 9.3 lists all the photo-electrode fabrication processes introduced above.

9.2.1.3 Fiber-type FDSCs

Most FDSCs are generally of the square type as with other photovoltaic devices, because of the area efficiency for a modular array structure. For a wearable or portable electronics field, the wire and fiber-types of FDSCs have been gradually improved. Recently, Zou and coworkers reported highly flexible and efficient fiber-type DSCs with 7.4% PCE (Figure 9.10) [37]. They employed atomic

Table 9.3 The PCE of FDSCs according to the photo-electrode preparation processes.

Photo-electrode preparation	Film deposition	Post-treatment	η (%)	References
Combined sol–gel and solvothermal process, followed by pre-sintering at 650 °C	Doctor blading	100 MPa cold isostatic pressing	7.5	[25]
Hydrothermal process	Doctor blading	120 °C, 80 MPa pressing	3.3	[27a]
Degussa P25 + Titanium(IV) isopropoxide (TTIP)	Doctor blading	UV-ozone treatment	3.3	[14]
Degussa P25	Aerosol spray	KrF excimer laser (248 nm) annealing	3.3	[33]
Degussa P25	Ultrasonic spray deposition	Cold isostatic pressing	4.3	[13]
Degussa P25	Electrospray	Sintering at 500 °C	5.6	[34]
Commercial ZnO powder (20 nm)	Electrophoretic deposition	100 MPa pressing + UV-ozone treatment	4.0	[35]
Zn powder + $AlCl_3 \cdot 6H_2O$ powder	Chemical vapor deposition at 700 °C (Ar/O_2, 100 sccm/ 8 sccm)	—	1.0	[32b]
Degussa P25	Doctor blading	UV irradiation + dipping in water 100 °C	6.8	[27c]
Hydrothermal process	Doctor blading	100 MPa pressing	7.0	[36]
Hydrothermal process	Doctor blading	100 MPa pressing	8.1	[29a]

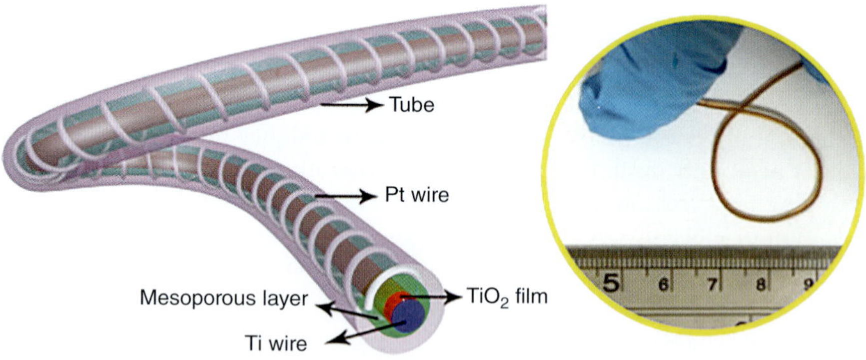

Figure 9.10 Schematic of the FDSC structure with the thin TiO_2 film deposited with ALD. *Source:* Song et al. 2016 [37]. Elsevier.

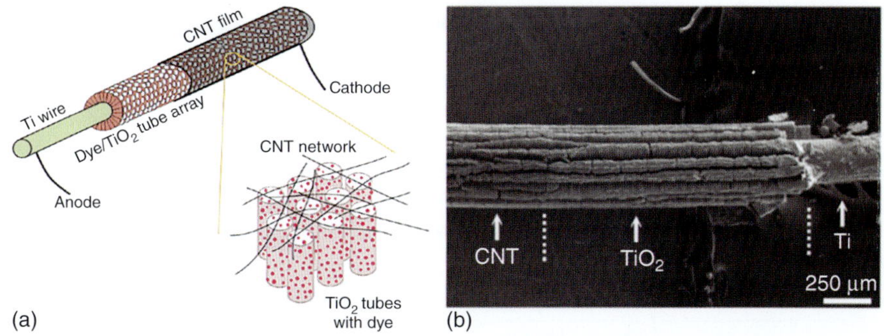

Figure 9.11 Schematic of the fiber DSC structure with (a) CNT counter-electrode and (b) SEM image of Ti–TiO_2-CNT-wire. *Source:* Zhang et al. 2011 [38]. American Chemical Society.

layer deposition (ALD) to form thin TiO_2 blocking layer on the Ti-wire. As a result, a pinhole-free and dense TiO_2 film enhanced the interconnection at the Ti-wire/mesoporous TiO_2 layer and suppressed the charge recombination at the Ti-wire/electrolyte.

A number of studies have focused on developing new electrode materials for replacing Ti/Pt wires, which are characterized by high cost and insufficient contact properties between Ti/Pt electrodes with/without bending, worsening the device performance. In addition, to retain stable performance, the Pt wires require encapsulation, which decreases their flexibility. In this regard, carbon-based materials, such as graphene and CNTs, have been considered as alternative electrode materials, owing to their good coverage, catalytic property, mechanical stability, and electrical conductivity [29b, 38]. As shown in Figure 9.11, CNT as a counter-electrode offers compact surface coverage on TiO_2 nanotube array and then minimal distance of the I_3^- ions diffusion. Other materials such as poly(3,4-ethylenedioxythiophene):poly(styrene sulfonate) (PEDOT:PSS) conducting polymers can also be introduced as counter-electrode materials in fiber-type FDSCs [39].

In summary, FDSCs face major challenges due to low-temperature processes and require new substrate materials to be developed. The fabrication processes are mainly responsible for enhancing device properties. The device performances that are comparable with those of conventional glass-based DSCs and commercial mass production can be realized by controlling the precursor colloidal properties and by developing new fabrication processes and electrode materials. Further, new types of device architecture allow for wider application, such as remote-controllable and portable electronic devices.

9.2.2 Counter-electrode

Flexible counter-electrodes should have both good catalytic activity toward the redox couples in the electrolyte and chemical stability with the electrolyte. In addition, high transmittance is needed for isotropic light harvesting. For this

Table 9.4 PCE of FDSCs with different counter-electrode preparation processes.

Counter-electrode (CE) catalyst/Substrate	Preparation method	R_{CT} (Ω)[a]	η (%)	References
Pt/ITO-PEN	Sputtering	—	8.1	[29a]
	Electrodeposition	2.2	6.5	[40]
	Electrophoretic deposition	2.0	5.8	[41]
	Atomic layer deposition	10.0	3.7	[42]
PEDOT/graphene/PET	Spin coating	—	6.3	[43]
Carbon/ITO-PET	Roll-to-roll	23.0	4.2	[44]
Carbon nanofiber/ITO-PET	Doctor blading	0.5	5.0	[45]
Activate carbon/graphite sheet	Doctor blading	1.2	6.5	[46]
PANI/flexible graphite	In situ chemical polymerization	3.9	7.4	[47]
PPy	Self-assembly	5.0	5.3	[48]
Pt/Ni sheet	Chemical deposition	0.7	7.3	[49]
Pt/Ti mesh	Electrodeposition	49.6	6.1	[50]
Vertical oriented graphene/stainless steel	Plasma-enhanced chemical vapor deposition	7.3×10^{-3}	5.4	[51]

a) Charge transfer resistance at the interface between the counter-electrode and the electrolyte.

purpose, numerous materials and catalysts have been developed, such as Pt-, carbonaceous-, polymer-, and metal-based ones. Table 9.4 summarizes some of the main research results about flexible counter-electrodes materials and fabrication methods.

Pt-based counter-electrodes are the most widely studied owing to their good catalytic properties. Thus far, sputtered Pt on the plastic substrate has often been used for flexible counter-electrodes. Yamaguchi et al. reported highly efficient FDSC by using a Pt sputtered-ITO/PEN with a 8.1% PCE [29a]. Fu et al. reported a highly transparent (over 75% average transmittance in 400–800 nm) and bendable Pt counter-electrode prepared via constant current electrodeposition and achieved 6.53% PCE under back illumination of 100 mW cm^{-2} [40]. Despite the excellent catalytic ability, their adverse price competitiveness makes a Pt electrode unsuitable for further large-scale research or mass production. Thus far, alternative catalytic materials that are inexpensive and show good catalytic properties, such as carbon and polymer composites have been proposed. Lee et al. demonstrated the Pt- and TCO-free counter-electrode that was composed of 70% transparent PEDOT, graphene, and PET substrate. They achieved 6.3% PCE using PEDOT/graphene/PET electrode, which was comparable to that using Pt/ITO/PET counterpart (6.7%) [43]. Figure 9.12 shows the fabrication process of PEDOT/Graphene/PET electrode. Miettunen et al. used carbon gel paste-deposited ITO-PEN counter-electrode and achieved only 10% lower

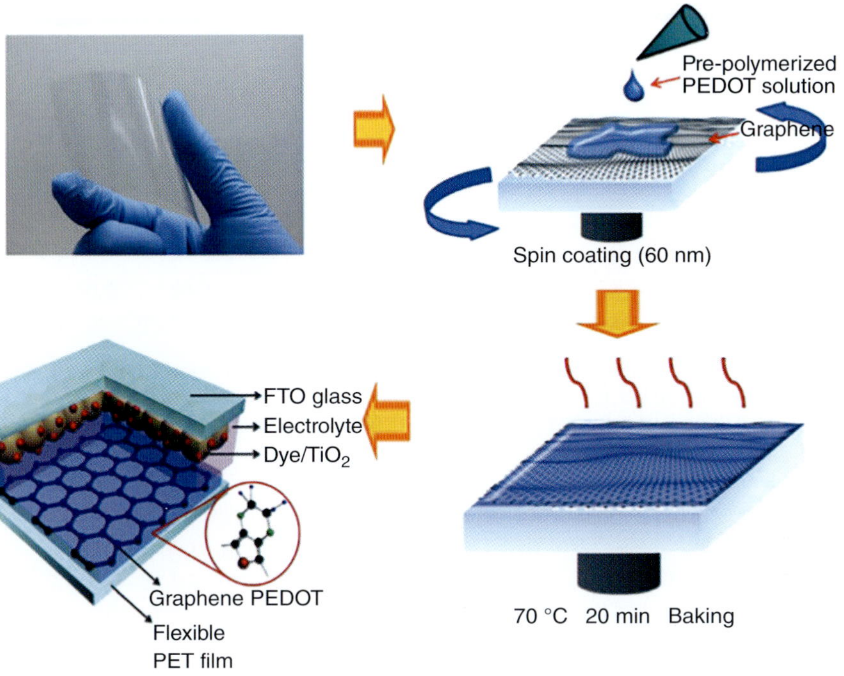

Figure 9.12 Schematic diagram of the fabrication of PEDOT/graphene/PET electrode. *Source:* Lee et al. 2012 [43]. John Wiley & Sons.

PCE (4.2%) than conventional DSCs with Pt/FTO glass counter-electrodes, suggesting the probability of application to roll-to-roll production [44].

Owing to stability and cost limitations, TCO and plastic-free counter-electrodes have been investigated. Sun et al. reported 7.4% PCE with a polymer composite counter-electrode made of conducting polyaniline (PANI) and flexible graphite as the substrate [47]. Peng et al. developed a paper-like polypyrrole (PPy) membrane via self-assembly from PPy nanotubes under vacuum conditions. The resulting PPy membrane well performed as the counter-electrode in DSCs, achieving the comparable PCE (5.3%) to that by the Pt/FTO glass counterpart [48].

Metallic counter-electrodes have been widely investigated owing to their superior thermal stability. As mentioned in Section 9.2.1.1, stainless steel and Ti are preferred substrates owing to their chemical stability with iodine-based electrolytes. Xiao et al. reported 6.1% PCE under 55 mW cm^{-2} light irradiation with highly transparent Pt-coated Ti mesh counter-electrode fabricated via the electrodeposition method [50]. C.-M. Chen et al. deposited a Pt layer on either stainless steel or nickel (Ni) by the chemical deposition method. As a result, Pt-coated Ni electrode provided lower charge transfer resistance and thus higher PCE (7.3%) than stainless steel-based devices [49]. In addition, the structure of a catalytic layer is crucial to catalytic ability. Vertically oriented graphene (VG) provides much higher electrochemical activity due to its large surface area. Yu et al. prepared VG-coated stainless steel counter-electrodes via

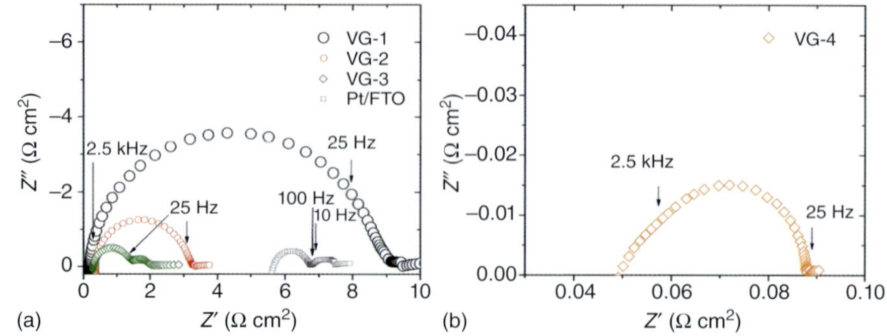

Figure 9.13 Nyquist plots of electrochemical impedance spectra measured on symmetrical dummy cells with (a) VG-1, VG-2, VG-3, Pt/FTO and (b) VG-4 (frequency sweep range: 100 kHz to 0.1 Hz, DC bias: 0 V). *Source:* Yu et al. 2013 [51]. Royal Society of Chemistry.

plasma-enhanced CVD. VG4 with the most moisture content showed extremely low charge transfer resistance ($7.3 \times 10^{-3}\ \Omega\,cm^{-2}$), corresponding to only 1% of platinized FTO glass ($5.9 \times 10^{-1}\ \Omega\,cm^{-2}$) as shown in Figure 9.13 [51].

In summary, considerable effort has been devoted toward realizing effective, low-cost alternative catalysts and substrates for FDSCs. However, the overall PCE has not reached that of conventional glass and Pt-based DSCs owing to less optimized conditions for the flexible system. For further improvement, other components such as the sensitizer and electrolyte should be investigated.

9.3 Sensitizers in FDSCs and Thin Photoactive Film DSCs

This section introduces various sensitizers that are applied to flexible and thin photoactive film-based DSCs. In addition, molecular design strategies of sensitizers for highly efficient FDSCs are discussed. A sensitizer is one of the most important parts of an FDSC because it plays a key role in generating electricity from photon energy. To enhance the performance of the sensitizer, we should design it with the following considerations: (i) absorption range, (ii) molar extinction coefficient, and (iii) systematic molecular structure. The absorption range and molar extinction coefficient are directly connected with the light harvesting ability of the sensitizer, which is especially an important factor in thin photoactive film DSCs. Moreover, the molecular structure profoundly affects the dye arrangement on the photoactive film as well as the charge recombination kinetics. Nevertheless, relatively less effort has been devoted to developing a sensitizer for FDSCs, which motivates us to examine this scarcely competitive market and raise the possibility of commercializing FDSCs.

9.3.1 State-of-the-Art Review of Sensitizers in FDSCs

Table 9.5 lists the photovoltaic parameters of flexible and conventional DSCs according to the sensitizers. Thus far, most FDSCs have employed Ru-based sensitizers because they have a broad absorption range (300–700 nm) and thus

Table 9.5 Photovoltaic parameters of sensitizers in flexible and conventional DSCs according to sensitizers.

Dye	Compound	Type	Thickness (μm)	J_{SC} (mA cm^{-2})	V_{OC} (V)	FF	η (%)	Electrolyte	References
N719	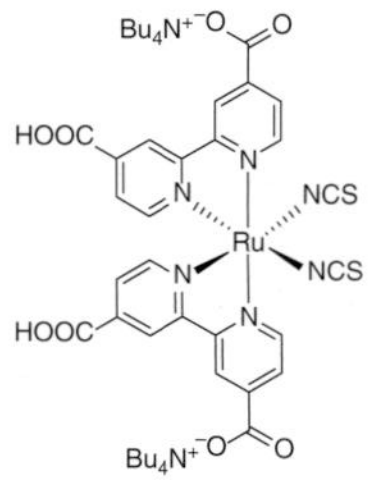	Flexible fiber-type	—	18.8	0.66	0.73	9.1	I^-/I_3^-	[52]
		Rigid	12+4	17.7	0.85	0.75	11.2	I^-/I_3^-	[53a]
Z907	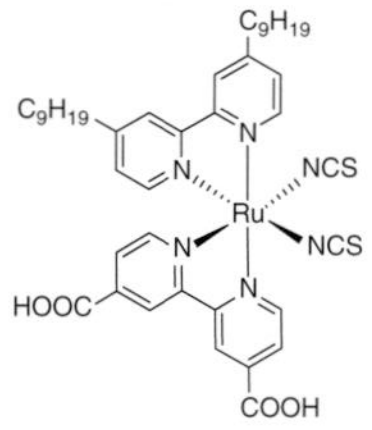	Flexible	14	16.1	0.72	0.66	7.7	I^-/I_3^-	[54]
		Rigid	26	18.2	0.75	0.73	9.9	I^-/I_3^-	[53b]
C106	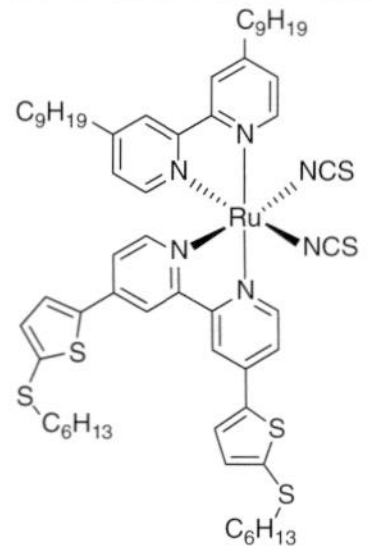	Flexible	8+1	10.9	0.66	0.69	5.0	I^-/I_3^-	[20]
		Rigid	10+5	18.3	0.75	0.77	10.6	I^-/I_3^-	[55]

(*continued*)

Table 9.5 (Continued)

Dye	Compound	Type	Thickness (μm)	J_{SC} (mA cm^{-2})	V_{OC} (V)	FF	η (%)	Electrolyte	References
D149	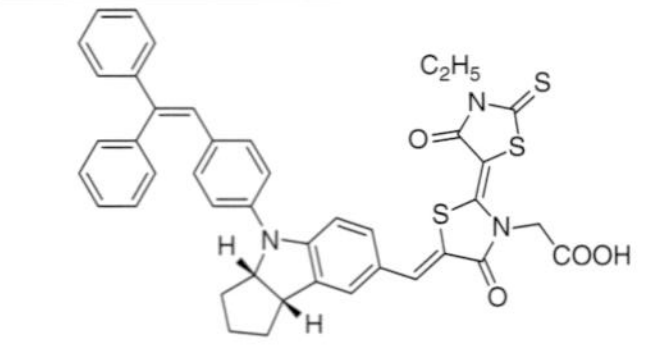	Flexible	5	8.7	0.66	0.70	4.0	I^-/I_3^-	[56]
		Rigid	14+5	19.9	0.64	0.69	8.9	I^-/I_3^-	[57]
D205	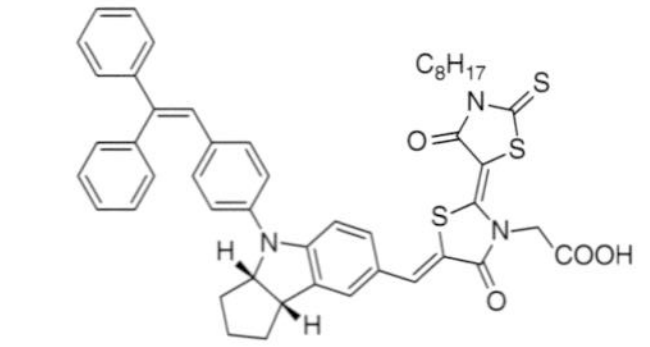	Flexible	11	6.9	0.71	0.68	3.3	Co^{2+}/Co^{3+}	[58]
		Rigid	14+5	18.7	0.71	0.71	9.4	I^-/I_3^-	[57]
MK-2	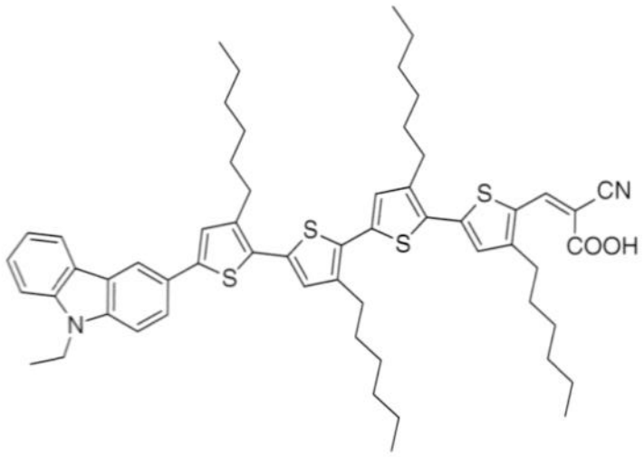	Flexible	1	8.5	0.66	0.72	4.0	IL	[20]
		Rigid	16	15.2	0.73	0.75	8.3	I^-/I_3^-	[59]

exhibit high PCE (over 10% for conventional glass-based DSCs). In particular, **N719** is a popular sensitizer for FDSCs. **N719** has been applied to various types of FDSCs including metal foil, plastic, fiber, and graphene-based types, which have recently shown high PCE (up to 9.1% for fiber-type FDSCs) [52]. For other Ru sensitizers, **Z907** has also shown a relatively high PCE (7.66%) with I^-/I_3^- couples in ITO/PEN-based FDSCs owing to the great catalytic ability of the nickel sulfide/PPy/Ti foil counter-electrode [54]. The improved electrochemical catalytic ability at the counter-electrode makes the redox circulation more active; thus, **Z907** generates more photoelectrons, thereby lowering the charge transfer resistance in the TiO_2 film. These results suggest a possibility that the PCE of FDSCs can reach that of conventional DSCs through extensive studies [53]. In contrast, **C106,** which was invented for enhancing the stability of the Ru sensitizer, showed significantly reduced PCE (5.0%) when applied to ITO/PEN-based FDSCs compared with the conventional **C106** DSCs (10.6%) [20, 55].

By reviewing the precedent studies on FDSCs with Ru-based sensitizers, we find that Ru sensitizers are not suitable for FDSCs because of the following issues. (i) They need thick TiO_2 film (>10-μm photoactive) owing to the low molar extinction coefficient ($\varepsilon_{max} = 1.30 \times 10^4\ M^{-1}\ cm^{-1}$ at 518 nm for **N719**) [60]. However, the use of thick film must be avoided for FDSCs, considering exfoliation of the photoactive film from the substrate during bending, which is related to the stability issue [61]. (ii) Thiocyanate ligands for anchoring to the metal oxide film are chemically unstable, causing degradation and desorption of sensitizers [62]. (iii) Ru is a rare metal, which fundamentally decreases the price competitiveness. Therefore, Ru sensitizers should be replaced with metal-free organic sensitizers, whose structure is easily tunable to overcome the disadvantages stated above. **D149** and **D205**, indoline-based sensitizers, were applied to FDSCs. A textile-type FDSC based on a ZnO nanorod/stainless steelwire exhibited good flexibility, whereas **D149** achieved low PCE of 0.95% [63]. For the case of ITO/PEN-based FDSC, **D149** showed a much higher PCE of 5.3% under the co-dyes system [56]. However, the PCE is still much lower than that of the conventional **D149** DSC (8.85%) because of the significantly reduced J_{SC} (19.86 to 13.16 $mA\ cm^{-2}$) [57]. The difference in PCE is attributed to the thin TiO_2 film and poor charge collection on the ITO/PEN substrate. Similarly, **D205** showed much lower PCE (4.0%) in FDSCs based on a stainless steel mesh compared to that of conventional **D205** DSCs (8.3%) because of the decreased J_{SC} [58]. In general, charge recombination is mainly responsible for low J_{SC} in solar cells. Han et al. developed a TiO_2 blocking layer on the ITO/PEN substrate using an ALD technique to reduce the charge recombination [20]. For testing the device performance, they employed **MK-2** based on the carbazole derivative and hexyl-thiophene units as a sensitizer, only 1-μm photoactive TiO_2 film prepared by cold isostatic press, and ionic liquid-based iodine electrolyte. The best result was 4.0% PCE, showing <10% drop after 1000 h light soaking at 60 °C. This was attributed to the high molar extinction coefficient ($\varepsilon_{max} = 3.84 \times 10^4\ M^{-1}\ cm^{-1}$ at 480 nm) of **MK-2** [59] and no use of volatile solvent.

As described above, most sensitizers having high PCE with thick TiO_2 film (>10-μm photoactive) do not maintain their performance in FDSCs with a thinner TiO_2 film because of the limitation of light harvesting. These results indicate the preconditions for organic sensitizers for FDSCs. First, they must be able to harvest photon energy efficiently even with a thin photoactive film, i.e. low dye amount. Sensitizers with poor molar absorptivity limit not only the power output but also the device architecture design. For example, the DSCs deployed on a curved substrate have a much smaller chance of facing the incident light. Second, generated excitons have to separate without exception, which is closely linked to the charge injection efficiency. One of the main obstacles to charge injection is intermolecular aggregation, causing exciton self-quenching through intermolecular new energy states [64]. Third, injected electrons must pass through the TiO_2 network safely without interception of oxidized redox species, which indicates slight charge loss. Introducing intrinsically photosensitive chromophores as donors is a fundamental way to increase molar absorptivity. Moreover, a planar molecular structure is favorable to light harvesting but causes dye aggregation at the same time. Adding suitable length of alkyl chains is helpful in avoiding this. It is also well known that bulky molecular units and alkyl chains introduced to the sensitizer retard the charge recombination because they act as a protective barrier against the oxidized redox mediators approaching the TiO_2 surface. However, few studies have investigated sensitizers showing high performance in FDSCs with thin photoactive film. We believe that it is possible to achieve highly efficient and bendable FDSCs by applying sensitizers that perform well in thin photoactive film.

9.3.2 Sensitizers in Thin Photoactive Film DSCs

As mentioned above, thick photoactive films are not suitable for FDSCs because they cannot withstand intense bending, leading to cracking and delamination [61]. Therefore, it is important to develop highly efficient sensitizers under thin photoactive film. Table 9.6 lists the photovoltaic parameters of various sensitizers in DSCs with thin photoactive films. To enhance the molar extinction coefficient of Ru-based sensitizer **Z907**, Cao et al. introduced triphenylamine electron donating antennas on the donor part of **Z907**, namely **KW1** and **KW2** according to conjugation length [65]. They exhibited improved PCE (5.0% for **KW1** and 6.3% for **KW2**) under TiO_2 film (2.3-μm photoactive) compared to that of **Z907** (4.1%). Chen et al. synthesized heteroleptic Ru sensitizer with high molar extinction coefficient ($\varepsilon_{max} = 2.42 \times 10^4\ M^{-1}\ cm^{-1}$ at 554 nm) by incorporating electron-rich hexylthio-bithiophene antennas, namely **CYC-B11**, showing high PCE values (8.2%) under 5.0-μm TiO_2 film without scattering film [66]. Hagberg et al. designed **L0-4** with triphenylamine- and thiophene-based simple structure with the aim of energy level tuning [67]. Depending on the π-conjugation length, their absorption peaks were gradually red-shifted and the molar extinction coefficient was remarkably enhanced. However, **L4** showed a deteriorated PCE (1.7%) with a low J_{SC} (4.56 mA cm^{-2}) and V_{OC} (0.58 V) in 3.0-μm photoactive TiO_2 film, despite enhanced molar absorptivity. This unexpected result can be explained by intermolecular aggregation on the TiO_2 surface because **L4** has a

Table 9.6 Photovoltaic parameters of sensitizers in DSCs with various film thicknesses.

Dye	Compound	Thickness (μm)	J_{SC} (mA cm^{-2})	V_{OC} (V)	FF	η (%)	Electrolyte	References
Z907	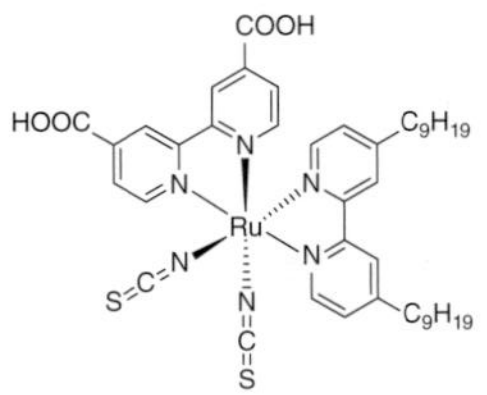	2.3	7.4	0.76	0.71	4.1	I^-/I_3^-	[65]
		8.5 + 5	16.7	0.73	0.72	8.9		
KW1	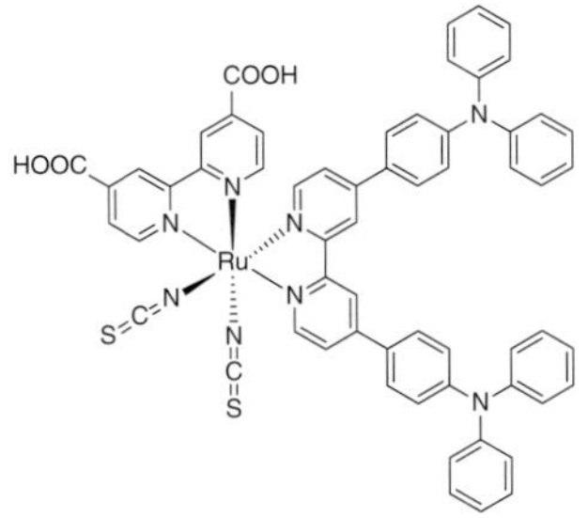	2.3	8.9	0.76	0.73	5.0	I^-/I_3^-	
		8.5 + 5	18.5	0.74	0.74	10.1		
KW2	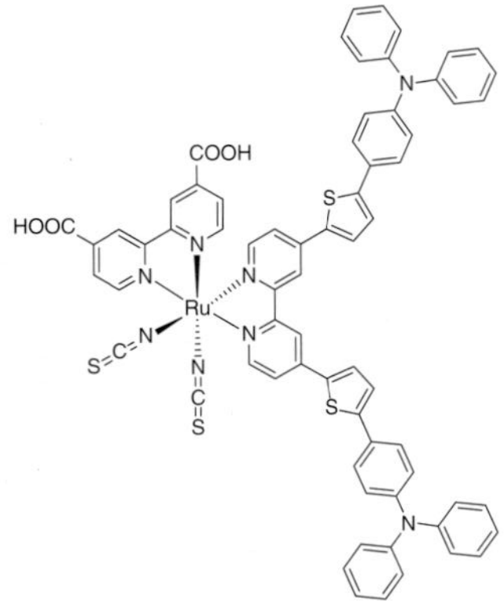	2.3	11.2	0.76	0.74	6.3	I^-/I_3^-	
		8.5 + 5	18.9	0.73	0.74	10.1		

(*continued*)

Table 9.6 (Continued)

Dye	Compound	Thickness (μm)	J_{SC} (mA cm^{-2})	V_{OC} (V)	FF	η (%)	Electrolyte	References
CYC-B11	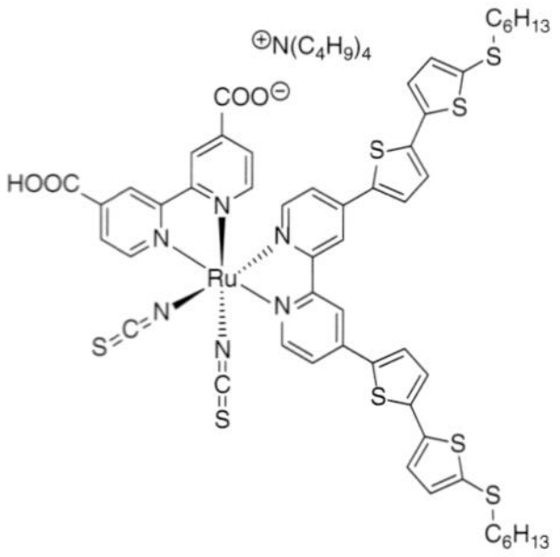	5	15.5	0.73	0.73	8.2	I^-/I_3^-	[66]
		8 + 5	18.3	0.71	0.73	9.4		
L0	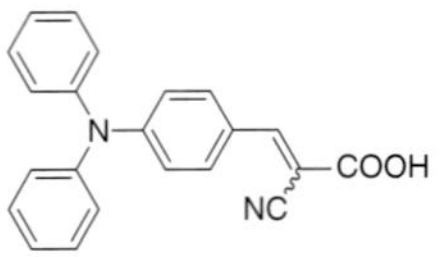	3	2.9	0.74	0.73	1.6	I^-/I_3^-	[67]
L1	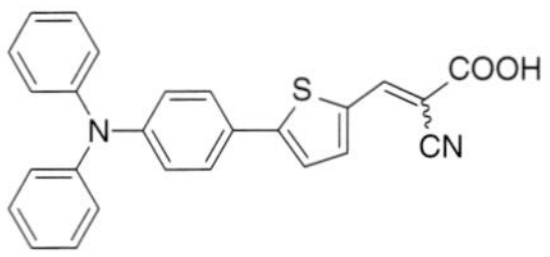	3	5.4	0.74	0.69	2.8		
L2	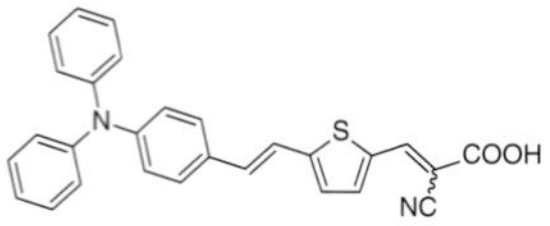	3	6.4	0.71	0.68	3.1		

L3	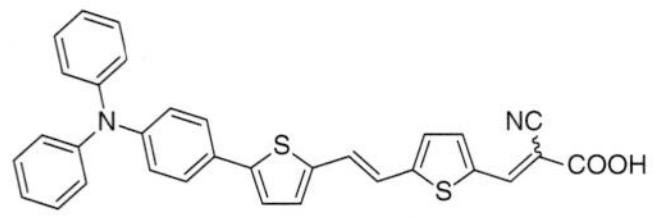	3	6.6	0.64	0.66	2.7		
L4	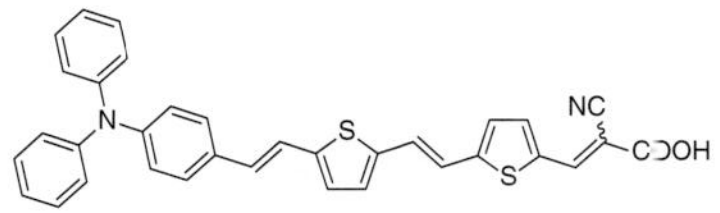	3	4.6	0.58	0.64	1.7		
D21L6	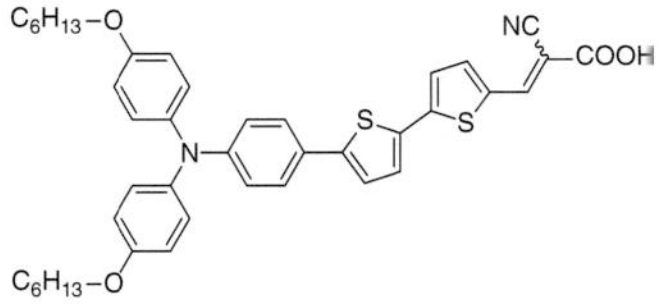	2.2	9.6	0.75	0.71	5.1	I^-/I_3^-	[68]
		7.0 + 5.0	13.8	0.71	0.72	7.1		
C218	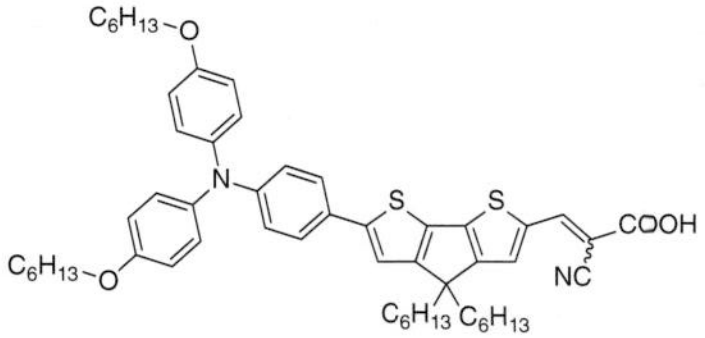	2.2	11.1	0.80	0.71	6.3		
		7.0 + 5.0	15.8	0.77	0.74	9.0		
Y123	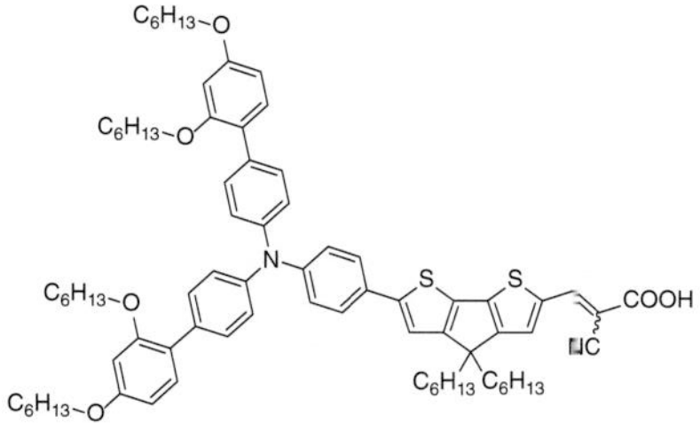	2 + 5	13.6	0.76	0.70	7.2	I^-/I_3^-	[69]
			14.6	0.86	0.70	8.8	Co^{2+}/Co^{3+}	
		1.3	8.5	0.94	0.73	5.8	Spiro-MeOTAD	[70d]

(*continued*)

Table 9.6 (Continued)

Dye	Compound	Thickness (μm)	J_{SC} (mA cm^{-2})	V_{OC} (V)	FF	(%)	Electrolyte	References
IQ4	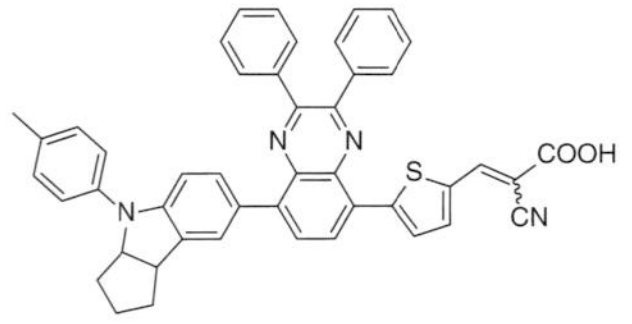	4 + 4	15.3	0.74	0.76	8.5	I^-/I_3^-	[71]
			14.7	0.77	0.69	7.8	Co^{2+}/Co^{3+}	
YA421	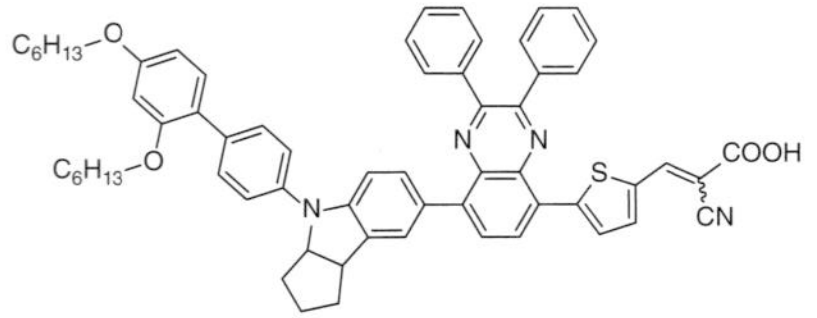	4 + 4	15.4	0.74	0.71	8.1	I^-/I_3^-	
			15.8	0.80	0.71	9.0	Co^{2+}/Co^{3+}	
YA422	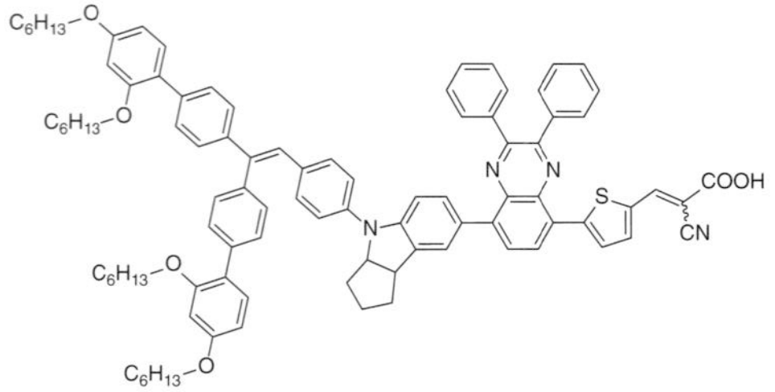	4 + 4	14.4	0.74	0.68	7.3	I^-/I_3^-	
			15.3	0.88	0.69	9.2	Co^{2+}/Co^{3+}	
C257	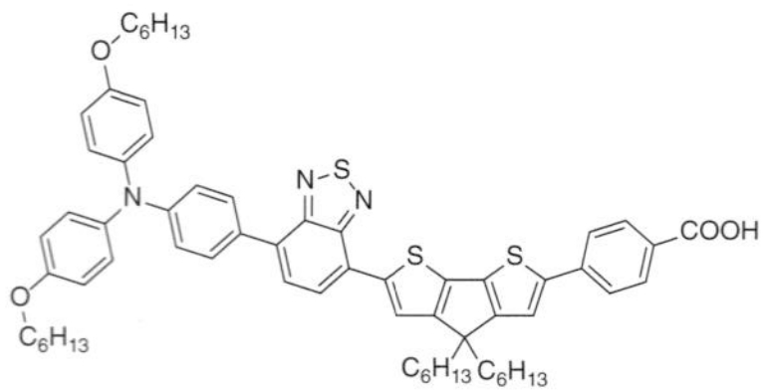	4.6 + 4.7	15.3	0.89	0.73	10.0	Co^{2+}/Co^{3+}	[72]

C258	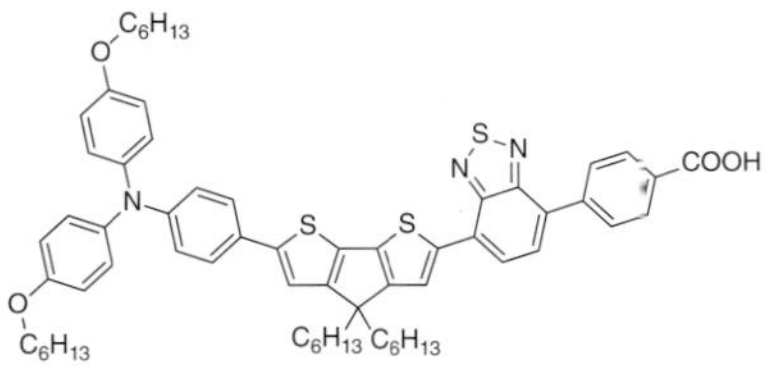	4.6 + 4.7	15.6	0.81	0.71	8.9		
YD2-o-C8	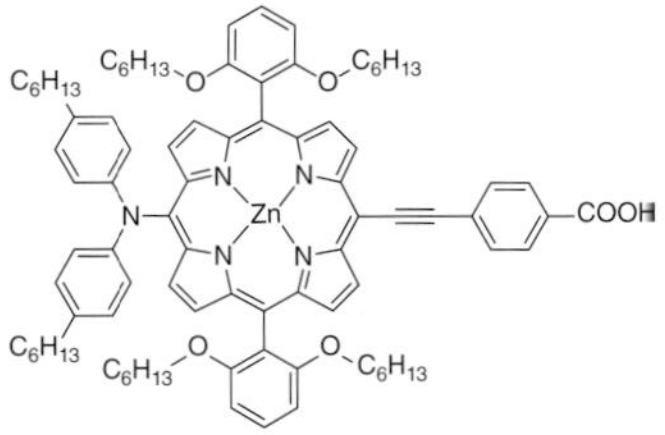	4	13.7	0.95	0.70	9.0		[73]
		4.0 + 5.0	16.3	0.87	0.74	10.5		[74]
Y350	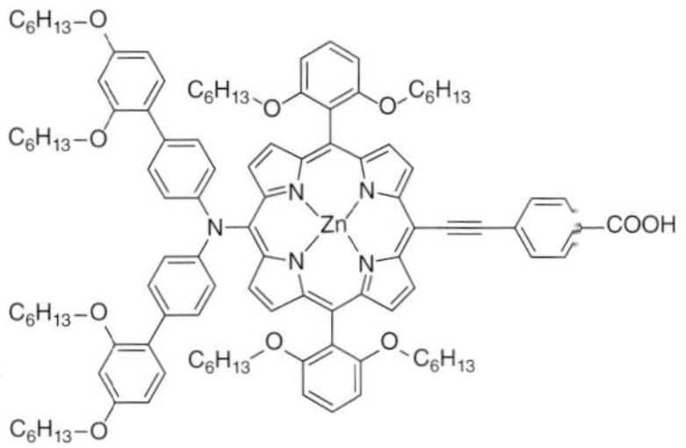	4	13.9	0.96	0.73	9.6	Co^{2+}/Co^{3+}	[73]
Y350-OC12	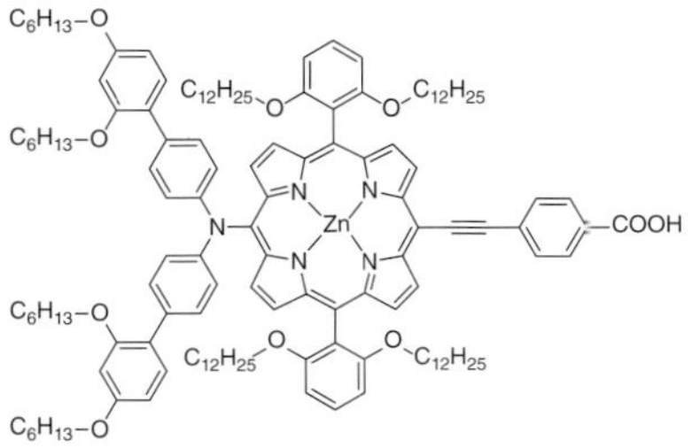	4 + 4	17.2	0.94	0.74	12.0		
		4	13.4	0.97	0.71	9.1		
		4 + 4	16.0	0.97	0.74	11.5		

(*continued*)

Table 9.6 (Continued)

Dye	Compound	Thickness (μm)	J_{SC} (mA cm^{-2})	V_{OC} (V)	FF	(%)	Electrolyte	References
C271	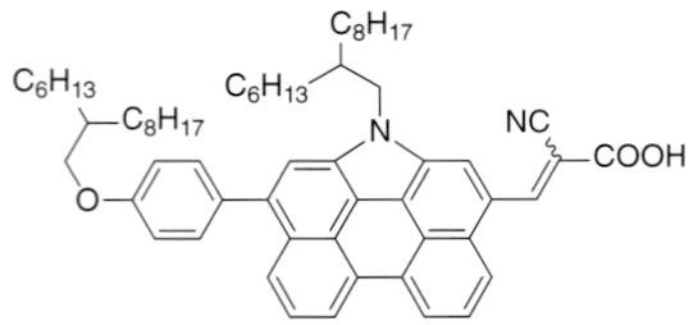	4.2 + 5.0	10.2	0.84	0.78	6.7	Co^{2+}/Co^{3+}	[75]
C272	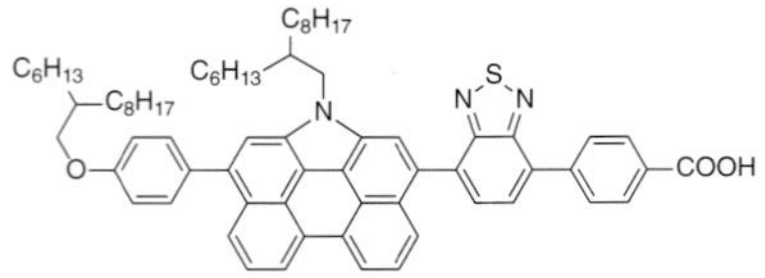	4.2 + 5.0	15.4	0.91	0.74	10.4		
C275	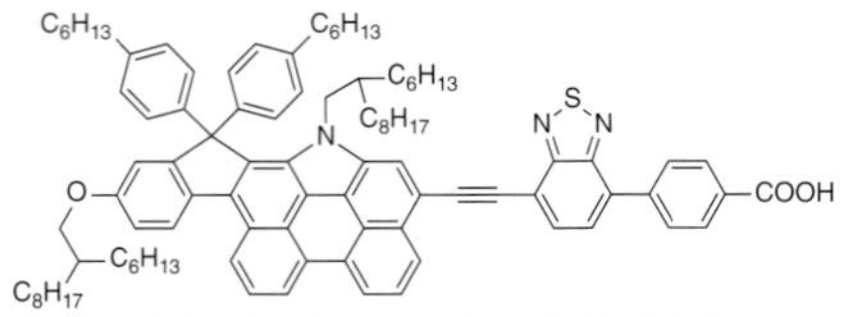	5.2 + 5.0	17.0	0.96	0.77	12.5	Co^{2+}/Co^{3+}	[76a]
C279	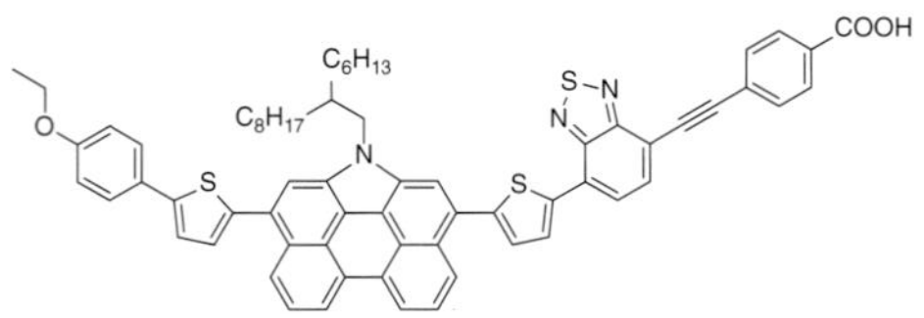	4.6 + 5.0	14.8	0.73	0.74	8.0	Co^{2+}/Co^{3+}	

C280	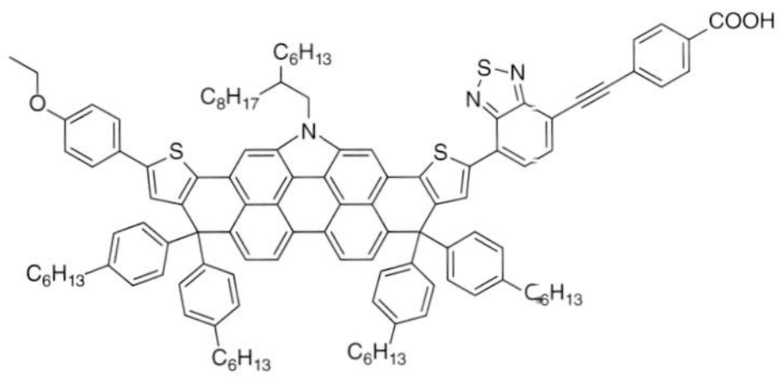	4.6 + 5.0	21.2	0.79	0.71	11.8	Co^{2+}/Co^{3+}	[76]
C281	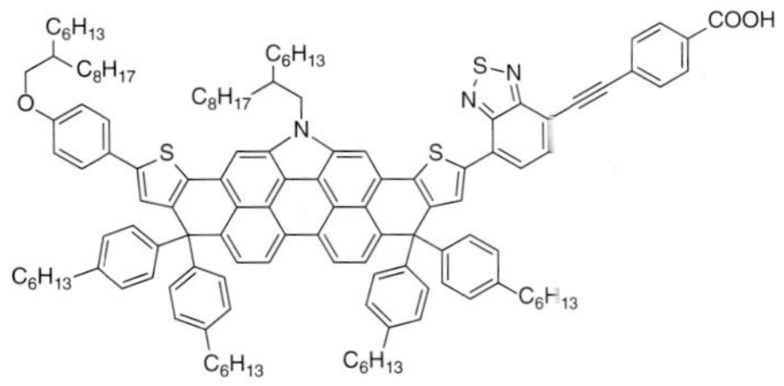	4.6 + 5.0	21.7	0.82	0.74	13.0	Co^{2+}/Co^{3+}	
BT-T	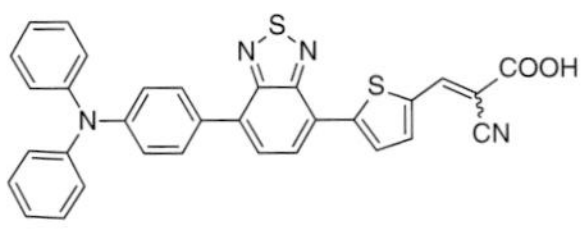	1.8 + 3.5	13.6	0.67	0.72	6.5	I^-/I_3^-	[70c]
		3.5 + 3.5	10.5	0.66	0.76	5.2		
		5.1 + 3.5	10.3	0.64	0.75	5.0		
		10 + 3.5	10.1	0.63	0.75	4.8		

(*continued*)

Table 9.6 (Continued)

Dye	Compound	Thickness (μm)	J_{SC} (mA cm^{-2})	V_{OC} (V)	FF	(%)	Electrolyte	References
Carbz-PAHTDTT	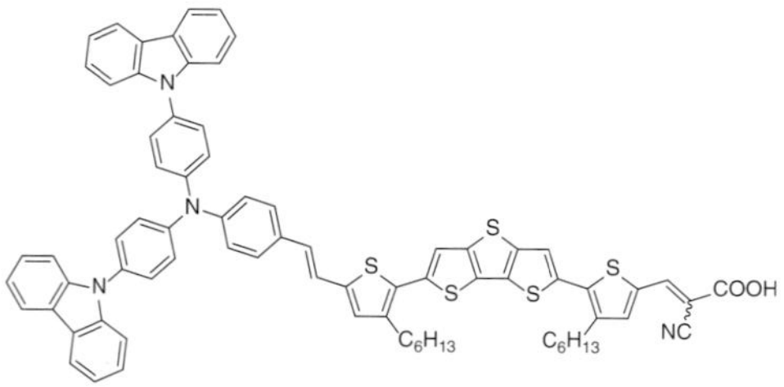	2.2 + 4.4	13.3	0.74	0.62	6.1	I^-/I_3^-	[70a]
		12 + 6	12.1	0.74	0.70	6.3		
		2.2 + 4.4	12.2	0.84	0.73	7.5	Fc/Fc^+	
RK 3	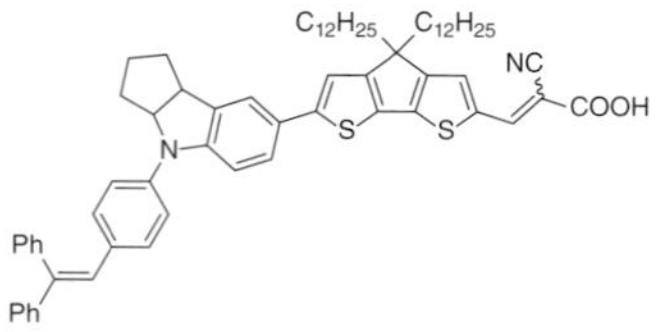	1.8 + 2.5	16.2	0.74	0.76	9.1	I^-/I_3^-	[70b]
		3.2 + 2.5	15.2	0.74	0.73	8.2		
		5.1 + 2.5	12.8	0.73	0.71	6.6		
MK-2	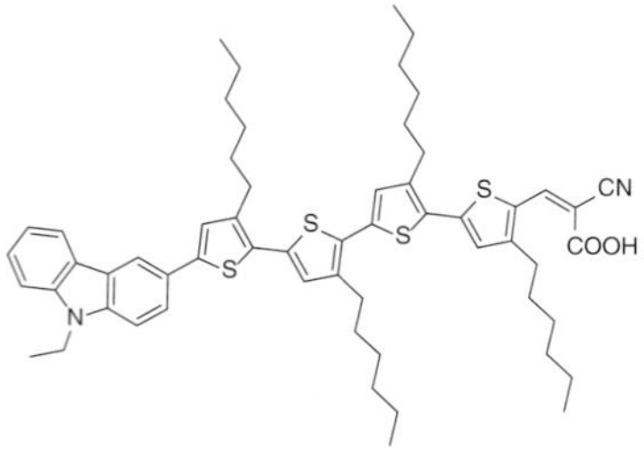	1 + 2	11.8	0.94	0.77	8.4	Co^{2+}/Co^{3+}	[77]

XY1	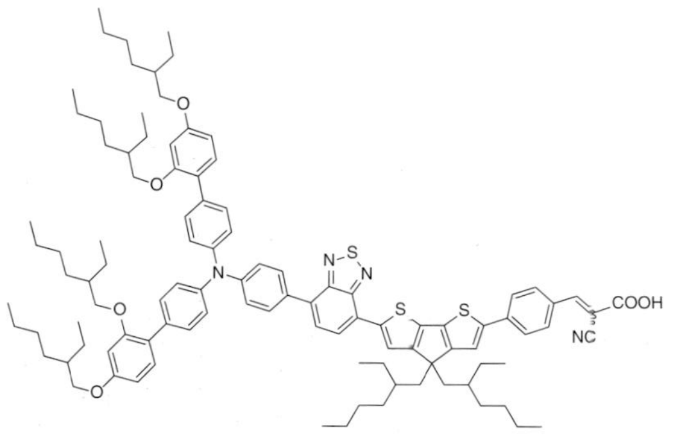	1.3	10.0	0.94	0.67	6.7	Spiro-OMeTAD	[70d]
XY2	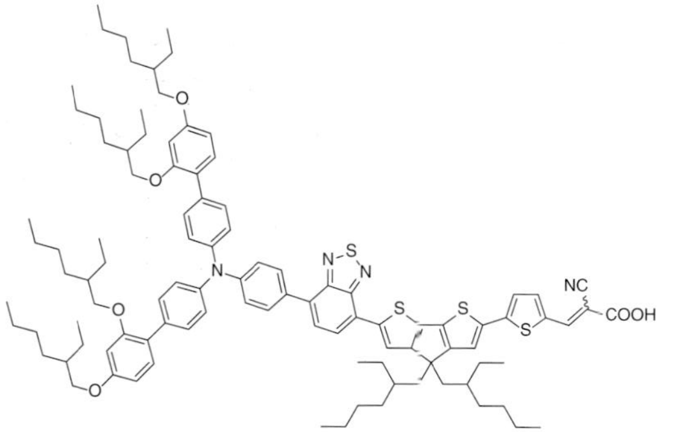	--	11.0	0.90	0.76	7.5	--	
XY3	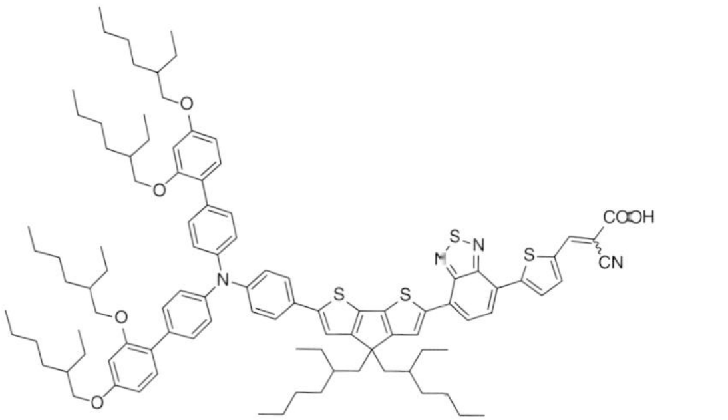	--	11.1	0.80	0.60	5.5	--	

Table 9.7 Summary of electrolyte composition for flexible dye-sensitized solar cells.

No	Solvent/gellator (phase)	Composition	Additives	η (%)	Counter-electrode	Stability test	References
1	ACN (liquid)	I_2 0.030 M PMII 0.60 M	LiI 0.06 M GuSCN 0.10 M tBP 0.50 M	8.5	Pt catalyst	None	[84a]
2	MPN/ACN 1 : 1 vol (liquid)	I_2 0.050 M DMPII 0.60 M	LiI 0.10 M tBP 0.50 M	[a] 8.3 [b] 8.2	[a] NGr/PEDOT [b] Pt catalyst	None	[84b]
3	MPN/ACN 2 : 8 vol (liquid)	I_2 0.035 M DMPII 1.20 M	GuSCN 0.10 M tBP 0.50 M	[a] 9.0 [b] 8.4	[a] CB-NPs/s-PT [b] Pt catalyst	None	[84d]
4	MPN/ACN 2 : 8 vol (liquid)	I_2 0.035 M DMPII 1.20 M	GuSCN 0.10 M tBP 0.50 M	[a] 8.9 [b] 8.2	[a] MOF-525/s-PT [b] Pt catalyst	None	[84e]
5	ACN/valeronitrile 85 : 15 vol (Liquid)	I_2 0.030 M PMII 0.70 M	GuSCN 0.050 M tBP 0.50 M	8.1	Pt/carbon catalyst	None	[84c]
6	ACN (liquid)	I_2 0.050 M TMAI 0.10 M TEAI 0.10 M TBAI 0.10 M	NaI 0.10 M KI 0.10 M LiI 0.10 M tBP 0.50 M	[a] 7.4 [b] 7.5	[a] Binary Ni/Se alloy [b] Binary Co/Se alloy	Storing in an etched groove on Ti foil [a] 99.8% after 600 min under irradiation [b] None	[85a] [85b]
7	[a] ACN [b] MPN (liquid)	[a] I_2 0.040 M PMII 0.80 M [b] I_2 0.10 M PMII 0.80 M	tBP 0.50M	[a] 7.1 [b] 5.2	Pt catalyst	[b] >80% after 1500 h under one sun (AM 1.5, 1000 W m^{-2}) at 50 °C	[86]

8	[a] ACN/MPN 1 : 1 vol [b] MPN (liquid)	I_2 0.040 M TBAI 0.10 M	LiI 0.04 M [a] NBB 0.30M [b] tBP 0.30M	[a] 5.8 [b] 5.2	Pt catalyst	None	[87]
9	MPN (liquid)	I_2 0.010 M BMII 0.060 M	LiI 0.10 M GuSCN 0.10 M tBP 0.50 M	5.4	Pt catalyst	90% after 130 days under dark at 23 °C	[88]
10	IL (liquid)	I_2 0.050 M PMII/EMITCB 65 : 35 vol	GuSCN 0.10 M tBP 0.50 M	3.6	Pt catalyst	None	[89]
11	IL (liquid)	I_2/DMII/EMII/EMITCB/NBB/GuSCN [1.67:12:12:16:3.33:0.67] mol ratio		[a] 6.5 [b] 6.5	[a] CoS Supersedes Pt [b] Pt catalyst	[a] 85% after 1000 h under irradiation at 60 °C	
12	ACN/PEO (quasi-solid)	DMPII 50 mg, KI 0.12 g in 2.5 ml ACN		4.3	Pt catalyst	None	
13	P25 TiO_2 nanoparticle [a] MPN (liquid) [b] IL (quasi-solid)	[a] I_2 0.10 M DMII 0.80 M [b] I_2 0.20 M DMII 0.80 M PMII/EMITCB 65 : 35 vol	[a] NMBI 0.45 M [b] NMBI 0.50 M	[a] 6.5 [b] 2.8	Pt catalyst	[a] 81% [b] 97% after 500 h under one sun (AM 1.5, 1000 W m^{-2}) at 60 °C	[90]

(continued)

Table 9.7 (Continued)

No	Solvent/gellator (phase)	Composition	Additives	η (%)	Counter-electrode	Stability test	References
14	ACN/PLGA (quasi-solid)	I_2 0.060 M BMII 0.50 M	KI 0.1 M GuSCN 0.10 M tBP 0.50 M	[a] 5.4 [b] 6.0	[a] Carbon-nanofiber [b] Pt catalyst	[a] 90% after 240 h in dark at room temperature	[45]
15	MPN/PVDF-HFP (quasi-solid)	I_2 0.010 M PMII 0.60 M	LiI 0.10 M NMBI 0.45 M TiO_2 P25	6.4	Pt catalyst	90% after 300 h at 65 °C	[91]
16	ACN/MPN 1 : 1 vol SiO_2 nanopaticles (quasi-solid)	I_2 0.04 M TBAI 0.40 M	LiI 0.10 M	5.5	Pt catalyst	None	[92]
17	CuI (solid)	—	—	1.4	Au-wire	None	[93a]
18	PEDOT (solid)	—	—	1.0	Pt catalyst	None	[94]
19	P3HT (solid)	—	—	1.9	Ag electrode	None	[95]

ACN, acetonitrile; MPN, 3-methoxypropionitrile; PMII, 1-methyl-3-propylimidazolium iodide; GuSCN, guanidinium thiocyanate; tBP, 4-*tert*-butylpyridine; NBB, *N*-butylbenzimidazole; DMPII, 1,2-dimethyl-3-propylimidazolium iodide; DMII, 1,3-dimethylimidazolium iodide; BMII, 1-butyl-3-methylimidazolium iodide; NMBI, *N*-methylbenzimidazole; TMAI, tetraethylammonium iodide; TEAI, tetramethylammonium iodide; TBAI, tetrabutylammonium iodide; EMII, 1-ethyl-3-methylimidazolium iodide; EMITCB, 1-ethyl-3-methylimidazolium tetracyanoborate; PEO, poly(ethylene oxide); PLGA, poly(lactic-*co*-glycolic acid); PVDF-HFP, poly(vinylidene fluoride-*co*-hexafluoropropylene); P3HT, poly(3-hexylthiophene-2,5-diyl); PEDOT, poly(3,4-ethylenedioxythiophene); NGr, nitrogen-dopedgraphene; s-PT, sulfonated-poly(thiophene-3-[2-(2-methoxyethoxy)ethoxy]-2,5-diyl); MOF, metal-organic framework.

maintained performance was realized after 1500 h under one sun (AM 1.5, 1000 W m^{-2}) light exposure at 60 °C [86]. MPN-based electrolyte generally gives lower PCE than ACN-based ones owing to the higher viscosity, but it has moderate durability. In addition to the modification of electrolytes, rigid encapsulation to seal the liquid electrolyte has also been suggested. The photo-electrode was prepared by etching Ti foils so that volatile liquid electrolytes are stored in a groove on the Ti foil, and it achieved 7.4% PCE with 600 min durability [85]. However, sustainable encapsulation and developing sealing materials that can be applied at low temperatures and that have high durability against corrosive materials in the electrolyte, remain a challenge in DSCs [6a, 96].

With this background, nonvolatile solvents have been studied as alternatives to volatile organic solvents. These electrolytes have been used by dissolving high concentration I^- for compensating the slow diffusion and thus achieving sufficient dye regeneration. In this system, direct ion diffusion does not occur; instead, ion exchange between compounds occurs, which is popularly known as the *Grotthus mechanism* [98]. However, high I^- concentration causes rapid recombination of injected electrons in the TiO_2 film, leading to poorer performance than that of liquid-based electrolytes. An ionic liquid (IL) that is a eutectic salt with moderate conductivity, negligible vapor pressure, and good chemical stability is a promising alternative solvent. For instance, imidazolium units, typical ILs in DSCs, are widely used as solvents. Wang et al. used a mixture of ILs as solvents of electrolytes for FDSCs and achieved 6.5% PCE with 85% maintained performance after light exposure for 1000 h at 60 °C [99]. Based on the IL system, the quasi-solid-state could be possible with high melting point ionic salts or gelling agents such as silica nanoparticles. Although the durability of IL electrolytes is promising, thus far, the PCE of IL electrolyte has been lower than that of organic solvent-based electrolytes because of the high viscosity.

To prevent evaporation or leakage of organic solvents, polymer and gel electrolytes have been chosen as alternatives to liquid electrolytes. In these types, both organic solvents and IL can be converted to the quasi-solid-state by the gelling agent or by polymerizing with polymeric materials. Fan et al. prepared quasi-solid-state FDSCs by mixing poly-ethylene oxide (PEO) in electrolyte. They employed hierarchical TiO_2 microsphere instead of normal TiO_2 nanoparticles for smooth penetration of quasi-solid-state electrolyte and achieved enhanced PCE (4.3%) compared with that using commercial P25 TiO_2 nanoparticles [100]. Dkhissi et al. reported 2.8% PCE with quasi-solid electrolyte combined with poly-(vinylidenefluoride-*co*-hexafluoropropylene)-based (PVDF-HFP) gel electrolytes [91]. Chen et al. proposed that mesoporous SiO_2 nanoparticles (MSNs) can be good gelling agents. MSNs facilitate ionic diffusion and light-scattering effect in FDSCs, resulting in 5.5% PCE [92]. Typically, these quasi-solid electrolytes show lower PCE than the host liquid electrolytes owing to low mobility in the mediator. In addition, when quasi-solid-state electrolytes are applied, photoactive films that are thinner than those of conventional DSCs are recommended to avoid pore filling problems.

Instead of I^-/I_3^- redox mediators, various solid-state hole-transporting materials (HTMs) have been applied to DSCs. Reported HTMs include: CuI [93], $CsSnI_3$ [101], 2,2′,7,7′-tetrakis-(*N*,*N*-di-*p*-methoxyphenyl-amine)-9,9′-spirobifluorene (spiro-MeOTAD) [102] as well as some conjugated polymers such as PPy [103], poly(3-hexylthiophene-2,5-diyl) (P3HT) [104], and PEDOT [105]. In FDSCs, a few studies have employed solid-state hole conductors. Long fabricated solid-state FDSCs based on TiO_2 nanowire-coated ITO/PEN substrate using photoelectropolymerized PEDOT as a hole conductor for the first time, obtaining 1.0% PCE [94]. Xue et al. also reported solid-state FDSCs using P3HT as a hole conductor. They employed ALD and EPD techniques to deposit dense and mesoporous TiO_2 films on ITO/PEN substrate at low temperatures. Pinhole-free thin TiO_2 films acted as rectifier well and resulted in 1.9% PCE [95]. In some cases, as a strategy for enhancing the charge collection, p-type HTMs were constructed at the interface between the metal substrate and photoactive TiO_2 film to form the p–n junction [106]. They formed a single directional pathway for charge transport from the n-type semiconductor to the p-type one and reduced the charge recombination loss. Xiao et al. introduced the novel structure of a flexible working electrode, TiO_2/PEDOT/Ti mesh, in the large-area (10 cm^2) FDSCs. The p–n junction of TiO_2/PEDOT brought higher PCE (6.5%) compared to that (4.9%) without PEDOT due to the rectifying effect on charge transport [106a].

However, flexible types with solid-state hole conductors show much lower performance than glass-based DSCs and even quasi-solid-state FDSCs. Although the PCEs are not as high as those with a liquid-type electrolyte, the higher V_{OC} and high transmittance in the visible region present their potential for further improvement in near future.

9.5 Conclusion and Outlook

PV technologies emerging in the market have generally focused on major parameters such as PCE, stability, and cost. Silicon cells that occupy over 80% of the current market meet such conditions and are still considered as the benchmark for the PV industry [107]. Nevertheless, consumer demands are increasing, suggesting a new roadmap for the development of PV technology. The recent roadmap for PV technology strongly depends on the functionalities for broad application, e.g. flexibility, transparency, and esthetics. DSCs deliver such functionalities that silicon cells do not meet, and this is the reason why FDSCs should be developed [108]. DSCs have shown power output that have outperformed their competitors, such as silicon cells, under indoor lighting conditions. In addition, their reasonable lifetime in less harsh indoor environments is commercially acceptable [109].

A number of studies have investigated the realization of FDSCs. To replace the rigid glass substrate, thermostable metal-based substrates have been employed as conductive substrates in FDSCs. However, they face constraints originating from opacity, corrosion, and difficulty in assembling them into tandem and module cells. Even though plastic substrates overcome such problems, some obstacles including weak thermal stability and exposure to water and oxygen penetration have motivated further studies to overcome them. Graphene-based conducting film can be easily formed on transparent plastic substrates through low-temperature methods, e.g. roll-to-roll process and CVD. Although few studies have investigated FDSCs using graphene-based substrates, their ideal properties suggest that it is a promising material as a transparent conducting substrate for FDSCs. Several methods for fabricating photo- and counter-electrodes have been reported. According to the substrate used, they are classified into two groups: high- and low-temperature processes. High-temperature sintering processes generally guarantee good device performance because of the well-connected TiO_2 nanoparticle network and enhanced crystallinity. However, they are not compatible with transparent plastic substrates, such as ITO/PET and ITO/PEN. To compensate the deficient activation energy in order to form the TiO_2 network under low temperature, binder-free TiO_2 paste goes through a chemical linking process or is treated with an intermediate connector, e.g. (Sb,In)SnO_2. Applying other types of energy, e.g. physical pressure, electric field, or microwave irradiation, is also helpful for good connection. Recently, new types of TCO-free architectures have been actively researched. In particular, fiber-type FDSCs are future-oriented because of their multifunctional properties, e.g. great bendability, isotropic light harvesting, and spatial efficiency. Along with the architecture and electrode materials, well-matched sensitizers and electrolytes are required. Both sensitizers and electrolytes are mainly responsible not only for the device performance but also for longevity. Nevertheless, relatively few studies have investigated sensitizers for FDSCs. For assigning sufficient mechanical stability, the photoactive film should be thin (<2 μm). In this case, the functionality of the sensitizer becomes more crucial for the device performance. As for the essential factors for a sensitizer suitable for thin photoactive film DSCs, a high molar extinction coefficient and broad absorption range are first required for efficient light harvesting ability, and an appropriate molecular structure is required according to redox mediators to reduce charge losses. The redox mediator in the electrolyte is another key factor for device performance and stability. Numerous electrolyte systems that have been used for FDSCs are limited to iodine-based ones, which are corrosive in the case of most metals and semiconductors used in DSCs, and they cause massive internal potential loss. Therefore, it is imperative to develop an alternative redox mediator to I^-/I_3^- couples. Inorganic and polymer hole conductors that have been rarely studied will be good alternatives considering their guaranteed stability.

Ultimately, maintaining the device performance in the long term similarly to that of initial-state cells from industrial manufacturing systems is a major challenge for the practical application of FDSCs, regardless of their environment. By reviewing the recent progress in FDSCs, we realize that little attention has been paid to the stability of the devices, while much effort has been devoted to fundamental studies. More intensive studies on device durability will accelerate the application of FDSCs to not only indoor systems but also outdoor to systems, such as BIPV and AIPV, which are the ultimate goals of the commercialization of FDSCs.

References

1 (a) Goncalves, L.M., Bermudez, V.D., Ribeiro, H.A., and Mendes, A.M. (2008). *Energy Environ. Sci.* 1: 655. (b) OECD-OCDE (2015). *Medium-Term Gas Market Report*, 2015. Paris, France: OECD Publishing; International Energy Agency.
2 Goldemberg, J. (2000). *World Energy Assessment: Energy and the Challenge of Sustainability*. New York, USA: United Nations Publications.
3 Shockley, W. and Queisser, H.J. (1961). *J. Appl. Phys.* 32: 510.
4 O'Regan, B. and Grätzel, M. (1991). *Nature* 353: 737.
5 Mathew, S., Yella, A., Gao, P. et al. (2014). *Nat. Chem.* 6: 242.
6 (a) Brown, T.M., De Rossi, F., Di Giacomo, F. et al. (2014). *J. Mater. Chem. A* 2: 10788. (b) Pugliese, D. (2014). New insights in dye-sensitized solar cells: novel nanostructured photoanodes, metal-free dye, quasi-solid electrolytes and physics-based modeling. PhD thesis. Politecnico di Torino, Italy; (c) Gratzel, M. (2005). *Inorg. Chem.* 44: 6841.
7 (a) Nasiri, A., Zabalawi, S.A., and Mandic, G. (2009). *IEEE Trans. Ind. Electron.* 56: 4502. (b) Seng, S. D. (2013). Fabrication and optimization of flexible dye sensitized solar cells. PhD thesis. National University of Singapore, Singapore; (c) Fakharuddin, A., Jose, R., Brown, T.M. et al. (2014). *Energy Environ. Sci.* 7: 3952.
8 (a) Roy-Mayhew, J.D. and Aksay, I.A. (2014). *Chem. Rev.* 114: 6323. (b) Hagfeldt, A., Boschloo, G., Sun, L.C. et al. (2010). *Chem. Rev.* 110: 6595.
9 Boschloo, G. and Hagfeldt, A. (2009). *Acc. Chem. Res.* 42: 1819.
10 Barnes, P.R.F., Miettunen, K., Li, X.E. et al. (2013). *Adv. Mater.* 25: 1881.
11 Papageorgiou, N., Athanassov, Y., Armand, M. et al. (1996). *J. Electrochem. Soc.* 143: 3099.
12 (a) Bae, S., Kim, H., Lee, Y. et al. (2010). *Nat. Nanotechnol.* 5: 574. (b) Bonaccorso, F., Sun, Z., Hasan, T., and Ferrari, A. (2010). *Nat. Photonics* 4: 611.
13 Han, H.-G., Weerasinghe, H.C., Min Kim, K. et al. (2015). *Sci. Rep.* 5: 14645.
14 Zhang, D., Yoshida, T., Oekermann, T. et al. (2006). *Adv. Funct. Mater.* 16: 1228.
15 (a) Kang, M.G., Park, N.-G., Ryu, K.S. et al. (2006). *Sol. Energy Mater. Sol. Cells* 90: 574. (b) Park, J.H., Jun, Y., Yun, H.-G. et al. (2008). *J. Electrochem. Soc.* 155: F145.
16 Tsai, T.-Y., Chen, C.-M., Cherng, S.-J., and Suen, S.-Y. (2013). *Prog. Photovolt.* 21: 226.
17 Linnemann, J., Giorgio, J., Wagner, K. et al. (2015). *J. Mater. Chem. A* 3: 3266.
18 Fan, X., Wang, F., Chu, Z. et al. (2007). *Appl. Phys. Lett.* 90: 073501.
19 Bella, F., Lamberti, A., Bianco, S. et al. (2016). *Adv. Mater. Technol.* 1: 1600002.
20 Han, Y., Pringle, J.M., and Cheng, Y.-B. (2015). *Photochem. Photobiol.* 91: 315.
21 Chen, H.-W., Hsu, C.-Y., Chen, J.-G. et al. (2010). *J. Power Sources* 195: 6225.
22 Gratzel, M. (2001). *Nature* 414: 338.

23 Ito, S., Chen, P., Comte, P. et al. (2007). *Prog. Photovolt.* 15: 603.
24 Zardetto, V., Brown, T.M., Reale, A., and Di Carlo, A. (2011). *J. Polym. Sci., Part B: Polym. Phys.* 49: 638.
25 Huang, F., Chen, D., Li, Q. et al. (2012). *Appl. Phys. Lett.* 100: 123102.
26 Hashmi, S.G., Halme, J., Saukkonen, T. et al. (2013). *Phys. Chem. Chem. Phys.* 15: 17689.
27 (a) Li, Y., Yoo, K., Lee, D.-K. et al. (2010). *Curr. Appl. Phys.* 10: e171. (b) Li, Y., Yoo, K., Lee, D.-K. et al. (2013). *Nanoscale* 5: 4711. (c) Fu, N.-Q., Duan, Y.-D., Fang, Y.-Y. et al. (2013). *Electrochem. Commun.* 34: 254.
28 Weerasinghe, H.C., Franks, G.V., Plessis, J.D. et al. (2010). *J. Mater. Chem.* 20: 9954.
29 (a) Yamaguchi, T., Tobe, N., Matsumoto, D. et al. (2010). *Sol. Energy Mater. Sol. Cells* 94: 812. (b) Weerasinghe, H.C., Sirimanne, P.M., Simon, G.P., and Cheng, Y.B. (2012). *Prog. Photovolt.* 20: 321. (c) Zardetto, V., Di Giacomo, F., Garcia-Alonso, D. et al. (2013). *Adv. Energy Mater.* 3: 1292. (d) Dang, B.H.Q., MacElroy, D., and Dowling, D.P. (2013). *Appl. Surf. Sci.* 275: 289.
30 Grinis, L., Dor, S., Ofir, A., and Zaban, A. (2008). *J. Photochem. Photobiol., A* 198: 52.
31 Lee, H., Hwang, D., Jo, S.M. et al. (2012). *ACS Appl. Mater. Interfaces* 4: 3308.
32 (a) Murakami, T.N., Kijitori, Y., Kawashima, N., and Miyasaka, T. (2003). *Chem. Lett.* 32: 1076. (b) Wang, W., Zhao, Q., Li, H. et al. (2012). *Adv. Funct. Mater.* 22: 2775.
33 Pan, H., Ko, S.H., Misra, N., and Grigoropoulos, C.P. (2009). *Appl. Phys. Lett.* 94: 071117.
34 Hwang, D., Lee, H., Jang, S.-Y. et al. (2011). *ACS Appl. Mater. Interfaces* 3: 2719.
35 Chen, H.-W., Lin, C.-Y., Lai, Y.-H. et al. (2011). *J. Power Sources* 196: 4859.
36 Fu, N.-Q., Fang, Y.-Y., Duan, Y.-D. et al. (2012). *ACS Nano* 6: 9596.
37 Song, W., Wang, H., Liu, G. et al. (2016). *Nano Energy* 19: 1.
38 Zhang, S., Ji, C., Bian, Z. et al. (2011). *Nano Lett.* 11: 3383.
39 Abid, A., Syed Mujtaba, S., Sinem, B. et al. (2016). *Nanotechnology* 27: 384003.
40 Fu, N., Xiao, X., Zhou, X. et al. (2012). *J. Phys. Chem. C* 116: 2850.
41 Yin, X.O., Xue, Z.S., and Liu, B. (2011). *J. Power Sources* 196: 2422.
42 Garcia-Alonso, D., Zardetto, V., Mackus, A.J.M. et al. (2014). *Adv. Energy Mater.* 4: 1300831.
43 Lee, K.S., Lee, Y., Lee, J.Y. et al. (2012). *ChemSusChem* 5: 379.
44 Miettunen, K., Toivola, M., Hashmi, G. et al. (2011). *Carbon* 49: 528.
45 Veerappan, G., Kwon, W., and Rhee, S.-W. (2011). *J. Power Sources* 196: 10798.
46 Chen, J., Li, K., Luo, Y. et al. (2009). *Carbon* 47: 2704.
47 Sun, H., Luo, Y., Zhang, Y. et al. (2010). *J. Phys. Chem. C* 114: 11673.
48 Peng, T., Sun, W., Huang, C. et al. (2014). *ACS Appl. Mater. Interfaces* 6: 14.
49 Chen, C.-M., Chen, C.-H., and Wei, T.-C. (2010). *Electrochim. Acta* 55: 1687.
50 Xiao, Y.M., Wu, J.H., Yue, G.T. et al. (2011). *Electrochim. Acta* 58: 621.
51 Yu, K., Wen, Z., Pu, H. et al. (2013). *J. Mater. Chem. A* 1: 188.

52 Liang, J., Zhang, G., Sun, W., and Dong, P. (2015). *Nano Energy* 12: 501.
53 (a) Nazeeruddin, M.K., De Angelis, F., Fantacci, S. et al. (2005). *J. Am. Chem. Soc.* 127: 16835. (b) Shiu, J.-W., Lan, Z.-J., Chan, C.-Y. et al. (2014). *J. Mater. Chem. A* 2: 8749.
54 Yue, G., Li, P., Li, F., and Chen, C. (2016). *RSC Adv.* 6: 61278.
55 Cao, Y., Bai, Y., Yu, Q. et al. (2009). *J. Phys. Chem. C* 113: 6290.
56 Syu, Y.-K., Tingare, Y., Yeh, C.-Y. et al. (2016). *RSC Adv.* 6: 59273.
57 Ito, S., Miura, H., Uchida, S. et al. (2008). *Chem. Commun.* 5194.
58 Molla, M.Z., Mizukoshi, N., Furukawa, H. et al. (2015). *Prog. Photovolt.* 23: 1100.
59 Wang, Z.-S., Koumura, N., Cui, Y. et al. (2008). *Chem. Mater.* 20: 3993.
60 Nazeeruddin, M.K., Zakeeruddin, S., Humphry-Baker, R. et al. (1999). *Inorg. Chem.* 38: 6298.
61 Miyasaka, T. and Ikegami, M. (2010). *J. Photopolym. Sci. Technol.* 23: 269.
62 Tuyet Nguyen, P., Degn, R., Thai Nguyen, H., and Lund, T. (2009). *Sol. Energy Mater. Sol. Cells* 93: 1939.
63 Chae, Y., Kim, S.J., Kim, J.H., and Kim, E. (2015). *Dyes Pigm.* 113: 378.
64 (a) Katoh, R. and Furube, A. (2014). *J. Photochem. Photobiol. C* 20: 1. (b) Kim, B.-M., Han, H.-G., Kim, J.S. et al. (2017). *ACS Appl. Mater. Interfaces* 9: 2572.
65 Cao, K., Lu, J., Cui, J. et al. (2014). *J. Mater. Chem. A* 2: 4945.
66 Chen, C.-Y., Wang, M., Li, J.-Y. et al. (2009). *ACS Nano* 3: 3103.
67 Hagberg, D.P., Marinado, T., Karlsson, K.M. et al. (2007). *J. Org. Chem.* 72: 9550.
68 Li, R., Liu, J., Cai, N. et al. (2010). *J. Phys. Chem. B* 114: 4461.
69 Tsao, H.N., Yi, C., Moehl, T. et al. (2011). *ChemSusChem* 4: 591.
70 (a) Daeneke, T., Kwon, T.-H., Holmes, A.B. et al. (2011). *Nat. Chem.* 3: 211. (b) Jin, M.Y., Kim, B.-M., Jung, H.S. et al. (2016). *Adv. Funct. Mater.* 26: 6876. (c) Roh, D.-H., Kim, K.M., Nam, J.S. et al. (2016). *J. Phys. Chem. C* 120: 24655. (d) Zhang, X., Xu, Y., Giordano, F. et al. (2016). *J. Am. Chem. Soc.* 138: 10742.
71 Yang, J., Ganesan, P., Teuscher, J. et al. (2014). *J. Am. Chem. Soc.* 136: 5722.
72 Zhang, M., Wang, Y., Xu, M. et al. (2013). *Energy Environ. Sci.* 6: 2944.
73 Yi, C., Giordano, F., Cevey-Ha, N.-L. et al. (2014). *ChemSusChem* 7: 1107.
74 Luo, J., Xu, M., Li, R. et al. (2014). *J. Am. Chem. Soc.* 136: 265.
75 Yao, Z., Wu, H., Ren, Y. et al. (2015). *Energy Environ. Sci.* 8: 1438.
76 (a) Yao, Z., Zhang, M., Wu, H. et al. (2015). *J. Am. Chem. Soc.* 137: 3799. (b) Yao, Z., Wu, H., Li, Y. et al. (2015). *Energy Environ. Sci.* 8: 3192.
77 Kashif, M.K., Axelson, J.C., Duffy, N.W. et al. (2012). *J. Am. Chem. Soc.* 134: 16646.
78 Yella, A., Lee, H.-W., Tsao, H.N. et al. (2011). *Science* 334: 629.
79 Yum, J.-H., Baranoff, E., Kessler, F. et al. (2012). *Nat. Commun.* 3: 631.
80 Feldt, S.M., Gibson, E.A., Gabrielsson, E. et al. (2010). *J. Am. Chem. Soc.* 132: 16714.
81 Wu, J.H., Lan, Z., Lin, J.M. et al. (2015). *Chem. Rev.* 115: 2136.
82 Ikegami, M., Suzuki, J., Teshima, K. et al. (2009). *Sol. Energy Mater. Sol. Cells* 93: 836.
83 Alebbi, M., Bignozzi, C.A., Heimer, T.A. et al. (1998). *J. Phys. Chem. B* 102: 7577.

84 (a) An, J., Guo, W., and Ma, T.L. (2012). *Small* 8: 3427. (b) Chen, P.Y., Li, C.T., Lee, C.P. et al. (2015). *Nano Energy* 12: 374. (c) Yoo, K., Kim, J.Y., Lee, J.A. et al. (2015). *ACS Nano* 9: 3760. (d) Li, C.T., Lee, C.T., Li, S.R. et al. (2016). *J. Power Sources* 302: 155. (e) Chen, T.Y., Huang, Y.J., Li, C.T. et al. (2017). *Nano Energy* 32: 19.

85 (a) Wang, X.P., Tang, Q.W., He, B.L. et al. (2015). *Chem. Commun.* 51: 491. (b) Duan, Y.Y., Tang, Q.W., Li, R. et al. (2015). *J. Power Sources* 274: 304.

86 Lee, K.M., Chen, C.Y., Tsai, Y.T. et al. (2013). *RSC Adv.* 3: 9994.

87 Kijitori, Y., Ikegami, M., and Miyasaka, T. (2007). *Chem. Lett.* 36: 190.

88 Jen, H.P., Lin, M.H., Li, L.L. et al. (2013). *ACS Appl. Mater. Interfaces* 5: 10098.

89 Kuang, D., Brillet, J., Chen, P. et al. (2008). *ACS Nano* 2: 1113.

90 Lee, H.F., Chua, Y.T., Yang, S.M. et al. (2013). *Thin Solid Films* 544: 301.

91 Dkhissi, Y., Huang, F.Z., Cheng, Y.B., and Caruso, R.A. (2014). *J. Phys. Chem. C* 118: 16366.

92 Chen, H.W., Chiang, Y.D., Kung, C.W. et al. (2014). *J. Power Sources* 245: 411.

93 (a) Wang, D., Hou, S.C., Wu, H.W. et al. (2011). *J. Mater. Chem.* 21: 6383. (b) Meng, Q.-B., Takahashi, K., Zhang, X.-T. et al. (2003). *Langmuir* 19: 3572.

94 Long, W. (2013). Flexible dye sensitized solar cells based on TiO_2 photoanode. PhD thesis. National University of Singapore, Singapore.

95 Xue, Z.S., Jiang, C.Y., Wang, L. et al. (2014). *J. Phys. Chem. C* 118: 16352.

96 Weerasinghe, H.C., Huang, F., and Cheng, Y.-B. (2013). *Nano Energy* 2: 174.

97 Yin, X., Wu, F., Fu, N.Q. et al. (2013). *ACS Appl. Mater. Interfaces* 5: 8423.

98 Bai, Y., Cao, Y.M., Zhang, J. et al. (2008). *Nat. Mater.* 7: 626.

99 Wang, M.K., Anghel, A.M., Marsan, B. et al. (2009). *J. Am. Chem. Soc.* 131: 15976.

100 Fan, K., Peng, T.Y., Chen, J.N. et al. (2013). *J. Power Sources* 222: 38.

101 Chung, I., Lee, B., He, J. et al. (2012). *Nature* 485: 486.

102 Bach, U., Lupo, D., Comte, P. et al. (1998). *Nature* 395: 583.

103 Murakoshi, K., Kogure, R., Wada, Y., and Yanagida, S. (1997). *Chem. Lett.* 26: 471.

104 (a) Yang, L., Cappel, U.B., Unger, E.L. et al. (2012). *Phys. Chem. Chem. Phys.* 14: 779. (b) Zhang, W., Zhu, R., Li, F. et al. (2011). *J. Phys. Chem. C* 115: 7038.

105 (a) Zhang, J., Häggman, L., Jouini, M. et al. (2014). *ChemPhysChem* 15: 1043. (b) Xia, J., Masaki, N., Lira-Cantu, M. et al. (2008). *J. Am. Chem. Soc.* 130: 1258.

106 (a) Xiao, Y., Wu, J., Lin, J. et al. (2014). *Electrochim. Acta* 117: 1. (b) Heng, L., Wang, X., Yang, N. et al. (2010). *Adv. Funct. Mater.* 20: 266.

107 Avrutin, V., Izyumskaya, N., and Morkoç, H. (2011). *Superlattices Microstruct.* 49: 337.

108 Yugis, A.R., Mansa, R.F., and Sipaut, C.S. (2015). *IOP Conf. Ser. Mater. Sci. Eng.* 78: 012003.

109 De Rossi, F., Pontecorvo, T., and Brown, T.M. (2015). *Appl. Energy* 156: 413.

10

Self-assembly in Fabrication of Semitransparent and Meso–Planar Hybrid Perovskite Photovoltaic Devices

Ravi K. Misra, Sigalit Aharon, Michael Layani, Shlomo Magdassi, and Lioz Etgar

The Hebrew University of Jerusalem, The Institute of Chemistry, Edmond J. Safra Campus, Givat Ram,, Jerusalem 91904, Israel

10.1 Introduction

Semitransparent photovoltaic (PV) devices are one of the most studied class of solar cells due to their potential to be used in building-integrated applications, as they offer harvesting of solar energy over wider surfaces leading to construction of self-sustaining buildings at low cost. Organic and inorganic materials have been widely used for the development of semitransparent solar cells [1–8], and have been suggested for integration in office buildings and replacement of windows.

The past few years have witnessed tremendous advancement in the field of perovskite solar cells (PSCs), since their first report as a visible light sensitizer in 2009 [9]. PSCs are currently approaching high power conversion efficiency (PCE) of 22.1% (https://www.nrel.gov/pv/assets/images/efficiency-chart.png) within few years of their development. Properties such as simple, low-cost deposition methods with high PCEs, have made them one of the promising candidates suitable for low-cost solar energy generation [10–12]. The methodologies for the development of perovskite-based semitransparent cells, is still less explored and only few reports on the topic are available so far. In order to achieve high transmittance, Eperon et al. [11, 12] utilized the de-wetting of the perovskite film to create "perovskite islands" utilizing the difference in solvent evaporation rates, which resulted in reduced overall PCE due to the formation of voids within the active layer. However, the precise control of the transparency of the cell is difficult using this method. Recent reports [13, 14] are based on fabrication of semitransparent devices using evaporation of thin perovskite films. However, the evaporation technique is an expensive process that needs high capital investments and is associated with highly complicated steps for scale-up to render the devices suitable for industrial applications. Top electrode made up of silver nanowires in semitransparent perovskite-based solar cells has also been introduced recently [15]. The average transparency of such cells depends on the top electrode transparency in place of transparency of the perovskite film. The use of such transparent top electrode has the potential to replace the requirement of evaporation of expensive

Flexible Energy Conversion and Storage Devices, First Edition. Edited by Chunyi Zhi and Liming Dai.

gold top electrode as well as to minimize the need of an expensive and complicated evaporation process.

In our recent work, we introduced a unique, simple, wet deposition process for the development of semitransparent perovskite devices, based on the mesh-assisted self-assembly of perovskite layer. We successfully demonstrated perovskite-based grid patterns, in which the perovskite is present in thin lines surrounding 2D voids using the one-step [16] perovskite deposition technique. In addition, we also developed a novel approach of mesoporous–planar hybrid PSC architecture using mesh-assisted self-assembly of *meso*-TiO_2; the cells have the planar and mesoporous components on a single device [17].

The self-assembly approach is basically inspired from the already demonstrated grid patterns for obtaining transparent electrodes that are made of silver nanoparticles [18, 19]. These grids have been deposited through self-assembly of silver nanoparticles from aqueous dispersions through an evaporative lithography process that is performed directly onto the required substrate, including plastics. As shown schematically (Figure 10.1), in essence, a droplet containing silver nanoparticles is placed on top of a metallic mesh, which instantaneously spreads over the mesh and the plastic substrate; following the

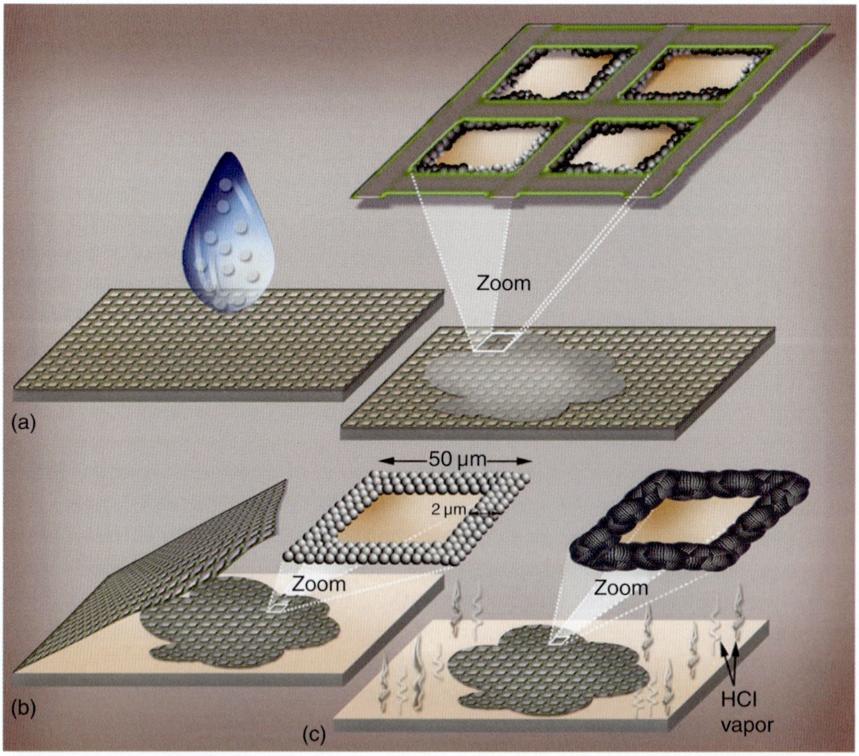

Figure 10.1 Schematic presentation of the fabrication process to obtain a transparent and conductive metallic grid.

flow of the dispersion toward the wires of the mesh and drying, a transparent grid composed of the nanoparticles is formed.

The silver nanoparticles are allowed to self-sinter at low temperature, upon short exposure to HCl vapors, due to the presence of polyacrylic acid salt on the surface of the particles. Therefore, immediate sintering of the silver nanoparticles in the thin lines of the grid occurs even at room temperature, enabling formation of a transparent, flexible, and conductive grid on heat-sensitive substrates. The process yielded a conductive array having a very low sheet resistance of 9–0.8 Ω/sq. and a transparency above 75%.

This chapter discusses the utilization of a unique, simple wet deposition approach namely, mesh-assisted self-assembly technique, for the fabrication of semitransparent perovskite-based solar cells on one hand, and the fabrication and characterization of novel mesoporous–planar hybrid perovskite cells on the other. The deposition method, used for the fabrication of semitransparent perovskite cells, is fundamentally different from previously reported deposition methods used for $CH_3NH_3PbI_3$ ($MAPbI_3$) perovskite. The film formation in this method is achieved by the mesh-assisted assembly of the perovskite solution through wetting along the wall of a conventional screen printing mesh, as described earlier [18]. The development of a novel architecture of mesoporous–planar hybrid architecture of PSCs, which relies on the self-assembling of *meso*-TiO_2 layer as a support for the perovskite-based cells, is also under the scope of this chapter. The chapter is hence subdivided into two sections, namely:

A) Semitransparent perovskite devices through self-assembly of perovskite
B) Mesoporous–planar hybrid perovskite devices through mesh-assisted self-assembly of mesoporous-TiO_2.

This is followed by a general summary as well as a future perspective, enabling the readers to get updated information on the topic and enhance their understanding toward the utilization of commonly used printing tools, i.e. mesh-assisted well-ordered, precisely controllable self-assembly for achieving the goals of semitransparent to meso–planar hybrid perovskite cells.

10.1.1 Semitransparent Perovskite Solar Cells Through Self-assembly of Perovskite in One Step

This section describes the fabrication of semitransparent perovskite devices using mesh-assisted self-assembly of perovskite layer on top of already prepared photo-anodes with Glass/FTO/BL/mesoporous TiO_2 structure. The well-aligned semitransparent perovskite layer was deposited along a *controlled pattern* through a *solution-process* and *in ambient conditions*, by using mesh-assisted assembly and subsequent annealing step; this was followed by the deposition of hole transporting material (HTM), specifically 2,2′,7,7′-Tetrakis-(*N*,*N*-di-4-methoxyphenylamino)-9,9′-spirobifluorene (spiro-OMeTAD). The cells were completed with the deposition of gold back contact. The HTM-free PSCs (without spiro-OMeTAD) have also been fabricated in order to compare the performance and device operation mechanism.

Nontransparent and HTM-free PSCs have already been reported to show 11.2% PCE [20–22]. The solution concentrations (wt%) and mesh openings of the printing screen were varied in order to achieve a precise control over the transparency. Semitransparent perovskite cells with 20–70% transparency have been developed and characterized for their morphology, layer thickness, average transparency, and PV performance. Ultrahigh resolution scanning electron microscopy (UHR-SEM), optical microscope, and a profilometer were used to characterize the perovskite grids. Intensity-modulated photovoltage spectroscopy (IMVS) was performed for the analysis of the recombination processes occurring in this unique structure of semitransparent PSCs and for comparing it with that of HTM-free cells.

10.1.1.1 Cell Architecture and Morphology

This work describes the preparation of semitransparent perovskite-based solar cells. The one-step deposition solution was used here for the deposition of perovskite layer using simple, low-cost mesh-assisted assembly deposition method (followed by annealing process) under ambient conditions. A strong correlation between the photovoltaic performances of the cells and its average transparency in entire visible range, i.e. 400–800 nm, has been observed in transparency measurements and photovoltaic characterizations of devices. The photovoltaic performance and transparency of cells have been found to vary with the change in the parameters of the perovskite deposition process (mesh opening and solution concentration).

The complete structure of the perovskite-based solar cells follows the order: FTO glass/compact TiO_2/mesoporous TiO_2/perovskite grid/spiro-OMeTAD/thin film of gold (as shown in Figure 10.2a); whereas the structure was similar, excluding the spiro-OMeTAD layer in the case of the HTM-free semitransparent PSCs (as shown in Figure 10.2b). The controlled voids in the perovskite grid, which introduce the transparency in these solar cells, are found to have a direct contact between the HTM and the mesoporous TiO_2 film in case of the cells with HTM and between the gold back contact and the mesoporous TiO_2 film in the case of HTM-free solar cells. Interestingly, despite those voids, the semitransparent devices with and without HTM showed PCE of 4.98% and 2.55%, respectively with 19% average transparency. The perovskite grid was formed by the wetting of the perovskite precursor solution along the walls of a screen printing mesh by placing it over an already deposited mesoporous TiO_2 layer. The uniqueness of the method is a simple control over the transparency of the cell by varying two parameters, (i) mesh opening and (ii) the precursor solution concentration.

A solution of PbI_2 and MAI in dimethyl formamide (DMF) was dropped over the screen printing mesh, after placing it carefully upon already deposited layers of the compact and mesoporous TiO_2. The precursor solution immediately wets the surface and fills the gap between the screen printing mesh and the mesoporous TiO_2 layer, due to the presence of the wetting agent. During solvent evaporation, the PbI_2 and MAI molecules align along the mesh wires due to capillary forces. The screen printing mesh was removed after the complete evaporation of

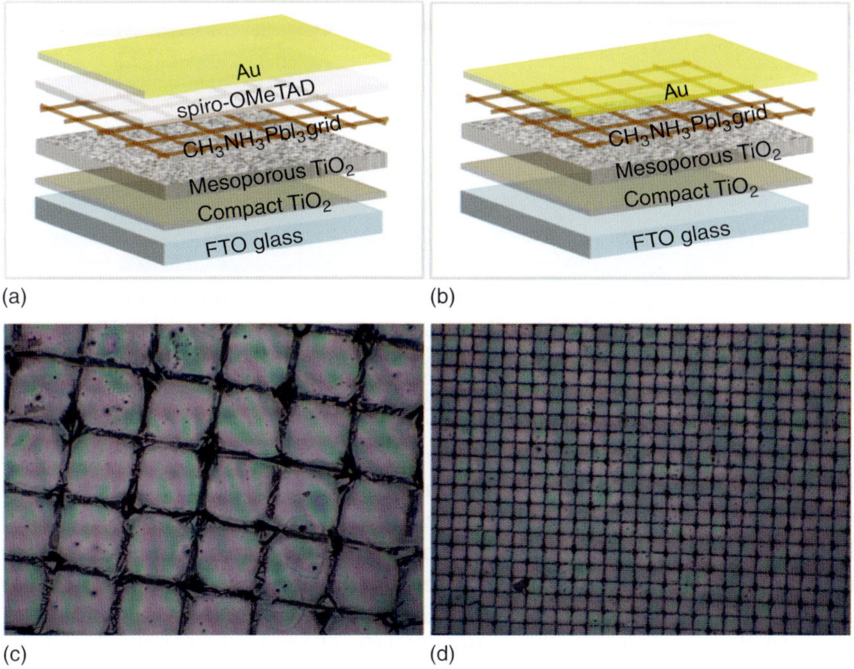

Figure 10.2 (a) The structure of the semitransparent cell: FTO glass/TiO_2 compact layer/mesoporous TiO_2/grid of $CH_3NH_3PbI_3$/spiro-OMeTAD/gold. (b) The structure of the HTM-free semitransparent cell: FTO glass/TiO_2 compact layer/mesoporous TiO_2/grid of $CH_3NH_3PbI_3$/gold. (c) Optical microscopic image of the wide grid: opening of ~200 μm. (d) Optical microscopic image of the dense grid: opening of ~60 μm.

the solvent (DMF), and a grid pattern was obtained. The walls of these grids are formed of crystalline perovskite synthesized from one-step solution of PbI_2 and MAI. An additional annealing step is necessary to complete the crystallization.

The optical-microscopic images of two different $CH_3NH_3PbI_3$ perovskite grids fabricated on top of the mesoporous TiO_2 surface, are shown in Figure 10.2c,d, using meshes with different openings: wide opening (Figure 10.2c, grid opening of ~200 μm), and narrow opening (Figure 10.2d, grid opening of ~60 μm). One of the advantages of this process is the easy control over the grid opening size and thus an actual control over the transparency of the perovskite film (which is actually empty squares surrounded by perovskite crystals) by changing the mesh opening size. Hence, the PCE is directly affected by the grid opening (or squares of perovskite crystals along the mesh wall), i.e. when the grid opening is large, the PCE decreases since the active area is smaller and there are more contact points between the HTM and the mesoporous TiO_2. Moreover, there are no perovskite crystals present inside the open areas of the squares in the perovskite grid as shown in Figures 10.2c,d and 10.3a. The efficient control over the final assembly of the perovskite attained by using this deposition method is evident from the absence of perovskite crystals in the voids.

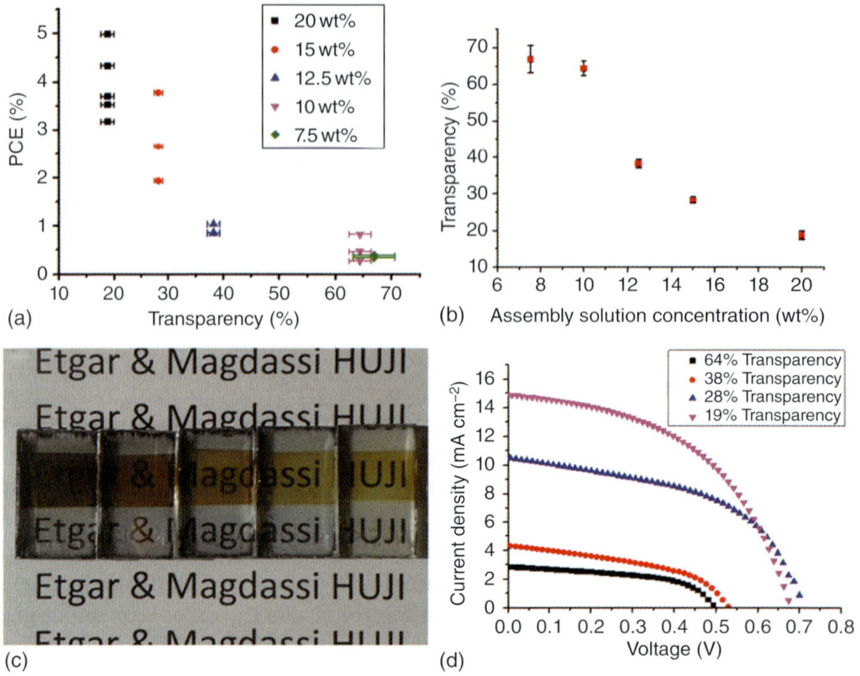

Figure 10.5 (a) The PCE of the semitransparent cells vs their average transparency calculated in the range of 400–800 nm wavelengths (the different colors represent the different solution concentrations). (b) The average transparency of the perovskite solar cells vs the concentration (wt%) of the assembly solution of the precursors. (c) Images of the semitransparent cells with different average transparencies from left to right, 19%, 28%, 38%, 64%, and 67%. (d) The current–voltage curves of the semitransparent cells with different average transparencies. The PCEs of the cells are 0.83%, 1.04%, 3.79%, and 4.98% for the cells of 64%, 38%, 28%, and 19% transparency, respectively.

Table 10.2 Photovoltaic parameters of the various HTM-free semitransparent cells, resulted from similar assembly-solution concentration.

Assembly-solution concentration (%)	V_{oc} (V)	J_{sc} (mA cm^{-2})	FF (%)	Efficiency (%)
20	0.54	11.70	40	2.55
15	0.62	8.94	45	2.51
12.5	0.54	4.17	39	0.89
10	0.23	1.25	31	0.09
7.5	0.20	0.10	31	0.01

accurately design the required coverage and transparency of the solar cell, using this unique mesh-assisted assembly deposition method. The coverage of the cell is the fraction of area that is covered by the perovskite grid lines, without the grid voids (in which there is no perovskite). It was observed that the coverage of the PSCs prepared using assembly solution concentrations of 7.5, 10, 12.5, and 15 wt% was similar (~35 ±3% of the area is covered by the perovskite grid lines), whereas the coverage of the cells that were prepared when using assembly solution concentration of 20 wt% was different (~50%). The increase in the coverage for the 20 wt% solution concentration can be expected already from the profile measurements (Figure 10.3a) in which the width of the grid lines displayed about the same average values in the case of the lower concentrations (till 15 wt%). However, despite the fact that the coverage of the cells in the case of the lower assembly solution concentrations was similar, the PV performance of the cells was different, as shown in Table 10.1. The reason for the difference in PV performance might be because the coverage calculations do not take into account the thickness of the perovskite lines, although it drastically affects the photovoltaic performance of the device.

10.1.1.3 Recombination Behavior of the Charges in Cells

We highlight an interesting outcome that is related to the unique structure of the semitransparent PSCs based on less coverage of substrate surface: there is an intrinsic direct contact between the HTM and the mesoporous TiO_2 film (or between the gold back contact and the mesoporous TiO_2 film, in case of HTM-free perovskite cells), resulting in the formation of a large number of recombination centers. However, a reasonable PV performance was evident. For example, in the case of 50% coverage, J_{sc} of 14.87 mA cm^{-2} with V_{oc} of 0.68 V were achieved. The smooth working of these cells, despite the presence of the large recombination centers, can be attributed to the large diffusion length [20] and absorption coefficient [23] of $MAPbI_3$. Moreover, in the absence of HTM, a J_{sc} of 9 mA cm^{-2} with V_{oc} of 0.63 V was achieved, which supports our previous reports on the possibility to avoid the need for HTM in the PSCs and demonstrate the superior properties of the perovskite [20, 21, 24].

In case of semitransparent solar cells that include the HTM layer, the recombination centers mainly lie at the interface between spiro-OMeTAD/mesoporous TiO_2 (please refer to Figure 10.3), whereas most of the recombination may occur at the interface mesoporous TiO_2/Au (please refer to Figure 10.2b) in the HTM-free semitransparent cells. It can be further explained with the help of IMVS characterization of cells. The IMVS measurements performed on the semitransparent cells with and without HTM layer, provided insight on the recombination life time [25–27] at these interfaces, as shown in Figure 10.6.

The drastic increase in recombination lifetime of HTM-containing cells as compared to that of HTM-free cells is evident from Figure 10.6. Interestingly, this main recombination channel has almost constant recombination lifetime, with a slight decrease on increasing the voltage, even in the presence of a direct contact of the HTM and the mesoporous TiO_2 interface.

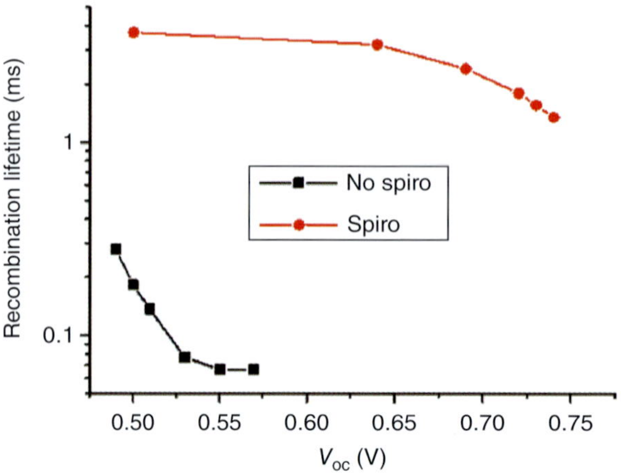

Figure 10.6 The recombination lifetimes that were extracted from the IMVS measurement for the cells with and without HTM vs the V_{OC}.

Conclusions Semitransparent perovskite-based solar cells were fabricated using simple mesh-assisted deposition method. The transparency of the cells was controlled by varying the mesh opening used for the perovskite grid assembly and by changing the concentrations of the perovskite precursor solutions. The well confined perovskite film within the grid lines are confirmed with the help of optical microscopy, UHR–SEM, and profilometer. The best PV performance for the semitransparent solar cells showed PCE of 4.98% with transparency of 19%, which is comparable to a recent report on semitransparent perovskite-based solar cell [16]. Possibility exists for further improvement by better control of additional parameters of the grid formation, such as the solvent evaporation rate, the wetting agent type, and concentration, which could result in even lesser defects in the perovskite grid and could improve the cell performance.

10.1.2 Mesoporous–Planar Hybrid Perovskite Devices Through Mesh-assisted Self-assembly of Mesoporous-TiO_2

In this section, realization of a unique architecture, i.e. mesoporous–planar hybrid PSCs through self-assembling of *meso*-TiO_2 followed by perovskite deposition is described. The uniqueness of this architecture lies in the presence of both the planar and mesoporous parts in a single cell (e.g. meso–planar hybrid architecture of PSCs), making them a potential structure to see the effects of meso-titania on a perovskite-based solar cell.

The most common architectures of PSCs are namely, mesoporous (meso) and planar device architectures [5]. The *meso*-TiO_2 layer is an essential component of mesoporous cells providing a porous support for the deposited perovskite layer and its infiltration. Most of the cells that report more than 20% PCE are based on mesoporous device architectures (http://www.nrel.gov/ncpv/images/efficiency_chart.jpg) [28, 29] making the mesoporous architecture a preferred choice, despite the fact that the exact role of *meso*-TiO_2 in these cells is yet to be well established [22, 30, 31]. An excellent PCE comes with an additional cost of the meso layer, making the fabrication more complex and also the phenomenon of hysteresis, which is quite common in *meso*-TiO_2 based PSCs. The realization of planar cells is a more generic and simple approach, as the perovskite layer is sandwiched between a hole blocking layer of compact titania, and HTM in these cells [32–36]. The demonstration of significant PCE even in the absence of a *meso*-TiO_2 layer is one of the major advantages that establish the popularity of the planar cell architecture of PSCs [32–39]. The planar cells are preferred in most of the studies, as they need less fabrication cost and eliminate the TiO_2-sintering step at high temperature. Planar cells with flat and inverted structures (based on the sequence of the layers on the device substrate) using organic electron transporting material (ETM/HTM) are seeking more attention nowadays because they are associated with less or almost negligible hysteresis as compared to their meso counterparts [37–39].

Combining mesoporous and planar architecture in a single cell to utilize the benefits of both the structures and getting an intermediate architecture for detailed study of the behavior of *meso*-TiO_2 in a PSC motivated us for the extension of our work in this direction. A mesh-assisted, directional self-assembly of TiO_2 nanoparticle layer (*meso*-TiO_2 layer) and subsequent perovskite deposition through a two-step method made the hybrid architecture possible. The deposition of HTM, specifically 2,2′,7,7′-Tetrakis-(*N*,*N*-di-4-methoxyphenylamino)-9,9′-spirobifluorene (spiro-OMeTAD) was deposited over the perovskite layer in these cells. The cells were then completed with the deposition of gold back contact.

In this work, the fabrication and characterization of novel meso–planar hybrid architecture of PSCs have been discussed in detail. High resolution scanning electron microscopy (HR-SEM) and optical microscopy have been used for the morphological characterization of the meso–planar hybrid layer and the fabricated cells. Additionally, the PV performance of these cells was compared with those of traditional meso- and planar-architectures. The effect of varying the meso component in the cells by changing the areal ratio of the meso/planar component on the performance of these cells has also been evaluated. The electronic properties of the hybrid architecture cells were analyzed through charge extraction measurements and compared with their meso and planar counterparts. The meso–planar hybrid cells provide a mechanism to compare the electronic properties of different architectures and to study the effect of the presence or absence of *meso*-TiO_2 on the photovoltaic parameters of the cells.

10.1.2.1 Cell Architecture and Morphology

A schematic representation of the different architectures of PSCs, namely the mesoporous, the meso–planar hybrid, and the planar PSCs are shown in Figure 10.7a–c, respectively. The arrows between the figures show the intermediate structure, i.e. meso–planar hybrid cells.

The common printing mesh-assisted self-assembly of the TiO_2 nanoparticles (NPs) form a *meso*-TiO_2 grid, which has been realized to serve as a base for the meso–planar hybrid PSCs. This TiO_2 layer was deposited on already etched and cleaned electrodes, containing a thin hole blocking layer of compact-TiO_2. The meso-titania's self-assembly took place along the mesh walls, which resulted in the formation of junction points as a hill and centers of the mesh as a valley without the presence of meso-titania. The technique has been already implemented

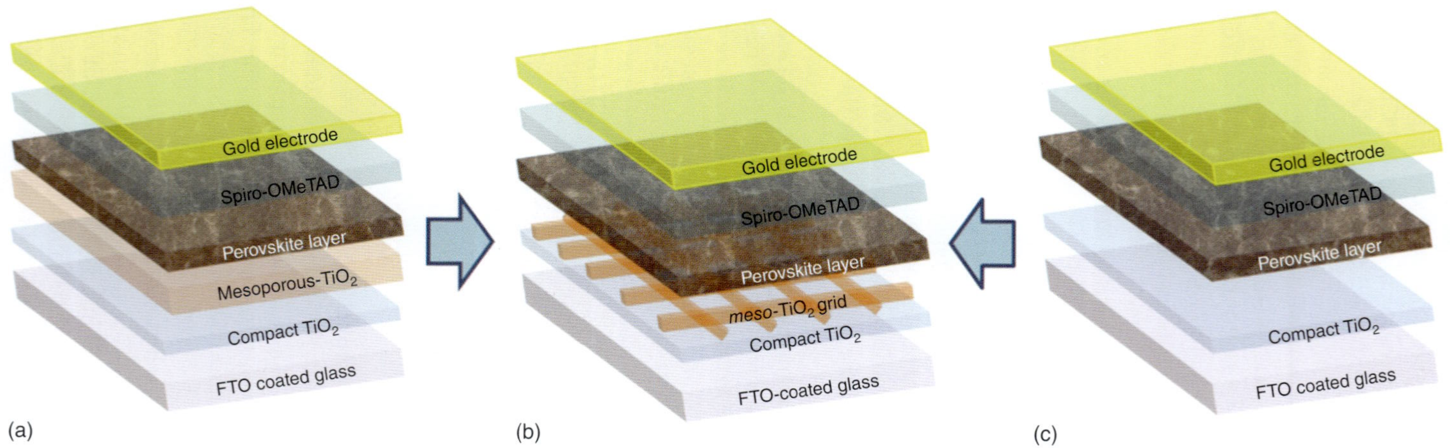

Figure 10.7 Schematic diagram of (a) mesoporous structure; (b) meso–planar hybrid structure; and (c) planar structure, of perovskite solar cells; the arrows show that the hybrid structure falls in between these two designs. *Source:* Misra et al. 2016 [17]. Reproduced with permission from Royal Society of Chemistry.

successfully by our group to realize the semitransparent PSCs as discussed in the previous section, and as also in the previous work on conducting grid of silver nanoparticles (Ag NPs) by our group and transparent conducting grid deposition by Higashitani et al. [16, 40, 41] In this study, we used metallic meshes with different mesh sizes of 50 and 72 μm for *meso*-TiO_2 grid deposition, to achieve areal variation (i.e. to vary the ratio between the meso and planar part of the cells). A methyl ammonium lead iodide ($MAPbI_3$) perovskite layer was deposited subsequently using the two-step process after the formation of a *meso*-TiO_2 grid. The cells were then completed by the deposition of spiro-OMeTAD, the HTM, and evaporation of gold back contact. The hybrid meso–planar cells have areas with and without mesoporous TiO_2 as shown in Figure 10.7b, enabling them to work as a hybrid between the mesoporous and planar architecture of PSCs.

Figure 10.8a,b shows the top view optical images of self-assembled *meso*-TiO_2 grid (with 50 μm mesh opening) and a completed meso–planar hybrid PSC. The preferential self-assembly of *meso*-TiO_2 along the wires of the mesh, due to the capillary forces and coffee-ring effect is evident from Figure 10.8a. Figure 10.8b shows the TiO_2 grid covered with the perovskite film on top of it; the dense perovskite coverage can be seen in the form of circles with dark color, whereas the area with the lighter part of the image shows the least perovskite covered parts.

HR-SEM top view image of the same grid is shown in Figure 10.8c. The dark grey junction lines are due to the higher density of TiO_2 alongwith the mesh wires, whereas decrease in the density of the junction while moving from the

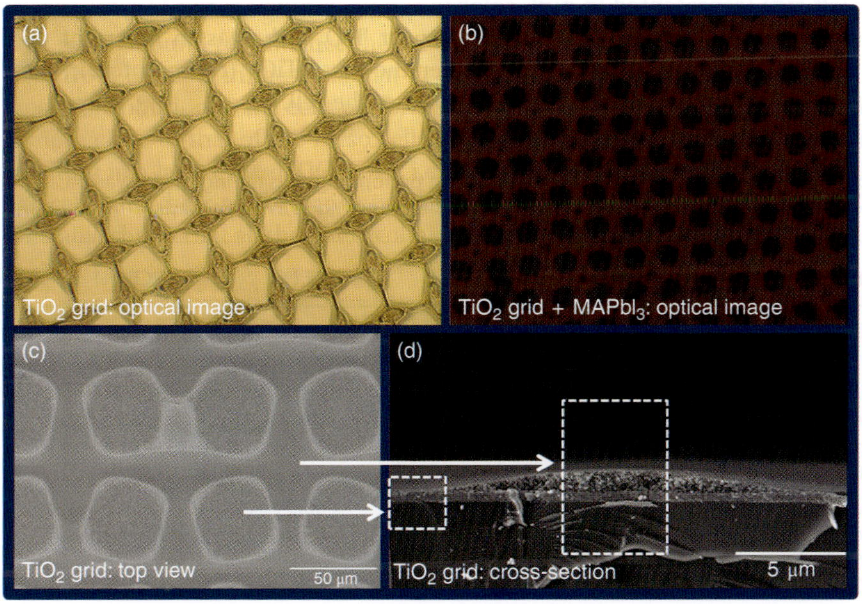

Figure 10.8 Optical image of the (a) mesoporous-titania grid deposited with the help of a mesh (mesh size 50 μm), (b) perovskite deposited on TiO_2 grid; (c) top view SEM image, (d) cross-sectional SEM image of the grid. *Source:* Misra et al. 2016 [17]. Reproduced with permission from Royal Society of Chemistry.

of the cells decrease while moving from meso to planar through meso–planar hybrid PSCs with increased planar behavior. Although higher V_{oc} of the planar cell was observed for the best cell fabricated, the average V_{oc} value follows the similar trend of decrease in these parameters with the increase in planarity. The highest FF of about 68.9% for the hybrid cells with 50 μm grid size is found to be the best amongst all other solar cells and can be attributed to the more efficient charge transportation in the presence of meso-titania even at low content, which is in good agreement with the outcome of the charge extraction measurements.

The *J*–*V* curves for representative cells with different architectures are presented in Figure 10.10a. The better short circuit current density J_{sc} and open circuit voltage V_{oc}, reaching up to 20.28 mA cm^{-2} and 0.99 V, can be observed for the mesoporous PSCs followed by the intermediate values for meso–planar hybrid cells and the least for planar cells. The possibility to conserve the current density even with the lesser meso character in hybrid PSCs with 1.08 meso/planar ratio indicates that the presence of *meso*-TiO_2 may not be an essential requirement to achieve high current in PSCs; hence, lesser amount of meso-titania in these cells is sufficient to provide good charge transportation and high current density.

Figure 10.10b shows the comparative EQE graphs for these cells. Better photon absorption and charge collection in the entire MAPbI3 absorption range for both the mesoporous and meso–planar hybrid cells are evident from the figure. This may be further confirmed by the better matching between the J_{sc} taken from the *J*–*V* measurements and is calculated by the integration of EQE values for both the meso and meso–planar hybrid architectures. However, the lowering in EQE has been observed in the case of planar cells, especially while moving toward the red region of the spectrum. The integrated value of the J_{sc} from EQE measurements is in very good agreement with J_{sc} observed during the *J*–*V* measurement of the meso–planar hybrid (50 μm with 1.08 meso/planar ratio) cell and was found to be 17.1 and 18.3 mA cm^{-2}, respectively. The trend was followed by the meso cells with calculated J_{sc} value of 15.7 mA cm^{-2} compared to 19.1 mA cm^{-2} measured under the solar simulator through *J*–*V* measurements. It is clear from the J_{sc} data in Table 10.1, that the meso–planar hybrid cell with 0.77 meso/planar ratio behaves similarly to the planar cell, and the short circuit current density for both of them is almost equal, i.e. 14.99 and 14.67 for the meso–planar hybrid (72 μm grid) and planar, respectively. The high degree of disagreement in current densities calculated from EQE data and taken from the *J*–*V* scan has been observed in case of planar cells as can be seen in figure 15b; and may be attributed to the measurement of EQE without any pretreatment of the cell architecture, including the planar one. The agreement between the EQE and the solar simulator measurements strengthens after the pretreatment of planar configuration as evidenced in the literature [13].

The lower current density in planar cells can be attributed to less photon absorption or to a difference in the thickness of the perovskite layer in the planar cells, due to the absence of *meso*-TiO_2 layer. This can be supported by the absorption profile of meso and planar solar cells as the planar cell has less absorption in the entire $MAPbI_3$ absorption range, as evident from the Figure 10.10, due to the thinner perovskite layer.

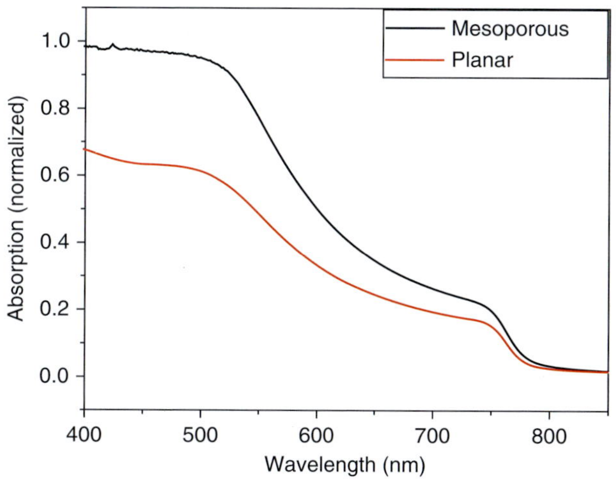

Figure 10.11 Absorption spectra of mesoporous and planar cells, showing normalized spectra in the range of 400–850 nm. *Source:* Misra et al. 2016 [17]. Reproduced with permission from Royal Society of Chemistry.

Figure 10.11a–d shows the top view scanning electron microscopy (SEM) images of the *meso*-TiO_2 grid, and the different densities of perovskite at different places are evident from the figure. The decrease in V_{oc} of meso–planar hybrid cells as compared to that of meso cells has been observed and can be attributed to the presence of recombination centers on the peaks (junction points, please refer to Figure 10.11b,d) of *meso*-TiO_2 in the grid structure, where, in these peaks, the perovskite is either absent or is at low coverage due to the height of the TiO_2. The direct contact between the spiro-OMeTAD and the *meso*-TiO_2, which increases the probability of recombination on these points is responsible for the decreasing V_{oc} in this architecture. The square nature of the J–V curve has been found to improve as results of lesser series and shunt resistance might be the reason for the highest FF value in hybrid PSCs. After improvement in perovskite coverage on less covered junction points, further enhancement in the performance of hybrid PSCs is expected, as a result of the disappearance of recombination centers (Figure 10.12a–d).

The cells were found to behave differently during forward and backward J–V scans, i.e. from J_{sc} to V_{oc} and V_{oc} to J_{sc}, respectively, resulting in the nonoverlapping J–V curves. This means that the cells are not exempt from the commonly occurring issue of hysteresis in PSCs. This behavior is most commonly observed in mesoporous perovskite cells [42], and can be overcome by using organic ETL in planar cells [35, 36]. The extent of hysteresis was different for different architectures, which means it is more related with the interfaces involved. The extent of hysteresis was found to increase while moving from meso to planar cells through hybrid meso–planar cells. This phenomenon may be explained on the basis of the structure of the cells: in meso architecture both the J_{sc} and V_{oc} has been found to be affected with the direction of bias without changing the shape of the curve. On the other hand the current remains unaffected in case of meso–planar hybrid

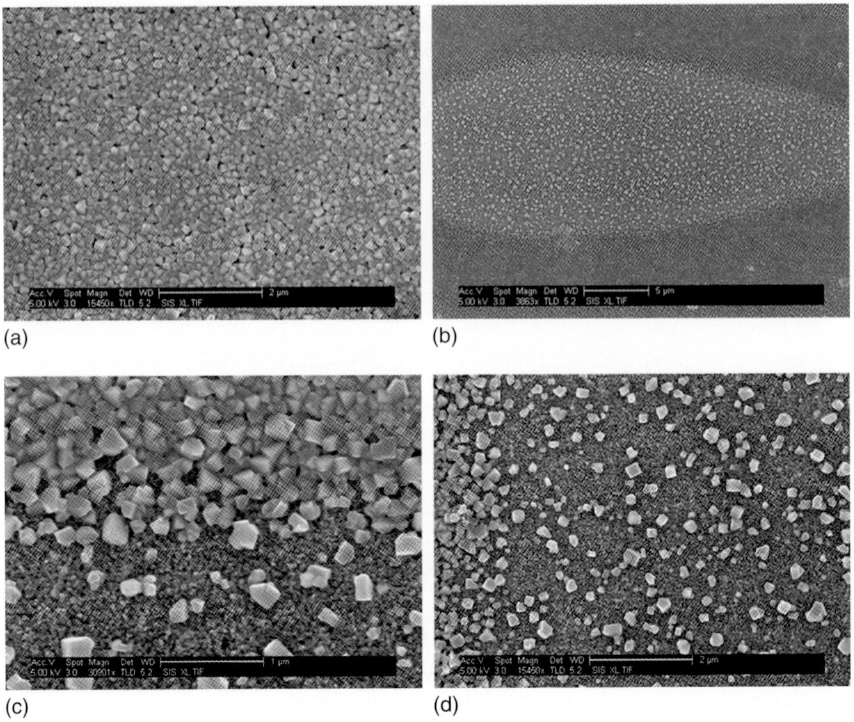

Figure 10.12 (a–d) show the perovskite density at different locations of grid cells, the much dense perovskite coverage is evident in image (a), the valley region of the titania grid, whereas (b–d) show the enlarged view of the junction of meso–planar area and peaks of titania at mesh junction points. These recombination centers are responsible for some of the performance loss of meso–planar hybrid cell structures. *Source:* Misra et al. 2016 [17]. Reproduced with permission from Royal Society of Chemistry.

cells with variation of voltage and introduction of "s" shape in the J–V curve, indicating the charge accumulation during the backward scan in this architecture. Interestingly, the "s" shape in the J–V curve of the backward scan is pronounced in the case of the planar structure compared to the hybrid structure, which further supports that the meso–planar hybrid structure is an intermediate structure between meso and planar architectures.

10.1.2.3 Study of Recombination Behavior Through Charge Extraction

Charge extraction measurement is a well-established and recognized tool for investigating electron transport and interfacial charge transfer in sensitized solar cells [43]. The different steps involved in a typical charge extraction (CE), also illustrated in Figure 10.13a, are as below:

i) A 2 s step in which the cell is being discharge at the dark;
ii) The cell is then being disconnected and illuminated for 2 s (illumination time);
iii) The light is then switched off and the system waits for a certain time called delay time. In this step a charge recombination occurs inside the cell;

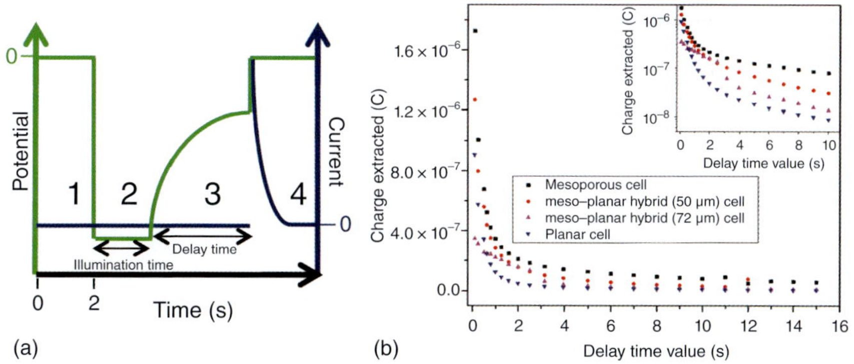

Figure 10.13 (a) The steps involved in the charge extraction measurements. (b) Charge extraction of different types of cells using white LED, inset shows semilogarithmic graph of the charge extraction, presenting the difference between the cells structure. *Source:* Misra et al. 2016 [17]. Reproduced with permission from Royal Society of Chemistry.

iv) The cell is reconnected and the charges which were left and did not recombine are extracted and measured. This process is being repeated for different delay times, from 0.5 to 15 s.

The charge extraction of the different cell architectures vs increasing delay time is shown in Figure 10.13b. The amount of charge extracted is found to follow the order of the meso to meso–planar hybrid followed by planar cells. The CE graphs show that in the case of the meso structure based cells, more charges are left behind to extract after a certain delay time compared to the other architectures; meaning that the rate of recombination in the meso cells is least amongst the other architectures of PSCs. These results are in good agreement with the J–V curves (Figure 10.10a), where the highest voltage has been obtained for mesoporous cells. The intermediate values of CE curves of the meso–planar hybrid cell (meso/planar ratio 1.08 with 50 μm grid and meso/planar ratio 0.77 with 72 μm grid) between the meso and the planar structures suggest that the rate of recombination in the hybrid cells is higher than the meso but lower than the planar structure. This can be attributed to the high junctions of the TiO_2 grid which possess a direct contact between the TiO_2 and the spiro, resulting in recombination centers. These recombination centers are the major cause of reduced voltage for the hybrid architectures, as also supported by the J–V measurements. The semi-logarithmic plot of the CE curves at the inset of Figure 10.13b clearly shows the closeness of these hybrid architectures with the mesoporous cells as compared to the planar structure, which means that the presence of *meso*-TiO_2, even in lesser coverage of substrate, decreases the rate of recombination in a cell architecture.

Conclusions To summarize, we have successfully fabricated and characterized a novel architecture namely, meso–planar hybrid perovskite-based solar cells. The fabrication of these hybrid cells is done via mesh-assisted self-assembly of a *meso*-TiO_2 layer, followed by perovskite deposition. The photovoltaic performance of these cells has been evaluated and compared with that of mesoporous and planar architectures. Interestingly, the cells were found to conserve the short circuit current density, as compared to their meso counterparts. The decrease in V_{oc} observed can be attributed to the complex structure of the cell with and without *meso*-TiO_2 at different places, resulting in two types of interfaces between the electron transporting layer and perovskite, specifically, ETL/perovskite (from the planar region) and ETL/*meso*-TiO_2/perovskite (in the meso region) of the cell. The intermediate photovoltaic properties between the meso and planar architectures have been observed for these hybrid cells. The effect of the architecture on the photovoltaic parameters of the cells has also been studied through variation of meso/planar ratio. Charge extraction measurements reveal the recombination rate in these different cell architectures. These cells were successfully demonstrated to function as an intermediate between meso and planar architectures and might be of potential importance during the investigation of working mechanisms of different perovskite cells under operational conditions.

10.2 Summary and Future Perspective

The mesh-assisted self-assembly technique for the deposition of different layers of perovskite-based solar cells for the fabrication of semitransparent solar cells has been successfully demonstrated. The cells with PCE up to 7% efficiency and 20% average transmittance have been demonstrated and characterized thoroughly.

The similar self-assembly technique is utilized for realization of a novel mesoporous–planar hybrid architecture of perovskite cells in which a grid of meso-titania was fabricated through mesh-assisted assembly, and the device was completed subsequently through two-step perovskite deposition. The cells were characterized extensively, and compared with their meso and planar counterparts. These hybrid cells may serve as an interesting tool to investigate the internal device mechanisms during standard operational conditions.

The utilization of the self-assembly technique, during different steps of perovskite cell fabrication, is unique of its type. The involvement of common printing tools in PSC fabrication is of great importance, and opens new avenues for fully printed perovskite devices.

In addition, the mesh-assisted technique provides a possible way for the scaling up of these perovskites-based solar cells, which is certainly an important step toward their commercialization.

References

1 Lim, J.W., Lee, S.H., Lee, D.J. et al. (2013). *Thin Solid Films* 547: 212–215.
2 Chae, Y.T., Kim, J., Park, H., and Shin, B. (2014). *Appl. Energy* 129: 217–227.
3 Lin, H.W., Chen, Y.H., Huang, Z.Y. et al. (2012). *Org. Electron.* 13: 1722–1728.
4 Reinhard, M., Eckstein, R., Slobodsky, A. et al. (2013). *Org. Electron.* 14: 273–277.
5 Mescher, J., Kettlitz, S.W., Christ, N. et al. (2014). *Org. Electron.* 15: 1476–1480.

6 Czolk, J., Puetz, A., Kutsarov, D. et al. (2013). *Adv. Energy Matter* 3: 386–390.
7 Park, N.G. and Kim, K. (2008). *Phys. Status Solidi A* 205 (8): 1895–1904.
8 Pandey, A.K. and Samuel, I.D.W. (2010). *IEEE J. Sel. Top. Quantum Electron.* 16 (6): 1560–1564.
9 Kojima, A., Teshima, K., Shirai, Y., and Miyasaka, T. (2009). *J. Am. Chem. Soc.* 131: 6050–6051.
10 Malinkiewics, O., Yella, A., Lee, Y.H. et al. (2014). *Nat. Photonics* 8: 128–132.
11 Eperon, G.E., Burlakov, V.M., Docampo, B. et al. (2014). *Adv. Funct. Mater.* 24: 151–157.
12 Eperon, G.E., Burlakov, V.M., Goriely, A., and Snaith, H.J. (2014). *ACS Nano* 8 (1): 591–598.
13 Roldan, C., Malinkiewicz, O., Betancur, R. et al. (2014). *Energy Environ. Sci.* 7: 2968–2973.
14 Ono, L.K., Wang, S., Kato, Y. et al. (2014). *Energy Environ. Sci.* 7: 3989–3993.
15 Liu, Y., Zhou, H., Hong, Z. et al. (2014). *J. Am. Chem. Soc.* 136: 622–625.
16 Aharon, S., Layani, M., Cohen, B.-E. et al. (2015). *Adv. Mater. Interfaces* 1500118.
17 Misra, R.K., Aharon, S., Layani, M. et al. (2016). *J. Mater. Chem. A* 4: 14423–14429.
18 Layani, M. and Magdassi, S. (2011). *J. Mater. Chem.* 21: 15378–15382.
19 Guo, F., Azimi, H., Hou, Y. et al. (2015). *Nanoscale* 7: 1642–1649.
20 Aharon, S., Cohen, B.E., and Etgar, L. (2014). *J. Phys. Chem. C* 118: 17160–17165.
21 Abu-Laban, W. and Etgar, L. (2013). *Energy Environ. Sci.* 6: 3249–3253.
22 Stranks, S.D., Eperon, G.E., Grancini, G. et al. (2013). *Science* 342: 341–344.
23 Eperon, G.E., Bryant, D., Troughton, J. et al. (2015). *J. Phys. Chem. Lett.* 6: 129–138.
24 Cohen, B.E., Gamliel, S., and Etgar, L. (2014). *APL Mater.* 2 (1–8): 081502.
25 Guillen, E., Ramos, F.J., Anta, J.A., and Ahmad, S. (2014). *J. Phys. Chem. C* 118: 22913–22922.
26 Zaoh, Y., Nardes, A.M., and Zhu, K. (2014). *J. Phys. Chem. Lett.* 5: 490–494.
27 Zaoh, Y. and Zhu, K. (2013). *J. Phys. Chem. Lett.* 4: 2880–2884.
28 Saliba, M., Matsui, T., Seo, J.-Y. et al. (2016). *Energy Environ. Sci.* 9: 1989–1997.
29 Saliba, M., Orlandi, S., Matsui, T. et al. (2016). *Nat. Energy* 15017. doi: 10.1038/NENERGY.2015.17.
30 Xing, G., Mathews, N., Sun, S. et al. (2013). *Science* 342: 344–347.
31 Kim, H.S., Mora-Sero, I., Gonzalez-Pedro, V. et al. *Nat. Commun.* 4: 2242.
32 Hou, F., Su, Z., Jin, F. et al. (2015). *Nanoscale* 7: 9427–9432.
33 Sum, T.C. and Mathews, N. (2014). *Energy Environ. Sci.* 7: 2518–2534.
34 Wang, K., Liu, C., Du, P. et al. (2015). *Energy Environ. Sci.* 8: 1245–1255.
35 Xiao, M., Huang, F., Huang, W. et al. (2014). *Angew. Chem. Int. Ed.* 53: 9898–9903.
36 Ye, S., Sun, W., Li, Y. et al. (2015). *Nano Lett.* 15: 3723–3728.
37 Wu, C.G., Chiang, C.H., Tseng, Z.L. et al. (2015). *Energy Environ. Sci.* 8: 2725–2733.
38 Heo, J.H., Han, H.J., Kim, D. et al. (2015). *Energy Environ. Sci.* 8: 1602–1608.

39 Kim, S.S., Bae, S., and Jo, W.H. (2015). *Chem. Commun.* 51: 17413–17416.

40 Layani, M., Darmawan, P., Foo, W.L. et al. (2014). *Nanoscale* 6: 4572.

41 Higashitani, K., McNamee, C.E., and Nakayama, M. (2011). *Langmuir* 2080–2083.

42 Jeon, N.J., Noh, J.H., Kim, Y.C. et al. (2014). *Nat. Mater.* 13: 897–903.

43 Duffy, N.W., Peter, L.M., Rajpakshe, R.M.G., and Wijayantha, K.G.U. (2000). *Electrochem. Commun.* 2: 658–662.

11

Flexible Organic Solar Cells

Lin Hu, Youyu Jiang, and Yinhua Zhou

Huazhong University of Science and Technology, School of Optical and Electronic Information, Wuhan National Laboratory for Optoelectronics, 1037 Luoyu Road, Wuhan, Hubei 430074, China

11.1 Introduction

Clean and renewable energy are highly desirable for sustainable development. Solar energy is clean, renewable, and is available in huge amounts, and is an ideal energy source to meet the world's energy demands. Photovoltaic (PV) cells or solar cells, which directly convert solar energy into electricity, are one of the most promising and efficient devices to harvest solar energy. In recent years, the need to continually increase the power conversion efficiency (PCE) and decrease the cost of PV systems has pushed PV technology into the electricity markets and enter the multi-terawatt regime [1–3].

To date, inorganic solar cells based on materials such as crystalline silicon, cadmium telluride, or copper indium gallium diselenide (CIGS) have been mature devices exhibiting high solar energy conversion efficiencies of around 20–25%, and, have dominated the photovoltaic market [4, 5]. However, the production of these inorganic solar cells continues to encounter problems such as huge energy consumption and environmental pollution. The current cost of a silicon solar cell panel is about 0.5 US$ W^{-1}, which is about five times the cost of coal electricity. This enormous cost has spurred research to explore alternative materials and technologies to reduce the cost of devices for solar energy utilization. In recent years, several emerging technologies, i.e. organic solar cells (OSCs), perovskite solar cells, and dye-sensitized solar cells have been rapidly developed. In this chapter, we will focus on OSCs that are composed of organic semiconductors as the active layers for light harvesting. The OSCs can be fully solution-processed from inks, (by printing, slot-die coating, or others) are light-weight and mechanically flexible, and are easy to be produced as semitransparent organic cells and in vivid colors. Organic semiconductors intrinsically have larger exciton binding energy and shorter exciton diffusion length compared to their inorganic counterparts. These properties require the particular device architecture to achieve high PCE. This chapter will discuss most of the important features of (flexible) OSCs as well as their working principles that include device structure, active layers, electrodes, interfaces, tandem structure, and fabrication.

Flexible Energy Conversion and Storage Devices, First Edition. Edited by Chunyi Zhi and Liming Dai.

11.1.1 Working Principle

The first successful OSC was reported by Tang [6]. He fabricated the configuration by constructing a bilayer thin-film heterojunction. The PCE was only about 1%. For the bilayer heterojunction architecture, the limited donor–acceptor (D–A) interface for exciton dissociation could influence the charge carrier generation efficiency because of the small exciton diffusion length (about 10 nm). Another type of device architecture is "bulk heterojunction (BHJ)" that was introduced by Yu et al. [7]. The BHJ active layer can be achieved by blending the donor and the acceptor materials together in solution and casting to film. A bicontinuous interpenetrating network with a large D–A interface is achievable. In this way, the dissociation, transportation, and collection of the excitons became more efficient, eventually leading to a significant enhanced conversion efficiency [8, 9]. Since then, BHJ OSCs based on blends of organic semiconductor and fullerene derivatives have become the main devices to be adopted by the community [3, 10]. Nowadays, PCEs of 11% have been achieved in single-heterojunction PSCs based on both fullerene and non-fullerene acceptors in small active area (<1 cm^2) [11, 12].

As shown in Figure 11.1, the fundamental mechanism of the conversion from photons to electrons can be classified into four main steps in an OSC device [13, 14]. First, the active layers absorb the light energy to generate excitons (bound electron–hole pairs) when the OSC device is illuminated by the Sun or other light sources (Step 1). Then these bound electron–hole pairs diffuse to the donor and acceptor interface (Step 2). Some excitons may recombine and release the energy to the surrounding (nonradiative recombination) and some may recombine and emit photons (radiative recombination) during the diffusion process. The others may dissociate into free electrons and holes at the interfaces due to the lowest unoccupied molecular orbitals (LUMOs) offset between the donor and acceptor as well as a strong internal electric field formed by the work function difference between two electrodes (Step 3). Finally, free charge carriers (electrons and holes) drift to the respective electrodes in the opposite directions with the aid of the internal electric field, thereby generating photocurrent and photovoltage (Step 4).

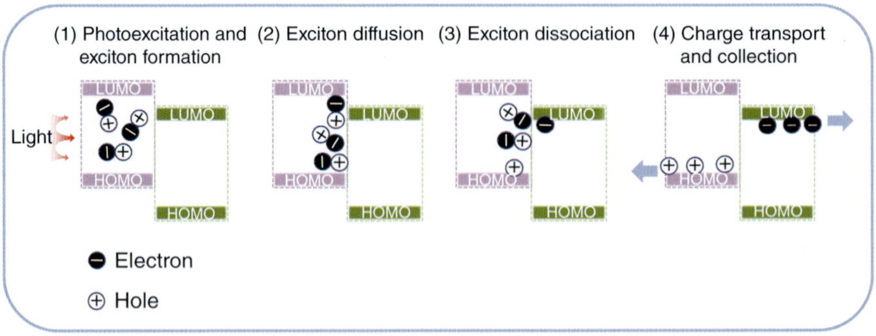

Figure 11.1 Schematic outline of the conversion from photons to electrons in OSCs.

11.1.2 Performance Characterization of OSCs

OSCs are evaluated in the same way as other types of solar cells. The electric power output is the product of photo-induced current (I) and voltage (V). The PCE is the ratio of the maximum power produced by the cell and the incident light power. For laboratory purposes, the incident light source is a simulated sunlight source with a power density P_{in} of 100 mW cm^{-2}. For determining the power density output, current density–voltage ($J-V$) curves are measured. At a certain voltage (V_{max}), corresponding to a certain current density (J_{max}), the power density output reaches a maximum (P_{max}) (Figures 11.2 and 11.3b). The PCE is calculated as P_{max}/P_{in}.

$$\mathrm{PCE}_{max} = \frac{P_{max}}{P_{in}} = \frac{(JV)_{max}}{P_{in}} = \mathrm{FF}\frac{J_{sc}V_{oc}}{P_{in}} \tag{11.1}$$

where J_{sc} is the short-circuit current density, V_{oc} is the open-circuit voltage, and the ratio $P_{max}/J_{sc}V_{oc}$ is called the fill factor (FF).

In addition, to evaluate the photon-to-electron conversion at each wavelength, external quantum efficiency (EQE) are essential. EQE is defined as the number of

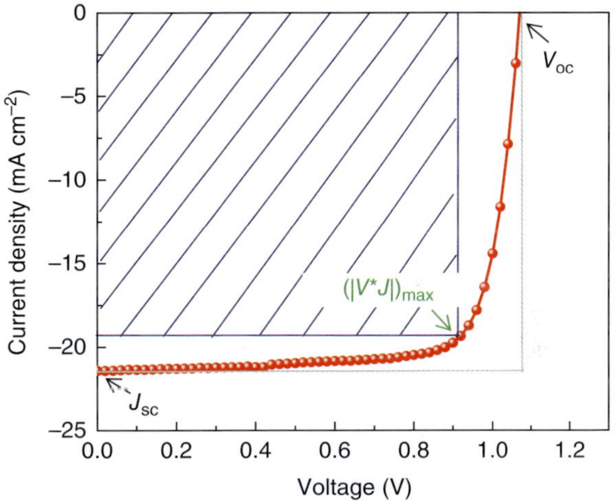

Figure 11.2 Typical *J–V* characteristics of a polymer solar cell.

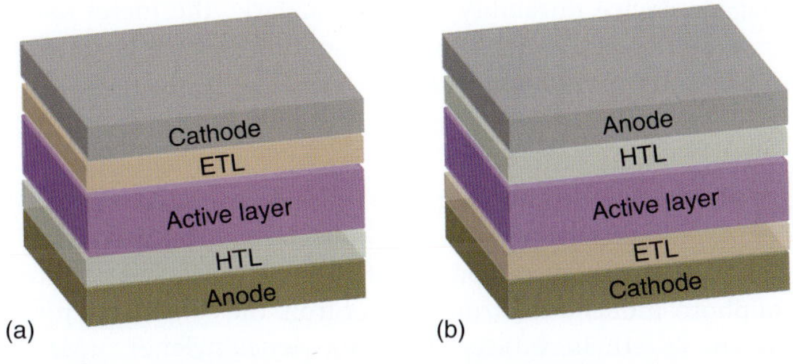

Figure 11.3 Schematic diagram of (a) a conventional device structure and (b) an inverted device structure.

extracted electrons from an OSC illuminated by a monochromatic light divided by the number of incident photons to the OSC (Equation (11.2)) where the J_{ph} is photo-induced current density, λ is the wavelength of the monochromatic light, and P_{in} is the power density input of the monochromatic light.

$$\mathrm{EQE} = \frac{\text{number of extracted electrons}}{\text{number of incident photons}} = 1240\frac{J_{ph}}{\lambda P_{in}} \tag{11.2}$$

11.1.3 Device Structure

11.1.3.1 Conventional Device Structure

A conventional OSC device is usually fabricated by subsequently depositing a hole transporting layer (HTL), an active layer, an electron transporting layer (ETL), and a metal layer on tin-doped indium oxide (ITO) or fluorine-doped tin oxide (FTO) electrodes covered glass substrate as shown in Figure 11.3a [15–17]. Typically, the p-doped conducting polymer poly(3,4-ethylenedioxythiophene):poly(styrenesulfonate) (PEDOT:PSS) is used as HTL to extract and transport holes. On the other hand, low-function metals LiF/Al and Ca/Al are often used as the top cathode to ensure effective electron extraction. The low work function metal is vulnerable to oxidation under ambient conditions, which would result in device degradation [18]. Poor device stability is a major issue in a conventional OSC device.

11.1.3.2 Inverted Device Structure

Inverted structure is another type of architecture of OSCs, as shown in Figure 11.3b [19, 20]. In an inverted structure, the bottom transparent ITO or FTO electrode serves as the cathode whereas the top high work function electrode is used as the anode. N-type metal oxides (such as ZnO or TiO_2) are usually inserted between the bottom cathode and organic active layer as the ETL to selectively collect electrons [21–23]. On the other hand, high work function metal oxides (such as MoO_3, WO_3, or NiO_x) serve as the HTL between the active layer and the top anode to selectively collect holes [24]. The polarity of charge collection is opposite in the inverted structure compared to that of the aforementioned conventional structure. In the inverted devices, high work function metals (such as Au or Ag) as the top anode with good air stability are beneficial for device durability [25]. Particularly, the metal oxide transporting layers can be deposited from colloidal solution, which is suitable for an all-solution roll-to-roll processing technology [26, 27].

11.2 Active Layer

A major breakthrough in the rapid development of BHJ solar cells has resulted from the discovery of photo-induced electron transfer from the conjugated polymer poly(2-methoxy-5-((2′-ethylhexyl)oxy)-1,4-phen-ylenevinylene) (MEH-PPV) to buckminsterfullerene. In a typical BHJ OSC device, the active layers are often composed of p-type and n-type organic semiconductors. The p-type organic semiconductors include conjugated polymers and small molecules, referred as electron donors (D). The n-type organic semiconductors are a series of fullerene derivatives and recently developed non-fullerene structures, referred as electron acceptors (A). To efficiently harvest the photo energy, the absorption spectra of the BHJ active layer should have large overlaps with the solar spectrum in the visible as well as in the near-infrared region, as shown in Figure 11.4. So, it is desirable to design and synthesize novel p-type donors with broader absorption capabilities through narrowing their optical bandgap. It is clear that raising the highest occupied molecular orbital (HOMO) or lowering the LUMO level of the donor or moving the two levels closer simultaneously can narrow the bandgap. Furthermore, a band offset between the donor LUMO level and the acceptor LUMO level is needed to guarantee an efficient exciton separation at the D/A interface as shown in Figure 11.1. In addition, narrowing the bandgap of the donor would require a lift-up of its HOMO level, which results in a decrease of V_{oc}. The V_{oc} in the BHJ OSCs is believed to be in correlation with the energy level difference between the HOMO of the donor and the LUMO of the acceptor [28, 29]. Therefore, it is important to balance the trade-off between the bandgap and the positions of the HOMO and LUMO energy levels of the active layers. Besides the bandgap and energy levels, other properties of active materials are also desired for high-performance OSCs, including: (i) good solubility and miscibility between donor and acceptor; (ii) high and balanced hole/electron

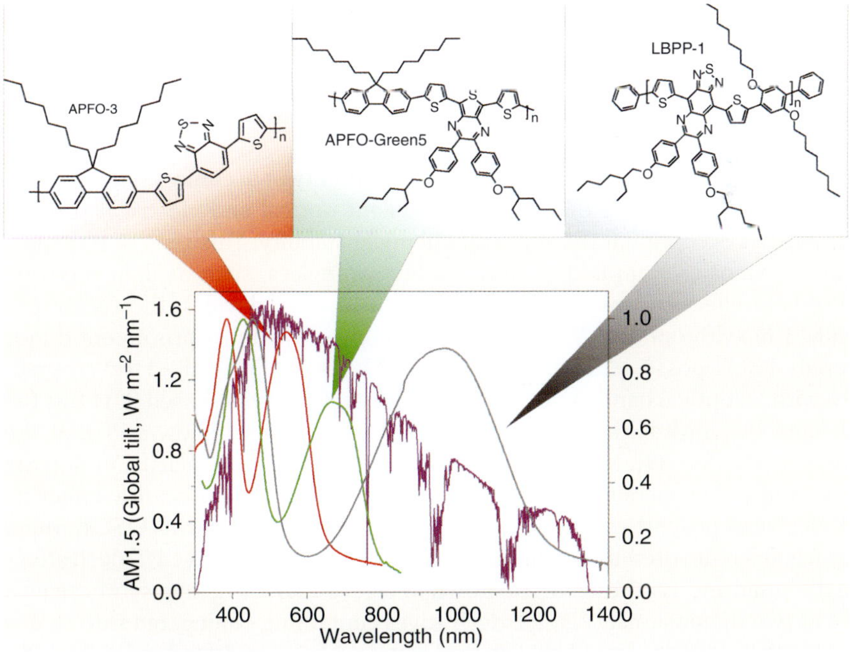

Figure 11.4 Absorption spectra and molecular structures of three representative polymers with different optical bandgaps. *Source:* Zhang et al. 2016 [9]. Copyright 2016. Reproduced with permission from OUP.

carrier mobility to enhance charge transport; and (iii) optimal morphology to promote charge separation and transport [13]. All these properties influence each other. Therefore, efficient active materials should be rationally designed by molecular engineering to compromise these requirements. During the past decades, a great number of donors and acceptors have been synthesized. This section will discuss some of the representative examples of polymers and small molecules donors and fullerene and non-fullerene acceptors.

11.2.1 Donor Materials

11.2.1.1 Poly(Phenylenevinylene) (PPV) and Polythiophene (PT) Derivatives

Conjugated poly(phenylenevinylene)s (PPVs) have attracted great attention in their early stages due to easy synthesis and good photoluminescence properties. However, the initially developed PPV without side chains was insoluble and infusible with very low molecular weight, which restricted its processing as well as application [30]. To circumvent these problems, Gilch incorporated long alkyl or alkoxy chains into the phenylene ring before polymerization, which ensured the solubility of the products and realized processable PPVs and its derivatives [31–33]. The representative solution-processable MEH-PPV with high molecular weight was successfully synthesized in this way. BHJ OSCs with MEH-PPV as donor and [6,6]-phenyl-C_{61}-butyric acid methyl ester ($PC_{61}BM$) as acceptor were fabricated and showed a PCE in the range of 1–2% [23]. Following this, another derivative poly(2-methoxy-5-((3′,7′-dimethyloctyl)oxy)-1,4-phenylenevinylene) (MDMO-PPV), with better processability was developed. The efficiency of devices-based MDMO-PPV/$PC_{61}BM$ active layer was about 3% [34]. As compared with MEH-PPV, the improvement of the PCE was mainly attributed to a more desirable morphology caused by the enhanced solubility and miscibility with $PC_{61}BM$.

Polythiophene (PT) and its derivatives are another type of conjugated polymers that are widely employed due to their excellent optical and electrical properties as well as exceptional thermal and chemical stability. This type of polymers has been extensively applied in semiconductor devices, such as light emitting diodes, field-effect transistors, OSCs, etc. [8]. In OSC devices, the regioregular poly(3-hexylthiophene-2,5-diyl) (P3HT) is one of the most prominent donor materials. P3HT possesses HOMO and LUMO levels at −5.0 and −3.0 eV, respectively, with an optical bandgap of c. 2.0 eV. To date, the device-based P3HT as the donor and $PC_{61}BM$ as the acceptor in the BHJ active layer show a PCE at the range of 3–5% [35–39]. It has been reported that the regioregularity, molecular weight, and polydispersity of P3HT have been shown to significantly affect its opto-electrical properties. In order to realize higher efficiency in OSCs, much effort has been devoted to ameliorating the PT. A series of other PT derivatives were designed and synthesized. As shown in Figure 11.5, Li's group synthesized a series of two dimensional (2D) PT derivatives containing conjugated side chains [40, 41]. They demonstrated that incorporation of a conjugated side chain on the PT main chain can effectively extend the degree of conjugation, and hence the absorption spectra of conjugated polymers in ultraviolet (UV) region were enhanced and red-shifted. By controlling the ratio between the thiophene units in

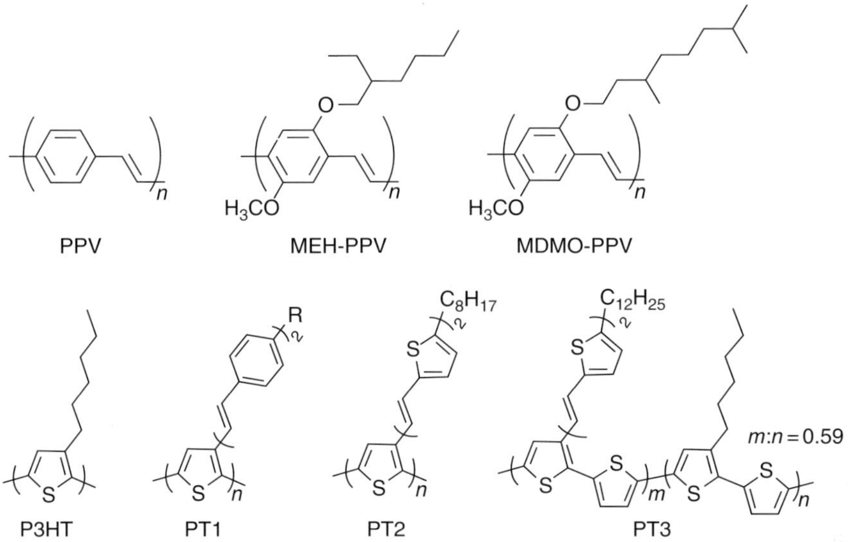

Figure 11.5 Structures of PPV and PT and their derivatives.

main chain and the conjugated side chains, the OSC devices based PT3:$PC_{61}BM$ delivered a PCE of 3.18% with a higher V_{oc} than the P3HT donor.

11.2.1.2 D–A Conjugated Polymers

The development of D–A copolymers brings a further breakthrough in donor materials, which leads to highly efficient BHJ OSCs [42–45]. Unlike homopolymer P3HT, D–A copolymers incorporate one electron-rich moiety (donor) and one electron-deficient moiety (acceptor). In these low bandgap polymers; the energy levels and molecular structures can be optimized via molecular engineering. Their HOMO and LUMO energy levels are largely determined by the HOMO energy level of the donor and the LUMO energy level of the acceptor, respectively [46]. As a result, the bandgap and energy level of the polymers could be individually tuned by engineering the donor and acceptor units. Over the past decades, a variety of D–A copolymers have been developed. In these D–A copolymers, the representative electron-rich moieties include fluorene, cyclopentadithiophene (CPDT), thiophene, benzodithiophene (BDT), indacenodithiophene (IDT), etc. The representative electron-deficient moieties include benzothiadiazole (BT), diketopyrrolopyrrole (DPP), thieno[3,4-*b*]thiophene (TT), thienopyrroledione (TPD), isoindigo (IID), etc. (Figure 11.6).

In 2003, Andersson first reported an alternating copolymer (poly{[2,7-(9,9-bis-(2-octyl)-fluorene)]-alt-[5,5-(4,7-di-20-thienyl-2,1,3-benzothiadiazole)]} (PFDTBT)) based on benzothiadiazole (A) and fluorine (D) with a considerable PCE of 2.2% [47]. After incorporating octyloxy chains on benzothiadiazole rings, the resulting copolymer showed an improved PCE of 3.1% [48]. Furthermore, changing the bridge atom of the fluorene from carbon to silicon enhances their electron-donating ability. The obtained silafluorene-based polymers PSiF-DBT realized a PCE over 5% [49]. When substituting the center carbon in the

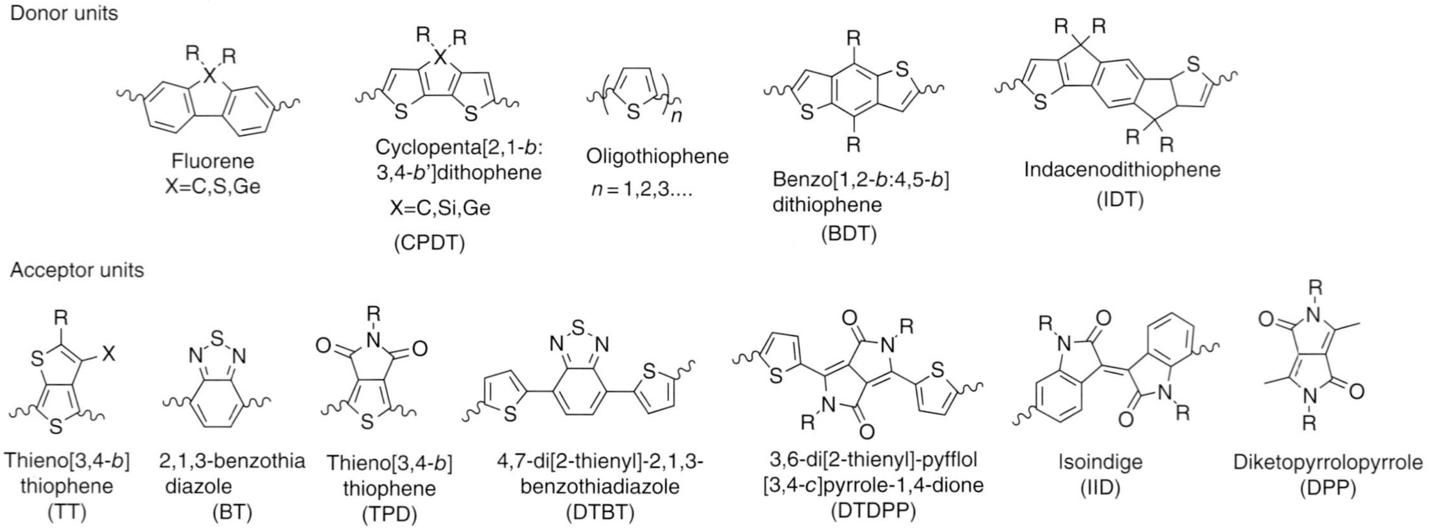

Figure 11.6 Structures of some widely used donor and acceptor fragments.

fluorene unit with a nitrogen (i.e. converting the fluorene into the carbazole), the carbazole-based polymers show a lower lying HOMO energy level. The carbazole-based polymer (PCDTBT) based OSCs demonstrated a V_{oc} of 0.88 V, a J_{sc} of 10.6 mA cm^{-2}, and a high FF of 66%, leading to an overall PCE over 6% in their BHJ cells [50]. The above-referred structures are all shown in Figure 11.7.

BDT derivatives are another popular class of donor units in constructing D–A copolymers. In 2008, Hou et al. first introduced BDT units in the design of photovoltaic polymers [51]. When the new BDT-containing polymers were combined with different acceptors, such as thiophene, benzothiadiazole, thieno[3,4-*b*]pyrazine (TPz), TT, etc. the bandgaps and molecular energy levels of these BDT-based polymers were tuned effectively, suggesting that BDT was a promising unit for conjugated photovoltaic polymers. For example, Jones and coworkers reported on the copolymer P2 based BDT and BT units in 2014, and the influence of the molecular weight on the photovoltaic performance was studied. The device displayed a high PCE of 8.5%, with a V_{oc} of 0.92 V, a J_{sc} of 14.5 mA cm^{-2}, and an FF of 64% in an inverted OSC. Then, a higher PCE of 9.4% was realized when a thin layer of fullerene derivative was employed to modify the ZnO as the electron-transporting layer in the inverted device [52]. In addition to the copolymer P10, Figure 11.7 lists several effective copolymers based on BDT units and some other acceptor units. Note that, the TT units are special donor fragments for their stabilized quinoidal structures. A significant breakthrough was achieved when the PTB polymers, which consisted of alternating electron-rich BDT units and electron-deficient TT units, were developed [53, 54]. PTB7 was one of the most studied donor materials that were developed in 2010. PTB7 shows a strong and broad absorption from 550 to 750 nm, accompanied by good solubility for solar cell fabrication. The conventional solar cell devices with an ITO/PEDOT:PSS/PTB7:PC$_{71}$BM/Ca/Al and chlorobenzene (CB)/diiodooctane (DIO) as processing solvents yielded an impressive PCE of 7.4% with a V_{oc} of 0.74 V, a J_{sc} of 14.50 mA cm^{-2}, and an FF of 68.97% [45]. Later on, Cao and coworkers improved the PCE of the PTB7-based device to 9.2% by applying an inverted device structure and new interface material [55]. Following the success of PTB7, Chen and coworkers replaced the alkoxyl side chains on BDT with 2D conjugated thienyl side chains on the basic structure of the polymer PTB7 to improve the coplanar character of the polymer backbone. The resulting polymer PTB7-Th revealed improved π-stacking and charge-transport properties, and showed very impressive photovoltaic performance [56]. Very recently, Cao and coworkers achieved a notable 9.94% PCE in inverted PTB7-Th devices [57].

11.2.2 Acceptor Materials

11.2.2.1 Fullerene Derivatives

Fullerenes are the predominant acceptor materials for OSCs because of their high electron mobility, appropriate energy levels, desirable processability, and efficient charge transfer from donors to the fullerenes. The pristine fullerene C_{60} possesses poor solubility in organic solvents and tends to crystallize, which hinders its application in OSCs. So, it is important to develop fullerene derivatives that

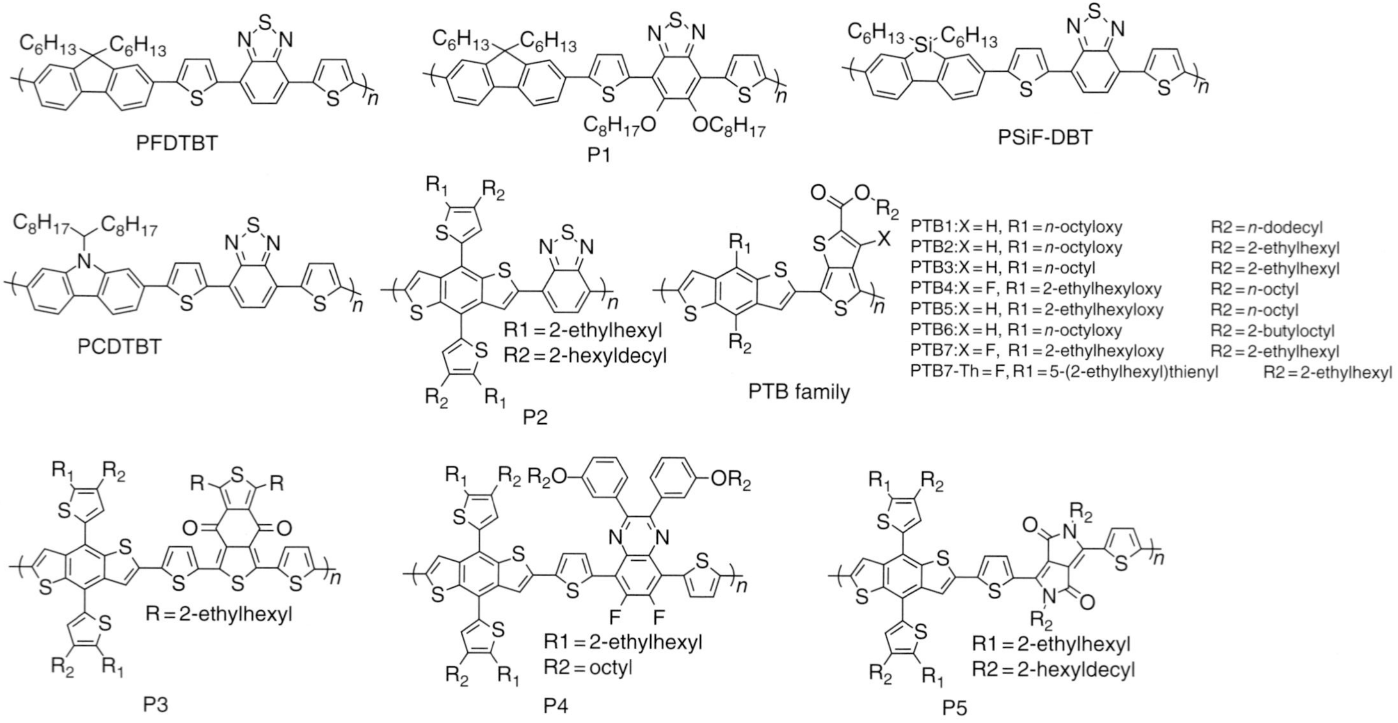

Figure 11.7 Structure of the representative D–A copolymer.

can endow appropriate energy levels as well as good solubility in common organic solvents, such as chloroform, toluene, chlorobenzene, and *o*-dichlorobenzene.

To date, soluble fullerene derivatives based on $PC_{61}BM$ and its analogue $PC_{71}BM$ (Figure 11.8) are considered as the most advanced acceptor materials and are widely used for BHJ OSCs. $PC_{61}BM$ was first synthesized by Hummelen and Wudl [58]. The solubility of $PC_{61}BM$ in toluene is 35 $mg\,ml^{-1}$, which is higher than that of C_{60} (2.15 $mg\,ml^{-1}$ in toluene) [59]. The electron mobility of $PC_{61}BM$ is c. 2×10^{-3} $cm^2\,V^{-1}$ [60]. The LUMO and HOMO levels of $PC_{61}BM$ reported by Li and coworkers are −3.91 and −5.93 eV, respectively [61]. The PCE of OSCs reached 4.37% depending on the most representative active layer P3HT/$PC_{61}BM$ [37]. Considering the fact that C_{70} exhibits significantly extended absorption than C_{60} in the visible part of the spectrum, Janssen and coworkers developed a $PC_{61}BM$ analogue (namely $PC_{71}BM$) to improve the light absorption in the visible region [62]. So far, the PCEs of single-junction PSCs based on $PC_{71}BM$ as an acceptor have broken the 10% efficiency barrier [63–65]. In addition, it was found that second functionalization on the core structure of the monosubstituted fullerene could further reduce the π-conjugation and electron delocalization in the cage. The bisadduct fullerene derivatives have large electrochemical reduction potentials, and, thus, higher lying LUMO levels [66]. Benefiting from the up-shifted LUMO level of Indene-C_{60} bisadduct ($IC_{60}BA$), the devices based on P3HT:$IC_{60}BA$ displayed an encouraging PCE of 5.44% and a high V_{oc} of 0.84 V, which is 0.26 V higher than that of P3HT:$PC_{61}BM$ based device. Similarly, Indene-C_{70} bisadduct ($IC_{70}BA$) possesses 0.19 eV LUMO which is higher than that of $PC_{71}BM$, which leads to a promising PCE of 5.79% in the OSCs based on P3HT:$IC_{70}BA$ [67].

11.2.2.2 Non-fullerene Acceptors

Fullerene derivatives ([6,6]-Phenyl-C61-butyric acid methyl ester (PCBM) and indene-C60 bisadduct (ICBA)) are the most common and successful acceptor materials in OSCs. However, the fullerene-based materials have some drawbacks, such as weak absorption in the visible spectral region, limited spectral breadth, and difficulty in variably tuning the band gap [68]. It is desirable to explore non-fullerene electron acceptors that will not only retain the favorable electron-accepting and transporting properties of fullerenes but also overcome their insufficiencies. After a decade of mediocrity, the PCE of fullerene-free OSCs has been improved to over 12%, almost approaching the best results of its fullerene counterparts. The rapid development of non-fullerene OSCs has opened a new avenue for the fundamental study of organic photovoltaics.

The most widely investigated non-fullerene acceptors to date have been based on the perylene diimide (PDI) core unit, which has shown to possess many desirable features as electron acceptors for PSCs, such as high electron mobility and large electron affinity. However, PDIs form micrometer-sized crystallites easily during solution processing due to their strong π–π stacking tendency. The large crystallites prevent forming a sufficiently large donor–acceptor interfacial area for efficient exciton splitting. A twisted PDI dimer (SdiPBI-S, Figure 11.8) with sulfur bridges in the bay positions was designed to solve the problems [69]. Owing to a more twisted molecular configuration, the molecule possesses a slightly lower high-conductivity PEDOT:PSS as an electrode to replace ITO that demonstrates the possibility of conducting polymer acting as the transparent electrodes for OSCs [75]. Subsequently, Zhou et al. reported the high electrical conducting PEDOT:PSS (PH500) as the anode electrode in all-solution-processed transparent inverted OSCs. But a relatively low PCE of 0.7% was observed, which was attributed to the less efficient absorption and a poor contact with the cathode [74]. Afterward, different strategies, such as adding high boiling polar solvents (dimethyl sulfone, ethylene glycol), post treatment with acid (sulfuric acid, phosphoric acid, or organic acids) or polar solvents have been used for enhancing its conductivity, affinity, etc. [76–79].

When used as the electrodes, an accurate pattern of the solution-processed PEDOT:PSS films is important. The metal electrodes are generally patterned via shadow masking. However, PEDOT:PSS films are typically prepared by spin coating, which makes fine patterning difficult. Film transfer printing (or transfer lamination) has been shown to be an effective technique to pattern PEDOT:PSS electrodes [80]. PEDOT:PSS is first deposited on the transfer medium (such as polydimethylsiloxane) and is then transferred onto the target surface. Before transfer printing, the PEDOT:PSS film is cut into the desired shape and size. The interplay between the adhesion of the film to the transfer medium and target surface is the key for the success of transfer printing. The yield of the transfer printing onto the hydrophilic surface is high because of the hydrophilic nature of the PEDOT:PSS. Mild treatment on the surface of the active layer with oxygen plasma before transfer printing, turning the surface hydrophilic, can enhance the transfer yield of PEDOT:PSS films. Another advantage of the transfer printing technique is that the PEDOT:PSS is deposited dry onto the target surface alleviating damage to the active layer or interfaces by water [81]. Owing to the fact that PEDOT:PSS electrodes are solution-processable, vacuum-free, optically transparent, possess tunable work function, and can be patterned with the transfer printing technique, both ITO electrode and metal electrode can be replaced by PEDOT:PSS films. Vacuum-free inverted PSCs including all-plastic solar cells with a plastic substrates/PEDOT:PSS/PEI/active layer/PEDOT:PSS architecture can be fabricated (Figure 11.9) [82].

11.3.2 Metal Nanowires and Grids

Other efficient emerging electrodes are metal nanowires and grids. Several studies reported solution-processed Ag NWs that had a sheet resistance and transmittance similar to those of ITO (10–20 Ω/sq. with 80% transmittance), together with a relatively high work function around 4.5 eV [83–86]. These promising characteristics enable its successful use as electrodes in OSCs. In 2008, Ag NWs were first investigated by Peumans et al., as a substitution for the ITO electrode [85]. The device delivered a PCE of only 0.38%, which was due to current leakage caused by the extremely rough Ag NW electrode with a low shunt resistance of less than 1 $k\Omega\,cm^{-2}$. To address this issue, an inorganic–organic composite consisting of Ag NWs and PEDOT:PSS was developed [87]. The conductive PEDOT:PSS could fill the gaps among Ag NWs, which would not only smoothen the film surface, but also increase the electrical conductivity of Ag NWs. Later

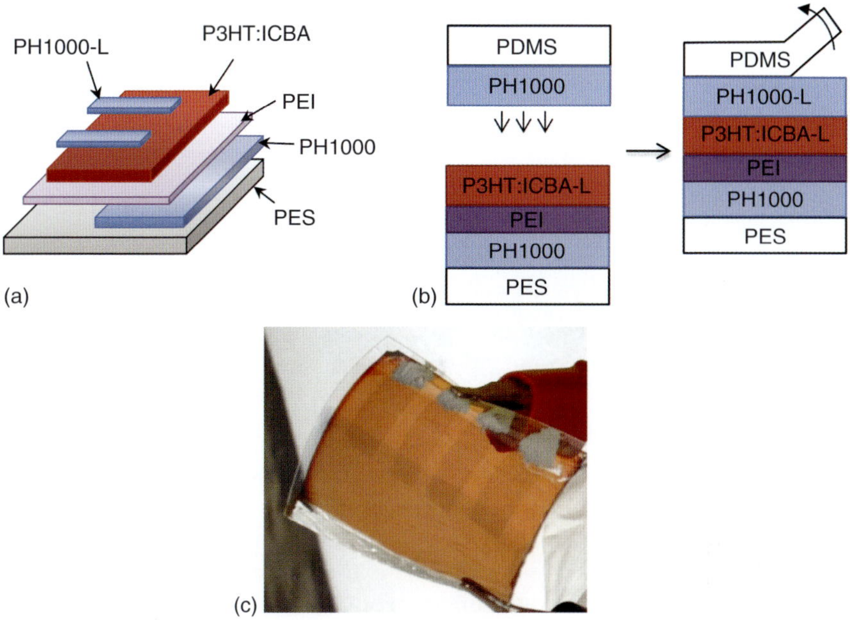

Figure 11.9 (a) Device structure of an all-plastic solar cell; (b) transfer printing of PEDOT:PSS PH1000 electrode; and (c) photographic picture of a fabricated all-plastic solar cell. *Source:* Zhou et al. 2014 [82]. Copyright 2014. Reproduced with permission from Royal Society of Chemistry.

on, Yang's group employed such a composite in visible-light transparent inverted OPVs with an all-solution process for high performance[88]. Recently, Guo et al. reported efficient, fully printed tandem PSCs using Ag NWs as the transparent top electrode, which showed efficiencies of 5.81% (on glass) and 4.85% (on flexible substrate) without open-circuit voltage losses [89].

11.3.3 Hybrid Carbon Material

Several nanomaterials including nanosized carbon, graphene, etc., attracted much attention in transparent electrodes, due to their fascinating properties, including high carrier mobility, good optical transparency, and low resistivity as well as excellent mechanical flexibility. In 1999, Ago et al. initially demonstrated the use of multiwalled carbon nanotubes (MWCNTs) as an opaque hole-collecting electrode to fabricate a photovoltaic device, which showed higher quantum efficiency than that of a controlled ITO device [90]. Since then, considerable effort has been devoted to developing higher transparent and conducting carbon nanotube electrodes for solar cells. Recently, Hu et al. reported that the PEDOT:PSS:CNTs hybrid film as electrode achieved similar efficiency of 7.47% with 7.69% of the ITO-controlled device, and also displayed promising application of the PEDOT:PSS:CNTs ink in the roll-to-roll process for large-scale flexible electronics [91]. In addition to the CNTs, graphene is another emerging carbon nanomaterial. Liu et al. used graphene-doped transparent electrodes as both cathode and anode to fabricate semitransparent OSCs with a PCE upto 3.4%, which also shows potential applications for OSCs [92].

11.4 Interfacial Layer

As shown in Figure 11.2, high work function interface for hole collection and low work function interface for electron collection are both needed to achieve high-performance OSCs. Besides, interfaces with proper surface energy are also important to prevent interfacial dewetting and delamination for efficient and stable OSCs.

11.4.1 Hole Transporting Layer (HTL)

Among various HTLs used in OSCs, conducting polymer PEDOT:PSS is the most commonly used HTL for anode modification in the conventional structure because of the following reasons: (i) it has high work function of 5.0 eV and has been demonstrated as a very efficient hole extractor from most active layers; (ii) the films are easily processed from aqueous solutions by spin coating; (iii) the PEDOT:PSS dispersion is commercially available; (iv) the films have good air stability and reproducibility; and (v) the films are highly transparent and do not affect the transmittance of ITO [93, 94]. However, in order to expand the application in an inverted architecture, one challenge is to deposit PEDOT:PSS from aqueous dispersion onto the hydrophobic active layers. Current feasible deposition strategies include: (i) adding a small amount of surfactant (<1 wt%) into the PEDOT:PSS dispersion and (ii) turning the surface of the active layer hydrophilic by air or oxygen plasma treatment or UV-Ozone treatment [95–97]. Commercially available formulated PEDOT:PSS CPP containing surfactant can also wet the hydrophobic active layer [77, 98].

Besides PEDOT:PSS, semiconducting oxides such as MoO_3, WO_3, NiO_x, and V_2O_5, have also been used as HTLs [99–102]. These oxides can be prepared either by vacuum deposition or solution processing. For inverted solar cells, the vacuum-deposited oxides are widely used because the layers can be deposited in sequence with the top metal electrode without breaking the vacuum. The current status of the role of HTL in OSCs is that PEDOT:PSS is widely used in conventional solar cells by directly coating PEDOT:PSS onto the ITO; MoO_3 is widely employed in inverted solar cells by vacuum deposition of MoO_3 on the active layers together with the top metal electrode.

11.4.2 Electron Transporting Layer (ETL)

As compared with HTL, use of ETL is more challenging because it typically displays poor air stability. In a conventional device architecture, vacuum-deposited thin films of low work function metals such as Al, LiF/Al, or Ca/Al are often used for electron collection. However, these low work function metals are very reactive and easily oxidized in air by oxygen or humidity. Considering the technology of vacuum deposition with high energy consumption and high cost, it becomes necessary to optimize solution-processable ETLs for the ultimate roll-to-roll manufacturing of OSCs. The n-type semiconducting oxides such as ZnO and TiO_x, which could be processed from solutions, nanoparticle dispersions, or precursors, are suitable [21, 103, 104]. These oxides are transparent in the Vis–NIR spectrum. They possess a matched energy level with the conduction band extract electrons and valence band block holes effectively. Note that, the working of ZnO or TiO_x as ETL is strongly dependent on the oxides' quality and surface states. Furthermore, the conductivity of the oxides is sensitive to UV illumination.

Organic-based low work function modifiers are another class of promising materials to achieve efficient electron collection. In 2009, Zhou et al. employed solution-processed polyethylene oxide (PEO) in vacuum-free inverted semitransparent PSCs with the structure of ITO/PEO/active layer/PEDOT. Ultraviolet photoemission spectroscopy (UPS) measurements confirmed that the PEO could convert ITO to an electron collection electrode by reducing the work function of ITO up to 0.5 eV [74]. This shows that ITO can be potentially tuned as an electron-collecting electrode after modified by a polymer despite low efficiency of the cells (about 0.7%). Later on, Zhou et al. found that the polymers polyethylenimine (PEI) and polyethylenimine ethoxylated (PEIE) are highly effective to reduce the work function of most conductors, including metals, metal oxides, conducting polymers, graphene, and ITO by about 1 eV [105]. The polymer modifiers are physically adsorbed onto the conductors due to the partial charge transfer and electrostatic interaction, independent of the surface chemistry. The low work function modification can be easily done by spin coating, printing, or immersing the conductors into the solution. The produced low work function interfaces exhibit good air stability. PEI or PEIE have been extensively used in OSCs [80, 106], organic LEDs [81], organic field-effect transistors [107], perovskite solar cells [108], and inorganic quantum dot LEDs [109]. It is interesting to note that the PEIE could also realize a low work function of a conducting polymer. After modification by PEI or PEIE, the work function of PEDOT:PSS decreased from 5.0 eV to about 3.6 eV, which enables the possibility of fabricating all-plastic solar cells [82, 105].

Other polymers have also been demonstrated as modifiers to lower the work function of ITO or metal electrode for efficient electron collection in OSCs [55, 110, 111]. Polymers containing side chains terminated with polar moieties, such as alkylamines, amine salts, phosphates, etc., could function as low work function modifiers. One other prominent example is the poly [(9,9-bis(3′-(*N*,*N*-dimethylamino)propyl)-2,7-fluorene)-*alt*-2,7-(9,9–dioctylfluorene)] (PFN, Figure 11.10) that reduced the work function of ITO from 4.7 to 4.1 eV and yielded a high PCE of 9.2% in inverted architecture.

11.5 Tandem Organic Solar Cells

Building tandem solar cells is an effective way to increase PCE of OSCs because a broader spectrum of solar radiation can be harvested by stacked junctions. The

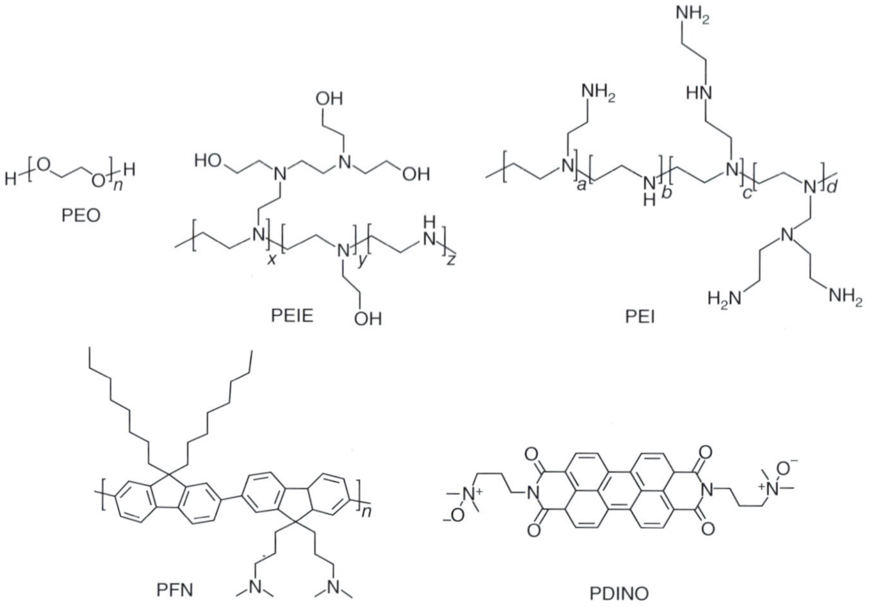

Figure 11.10 Chemical structures of representative materials to produce low work function interfaces.

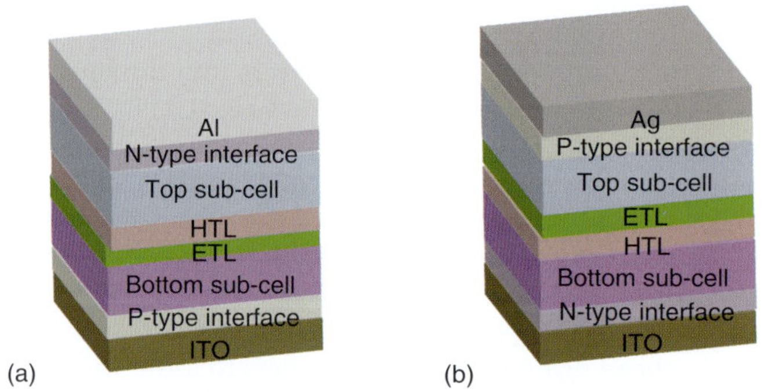

Figure 11.11 Schematic representation of an organic tandem device of an organic tandem solar cell in a normal (a) and an inverted configuration (b).

thermalization loss of photon energy can also be minimized. A typical structure of a tandem solar cell is shown in Figure 11.11. The double-junction tandem structure consists of a front cell with a high bandgap junction, an interconnecting layer (ICL), and a rear cell with an LBG junction.

11.5.1 Interconnecting Layer

In tandem OSCs, ICL, also known as charge recombination layer (CRL), is a critical component for realizing high device performance. The ICLs, generally consisting of a p-type HTL and an n-type ETL, function as the charge recombination zone to connect the two sub-cells. An ideal ICL is straightforward to engineer subject to the following requirements: (i) high optical transmittance to minimize light absorption losses; (ii) large work function contrast to efficiently collect electrons from one sub-cell and holes from the other sub-cell, respectively; (iii) a quasi-ohmic-contact between the ETL and HTL to ease bipolar recombination; and (iv) robust impermeability to protect the underlying layers of bottom cell from dissolution by the processing solvents of the top cell [112, 113].

In 2006, Dennler et al. stacked a vacuum-deposited ZnPc/C_{60} based cell on the top of a solution-processed P3HT:PC_{61}BM layer with a 1 nm thick Au as the intermediate recombination layer. The tandem devices exhibited a J_{sc} of 4.8 mA cm^{-2}, V_{oc} of 1.02 V, and FF of 0.45 [114]. Spectral photocurrent response region was effectively extended from about 650 nm (P3HT:PC_{61}BM) to about 800 nm contributed by the ZnPc/C_{60} junction. The vacuum-evaporated nanometer-scale metal ICLs did not yield high-performance tandem cells. Later on, researchers focused on the solution-processed ICL. In 2007, Kim et al. reported a solution-processed ICL TiO_x/PEDOT:PSS in the tandem cells, which employed P3HT:PC_{61}BM and PCPDTBT:PC_{70}BM as the active layers that have complementary optical absorption. The device achieved a record-high PCE of 6.5% at that time (Figure 11.12) [115]. Since then, solution-processed ZnO/PEDOT:PSS and TiO_x/PEDOT:PSS ICLs were widely studied and applied in tandem OSCs.

With the development of inverted single-junction OSCs, inverted tandem solar cells were also widely studied because of their better air stability as compared to conventional configuration.

In an inverted tandem device, the polarity of the ICL has to be reversed according to the efficient charge-carrier collection. In 2012, Dou et al. demonstrated a PCE of 8.6% in inverted tandem OSCs with PEDOT:PSS/ZnO as the ICL [116]. In the inverted device, another important ICL is PEDOT:PSS/PEI or PEDOT:PSS/PEIE (Figure 11.13) [117]. This ICL is developed based on the findings of the surface modification function of the PEIE or PEI by Zhou et al. [105] As mentioned earlier, the polymer surface modifier PEIE or PEI containing simple aliphatic amine groups can substantially reduce the work function of conductors including metals, transparent conductive metal oxides, conducting polymers, etc. They used this polymer modifier to develop an all-polymer ICL PEDOT:PSS/PEIE. This ICL shows very low optical absorption, high electrical conductivity, and a large work function contrast of 1.3 eV between its top and bottom interfaces. Simple polymeric composition as well as unprecedented performance made it a promising ICL for tandem solar cells. Recently, Tong et al. further optimized the conductivity of the ICL via modifying the PEDOT:PSS formulation and selecting the processing solvent of the PEI to avoid patterning and leakage [118]. With the optimized ICL, they firstly demonstrated all-solution-processed multijunction (up to seven junctions, 22 layers, Figure 11.13) all-plastic solar cells with high yields. These cells with high photovoltage (5.37 V for seven junctions) have been shown to meet the requirements of powering liquid-crystal displays and full-color light emitting diodes under low-intensity light conditions.

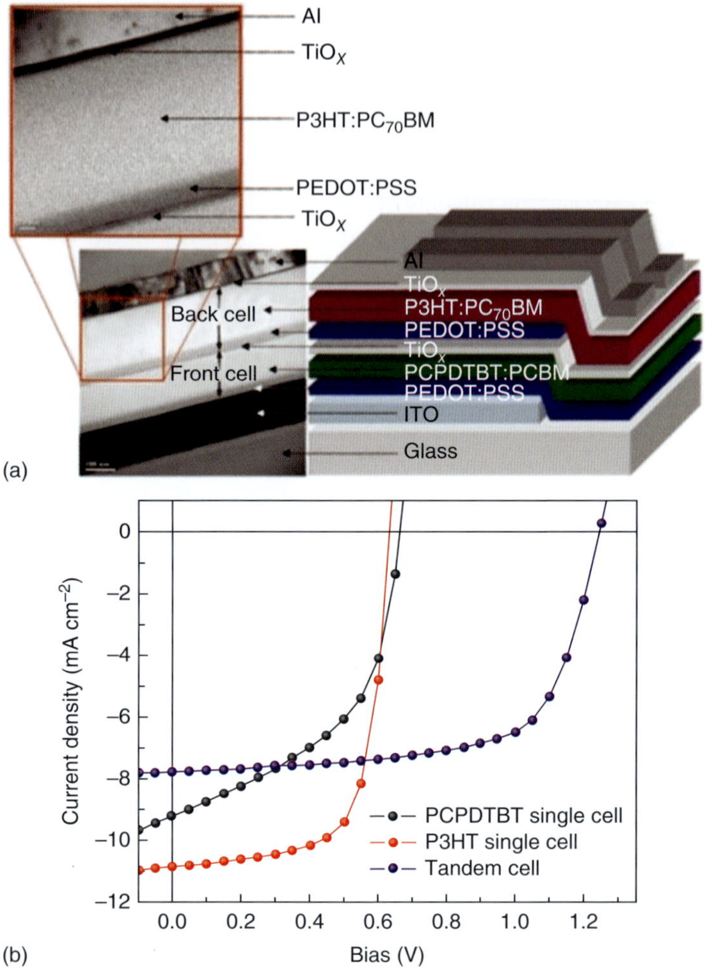

Figure 11.12 (a) Structure of organic tandem solar cells and (b) the *J–V* characteristics. *Source:* Reproduced with permission from Kim et al. [115] Copyright 2007, Science.

11.5.2 Low Bandgap Polymer Sub-cell

The advantage of the tandem solar cells is to extend the light harvest in a broad spectral region by stacking different junctions with different bandgaps. The typically used large bandgap junctions are P3HT:ICBA and PCDTBT:PC_{71}BM with a bandgap of ~2 eV. The two large bandgap junctions could deliver a J_{sc} about 10 mA cm^{-2} and an EQE of about 70% from 400 to 680 nm. Efficient LBG polymers with complementary spectral absorption region are desired, compared to the large bandgap polymers, to build efficient polymer tandem solar cells.

In recent years, progresses have been made in the development of novel high-efficiency LBG polymers. The PCE of the tandem solar cells is also rapidly enhanced. In 2012, You et al. designed a novel LBG D–A copolymer

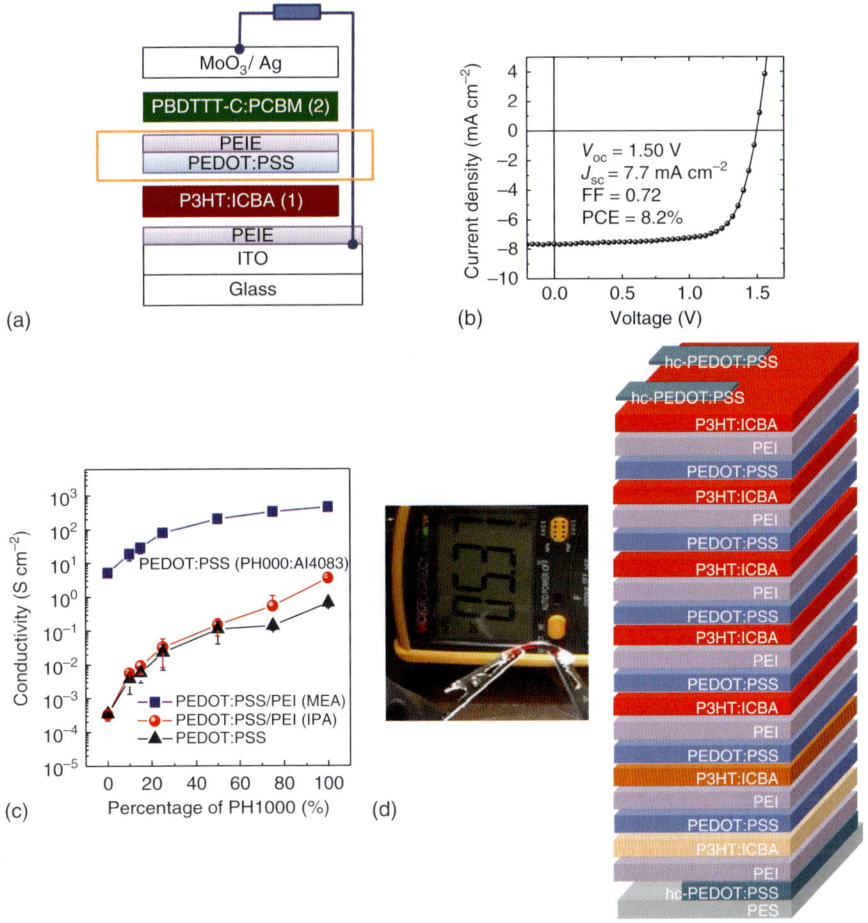

Figure 11.13 (a,b) Tandem solar cells with PEDOT:PSS/PEIE as the ICL and the *J–V* characteristics. *Source:* Zhou et al. 2012 [117]. Copyright 2012. Reproduced with permission from Royal Society of Chemistry. (c) Conductivity tuning of the PEDOT:PSS/PEI ICL; (d) All-solution-processed all-plastic multijunction (seven junctions, 22 layers) cells displayed V_{oc} of 5.37 V. *Source:* Tong et al. 2016 [118]. Copyright 2016. Reproduced with permission from Royal Society of Chemistry.

poly[2,7-(5,5-*bis*-(3,7-dimethyloctyl)-5*H*–dithieno[3,2-*b*:20,30-*d*]pyran)-*alt*-4,7-(5,6-difluoro-2,1,3-benzothiadiazole)] (PDTP-DFBT) based on a strong electron-withdrawing moiety difluorobenzothiadiazole (DFBT) and a strong electron-donating dithienopyran (DTP) unit. The copolymer PDTP-DFBT showed a bandgap as low as 1.38 eV, accompanied with a high hole mobility. With the combination of P3HT:ICBA as the large bandgap junction and PDTP-DFBT:PC_{71}BM as the LBG junction, tandem solar cells with a certified PCE of 10.6% have been achieved [119]. In 2013, Li et al. reported the design and synthesis of a new LBG semiconducting polymer, poly[[2,5-bis(2-hexyldecyl-2,3,5,6-tetrahydro-3,6-dioxopyrrolo[3,4 *c*]pyrrole-1,4-diyl] -*alt*-[3′,3″ -dimethyl-2,2′:5′,2″-terthiophene]-5,5″-diyl] (PMDPP3T, Figure 11.14) with a

2 Polman, A., Knight, M., Garnett, E.C. et al. (2016). Photovoltaic materials: present efficiencies and future challenges. *Science* 352: aad4424.

3 Lu, L., Zheng, T., Wu, Q. et al. (2015). Recent advances in bulk Heterojunction polymer solar cells. *Chem. Rev.* 115: 12666–12731.

4 Green, M.A. (2015). Corrigendum to 'solar cell efficiency tables (version 46)' [*Prog. Photovolt. Res. Appl.* 2015; 23: 805–812]. *Prog. Photovolt. Res. Appl.* 23: 1202.

5 Jackson, P., Hariskos, D., Wuerz, R. et al. (2015). Properties of Cu(In,Ga)Se_2 solar cells with new record efficiencies up to 21.7%. *Phys. Status Solidi RRL* 9: 28–31.

6 Tang, C.W. (1986). Two-layer organic photovoltaic cell. *Appl. Phys. Lett.* 48: 183–185.

7 Yu, G., Gao, J., Hummelen, J.C. et al. (1995). Polymer photovoltiac cells: enhanced efficiencies via a network of internal donor-acceptor heterojunctions. *Science* 270: 1789.

8 Cheng, Y.J., Yang, S.H., and Hsu, C.S. (2009). Synthesis of conjugated polymers for organic solar cell applications. *Chem. Rev.* 109: 5868–5923.

9 Zhang, F., Inganäs, O., Zhou, Y., and Vandewal, K. (2016). The development of polymer–fullerene solar cells. *Nat. Sci. Rev.* 3: doi: 10.1093/nsr/nww020.

10 Dou, L., You, J., Hong, Z. et al. (2013). 25th Anniversary article: a decade of organic/polymeric photovoltaic research. *Adv. Mater.* 25: 6642–6671.

11 Hu, H., Jiang, K., Yang, G. et al. (2015). Terthiophene-based D-A polymer with an asymmetric arrangement of alkyl chains that enables efficient polymer solar cells. *J. Am. Chem. Soc.* 137: 14149–14157.

12 Zhao, W., Qian, D., Zhang, S. et al. (2016). Fullerene-free polymer solar cells with over 11% efficiency and excellent thermal stability. *Adv. Mater.* 28: 4734–4739.

13 Thompson, B.C. and Frechet, J.M. (2008). Polymer-fullerene composite solar cells. *Angew. Chem.* 47: 58–77.

14 Yip, H.-L. and Jen, A.K.Y. (2012). Recent advances in solution-processed interfacial materials for efficient and stable polymer solar cells. *Energy Environ. Sci.* 5: 5994.

15 de Jong, M.P., van Ijzendoorn, L.J., and de Voigt, M.J.A. (2000). Stability of the interface between indium-tin-oxide and poly(3,4-ethylenedioxythiophene)/poly(styrenesulfonate) in polymer light-emitting diodes. *Appl. Phys. Lett.* 77: 2255–2257.

16 Wong, K.W., Yip, H.L., Luo, Y. et al. (2002). Blocking reactions between indium-tin oxide and poly(3,4-ethylene dioxythiophene):poly(styrene sulphonate) with a self-assembly monolayer. *Appl. Phys. Lett.* 80: 2788–2790.

17 Wang, K., Ren, H., Yi, C. et al. (2013). Solution-processed Fe_3O_4 magnetic nanoparticle thin film aligned by an external magnetostatic field as a hole extraction layer for polymer solar cells. *ACS Appl. Mater. Interfaces* 5: 10325–10330.

18 Jørgensen, M., Norrman, K., and Krebs, F.C. (2008). Stability/degradation of polymer solar cells. *Sol. Energy Mater. Sol. Cells* 92: 686–714.

19 Lloyd, M.T., Peters, C.H., Garcia, A. et al. (2011). Influence of the hole-transport layer on the initial behavior and lifetime of inverted organic photovoltaics. *Sol. Energy Mater. Sol. Cells* 95: 1382–1388.

20 Wang, K., Liu, C., Meng, T. et al. (2016). Inverted organic photovoltaic cells. *Chem. Soc. Rev.* 45: 2937–2975.

21 White, M.S., Olson, D.C., Shaheen, S.E. et al. (2006). Inverted bulk-heterojunction organic photovoltaic device using a solution-derived ZnO underlayer. *Appl. Phys. Lett.* 89: 143517.

22 Waldauf, C., Morana, M., Denk, P. et al. (2006). Highly efficient inverted organic photovoltaics using solution based titanium oxide as electron selective contact. *Appl. Phys. Lett.* 89: 233517.

23 Shrotriya, V., Wu, E.H.-E., Li, G. et al. (2006). Efficient light harvesting in multiple-device stacked structure for polymer solar cells. *Appl. Phys. Lett.* 88: 064104.

24 Choy, W.C. and Zhang, D. (2016). Solution-processed metal oxides as efficient carrier transport layers for organic photovoltaics. *Small* 12: 416–431.

25 Lloyd, M.T., Olson, D.C., Lu, P. et al. (2009). Impact of contact evolution on the shelf life of organic solar cells. *J. Mater. Chem.* 19: 7638.

26 Hau, S.K., Yip, H.-L., Leong, K., and Jen, A.K.Y. (2009). Spraycoating of silver nanoparticle electrodes for inverted polymer solar cells. *Org. Electron.* 10: 719–723.

27 Krebs, F.C., Gevorgyan, S.A., and Alstrup, J. (2009). A roll-to-roll process to flexible polymer solar cells: model studies, manufacture and operational stability studies. *J. Mater. Chem.* 19: 5442.

28 Gadisa, A., Svensson, M., Andersson, M.R., and Inganäs, O. (2004). Correlation between oxidation potential and open-circuit voltage of composite solar cells based on blends of polythiophenes/fullerene derivative. *Appl. Phys. Lett.* 84: 1609–1611.

29 Brabec, C.J., Cravino, A., Meissner, D. et al. (2001). Origin of the open circuit voltage of plastic solar cells. *Adv. Funct. Mater.* 11: 374–380.

30 McDonald, R.N. and Campbell, T.W. (1960). The wittig reaction as a polymerization Method1a. *J. Am. Chem. Soc.* 82: 4669–4671.

31 Braun, D. and Heeger, A.J. (1991). Visible light emission from semiconducting polymer diodes. *Appl. Phys. Lett.* 58: 1982–1984.

32 Wan, W.C., Antoniadis, H., Choong, V. et al. (1997). Halogen precursor route to poly[(2,3-diphenyl-*p*-phenylene) vinylene](DP-PPV): synthesis, photoluminescence, electroluminescence, and photoconductivity. *Macromolecules* 30: 6567–6574.

33 Gilch, H. and Wheelwright, W. (1966). Polymerization of α-halogenated p-xylenes with base. *J. Polym. Sci., Part A-1: Polym. Chem.* 4: 1337–1349.

34 Shaheen, S.E., Brabec, C.J., Sariciftci, N.S. et al. (2001). 2.5% Efficient organic plastic solar cells. *Appl. Phys. Lett.* 78: 841–843.

35 Berson, S., De Bettignies, R., Bailly, S., and Guillerez, S. (2007). Poly(3-hexylthiophene) fibers for photovoltaic applications. *Adv. Funct. Mater.* 17: 1377–1384.

36 Moulé, A.J. and Meerholz, K. (2008). Controlling morphology in polymer–fullerene mixtures. *Adv. Mater.* 20: 240–245.

37 Li, G., Shrotriya, V., Huang, J. et al. (2005). High-efficiency solution processable polymer photovoltaic cells by self-organization of polymer blends. *Nat. Mater.* 4: 864–868.

38 Erb, T., Zhokhavets, U., Gobsch, G. et al. (2005). Correlation between structural and optical properties of composite polymer/fullerene films for organic solar cells. *Adv. Funct. Mater.* 15: 1193–1196.

39 Ma, W., Yang, C., Gong, X. et al. (2005). Thermally stable, efficient polymer solar cells with nanoscale control of the interpenetrating network morphology. *Adv. Funct. Mater.* 15: 1617–1622.

40 Hou, J., Tan, Z.A., Yan, Y. et al. (2006). Synthesis and photovoltaic properties of two-dimensional conjugated polythiophenes with bi(thienylenevinylene) side chains. *J. Am. Chem. Soc.* 128: 4911–4916.

41 Hou, J., Huo, L., He, C. et al. (2006). Synthesis and absorption spectra of poly(3-(phenylenevinyl) thiophene)s with conjugated side chains. *Macromolecules* 39: 594–603.

42 Fang, L., Zhou, Y., Yao, Y.-X. et al. (2013). Side-chain engineering of isoindigo-containing conjugated polymers using polystyrene for high-performance bulk heterojunction solar cells. *Chem. Mater.* 25: 4874–4880.

43 Cabanetos, C., El Labban, A., Bartelt, J.A. et al. (2013). Linear side chains in benzo[1,2-*b*:4,5-*b*′]dithiophene-thieno[3,4-*c*]pyrrole-4,6-dione polymers direct self-assembly and solar cell performance. *J. Am. Chem. Soc.* 135: 4656–4659.

44 Price, S.C., Stuart, A.C., Yang, L. et al. (2011). Fluorine substituted conjugated polymer of medium band gap yields 7% efficiency in polymer-fullerene solar cells. *J. Am. Chem. Soc.* 133: 4625–4631.

45 Liang, Y., Xu, Z., Xia, J. et al. (2010). For the bright future-bulk heterojunction polymer solar cells with power conversion efficiency of 7.4%. *Adv. Mater.* 22: E135–E138.

46 Brédas, J.-L., Norton, J.E., Cornil, J., and Coropceanu, V. (2009). Molecular understanding of organic solar cells: the challenges. *Acc. Chem. Res.* 42: 1691–1699.

47 Svensson, M., Zhang, F., Veenstra, S.C. et al. (2003). High-performance polymer solar cells of an alternating polyfluorene copolymer and a fullerene derivative. *Adv. Mater.* 15: 988–991.

48 Li, W., Qin, R., Zhou, Y. et al. (2010). Tailoring side chains of low band gap polymers for high efficiency polymer solar cells. *Polymer* 51: 3031–3038.

49 Wang, E., Wang, L., Lan, L. et al. (2008). High-performance polymer heterojunction solar cells of a polysilafluorene derivative. *Appl. Phys. Lett.* 92: 033307.

50 Park, S.H., Roy, A., Beaupré, S. et al. (2009). Bulk heterojunction solar cells with internal quantum efficiency approaching 100%. *Nat. Photonics* 3: 297–302.

51 Hou, J., Park, M.-H., Zhang, S. et al. (2008). Bandgap and molecular energy level control of conjugated polymer photovoltaic materials based on benzo[1,2-*b*:4,5-*b*′]dithiophene. *Macromolecules* 41: 6012–6018.

52 Subbiah, J., Purushothaman, B., Chen, M. et al. (2015). Organic solar cells using a high-molecular-weight benzodithiophene–benzothiadiazole copolymer with an efficiency of 9.4%. *Adv. Mater.* 27: 702–705.
53 Chen, H.-Y., Hou, J., Zhang, S. et al. (2009). Polymer solar cells with enhanced open-circuit voltage and efficiency. *Nat. Photonics* 3: 649–653.
54 Liang, Y., Feng, D., Wu, Y. et al. (2009). Highly efficient solar cell polymers developed via fine-tuning of structural and electronic properties. *J. Am. Chem. Soc.* 131: 7792–7799.
55 He, Z., Zhong, C., Su, S. et al. (2012). Enhanced power-conversion efficiency in polymer solar cells using an inverted device structure. *Nat. Photonics* 6: 593–597.
56 Liao, S.H., Jhuo, H.J., Cheng, Y.S., and Chen, S.A. (2013). Fullerene derivative-doped zinc oxide nanofilm as the cathode of inverted polymer solar cells with low-bandgap polymer (PTB7-Th) for high performance. *Adv. Mater.* 25: 4766–4771.
57 He, Z., Xiao, B., Liu, F. et al. (2015). Single-junction polymer solar cells with high efficiency and photovoltage. *Nat. Photonics* 9: 174–179.
58 Hummelen, J.C., Knight, B.W., LePeq, F. et al. (1995). Preparation and characterization of fulleroid and methanofullerene derivatives. *J. Org. Chem.* 60: 532–538.
59 Sivaraman, N., Dhamodaran, R., Kaliappan, I. et al. (1992). Solubility of C60 in organic solvents. *J. Org. Chem.* 57: 6077–6079.
60 Mihailetchi, V.D., van Duren, J.K., Blom, P.W. et al. (2003). Electron transport in a methanofullerene. *Adv. Funct. Mater.* 13: 43–46.
61 He, Y. and Li, Y. (2011). Fullerene derivative acceptors for high performance polymer solar cells. *Phys. Chem. Chem. Phys.* 13: 1970–1983.
62 Wienk, M.M., Kroon, J.M., Verhees, W.J.H. et al. (2003). *Angew. Chem.* 115: 3493–3497.
63 Zhang, S., Ye, L., Zhao, W. et al. (2015). Realizing over 10% efficiency in polymer solar cell by device optimization. *Sci. China Chem.* 58: 248–256.
64 Chen, J.D., Cui, C., Li, Y.Q. et al. (2015). Single-junction polymer solar cells exceeding 10% power conversion efficiency. *Adv. Mater.* 27: 1035–1041.
65 Liu, Y., Zhao, J., Li, Z. et al. (2014). Aggregation and morphology control enables multiple cases of high-efficiency polymer solar cells. *Nat. Commun.* 5: 5293.
66 He, Y., Chen, H.-Y., Hou, J., and Li, Y. (2010). Indene–C60 bisadduct: a new acceptor for high-performance polymer solar cells. *J. Am. Chem. Soc.* 132: 1377–1382.
67 He, Y., Zhao, G., Peng, B., and Li, Y. (2010). High-yield synthesis and electrochemical and photovoltaic properties of indene-C70 bisadduct. *Adv. Funct. Mater.* 20: 3383–3389.
68 Lin, Y. and Zhan, X. (2014). Non-fullerene acceptors for organic photovoltaics: an emerging horizon. *Mater. Horiz.* 1: 470.
69 Sun, D., Meng, D., Cai, Y. et al. (2015). Non-fullerene-acceptor-based bulk-heterojunction organic solar cells with efficiency over 7. *J. Am. Chem. Soc.* 137: 11156–11162.

70 Meng, D., Sun, D., Zhong, C. et al. (2016). High-performance solution-processed non-fullerene organic solar cells based on selenophene-containing perylene bisimide acceptor. *J. Am. Chem. Soc.* 138: 375–380.

71 Lin, Y., Zhang, Z.-G., Bai, H. et al. (2015). High-performance fullerene-free polymer solar cells with 6.31% efficiency. *Energy Environ. Sci.* 8: 610–616.

72 Lin, Y., Wang, J., Zhang, Z.G. et al. (2015). An electron acceptor challenging fullerenes for efficient polymer solar cells. *Adv. Mater.* 27: 1170–1174.

73 Wang, M., Hu, X., Liu, P. et al. (2011). Donor-acceptor conjugated polymer based on naphtho[1,2-*c*:5,6-*c*]bis[1,2,5]thiadiazole for high-performance polymer solar cells. *J. Am. Chem. Soc.* 133: 9638–9641.

74 Zhou, Y., Li, F., Barrau, S. et al. (2009). Inverted and transparent polymer solar cells prepared with vacuum-free processing. *Sol. Energy Mater. Sol. Cells* 93: 497–500.

75 Zhang, F., Johansson, M., Andersson, M.R. et al. (2002). Polymer photovoltaic cells with conducting polymer anodes. *Adv. Mater.* 14: 662–665.

76 Zhou, Y., Cheun, H., Choi, S. et al. (2010). Indium tin oxide-free and metal-free semitransparent organic solar cells. *Appl. Phys. Lett.* 97: 153304.

77 Zhou, Y., Cheun, H., Choi, S. et al. (2011). Optimization of a polymer top electrode for inverted semitransparent organic solar cells. *Org. Electron.* 12: 827–831.

78 Admassie, S., Zhang, F., Manoj, A.G. et al. (2006). A polymer photodiode using vapour-phase polymerized PEDOT as an anode. *Sol. Energy Mater. Sol. Cells* 90: 133–141.

79 Dong, Q., Zhou, Y., Pei, J. et al. (2010). All-spin-coating vacuum-free processed semi-transparent inverted polymer solar cells with PEDOT:PSS anode and PAH-D interfacial layer. *Org. Electron.* 11: 1327–1331.

80 Kyaw, A.K., Wang, D.H., Gupta, V. et al. (2013). Efficient solution-processed small-molecule solar cells with inverted structure. *Adv. Mater.* 25: 2397–2402.

81 Hofle, S., Schienle, A., Bernhard, C. et al. (2014). Solution processed, white emitting tandem organic light-emitting diodes with inverted device architecture. *Adv. Mater.* 26: 5155–5159.

82 Zhou, Y., Khan, T.M., Shim, J.W. et al. (2014). All-plastic solar cells with a high photovoltaic dynamic range. *J. Mater. Chem. A* 2: 3492.

83 Han, B., Pei, K., Huang, Y. et al. (2014). Uniform self-forming metallic network as a high-performance transparent conductive electrode. *Adv. Mater.* 26: 873–877.

84 De, S., Higgins, T.M., Lyons, P.E. et al. (2009). Silver nanowire networks as flexible, transparent, conducting films: extremely high DC to optical conductivity ratios. *ACS Nano* 3: 1767–1774.

85 Lee, J.-Y., Connor, S.T., Cui, Y., and Peumans, P. (2008). Solution-processed metal nanowire mesh transparent electrodes. *Nano Lett.* 8: 689–692.

86 Fang, Y., Ding, K., Wu, Z. et al. (2016). Architectural engineering of nanowire network fine pattern for 30 μm wide flexible quantum dot light-emitting diode application. *Acs Nano* 10: 10023–10030.

87 Gaynor, W., Burkhard, G.F., McGehee, M.D., and Peumans, P. (2011). Smooth nanowire/polymer composite transparent electrodes. *Adv. Mater.* 23: 2905–2910.
88 Chen, C.-C., Dou, L., Zhu, R. et al. (2012). Visibly transparent polymer solar cells produced by solution processing. *ACS Nano* 6: 7185–7190.
89 Guo, F., Li, N., Radmilović, V.V. et al. (2015). Fully printed organic tandem solar cells using solution-processed silver nanowires and opaque silver as charge collecting electrodes. *Energy Environ. Sci.* 8: 1690–1697.
90 Ago, H., Petritsch, K., Shaffer, M.S. et al. (1999). Composites of carbon nanotubes and conjugated polymers for photovoltaic devices. *Adv. Mater.* 11: 1281–1285.
91 Hu, X., Chen, L., Zhang, Y. et al. (2014). Large-scale flexible and highly conductive carbon transparent electrodes via roll-to-roll process and its high performance lab-scale indium tin oxide-free polymer solar cells. *Chem. Mater.* 26: 6293–6302.
92 Liu, Z., Li, J., Sun, Z.-H. et al. (2011). The application of highly doped single-layer graphene as the top electrodes of semitransparent organic solar cells. *ACS Nano* 6: 810–818.
93 Roman, L.S., Berggren, M., and Inganäs, O. (1999). Polymer diodes with high rectification. *Appl. Phys. Lett.* 75: 3557–3559.
94 Zhang, F.L., Gadisa, A., Inganäs, O. et al. (2004). Influence of buffer layers on the performance of polymer solar cells. *Appl. Phys. Lett.* 84: 3906–3908.
95 Savva, A., Neophytou, M., Koutsides, C. et al. (2013). Synergistic effects of buffer layer processing additives for enhanced hole carrier selectivity in inverted organic photovoltaics. *Org. Electron.* 14: 3123–3130.
96 Li, Z., Meng, W., Tong, J. et al. (2015). A nonionic surfactant simultaneously enhancing wetting property and electrical conductivity of PEDOT:PSS for vacuum-free organic solar cells. *Sol. Energy Mater. Sol. Cells* 137: 311–318.
97 Baierl, D., Fabel, B., Gabos, P. et al. (2010). Solution-processable inverted organic photodetectors using oxygen plasma treatment. *Org. Electron.* 11: 1199–1206.
98 Glatthaar, M., Niggemann, M., Zimmermann, B. et al. (2005). Organic solar cells using inverted layer sequence. *Thin Solid Films* 491: 298–300.
99 Ratcliff, E.L., Meyer, J., Steirer, K.X. et al. (2012). Energy level alignment in PCDTBT:PC70BM solar cells: solution processed NiO_x for improved hole collection and efficiency. *Org. Electron.* 13: 744–749.
100 Bai, S., Cao, M., Jin, Y. et al. (2014). Low-temperature combustion-synthesized nickel oxide thin films as hole-transport interlayers for solution-processed optoelectronic devices. *Adv. Energy Mater.* 4: 1301460.
101 Terán-Escobar, G., Pampel, J., Caicedo, J.M., and Lira-Cantú, M. (2013). Low-temperature, solution-processed, layered V_2O_5 hydrate as the hole-transport layer for stable organic solar cells. *Energy Environ. Sci.* 6: 3088.
102 Tao, C., Ruan, S., Xie, G. et al. (2009). Role of tungsten oxide in inverted polymer solar cells. *Appl. Phys. Lett.* 94: 043311.

103 Hu, L., Chen, L., Hu, X., and Chen, Y. (2014). Solution processed and self-assembled polymerizable fullerenes/metal oxide as an interlayer for high efficient inverted polymer solar cells. *J. Mater. Chem. C* 2: 10282–10290.

104 Beek, W.J., Wienk, M.M., Kemerink, M. et al. (2005). Hybrid zinc oxide conjugated polymer bulk heterojunction solar cells. *J. Phys. Chem. B* 109: 9505–9516.

105 Zhou, Y., Fuentes-Hernandez, C., Shim, J. et al. (2012). A universal method to produce low–work function electrodes for organic electronics. *Science* 336: 327–332.

106 O'Connor, T.F., Zaretski, A.V., Shiravi, B.A. et al. (2014). Stretching and conformal bonding of organic solar cells to hemispherical surfaces. *Energy Environ. Sci.* 7: 370–378.

107 Weber, C., Oberberg, M., Weber, D. et al. (2014). Improved morphology and performance of solution-processed metal-oxide thin-film transistors due to a polymer based Interface modifier. *Adv. Mater. Interfaces* 1: 1400137.

108 Zhou, H., Chen, Q., Li, G. et al. (2014). Interface engineering of highly efficient perovskite solar cells. *Science* 345: 542–546.

109 Kim, H.H., Park, S., Yi, Y. et al. (2015). Inverted quantum dot light emitting diodes using polyethylenimine ethoxylated modified ZnO. *Sci. Rep.* 5: 8968.

110 Hu, L., Wu, F., Li, C. et al. (2015). Alcohol-soluble n-type conjugated polyelectrolyte as electron transport layer for polymer solar cells. *Macromolecules* 48: 5578–5586.

111 Tang, Z., Andersson, L.M., George, Z. et al. (2012). Interlayer for modified cathode in highly efficient inverted ITO-free organic solar cells. *Adv. Mater.* 24: 554–558.

112 Ameri, T., Dennler, G., Lungenschmied, C., and Brabec, C.J. (2009). Organic tandem solar cells: a review. *Energy Environ. Sci.* 2: 347.

113 Ameri, T., Li, N., and Brabec, C.J. (2013). Highly efficient organic tandem solar cells: a follow up review. *Energy Environ. Sci.* 6: 2390.

114 Dennler, G., Prall, H.-J., Koeppe, R. et al. (2006). Enhanced spectral coverage in tandem organic solar cells. *Appl. Phys. Lett.* 89: 073502.

115 Kim, J.Y., Lee, K., Coates, N.E. et al. (2007). Efficient tandem polymer solar cells fabricated by all-solution processing. *Science* 317: 222–225.

116 Dou, L., You, J., Yang, J. et al. (2012). Tandem polymer solar cells featuring a spectrally matched low-bandgap polymer. *Nat. Photonics* 6: 180–185.

117 Zhou, Y., Fuentes-Hernandez, C., Shim, J.W. et al. (2012). High performance polymeric charge recombination layer for organic tandem solar cells. *Energy Environ. Sci.* 5: 9827.

118 Tong, J., Xiong, S., Zhou, Y. et al. (2016). Flexible all-solution-processed all-plastic multijunction solar cells for powering electronic devices. *Mater. Horiz.* 3: 452–459.

119 You, J., Dou, L., Yoshimura, K. et al. (2013). A polymer tandem solar cell with 10.6% power conversion efficiency. *Nat. Commun.* 4: 1446.

120 Li, W., Furlan, A., Hendriks, K.H. et al. (2013). Efficient tandem and triple-junction polymer solar cells. *J. Am. Chem. Soc.* 135: 5529–5532.

121 Yusoff, A.R.B.M., Kim, D., Kim, H.P. et al. (2015). A high efficiency solution processed polymer inverted triple-junction solar cell exhibiting a power conversion efficiency of 11.83%. *Energy Environ. Sci.* 8: 303–316.
122 Krebs, F.C. (2009). Fabrication and processing of polymer solar cells: a review of printing and coating techniques. *Sol. Energy Mater. Sol. Cells* 93: 394–412.
123 Guo, F., Li, N., Fecher, F.W. et al. (2015). A generic concept to overcome bandgap limitations for designing highly efficient multi-junction photovoltaic cells. *Nat. Commun.* 6: 7730.
124 Shaheen, S.E., Radspinner, R., Peyghambarian, N., and Jabbour, G.E. (2001). Fabrication of bulk heterojunction plastic solar cells by screen printing. *Appl. Phys. Lett.* 79: 2996–2998.
125 Angmo, D., Sweelssen, J., Andriessen, R. et al. (2013). Inkjet printing of back electrodes for inverted polymer solar cells. *Adv. Energy Mater.* 3: 1230–1237.
126 Angmo, D., Larsen-Olsen, T.T., Jørgensen, M. et al. (2013). Roll-to-roll inkjet printing and photonic sintering of electrodes for ITO free polymer solar cell modules and facile product integration. *Adv. Energy Mater.* 3: 172–175.
127 Eggenhuisen, T.M., Galagan, Y., Biezemans, A.F.K.V. et al. (2015). High efficiency, fully inkjet printed organic solar cells with freedom of design. *J. Mater. Chem. A* 3: 7255–7262.
128 Søndergaard, R., Hösel, M., Angmo, D. et al. (2012). Roll-to-roll fabrication of polymer solar cells. *Mater. Today* 15: 36–49.
129 Carle, J.E., Helgesen, M., Madsen, M.V. et al. (2014). Upscaling from single cells to modules - fabrication of vacuum- and ITO-free polymer solar cells on flexible substrates with long lifetime. *J. Mater. Chem. C* 2: 1290–1297.
130 Angmo, D., Gevorgyan, S.A., Larsen-Olsen, T.T. et al. (2013). Scalability and stability of very thin, roll-to-roll processed, large area, indium-tin-oxide free polymer solar cell modules. *Org. Electron.* 14: 984–994.

12

Flexible Quantum Dot Sensitized Solar Cells

Yueli Liu [1], Keqiang Chen [1], Zhuoyin Peng [2], and Wen Chen [1]

[1] *Wuhan University of Technology, School of Materials Science and Engineering, State Key Laboratory of Silicate Materials for Architectures, No. 122, Luoshi Road, Wuhan 430070, China*
[2] *Changsha University of Science and Technology, School of Energy and Power Engineering, No. 960, Second Section, South of Wanjiali Road, Changsha 410111, China*

12.1 Introduction

Energy shortage and environmental pollution are two major problems that mankind faces in the twenty-first century. To date, the focus on energy resources in the world is still mainly on coal, petroleum, natural gas, hydropower, and nuclear energy. However, most of these resources are not inexhaustible and moreover, their use generates massive emission of CO_2 and NO_x and other contaminants such as hydrocarbons and sulfides, which leads to the greenhouse effect, acid rain, ozone depletion, and other types of environmental pollution. Considering these problems, the development of novel and clean energy sources has attracted more and more attention worldwide.

Photovoltaic (PV) devices are kinds of potential and promising choices among the group of energy conversion systems, wherein direct conversion from solar energy to electrical energy would be possible at lost cost. As an inexhaustible and clean energy that does not create any environmental pollution, many countries have now set up projects for the research and development of novel solar cells. To date, many kinds of solar cells have been proposed, such as silicon solar cells, multijunction solar cells, thin-film solar cells (CIGS, CdTe, etc.), and third generation solar cells, including dye-sensitized solar cells (DSSCs), quantum dots sensitized solar cells (QDSSCs), organic solar cells, perovskite solar cell, and so on [1, 2]. However, most of them are limited by their expensive outlay due to complicated manufacturing cost, which prevents their wide application.

Among them, QDSSCs are considered to be one kind of low-cost solar cells, which encompass devices whose maximum conversion efficiency is above the 32% limit for single-junction solar cell in air mass (AM) 1.5 sunlight [3]. Plenty of research work on QDSSCs has been blooming in the past 10 years because of their versatile properties, such as their narrow bandgaps, large absorption efficiencies, special photoelectric properties, and quantum size effects. Although a highest photovoltaic performance of over 12% has been reported recently, to meet the

Flexible Energy Conversion and Storage Devices, First Edition. Edited by Chunyi Zhi and Liming Dai.

Figure 12.1 Roll-to-roll fabrication of flexible photovoltaic cells. *Source:* Cheng et al. 2016 [5]. Copyright 2016, Nature Publishing Group.

demands of commercialization, there are still many questions to be solved, such as flexibility, roll-to-roll fabrication, utilization of the substrate at low cost, and so on [4]. Therefore, the development of flexible QDSSCs not only has to address the demanding application needs, but also consider the feasibility of roll-to-roll fabrication and cost-effectiveness in the manufacturing industry [5] (Figure 12.1).

Conventional solar cells with inflexible substrates are rigid and bulky, which is a limitation in many unique commercial applications. Flexible solar cells on plastic films and thin metal foils are already being made due to their advantages of lightweight, translucence, flexibility, and low cost. Compared with conventional solar cells, flexible solar cells possess more and wider application areas, for example, installing them in curved places to supply energy consumption; using little material and then generating electricity in low light, even indoors; and integrating them into wearable products of phones and watches as well as on walls and windows. All of the promising applications would transform the world's energy generation, reduce pollution, and mitigate climate change. Therefore, replacing the fluorine-doped tin oxide (FTO) glass substrates with flexible conductive substrates in QDSSCs not only reduces the weight and cost of the devices but also extends their applications.

12.2 Basic Concepts

12.2.1 Quantum Dots (QDs)

A bulk solid usually has stable physical and chemical properties. However, by tuning the physical size of materials to the nanoscale, a number of unique and interesting phenomena can be observed. Furthermore, decreasing the size of nanomaterials to less than their three-dimensional Bohr radius, some additional characteristics could appear [6, 7]. These kinds of nanoparticles are known as quantum dots (QDs). The particular physical size of QDs gives them some special properties, which are different from those of bulk and molecular systems.

12.2.1.1 Quantum Size Effect

As the size of QDs is smaller than their Bohr radius, which is commonly less than 10 nm, and the limited lattice structure is not able to build a continuous band structure similar to that of bulk materials, this would result in the energy level to be discrete (Figure 12.2) [6, 8]. The most obvious variation lies in the confinement of electron and hole wave functions, which leads to the varying of the bandgap with the size of QDs, known as the quantum size effect [9].

The bandgaps of QDs can be continuously regulated by altering the size of the QDs, which proposes a novel pathway to achieve QDs with various bandgaps, without changing their compositions and phases. This excellent characteristic makes it more convenient to fabricate different kinds of PVs. The variation of QD bandgaps can be observed from their optical absorption spectra. With the decreasing of QD sizes, their bandgaps increase dramatically, and the absorption onsets shift to the short wavelength range. Figure 12.3a shows the relationship between the QDs' size and their absorption properties [10]. A distinct red-shift of the absorption spectra can be clearly observed with the increasing size of $CuInS_2$ QDs. Moreover, the optical bandgaps vary with the QD sizes, 3.5-nm-sized $CuInS_2$ QDs have an E_g of ~2 eV, the value reduces to ~1.6 eV when the QD sizes reach up to 7.3 nm as shown in Figure 12.3b, which exhibits the obvious quantum size effect.

12.2.1.2 Multiple Exciton Generation

Enormous excitement has followed the prediction, discovery, and confirmation from the fact that several excitons can be produced from the absorption of a single

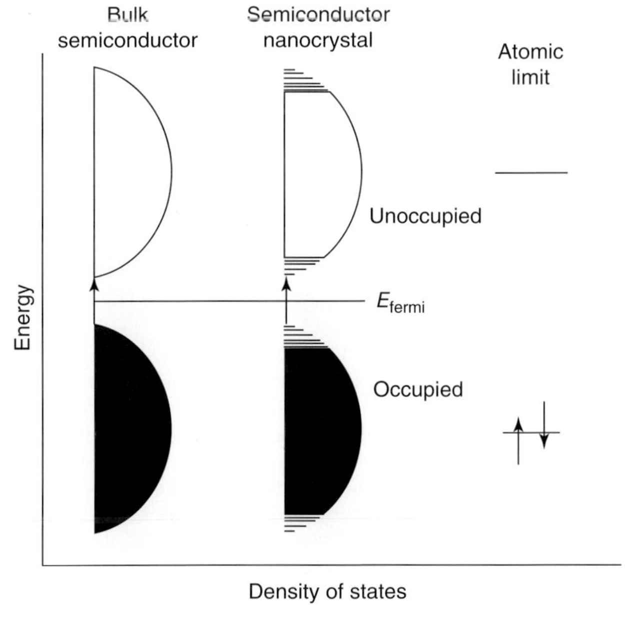

Figure 12.2 Density states in semiconductor nanocrystals. *Source:* Alivisatos 1996 [8]. Copyright 1996, American Chemical Society.

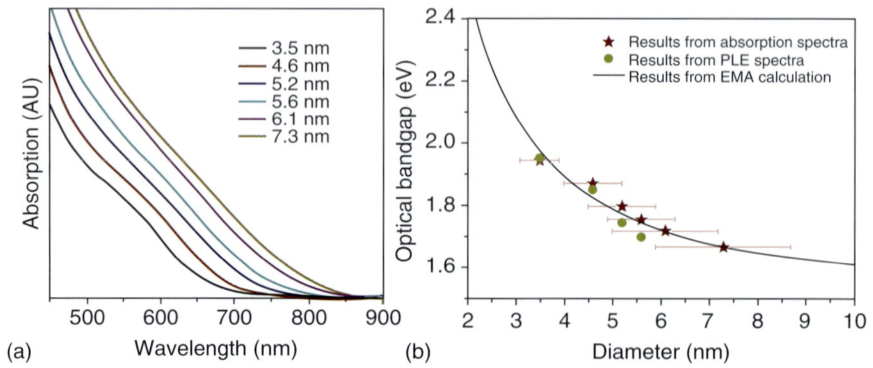

Figure 12.3 (a) Absorption spectra of $CuInS_2$ QDs with different average sizes and (b) size-dependent optical bandgaps of $CuInS_2$ QDs determined from absorption (red stars) and 2D photoluminescence excitation (PLE) spectra (green spheres). The solid line represents the calculation result from finite depth-well effective mass approximation (EMA) theory. *Source:* Zhong et al. 2010 [10]. Copyright 2010, American Chemical Society.

photon by very small QDs because their electronic properties are different from those of bulk-size materials due to the confinement of the electron–hole pairs produced by optical excitation [11–13]. This effect occurs if the photon energy is several times higher than the semiconductor band-gap. In a bulk semiconductor, the excess energy would be converted into heat, while a change in the selection rules allows several electron–hole pairs to be generated by a single photon in QDs (Figure 12.4).

While initial works have reported the generation of up to *seven* electron–hole pairs per photon in a PbSe nanocrystal through multiple exciton generation (MEG) [12], these values are changed in 2008. The generation of only 1.1–2.4 excitons per absorbed photon is evidenced depending on the materials and preparation. A recent contribution of Nozik's group has pointed out the very high sensitivity of the MEG efficiency on the nanocrystals' surface chemistry, originating from the difference in reproducibility [14].

12.2.1.3 Ultrafast Electron Transfer

In inorganic semiconductor based photoelectronic devices, one absorbed photon are presumed to generate two or more electron–hole pairs when the photon

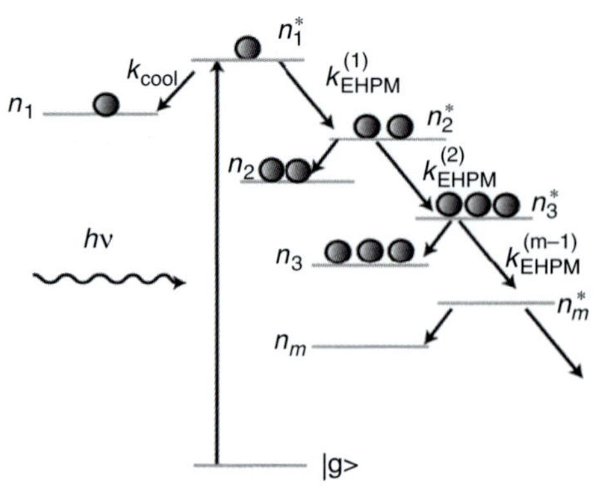

Figure 12.4 Illustration of cascade scheme for electron–hole pair multiplication. A high-energy photon creates an exciton with excess energy, n_1^*. The hot exciton can lose energy by cooling or electron–hole pair multiplication to form either n_1 or a hot biexciton, n_2^*, and so on. *Source:* Beard et al. 2010 [13]. Copyright 2010, American Chemical Society.

energy exceeds its bandgap by 2.7 times, which could sharply enhance the performances of devices [15, 16]. However, the as-generated electron–hole pairs could not be effectively separated quickly in the bulk inorganic semiconductor materials, which leads to a remarkable recombination process, and restricts the performances of the corresponding PV devices. In contrast, previous reports have suggested that the multiexcited electron–hole pairs could be ultrafast transferred from QDs to the electron acceptor. Huang et al. have achieved an exciton dissociation half-time of ~2.3 ps [17]. Moreover, Wachtveitl's group has covered the molecular electron acceptor methylviologen (MV^{2+}) onto the surface of CdSe QDs, and their outcomes indicate that the electron could be transferred from the conduction band of the CdSe QDs to MV^{2+} in a short time of ~70 fs (Figure 12.5) [18]. These works propose strong possibilities for obtaining high-performance PV devices by utilizing the MEG process.

12.2.1.4 Large Specific Surface Area

Besides the special physical properties mentioned above, QDs also have other distinguishing chemical properties compared with their bulk materials [19]. For any bulk material, there are only small amounts of surface atoms, which reveal that the internal atoms play a determinant role on the material property. With the decreasing of the material sizes, the proportion of the surface atoms increases, and they undertake more contributions to the material property.

When the substances show as QDs, the tiny sizes enlarge the concentration of surface atoms; moreover, the surface-to-volume ratios increase sorely. From Figure 12.6, one can clearly see that the surface-to-volume atom ratio increases rapidly with shrinking size [20]. When the crystal size of InP is 2 μm, the surface-to-volume atom ratio is about 0.06%. On decreasing the crystal size to 20 nm, the surface-to-volume atom ratio rises promptly to 5.9%. If the crystal size of InP is as small as 2 nm, the surface-to-volume atom ratio would be as large as 54.2%, which is outstanding. The high surface-to-volume atom ratio suggests that the chemical properties of QDs should be significantly different from that of their bulk materials. For example, the particular surface states reduce their stabilities, and surface ligands are necessary to prevent their aggregation, which have great influences on their solubility, optical and electronic characteristics [21]. Furthermore, the large amount of surface atoms can achieve several new

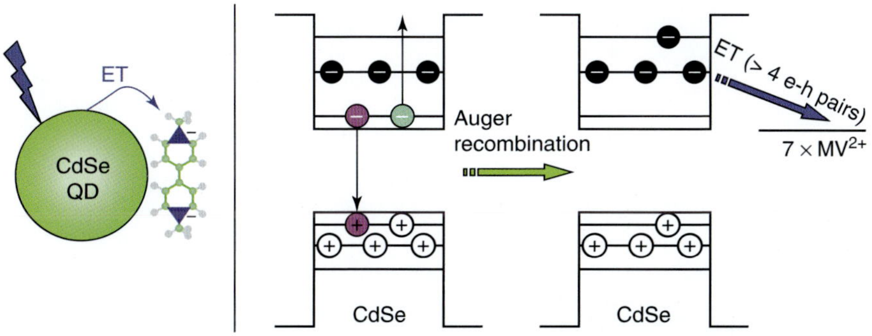

Figure 12.5 Ultrafast electron transfer from CdSe nanocrystals to methylviologen (MV^{2+}). *Source:* Matylitsky et al. 2009 [18]. Copyright 2009, American Chemical Society.

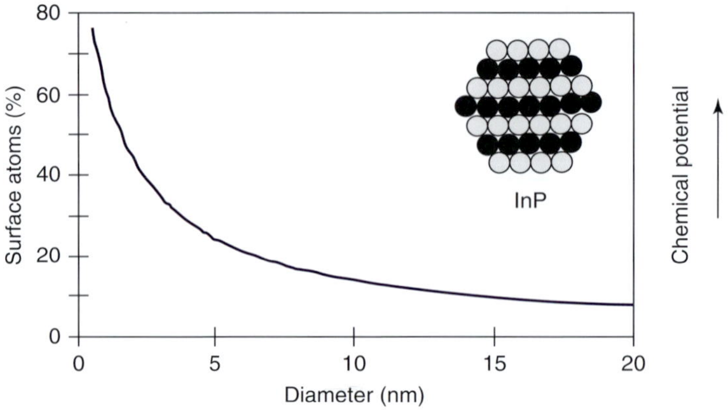

Figure 12.6 Surface-to-volume atom ratio and chemical potential of nanocrystals calculated using InP lattice parameters. *Source:* Peng 2009 [20]. Copyright 2009, Springer.

effects. Brittman et al. have fabricated a single-nanowire (NW) silicon solar cell device which is decorated with a silver nanocrystal, and their results reveal that the silver nanocrystal could enhance the absorption property due to surface plasmon resonance [22]. In addition to this, the special surface characteristic can induce size-dependent catalytic activity [23].

12.2.2 Quantum Dots Sensitized Solar Cells (QDSSCs)

12.2.2.1 Schematic of the Structure and Charge Circulation of QDSSCs

The special properties of QDs mentioned above reveal that QDs could be an ideal candidate for photo-electronic devices, and QDSSCs are one of the most important applications. In QDSSCs devices, QDs are regarded as absorber materials (sensitizers) owing to their high absorption efficiency and size-dependent bandgap, which can enhance the quantity of photo-excited electron–hole pairs and optimize the energy level alignment. Moreover, there are large numbers of QD materials for QDSSCs, which provide more choices for fabricating high-performance PV devices.

Figure 12.7 shows the schematic of QDSSCs [24]. Metal oxides with wide bandgap, such as TiO_2 and ZnO, are prepared as mesoporous films on

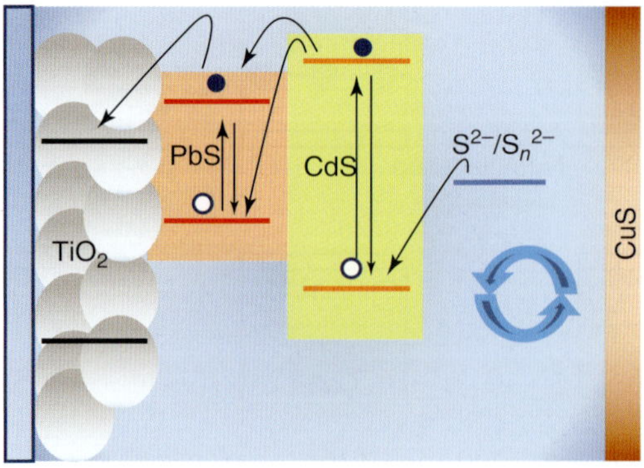

Figure 12.7 (a) Schematic of QDSSCs and (b) energy-band diagram of QDSSCs. *Source:* González-Pedro et al. 2013 [24]. Copyright 2013, Royal Society of Chemistry.

transparent conducting oxide (TCO) substrate, and these kinds of photo-anode electrodes are usually used to form a framework as electron transport materials for QDSSCs, which provide a tremendous surface area for absorbing a sufficient number of QDs [25]. After packaging with counter electrode (CE) (called as photo-cathode electrodes) and injecting electrolyte, sandwich-like structured QDSSCs are achieved.

For silicon solar cells, a built-in electric field will emerge at the interface of the p–n junction, which could separate the photo-excited electron–hole pairs and yield electric current. However, the QD layer in QDSSCs is too thin to form a space-charge region with photo-anodes, which means that there is no built-in electric field in the QDSSCs devices. Therefore, the separation mechanism of electron–hole pairs in QDSSCs and its working principle are sharply different with those of silicon solar cells. In the QDSSCs, QDs absorb the sunlight and generate electron–hole pairs, and the electrons in the conduction band of QDs shift to the conduction band of TiO_2 (or ZnO). The driving force should be attributed to the difference between the conduction bands of these two materials, and the electrons prefer to transfer from the high energy level to a low one. Similarly, the holes in the valence band of QDs will move to the electrolyte, which means that the oxidized QDs could be reduced by the electrolyte. The electrons obtained from the QDs will move to the TCO and external circuit to perform their role, and eventually return to the QDs and complete the circulation.

12.2.2.2 Evaluation of the Photovoltaic Performances of QDSSCs

The as-fabricated QDSSCs are commonly tested under AM 1.5 illumination to obtain the J–V characteristic curve (Figure 12.8). The current achieved at the short circuit conditions is called the short current (I_{sc}), which indicates that the voltage should be zero, and the short-circuit current density (J_{sc}) is obtained after the value is divided by the active area. Similarly, the open-circuit voltage (V_{oc}) is the value obtained under open-circuit conditions, the electron injection

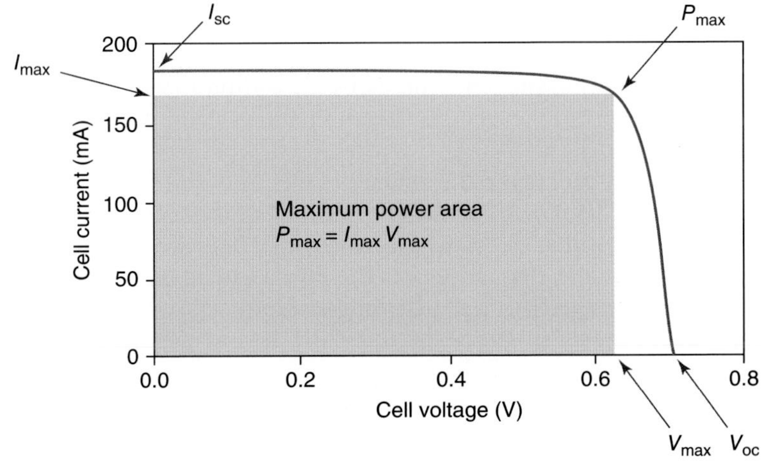

Figure 12.8 Typical I–V cure of QDSSCs.

is balanced by the recombination and there is no current. The output power of the cells is equal to the product of J and V, the higher one of them is the power conversion efficiency (PCE) (η). The J and V corresponding to the η are called as J_{max} and V_{max}, and the fill factor (FF) equals the ratio of $J_{max} \times V_{max}$ to $J_{sc} \times V_{oc}$.

For the QDSSCs, the V_{oc} is determined by the energy difference between the conduction band (E_{CB}) of photo-anodes and the energy level of the electrolyte. As the E_{CB} of photo-anodes (TiO_2 or ZnO) is substantially retained, the only pathway to alter the V_{oc} is to select various kinds of electrolytes. The S^{2-}/S_x^{2-} redox couple containing electrolyte is the most used one [26]. Researchers usually control the concentration of S^{2-}/S_x^{2-} redox couple and the pH value of aqueous electrolytes by tuning the quantities of different precursors, which can influence the energy level of the electrolyte to some extent, as well as affect the density of surface traps [27, 28]. However, the change in its energy level is not very significant, which indicates that the V_{oc} of QDSSCs is almost the fixed one. For TiO_2-based QDSSCs, the V_{oc} is about 500–700 mV.

The above statement suggests that enhancing the values of J_{sc} and FF is the practical way to achieve high-performance QDSSCs. The J_{sc} is closely associated with the light absorption property of the cells and therefore, the common approaches proposed to reach this goal, are structure optimization, increasing the QDs absorbance, and co-sensitization with different kinds of QDs. The photo-anodes usually have two layers: The first layer is of TiO_2 (or ZnO) nanocrystals with a diameter of 5–20 nm: the smaller size could provide larger surface area, which could promote the deposition of QDs; the second layer is called the scattering layer that consists of TiO_2 (or ZnO) nanocrystals with diameters of about 200–400 nm and is used to scatter the incident light for re-absorption [29]. Besides this, doping or designing the morphology of photo-anodes is an efficient route to optimize the structure of QDSSCs, too [30, 31].

Increasing the QDs' absorbance needs optimization of the structure of photo-anodes, which should be mesoporous. Moreover, many factors have a great influence on the adsorption ability of QDs, such as the surface states of photo-anodes, sizes, and surface ligand engineering of QDs [32–34]. However, excess QDs on the photo-anodes would lead to their aggregation and enlarge the amount of defect trap sites, and as a result, the recombination probability of the photo-generated electron–hole pairs increases. In addition, choosing QDs with wide light absorption range or co-sensitization shows good effectiveness for boosting the PCE of QDSSCs [35].

FF reflects the transport process of the generated electron–hole pairs. If the electron–hole pairs could be separated and transported to the external circuit as fast as possible, the FF value of the as-fabricated QDSSCs would be high. The most important parameter that influences the FF should be the resistance, and it contains series resistance (R_s) and parallel resistance (R_{sh}). R_s represents the charge transfer resistance in the solar cell. Low R_s is conducive to the transportation of electrons and can enhance the FF at last. Conversely, low R_{sh} would aggravate the recombination of electron–hole pairs. Suitable electrolyte and CE, especially the latter, could promote electron transportation. R_{sh} reflects the quantity of defect trap sites, QDs with high quality and optimal interface between the QDs and photo-anodes result in low defect trap sites and high R_{sh}, which prevent the electron–hole pairs from recombination and promote the enhancement of PCE.

12.3 Development of the Flexible QDSSCs

12.3.1 Choosing of the Types of QDs

Owing to the excellent properties of QDs, more and more works have been focusing on the preparation, characterization, and utilization of QDs. A large number of materials, including metal and semiconductor materials, have been synthesized as QDs. Among them, transition metal chalcogenide semiconductor QDs, especially the Cd-, Pb-, and Cu-based QDs, have extremely wide application in the field of QDSSCs.

12.3.1.1 Cd-based QDs

Cd-based QDs, such as CdS, CdSe, CdTe, CdS_xSe_{1-x}, $CdSe_xTe_{1-x}$, CdSe/CdS (core/shell), CdTe/CdSe, etc., are the most deeply studied for QDSSCs applications. All of them have direct bandgaps that vary from ~1.45 to ~2.45 eV. Moreover, their high absorption coefficients promote the effective absorption of sunlight, which results in the generation of a large number of electron–hole pairs. These properties suggest that Cd-based QDs might be suitable materials for solar cell application. In addition to these application potentials, the Cd-based QDs can be easily sensitized onto the surface of photo-anodes, which promotes their applications.

During the early research of QDSSCs, Kamat and coworkers have linked CdSe QDs onto mesoscopic TiO_2 films and fabricated a photoelectrochemical cell, which exhibits a photon-to-charge carrier generation efficiency of 12% [36]. This work proposes a possible way for the application of QDs in the field of QDs-based solar cells. Subsequently, further efforts have been put into effect. Lee et al. have synthesized colloidal CdSe QDs and treated it as the inorganic dyes to sensitize TiO_2 photo-anode films for QDSSCs, the as-fabricated device achieves a much higher photon-to-charge carrier generation efficiency of 36% and has a PCE of 1.7% at 0.1 full-sun intensity [37]. Lee et al. have prepared CdS/CdSe QDs co-sensitized solar cells and improved the efficiency up to 4.22% [38]. Recently, Zhong's group has employed the alloyed QDs as the sensitizers to fabricate QDSSCs, and the efficiency is as high as 8.55% [39]. Meng's group has prepared $CdSe_xTe_{1-x}$ QDSSCs with a PCE of 11.23% [40].

Moreover, the QDs doping has also been developed to enhance the performance of QDSSCs. Santra and Kamat has prepared Mn-doped-CdS/CdSe QDSSCs and received a PCE enhancement from 4.19% to 5.42% [41]. Yang et al. have fabricated QDSSCs with PCE exceeding 9% by using Mn-doped CdSeTe QDs as sensitizer [42]. The efficiency enhancement could be attributed to the Mn *d–d* transition ($^4T_1 - {}^6A_1$), which shows an emission at ~585 nm and promotes the absorption properties of as-fabricated solar cells. In addition, this transition could prolong the lifetime of generated exactions, which can be as long as several hundreds of microseconds [41]. Hence, the QDSSCs using Mn-doped (or $CuZnSnSe_2$) is another important photovoltaic material. However, it is usually fabricated to thin-film solar cells. The performances of QDSSCs based on $CuZnSnS_2$ QDs are relatively low [52]. However, several new materials, such as Cu–Sb–S [53] and $Cu_2GeS_{3-x}Se_x$ [54] have been reported as potential materials for QDSSCs.

Besides these common materials, InP [55], InAs [56], Bi_2S_3 [57], Sb_2S_3 [58], and Ag_2S [59] are also outstanding materials for photovoltaic materials, and they have been prepared as QDs and employed as sensitizers in QDSSCs.

12.3.2 Fabrication of the Flexible Photo-anode Films

Photo-anode is a significant part of QDSSCs, and some parameters such as the type of porous films, pore size distribution, surface electronic structure, film thickness, and particle size distribution in photo-anodes have important influences on the performance of QD-adsorption quantity in QDSSCs. Moreover, the interface between electron recombination, electron transport, and optical absorption efficiency of photo-anode electrodes will influence the photovoltaic conversion efficiency of QDSSCs to a large extent.

Recently, many metal oxides have been employed as basic materials of the photo-anodes in QDSSCs, which are introduced as the loading sites for QDs and as effective paths for carrier transfer. According to the working principle of QDSSCs, there are four photo-induced charge transfer processes to ensure the normal output current [60]. The composition and structure of metal oxides can influence the deposition of QDs, as well as the transfer and recombination of photo-induced charges. Therefore, the investigation of metal oxides becomes the important areas of focus for the enhancement of the photovoltaic performance in QDSSCs.

In order to obtain high photovoltaic conversion efficiency, the energy level and the structure of photo-anode films are the two important factors [61]. The transfer behavior of the photo-induced electron plays an important role in the photo-current generation process of QDSSCs. Photo-induced electron injection from QDs to metal oxides is the first step in photo-induced electron transfer, which has a significant influence on the subsequent electron transfer process. Therefore, the energy level engineering of QDs and metal oxides is important for the injection of photo-induced electrons from QDs to metal oxides. Furthermore, the different structures of oxide semiconductors have a critical effect on the charge separation, transfer, and collection. The higher surface ratio and porosity of nanoparticle (NP)-type metal oxides can ensure large adsorption ability of QDs, which can also be beneficial for the electrolyte infiltration. Moreover, with single-crystalline structure, the nanoarray-type metal oxides can obviously improve the ability of light scattering and distance of light transport. Additionally, it can also provide a direct pathway for the transfer of photo-induced electrons, which greatly enhances the separation efficiency of the photo-induced electron–hole pairs [62]. Therefore, the optimization of oxide semiconductors for the photo-anodes of QDSSCs has been widely investigated for parameters such as selection, preparation, phase, structure, etc., which enables the photovoltaic conversion efficiency enhancement of QDSSCs over 10% .

Actually, the metal oxides on flexible thin metal foils or plastic films are critical materials for the photo-anodes in flexible QDSSCs. As we know, traditional QDSSCs are prepared on the FTO glass substrates, and the photo-anode films need high temperature calcination at around 450 °C, which is required to remove the organic additives and promote chemical interconnection between the particles to establish their electrical connection [63, 64]. Although the flexible metal substrate can sustain high temperature, the light must be illuminated from the front, which inevitably causes optical absorption losses. The other flexible plastic conductive substrates can be illuminated from the back to improve the optical absorption property of solar cells, such as indium tin oxide/polyethylene naphthalene-2,6-dicarboxylate (ITO/PEN), indium tin oxide/polyethylene terephthalate (ITO/PET), polymer, etc. However, all of them cannot withstand high temperature over 200 °C due to their low melting point. Therefore, the low temperature processing of oxide semiconductor photo-anodes becomes crucial for flexible QDSSCs.

The selection and morphology of the photo-anode films are the key factors to achieve high optical absorption, fast charge transport, and low recombination properties in QDSSCs. Recently, many kinds of metal oxide nanomaterials with different morphologies have been prepared as discussed further.

12.3.3 TiO_2-Based Photo-anodes

In the past decade, many TiO_2 nanomaterials with different structures have been developed as the photo-anodes in QDSSCs, such as TiO_2 nanoparticles, nanorods (NRs), NWs, nanotubes (NTs), etc.

12.3.3.1 Photo-anodes of TiO_2 Nanoparticles

NPs-based thin film can provide large surface ratio for solar cells. As the commercial material of photocatalysis, P25 has become the original TiO_2 nanoparticle for the photo-anodes in QDSSCs, which can achieve a high photovoltaic conversion efficiency of 4.22% for CdS/CdSe QDSSCs reported by Lee et al. [38]. In order to eliminate the photovoltaic performance limitation of the rutile phase in P25, pure anatase TiO_2 nanoparticles have been employed to prepare photo-anodes to realize the efficiency enhancement of QDSSCs. Mali et al. have prepared PbS QDs sensitized anatase TiO_2 nanocrystal solar cells to improve the photoelectrochemical performance [65]. Furthermore, many novel highly active anatase TiO_2 nanoparticles have been prepared to improve the charge transfer in QDSSCs. Especially, TiO_2 nanosheets (NSs) with high-active crystal planes have also been developed to obtain higher photovoltaic efficiency from QDSSCs. Wu et al. have prepared TiO_2 NSs with NPs as photo-anodes of QDSSCs, which exhibit a high efficiency of 3.47% [66]. In order to fabricate flexible TiO_2-based solar cells, Weerasinghe et al have introduced a low-temperature process by employing ethanol, acid, and many other low-temperature binder solvents into TiO_2 NPs to fabricate TiO_2 nanoparticle photo-anodes, which achieves a photovoltaic efficiency of 7.4% in DSSCs [67]. It can also be employed for the preparation of TiO_2 photo-anodes in flexible QDSSCs.

photo-anodes, and the light-harvesting ability of conical structure is enhanced [80]. TiO_2 nanoarrays such as NWs and NTs have been prepared based on Ti foil or wire substrate, which are employed to fabricate flexible QDSSCs, achieving a photovoltaic efficiency over 3% [81].

12.3.3.3 Designing of Novel TiO_2 Architecture as Photo-anodes

Many other novel nanostructures have been prepared as the photo-anodes in QDDSCs, such as microflowers, nanospheres, etc. TiO_2 microflowers have been prepared to enhance photovoltaic performance [82]. Three-dimensional hierarchical TiO_2 hollow sphere has been prepared as photo-anodes in QDSSCs to achieve a high photovoltaic efficiency of 6.01% [83]. Bamboo-like TiO_2 NTs are fabricated by alternating the anodic oxidation process to improve the optical absorption of QDSSCs [84]. It is believed that these potential TiO_2 nanostructures will enhance the photovoltaic efficiency of QDSSCs in future applications.

12.3.4 ZnO based Photo-anodes

Although the efficiency is still lower than that of TiO_2-based QDSSCs, ZnO nanomaterials still exhibit large potential as photo-anodes in QDSSCs, which are considered as effective materials in QDSSCs investigations.

Owing to the effective charge transfer of nanoarray structures, ZnO NRs and NWs have been prepared by hydrothermal and electrodepostion processes for photo-anodes in QDSSCs, which can also achieve a desired photovoltaic performance for both general and flexible QDSSCs. Double-sided CdS and CdSe QDs have been co-sensitized on ZnO NRs by Wang, which exhibit a significant enhancement of photovoltaic performance [85]. Based on ZnO NRs photo-anodes, Raj et al. have prepared CdSe/CdS/PbS QDSSCs to achieve a photovoltaic efficiency of 2.35% [86]. Flexible CdSe QDSSCs have been also prepared based on ZnO NRs by Chen, which achieve an efficiency of 1% [87]. Similarly to TiO_2 NDs, hexagonal ZnO nanodisks have been prepared to improve the light scattering and charge collection in Figure 12.12a, exhibiting a high photovoltaic efficiency of 4.86% [88]. ZnO NWs and dense branched ZnO NWs have been prepared as photo-anodes in QDSSCs as shown in Figure 12.12b,c, which can enhance the optical absorption and charge transfer properties. Branched ZnO NWs based QDSSCs exhibit twice the efficiency than that of bare ZnO NWs [89]. Tian et al. have prepared ZnO NS/NWs nanostructures to fabricate QDSSCs, demonstrating a photovoltaic efficiency as high as 3.28% [91]. Three-dimensional ZnO microspheres have been prepared based on ZnO NWs as shown in Figure 12.12d, which can effectively increase the surface area of photo-anodes and exhibit a high J_{sc}, FF and efficiency [90]. Furthermore, TiO_2 and ZnO nanostructures are composited to prepare the photo-anodes of QDSSCs. ZnO NRs have been coated with TiO_2 NSs for QDSSCs to increase photocurrent response [92]. Feng et al. have prepared TiO_2 NWs/ZnO NS photo-anode structure to improve photovoltaic efficiency to 4.57% [93]. These different nanostructures provide excellent photovoltaic performance of QDSSCs.

Figure 12.12 Different ZnO structures: (a) Nanodisks. *Source:* Chetia et al. 2015 [88]. Copyright 2015, American Chemical Society. (b) Nanorods; (c) branched-nanorods. *Source:* (b,c) Lee et al. 2015 [89]. Copyright 2015, Elsevier. (d) Microspheres. *Source:* Chetia et al. 2014 [90]. Copyright 2014, Royal Society of Chemistry.

12.3.5 Other Metal Oxide Based Photo-anodes

In order to further improve the photovoltaic performance of QDSSCs, some other metal oxide nanomaterials, such as SnO_2, are also introduced to act as the photo-anodes. Employing different SnO_2 nanostructures for photo-anodes can provide efficient charge injection and collection in QDSSCs. $TiCl_4$-modified SnO_2 can enhance photovoltaic performance of PbS/CdS based photo-anodes to achieve efficiency of 1.6% [94]. Cauliflower-like SnO_2 hollow microspheres sensitized with CdS/CdSe QDs have a high surface area and good light scattering effect, which improve the photovoltaic efficiency to 2.5% [95]. SnO_2/ZnO based photo-anodes can enhance the electron transfer performance of CdS QDSSCs to achieve a photovoltaic efficiency of 3.41% [96]. SnO_2 photo-anodes with three-dimensional interconnected ordered porous structure can further enhance the photovoltaic conversion efficiency of 4.37% [97].

12.3.6 Development of the Sensitization Method

QD sensitizers are the key component of QDSSCs, as they harvest the whole ultraviolet and visible light and produce photo-excited electron–hole pairs. In order to obtain excellent photovoltaic performance QDSSCs, the QD sensitizers should have several requirements in the photo-anodes.

Firstly, they must have a great stability for the photo-anodes to ensure the electron transfer process because the electrons are photo-generated from QD are dropped onto the surface of an oxide semiconductor, which are spin coated without rinsing and drying, and this spin-coating cycle is also repeated several times to form the desired QD layer.

12.3.6.2 Ex situ Techniques

Contrary to the in situ process, QDs are separately prepared and further deposited on the metal oxide semiconductor-coated photo-anodes in the ex situ techniques. When compared with an in situ technique, the ex situ one has many advantages, such as high crystallization of QDs, controllable size, uniform size distribution, few surface defects, etc., which can improve the optical absorption property as well as electron generation, separation, and transfer ability. Generally, QDs are adsorbed on the surface of metal oxides by two major processes in ex situ techniques: linking with molecular linkers by various functional groups and the direct deposition without linker molecules.

Linking with Molecular Linkers In order to link the QDs and metal oxides, the linking agents are always employed by the double functional short molecular linkers, such as 3-mercaptopropionic acid (3-MPA) and mercaptoacetic acid (or thioglycolic acid (TGA)). Generally, the carboylic acid group in these double-functional molecular linkers is bonded to the surface of metal oxides and the thiol group in the other side of these double-functional molecular linkers is bonded on the surface of QDs through the QD preparation process. Therefore, employing this process to fabricate QDSSCs can provide a good connection between QDs and oxide semiconductors to obtain an excellent stability of the photo-anodes. At the same time, the length and electronic property of molecular linkers will influence the adsorption quantity of QDs, as well as charge separation and transfer performance. Recently, Kamat's group has investigated the effect of molecular linkers on the photo-anodes as shown in Figure 12.14a, and molecular linkers have been also introduced in the $CuInS_2$ QDSSCs by our group, which have obtained a desired stability and FF parameters in QDSSCs [104–106].

Direct Deposition Without Linker Molecules Normally, many kinds of QD sensitizers with different sizes and shapes are effectively prepared using the oil phase based capping agents, such as dodecanethiol (DDT), trioctylphosphine (TOP), trioctylphosphine oxide (TOPO), oleylamine (OLA), etc. [107, 108]. All of them do not have a molecular group to bond to the surface of oxide semiconductors, and these kinds of QD sensitizers are directly deposited on the surface of oxide semiconductor photo-anodes without linker molecules. There is a good dispersion for QDs colloidal solution under these kinds of capping agents, which can deposit more uniform and quantity of QDs in the photo-anodes. However, some of the capping agents have long chain molecules, which will reduce the efficiency of charge transfer between QDs and metal oxides, and further influence the photovoltaic conversion efficiency of the solar cells. In recent investigations, a high photovoltaic efficiency of 11.6% has been achieved by this direct deposition in QDSSCs, which can be considered viable for potential large-scale applications [51].

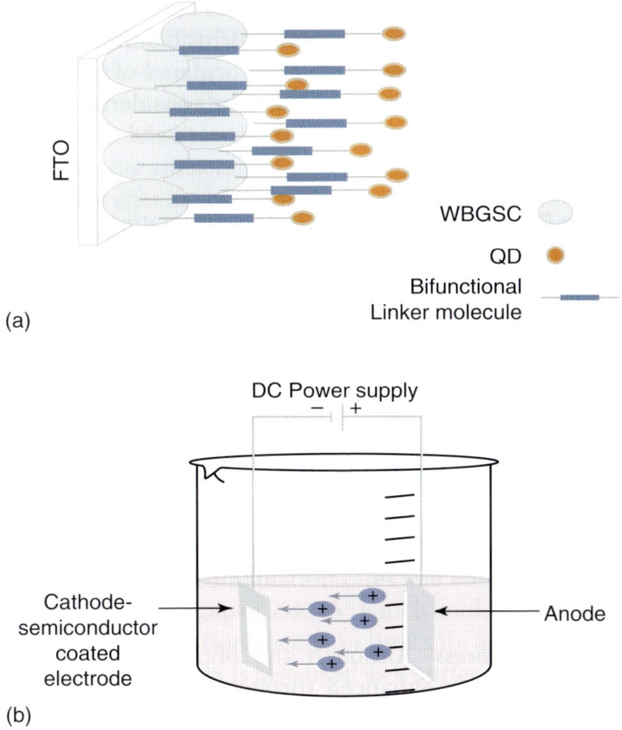

Figure 12.14 Schematic of ex situ sensitization techniques: (a) Molecular linker. *Source:* Pernik et al. 2011 [104]. Copyright 2011, American Chemical Society. (b) electrophoretic deposition. *Source:* Santra et al. 2013 [105]. Copyright 2013, American Chemical Society.

Other Techniques QDs are directly deposited on the surface of oxide semiconductors without linker molecules, which is a natural self-adsorption and assembly process. An electrophoretic deposition (EPD) process is developed in order to improve the interface connection between QDs and oxide semiconductors and reduce the time of QD sensitization in Figure 12.14b [105]. Typically, a positive terminal (oxide semiconductor photo-anodes) and a negative terminal (FTO glass substrate) are immersed into an electrophoretic bath with a desired distance, which contains the as-prepared QDs colloidal solution. A DC voltage is introduced for a desired value to deposit QDs on the surface of oxide semiconductors, which can provide a good interface connection between QDs and oxide semiconductors. The quantity and uniformity of QDs are controlled by the concentration of EPD solution and deposition time. It is considered to be an alternative route for different QD-deposition processes.

QDs can also be prepared by electro-spray process [109]. Typically, QDs colloidal is prepared and loaded in a syringe, which is fed through the needle at a constant flow rate by the syringe pump. QDs are sprayed onto the surface of metal oxides from the needle under high voltage.

which also suppresses the back electron transfer from TiO_2 photo-anodes to the electrolyte [120].

MgO is also used as the surface passivation layer of QDSSCs. MgO passivation layers are coated both above and below the CdS/CdSe interfaces by EPD, and it is found that the electron-recombination rates at the TiO_2/QDs, TiO_2/electrolyte, and QDs/electrolyte interfaces are significantly decreased, thereby leading to an increase in the photovoltaic efficiency by over 20% [121].

12.3.7.3 Surface Passivation by Molecular Dipoles

Besides large-bandgap semiconductors and metal oxides, small organic molecules with a dipolar moment modification are also considered to act as co-adsorbents for the surface passivation of QDs, as the negative dipoles containing the thiol or carboxylic acid groups may downshift the conduction band level of TiO_2 photo-anodes [122].

For QDs, the surface ligand exchanging is very important to modify their surface chemistry property, as the surface modification by the molecular dipoles may shift the energy levels. Bisquert's group has sensitized colloidal QDs onto the TiO_2 surface through a bifunctional molecular linker, and the linker introduces a dipole between the QDs and TiO_2 photo-anodes, which has an effective influence on the electron injection [123].

The energy levels of QD sensitizers using the SILAR or CBD methods may also be modulated by grafting dipolar molecules. The negative dipoles, such as 4-methoxy benzenethiol (BT) and 4-methyl BT, could upward shift the conduction band of QDs, while the positive dipoles, such as 4-fluoro BT, 4-chloro BT, and 4-nitro BT, could downward shift their conduction band. Therefore, an electrostatic field oriented toward/away from the QD surface is formed by the negative/positive dipole moments introduced, which influences the efficient electron injection [124].

12.3.7.4 Surface Passivation by Dye Molecules

QDSSCs are originated from DSSCs by replacing the sensitizers of organic dyes by QDs due to their excellent light harvesting ability. However, most of the common electrolytes in DSSCs, such as I^-/I_3^- and cobalt complexes, are not compatible in QDSSCs, as they undergo severe photocorrosion [55]. Therefore, the various dye molecules of ruthenium complexes can be used to modify and stabilize QDs.

QDs/dye systems have been explored in QDSSCs, and the improvement of the photovoltaic performances is clearly observed in various QDs/dye systems, such as PbS/N719, CdS/N3, CdSe/N3, and CdSe/Ru505 [125, 126]. As there is a significant overlapping range between the absorption spectra of QDs and ruthenium dyes, the light harvesting may be effectively enhanced, and the photons may be captured more selectively and complementarily.

12.3.7.5 Surface Passivation by Molecular Relays

The most commonly used liquid electrolyte in QDSSCs is a polysulfide aqueous solution, which exhibits excellent hole-extraction ability from the valence band of QDs, thereby stabilizing them in the liquid-junction devices. However, the V_{oc}

and FF of QDSSCs are much low, which is induced by the negative redox potential (−0.67 V vs Ag/AgCl) of polysulfide and its slow interfacial charge transfer at the CEs [127].

One possible solving route is to use the polysulfide with a redox mediator with a more positive redox potential, such as a cobalt–bipyridyl complex ($[Co(bpy)_3]^{2+/3+}$). To overcome the photocorrosion, Shen et al. have proposed that QDs are decorated with a molecular relay layer with a strong electron-donating property, such as 4′-(bis(4-(hexyloxy)phenyl)amino)-biphenyl-4-carboxylic acid (PAPC). Surface relay is considered to play the role of a charge relay, which could instantaneously intercept holes from the sensitizers of QDs (Figure 12.15) [128].

Time-resolved transient absorption reveals that the photo-generated holes are quickly removed by PAPC on the picosecond timescale while the photo-generated electrons are injected into TiO_2, which results in a rapid charge separation and stabilization of the QDs to avoid photocorrosion. The photovoltaic performance of the PAPC-modified QDSSCs is significantly improved, thereby yielding $V_{oc} = 858$ mV, $J_{sc} = 4.91$ mAcm^{-2}, FF = 64%, and PCE = 2.84%, which are among the highest values for CdS-sensitized TiO_2 solar cells.

Further studies have been undertaken on the influence of the chemical structure of relay molecules, and it is thought that long alkyl chains and benzoic acid groups may play a key role in slowing the recombination rate in QDSSCs.

12.3.7.6 Combined Interfacial Engineering Methods

From the above discussions, we can see that every interfacial engineering approach may only partially solve problems in QDSSCs. Hence, it would be more effective to combine several methods to obtain overall improved performance. Barea et al. have modified CdSe QD surfaces by depositing a conformal ZnS layer and subsequently grafting a BT monolayer, such as $BTOCH_3$, $BTCH_3$, BTF, and $BTNO_2$, which are used to reduce the recombination rate and enhance charge injection [129].

The combined use of amorphous TiO_2/dye, Al_2O_3/dye, and ZnS/dye has also been investigated, in which the large bandgap materials may suppress recombination while the dye molecules may enhance light harvesting and favor for charge separation [114, 130]. In addition, Zhong's group has prepared sequential ZnS/SiO_2 double layers on $CdSe_xTe_{1-x}$ sensitizers, which strongly inhibit the interfacial recombination to improve the charge-collection efficiency as well as the cell stability [39]. Recently, more improved interfacial engineering approaches have been reported, such as Mn-ZnS/SiO_2 and amorphous TiO_2/ZnS/SiO_2, where the photovoltaic efficiencies achieved are as high as 9.40% and 9.48%, respectively [131].

12.3.8 Optimization of the Counter Electrodes

CE is the part where the electrons flow back to the electrolyte from the external circuit in QDSSCs, and it may play a vital influence on the photovoltaic performance, which is mainly confirmed by its conductivity for the electron transfer as well as the catalytic ability for the generation of the reduced species (S^{2-}) in the electrolyte [132]. Therefore, the demand for efficient CEs in QDSSCs is based on

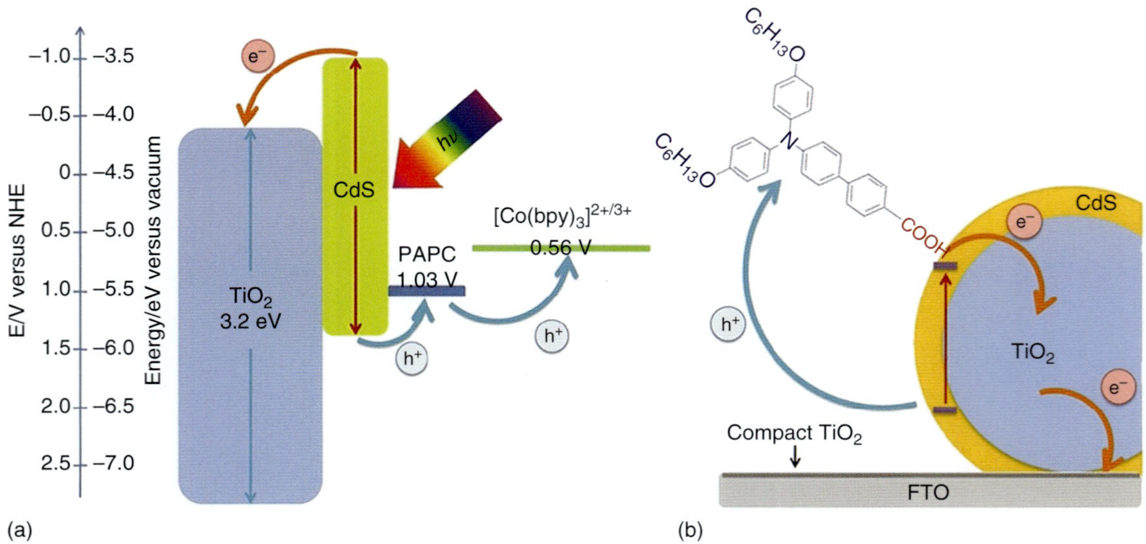

Figure 12.15 (a) Energy-band diagram and charge separation processes in the CdS/PAPC QDSC system and (b) schematic representation of a mesoscopic sensitized TiO_2 electrode grafted onto a PAPC molecular relay on a CdS surface. *Source:* Shen et al. 2015 [128]. Copyright 2015, American Chemical Society.

good charger transfer ability and catalytic property. Until now, a variety of CEs have been explored on the different materials and structures.

12.3.8.1 Noble Metal Counter Electrodes

Pt and Au noble metals have been widely used as CEs in QDSSCs, and normally they are used as a form of flatly deposited films on FTO or ITO glass. For a long time the researches have focused on the improvement of the performance by controlling their morphologies and surface properties.

Pt Counter Electrodes Pt is the most widely used CE material in DSSCs due to its highly catalytic, appreciable electrocatalytic ability toward I_3^- reduction and superior conductivity. During the early stages of the research on QDSSCs, the Pt CEs are introduced to catalyze the S^{2-}/S_n^{2-} redox couple, while the FF and short-circuit current density are quite poor. The reason lies in the fact that the sulfur-containing compounds in the polysulfide electrolytes adsorb preferably and strongly on the Pt surface, which reduces the stability and surface activity [132]. To solve the problem, Raj et al. have adopted CuS as the passivation layer and a catalytic assistant for Pt CEs, and CuS/Pt CEs in CdS/ZnS QDSSCs to achieve an enhanced current density in both the cathodic and anodic regions compared with the Pt CEs [133]. It is an attractive example for the surface-reinforcement of Pt CEs by the CuS passivation layer to improve electrocatalysis and surface stability in QDSSCs.

Au Counter Electrodes Au has been considered as an alternative CE in QDSSCs to functionalize the polysulfide electrolyte. Lee and Lo have proven that the S^{2-} ions are more weakly adsorbed on both Au surfaces as compared with Pt CEs, which induces Au CEs to work as better CEs in QDSSCs [38]. Kiyonaga et al. have examined the catalytic properties of Au NPs in the reduction of S_n^{2-} ions in a photoelectrochemical solar cell, which reveals that the size and crystal orientation control of Au nanoparticles have a great effect on the catalytic properties [134]. Zhu et al. have further prepared Au NPs in a graphene network to build a reduced graphene oxide (rGO)-Au NP composite film as the CEs in QDSSCs; both of the high catalytic properties of Au NPs and the conductivity of graphene are used to improve the photovoltaic performance [135]. Yoon et al. have coated Au on Pt NPs by an electro-deposition method to form composite CEs in QDSSCs, which possess the maximized photovoltaic performance [136].

12.3.8.2 Carbon Counter Electrodes

To meet up with the development of efficient and low-cost QDSSCs, carbon-derived materials would be remarkable candidates for CEs, as they have competitive prices and good electrocatalytic properties in comparison to other types of CEs. Various carbon CEs have been employed in QDSSCs, such as graphene, carbon nanotubes (CNTs), mesoporous carbon, fullerene, activated carbon, and so on.

Fan et al. have firstly used the activated carbon and hierarchical nanostructured spherical carbon with a hollow core/mesoporous shell (HCMS) as CE materials in QDSSCs [137]. The result shows that the carbon CEs possess better catalytic ability as compared to Pt CEs, and it also shows that the HCMS CEs may better functionalize the polysulfide electrolyte compared to active carbon, as the unique structure of the HCMS may enhance mass transportation and thus lead to more efficient diffusion of the reactants and products in the electrolyte's redox reactions. Ordered multimodal porous carbon(OMPC) CEs are further developed in QDSSCs, and enhanced photovoltaic performance of 4.36% is achieved due to the improvement in mass transportation resulting from the interconnected multimodal pore framework [138]. Lee et al. have prepared the mesocellular carbon foam (MMSU-F-C) as the CEs in QDSSCs to enlarge the surface area of the carbon electrodes, which show a relatively good electron transfer and an enhanced stability compared with Pt and Au CEs [139].

The various carbon structures mentioned above are all derivatives of intrinsic carbon. In QDSSCs, the research on extrinsic carbons is proposed by using nitrogen-doped hollow carbon nanoparticles (N-HCNPs) as an excellent CE, as the additional electrons provided by the N atoms play a critical role in improving their electrocatalytic properties [140]. N-HCNPs are synthesized by a detonation-assisted chemical vapor deposition process, which is proven to possess improved electrocatalytic ability, and their resistance is reduced by 35 times compared with traditional Pt CEs.

Carbon nanocomposites, such as CNTs and rGO, have the advantages of excellent electrical conductivity and large surface area, which are responsible for their excellent electrocatalytic characteristics. Giordano et al. have successfully prepared Mo-compound (Mo_2N, Mo_2C, and MoS_2)/CNT-rGO heterostructures by using a modified urea-glass route [141]. The photovoltaic performance of these novel CE-based QDSSCs is enhanced by 5%, which demonstrates the clearly enhanced electrocatalytic activities of the CEs. The reason lies in the fact that the combination of 1-D CNTs and 2-D rGO of the CNT/rGO complex provides a greatly enlarged surface area, which allows an increased loading amount on the Mo compound catalysts. Zeng et al. have prepared CZTSe NPs/multiwalled carbon nanotube (MWCNT) composites, and the advantages of both CZTSe and CNTs could create a synergistic effect for use as effective CEs [142].

Until now, many other carbon materials have also been developed to enrich the carbon family used as CEs in QDSSCs, including ordered mesocellular carbon foam, hollow carbon-mesoporous shell carbon, and vertically aligned single-walled CNTs, and the continuous researches may further widely spread the interest in carbon CEs, especially in a graphene based one [132].

12.3.8.3 Metallic Compound Counter Electrodes

Metallic compound CEs are considered as good alternatives to Pt CEs due to their stability to the polysulfide electrolyte, including sulfide CEs and metal selenides.

***Cu_2S* Counter Electrodes** As a p-type semiconductor with a bandgap of 1.1–1.14 eV, Cu_2S has a wide application as a CE in QDSSCs due to its improved catalytic properties. In general, although Cu_2S CEs are synthesized directly on the brass foil, the residual Cu in the brass substrate constantly reacts with the electrolyte, resulting in the ultimate disintegration of the CEs [143]. Moreover, the poor mechanical stability and relatively low surface area of the Cu_2S CEs also affect its photovoltaic performance in QDSSCs. To avoid the above mentioned problem, the electroplating approach is firstly used to deposit porous Cu_2S films on FTO conducting glass by replacing Cu brass as the substrate [144], this structure enables the Cu_2S films to have a superior catalytic performance, and the Cu_2S CEs in CdSSe and CdSeTe QDSSCs achieve enhanced photovoltaic performances of 5.41% and 6.36%, respectively.

Cu_2S CEs on FTO substrates provide good catalytic ability but an undesirable electrical conductivity due to the intrinsically poor conductivity of Cu_2S, and the formation of the composite materials with conductive species may be issues that have to be solved [74]. Jiang et al. have designed a core–shell structure, containing a tin-doped indium oxide nanowire (ITO) core as a three-dimensional conductive network as well as a p-type Cu_2S nanocrystal shell as a catalyst [145]. This core–shell configure the tunnel junction arrays and could facilitate the charge transfer processes, which decreases the charge transfer resistance and sheet resistance of the CEs, resulting in a 35% enhancement in the photovoltaic performance in QDSSC PCE compared to that of Cu_2S film. Kamat's group has developed another highly efficient CE structure by forming the composite of Cu_2S nanocrystals and rGO [143]. The high surface area of the 2-D rGO mat promotes a large number of Cu_2S reactive sites, and the conductivity of the composite material enables efficient electron shuttling, which induces a dramatic improvement of 75% for the FF of QDSSCs. The preferable stability and easy fabrication of these CEs enable the possibility of large-scale QDSSCs production. CdS/CdSe QDSSCs modules with the aperture area over 12 cm^2 have been fabricated, exhibiting a promising efficiency of 2.31% without cell performance degradation in 30 days as shown in Figure 12.16 [146].

Cu_xS Counter Electrodes Cu_2S is a relatively unstable phase in its stoichiometric member, and it is easily degraded into a Cu-deficient phase [132]. Therefore, the the attention of researchers has been focused entirely on the Cu_xS: from Cu_2S (chalcocite) to CuS (covellite), $Cu_{1.96}S$ (djurleite), $Cu_{1.8}S$ (digenite), $Cu_{1.75}S$ (anilite), and $Cu_{1.12}S$ (yarrowite).

As a p-type semiconductor, CuS possesses both electronic and ionic conductivity due to the presence of Cu vacancies, and its good stability and conductivity enable it to be applied as CEs in QDSSCs [147]. Wang et al. have synthesized a hierarchical CuS structure on a conductive substrate by a one-step EDC method, and the CuS CEs exhibit the best PV performance and stability in QDSSCs compared with those of brass-based Cu_2S and Pt CEs [148]. Savariraj et al. have prepared a knit coir mat structure of CuS film directly on a FTO substrate by an in situ CBD method, and found that the density of Cu vacancies increases by varying the deposition time, resulting in enhanced CE performance [149]. CuS nanoflakes, nanoplatelets, and nanopeas as well as the novel materials like Mn-doped CuS, CuS/carbon nanofiber, CuS/rGO nanocomposite, and CuS/porous conductive substrate have been also proposed to further promote the performance of CuS-based CEs [132].

Some other Cu_xS materials are also considered to be promising candidates in QDSSCs. Ye et al. have designed a skeletal $Cu_{1.75}S$ nanocage structure, and

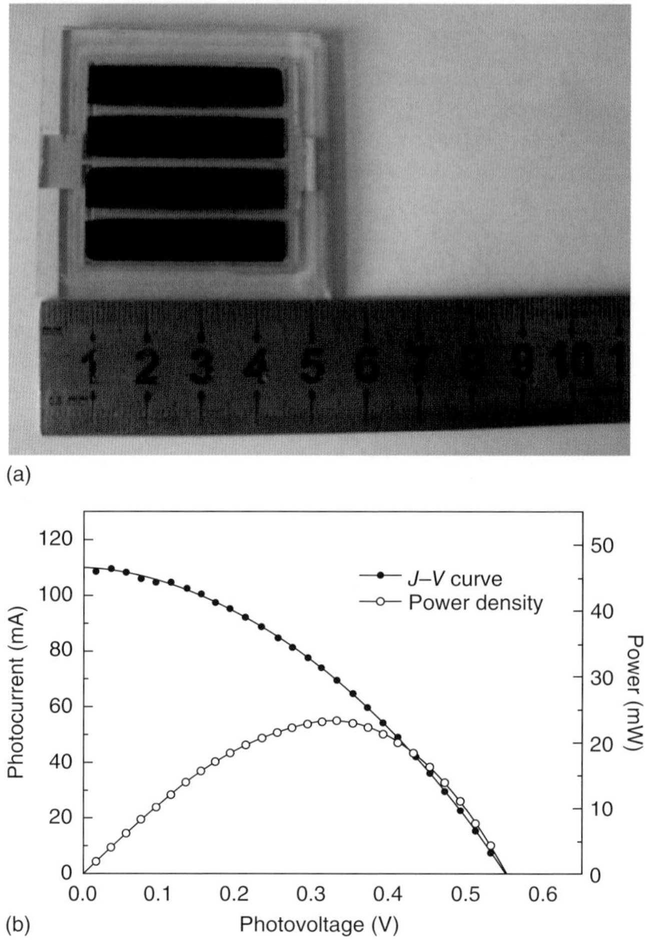

Figure 12.16 (a) Digital photograph and (b) *J–V* curve and power density (under AM 1.5 illumination) of the 12.97 cm^2 CdS/CdSe QDSSC module. *Source:* Li et al. 2014 [146]. Copyright 2014, Elsevier.

the 3-D hollow structure provides a large surface area to facilitate electrolyte infiltration/diffusion, leading to enhanced PV performance in QDSSCs [150]. Zhao et al. have deposited a CuS/$Cu_{1.8}S$ composite on FTO conducting glass, and obtained a high and reproducible efficiency of 6.28% by using these CEs in QDSSCs [151]. Kim et al. have developed a novel CBD method to deposit Cu_xS films with various stoichiometries by varying the concentration ratio of the precursors, and it is found that CuS CEs possess superior performance compared to other Cu_xS CEs in QDSSCs [152]. Moreover, Tafel polarization and Nyquist plots prove that the Cu-deficient films (CuS and $Cu_{1.12}S$) are more stable than the Cu-rich counterparts ($Cu_{1.75}S$ and $Cu_{1.8}S$) due to their greater resistance against surface oxidation.

Other Metal Sulfides Counter Electrodes Yang et al. have prepared low cost CoS CEs for CdS/CdSe QDSSCs by the CBD method, and found that the electrocatalytic activities of the sulfide electrodes follow the order of CuS > CoS > NiS > Pt, while CuS CEs possess an inferior stability [153]. Yuan et al. have developed an electrodeposition method to deposit the flexible CoS-containing CEs in flexible QDSSCs, including carbon fiber/CoS NT array, CoS NR array/graphite paper and CoS NWs/Au hybridized networks. These flexible CEs exhibit dramatic catalytic activity and superior mechanical strength, and no obvious changes in morphology and catalytic ability are observed after bending tests [154].

PbS CEs in QDSSCs are prepared by Tachan, and the charge transfer resistance of the PbS CEs in the aqueous polysulfide electrolyte is found to be quite well with good stability [155]. PbS composited with carbon black and graphene is prepared to obtain a more efficient and stable CEs, which is an order of magnitude lower than that of the pure PbS CEs [156]. Some other PbS composites are also designed as promising PbS-containing CEs in QDSSCs, including PbS/CuS [157] and PbS/ZnO NR core–shell materials [158].

Earth-abundant metal pyrite (FeS_2, CoS_2, NiS_2, and their alloys) thin films are also systemically investigated as alternative electrocatalysts for polysulfide reduction, and it is interesting to point out the high catalytic activity of the above pyrites in the polysulfide reduction and highlight the particularly high intrinsic activity of NiS_2 [159]. Further work should be focused on the alloying of different transition metal pyrites to introduce structural disorders, which may increase their real density of catalytic active sites as well as enhance the PV performance.

Owing to the bandgap energy of 1.5 eV and a considerably high absorption coefficient ($>10^{-4}$ cm^{-1}), $Cu_2ZnSn(S_{1-x}Se_x)_4$ (CZTS) has received much attention as one of the promising sensitizers in solar cells. Similarly to PbS and Cu_2S, the p-type sensitizer can also act as the CE in QDSSCs due to its good stability and catalytic ability toward the aqueous polysulfide electrolyte. Cao et al. have prepared the CZTS nanocrystals by a hot-injection method, and used further stoichiometric approaches to optimize the S/Se ratio in CZTS films, which illustrate that the CZTS CEs possess better electrocatalytic activity and enhanced current compared with conventional Pt CEs in QDSSCs [160].

Metal Selenides Counter Electrodes As an important class of chalcogenides, semiconducting selenides have drawn enormous attention due to their distinctive electronic properties and interesting chemical behaviors, and they can be applied in a wide variety of potential applications, such as optical recording materials, sensors, laser materials, and PV fields.

Recently, metal selenides have been regarded as the new class of CEs in QDSSCs. Choi et al. have conducted a systematic study on binary metal selenide materials as CEs in QDSSCs, and eight different types of binary selenides (MnSe, $CoSe_2$, $NiSe_2$, $Cu_{1.8}Se$, $MoSe_2$, WSe_2, PbSe, and Bi_2Se_3) are randomly selected as electrocatalysts and employed as CEs in QDSSCs [161]. It is found that the photovoltaic performances of $Cu_{1.8}Se$ and PbSe CEs are better than that of Pt CEs.

Bo et al. have developed a facile approach to fabricate copper selenide CEs, in which a red layer of Se is first deposited on the conductive substrate, followed by a simple Cu^{2+} treatment [162]. The method is also used to prepare other metal selenide functional CEs, such as PbSe, Ag_2Se, etc. Liu et al. have reported a low-temperature, solution-based method to deposit metal selenide films as efficient CEs in QDSSCs, and $FeSe_2$, $Cu_{1.8}S$, and CuSe films are also prepared by this method [163]. This facile method offers the possibility for low-cost and large-scale production of metal selenide materials in QDSSCs. Chen et al. have prepared $Cu_{2-x}Se$ NTs with hierarchical architecture as efficient CEs in QDSSCs; the one-dimensional hollow nanostructure consisting of a flower-like surface provides favorable catalytic ability and electrical conductivity, resulting in the photovoltaic performance as high as 6.25% [164].

12.3.8.4 Polymer Counter Electrodes

Conducting polymers such as poly(3, 4-ethylenedioxythiophene) (PEDOT), polythiophene (PT), polypyrrole (PPy), and polyaniline (PANI) are also considered to be good candidates as CEs in QDSSCs due to their superior electron conductivities, and have the advantages of low-cost preparation processes and flexibilities. Yeh et al. have reported the preparation of PEDOT, PT, and PPy on ITO glasses as CEs in QDSSCs through electrochemical polymerization, and the PEDOT CEs possess the best electrochemical and photovoltaic performance of 1.16%, which is comparable with that of Au CEs, while their cycling stability is better than that of Pt CEs [165].

Although the photovoltaic performance of the QDSSC using the PEDOT CEs is not very high, it still suggests the possible using of conducting polymers as CEs. However, few studies have been reported on the use of these conducting polymer CEs in QDSSCs until now. Further studies might be able to suggest new insights for novel and efficient CEs in the future.

12.4 Conclusion and Future Outlook

QDSSCs have become a promising subject of research in the field of solar energy conversion due to the excellent properties of QD photosensitizers and the photovoltaic performance of QDSSCs and have exhibited a steady growth over the past decade. However, it is still in the shadow compared with other new generation solar cells, with regard to relatively low PCE, less attention on the development of the flexible QDSSCs, etc.

The limitations of the photovoltaic performance of QDSSCs are mainly attributed to the narrow absorption range of the incident light, the interface combination, and the low FF value originating from unsuitable CEs. However, optimization of the construction of QD photosensitizers, CEs with superior catalytic activity as well as the redox couple electrolyte may further improve the performance of QDSSCs. The first possible route is to make full use of the excellent light absorption ability of the QD photosensitizers as well as the MEG effect, which will be effective in producing more photo-generated electrons under sunlight. Therefore, the construction of the QDs with various kinds and sizes as well as the

QDs passivated with organic and inorganic passivation agents may be adopted, as they can absorb the various sunlight ranges or excite more photo-generated electrons, which is expected to enhance the light absorption and electron excitation in the whole sun irradiation range. The development of judiciously engineered interfaces is crucial for the QDSSCs to passivate the surface state of QDs and form the well-combined interfaces, which is used to reduce the recombination of the photo-generated electrons and holes, and more efforts on the surface chemistry of QDs and the interface engineering should be carried out. Another critical factor to boost the QDSSCs efficiency is to design a compatible electrolyte–CE system, and the demands of high catalytic activity and good charge transfer for CEs in electrolyte are required to be met. It might be difficult to make a breakthrough by using one single component as the CEs, and the previous efforts on the heterostructured CEs may still remain the promising future for this great challenge.

Flexible devices have found wide application in the field of solar-energy conversion, and the continuously increasing flexible PV market also gives a huge motivation for research in domestic and industrial applications. However, the development of flexible QDSSCs has been extremely slow since their emergence, and there are great challenges as well as opportunities in this area. Fortunately, the huge success of flexible silicon solar cells has become a motivation for emerging technologies, and the manufacturers of silicon-based flexible PV devices share materials, equipment, and practices with sibling industries. The most difficult issue is to fabricate uniform printing layers on a large scale to facilitate mass production. With the success in the improved photovoltaic efficiency and the printing technology in the large-scale area, wide application of flexible QDSSCs will be not far away from our daily life.

Acknowledgments

The authors are supported by the National Nature Science Foundation of China (No. 11674258, 51506155), International Science & Technology Cooperation Program of China (No. 2013DFR50710), and Science and Technology Support Program of Hubei Province (No. 2014BAA096).

References

1 Brown, T., Rossi, F., Di Giacommo, F. et al. (2014). Progress in flexible solar cell materials, progresses and devices. *J. Mater. Chem. A* 2: 10788–108717.

2 Balsingam, S., Kang, M., and Jun, Y. (2013). Metal substrate based electrodes for flexible dye-sensitized solar cells: fabrication methods, progress and challenges. *Chem. Commun.* 49: 11457–11475.

3 Shockley, W. and Queisser, H. (1961). Detailed balance limit of efficiency of p–n junction solar cells. *J. Appl. Phys.* 32: 510–519.

4 Jiao, S., Du, J., Du, Z. et al. (2017). Nitrogen-doped mesoporous carbons as counter electrodes in quantum dot sensitized solar cells with a conversion efficiency exceeding 12%. *J. Phys. Chem. Lett.* 8: 559–564.

5 Cheng, Y., Pascoe, A., Huang, F., and Peng, Y. (2016). Print flexible solar cells. *Nature* 539: 488–489.
6 Alivisatos, A.P. (1996). Semiconductor clusters, nanocrystals, and quantum dots. *Science* 271: 933.
7 Tang, J. and Sargent, E.H. (2011). Infrared colloidal quantum dots for photovoltaics: fundamentals and recent progress. *Adv. Mater.* 23: 12–29.
8 Alivisatos, A.P. (1996). Perspectives on the physical chemistry of semiconductor nanocrystals. *J. Phys. Chem.* 100: 13226–13239.
9 Sugimoto, T. (1987). Preparation of monodispersed colloidal particles. *Adv. Colloid Interface Sci.* 28: 65–108.
10 Zhong, H., Lo, S.S., Mirkovic, T. et al. (2010). Noninjection gram-scale synthesis of monodisperse pyramidal $CuInS_2$ nanocrystals and their size-dependent properties. *ACS Nano* 4: 5253–5262.
11 Nozik, A. (2002). Quantum dot solar cells. *Physica E* 14: 115–120.
12 Schaller, R. and Klimov, V. (2004). High efficiency carrier multiplication in PbSe nanocrystals: implications for solar energy conversion. *Phys. Rev. Lett.* 92: 186601.
13 Beard, M.C., Midgett, A.G., Hanna, M.C. et al. (2010). Comparing multiple exciton generation in quantum dots to impact ionization in bulk semiconductors: implications for enhancement of solar energy conversion. *Nano Lett.* 10: 3019–3027.
14 Beard, M., Midgett, A., Law, M. et al. (2009). Variations in the quantum efficiency of multiple exciton generation for a series of chemically treated PbSe nanocrystal films. *Nano Lett.* 9 (2): 836–845.
15 Sukhovatkin, V., Hinds, S., Brzozowski, L. et al. (2009). Colloidal quantum-dot photodetectors exploiting multiexciton generation. *Science* 324: 1542–1544.
16 Semonin, O.E., Luther, J.M., Choi, S. et al. (2011). Peak external photocurrent quantum efficiency exceeding 100% via MEG in a quantum dot solar cell. *Science* 334: 1530–1533.
17 Huang, J., Stockwell, D., Huang, Z. et al. (2008). Photoinduced ultrafast electron transfer from CdSe quantum dots to re-bipyridyl complexes. *J. Am. Chem. Soc.* 130: 5632–5633.
18 Matylitsky, V.V., Dworak, L., Breus, V.V. et al. (2009). Ultrafast charge separation in multiexcited CdSe quantum dots mediated by adsorbed electron acceptors. *J. Am. Chem. Soc.* 131: 2424–2425.
19 Boles, M.A., Ling, D., Hyeon, T. et al. (2016). The surface science of nanocrystals. *Nat. Mater.* 15: 141–153.
20 Peng, X. (2009). An essay on synthetic chemistry of colloidal nanocrystals. *Nano Res.* 2: 425–447.
21 Liu, P., Qin, R., Fu, G. et al. (2017). Surface coordination chemistry of metal nanomaterials. *J. Am. Chem. Soc.* 139 (6): 2122–2131.
22 Brittman, S., Gao, H., Garnett, E.C. et al. (2011). Absorption of light in a single-nanowire silicon solar cell decorated with an octahedral silver nanocrystal. *Nano Lett.* 11: 5189–5195.
23 Bell, A.T. (2003). The impact of nanoscience on heterogeneous catalysis. *Science* 299: 1688–1691.

24 González-Pedro, V., Sima, C., Marzari, G. et al. (2013). High performance PbS quantum dot sensitized solar cells exceeding 4% efficiency: the role of metal precursors in the electron injection and charge separation. *Phys. Chem. Chem. Phys.* 15: 13835–13843.

25 Chen, X. and Mao, S.S. (2007). Titanium dioxide nanomaterials: synthesis, properties, modifications, and applications. *Chem. Rev.* 107: 2891–2959.

26 Hodes, G., Manassen, J., and Cahen, D. (1980). Electrocatalytic electrodes for the polysulfide redox system. *J. Electrochem. Soc.* 127: 544–549.

27 Du, Z., Zhang, H., Bao, H., and Zhong, X. (2014). Optimization of TiO_2 photoanode film for high efficiency quantum dot sensitized solar cells. *J. Mater. Chem. A* 2: 13033–13040.

28 Tian, J., Lv, L., Fei, C. et al. (2014). A highly efficient (>6%) $Cd_{1-x}Mn_xSe$ quantum dot sensitized solar cell. *J. Mater. Chem. A* 2: 19653–19659.

29 Ren, Z., Wang, J., Pan, Z. et al. (2015). Amorphous TiO_2 buffer layer boosts efficiency of quantum dot sensitized solar cells to over 9%. *Chem. Mater.* 27: 8398–8405.

30 Li, L., Yang, X., Zhang, W. et al. (2014). Boron and sulfur co-doped TiO_2 nanofilm as effective photoanode for high efficiency CdS quantum-dot-sensitized solar cells. *J. Power Sources* 272: 508–512.

31 Gakhar, R., Merwin, A., Summers, K. et al. (2014). Application of $Zn_xCd_{1-x}Se$-sensitized TiO_2 nanotube arrays as photoanodes for solar cells. *J. Mater. Chem. A* 2: 10116–10125.

32 Zhang, H., Cheng, K., Hou, Y.M. et al. (2012). Efficient CdSe quantum dot-sensitized solar cells prepared by a postsynthesis assembly approach. *Chem. Commun.* 48: 11235–11237.

33 Guijarro, N., Lana-Villarreal, T., Mora-Seró, I. et al. (2009). CdSe quantum dot-sensitized TiO_2 electrodes: effect of quantum dot coverage and mode of attachment. *J. Phys. Chem. C* 113: 4208–4214.

34 Salant, A., Shalom, M., Hod, I. et al. (2010). Quantum dot sensitized solar cells with improved efficiency prepared using electrophoretic deposition. *ACS Nano* 4: 5962–5968.

35 Pan, Z., Zhao, K., Wang, J. et al. (2013). Near infrared absorption of $CdSe_xTe_{1-x}$ alloyed quantum dot sensitized solar cells with more than 6% efficiency and high stability. *ACS Nano* 7: 5215–5222.

36 Robel, I., Subramanian, V., Kuno, M. et al. (2006). Quantum dot solar cells. Harvesting light energy with CdSe nanocrystals molecularly linked to mesoscopic TiO_2 films. *J. Am. Chem. Soc.* 128: 2385–2393.

37 Lee, H.J., Yum, J.-H., Leventis, H.C. et al. (2008). CdSe quantum dot-sensitized solar cells exceeding efficiency 1% at full-sun intensity. *J. Phys. Chem. C* 112: 11600–11608.

38 Lee, Y.-L. and Lo, Y.-S. (2009). Highly efficient quantum-dot-sensitized solar cell based on co-sensitization of CdS/CdSe. *Adv. Funct. Mater.* 19: 604–609.

39 Zhao, K., Pan, Z., Mora-Seró, I. et al. (2015). Boosting power conversion efficiencies of quantum-dot-sensitized solar cells beyond 8% by recombination control. *J. Am. Chem. Soc.* 137: 5602–5609.

40 Wei, H., Wang, G., Shi, J. et al. (2016). Fumed SiO_2 modified electrolytes for quantum dot sensitized solar cells with efficiency exceeding 11% and better stability. *J. Mater. Chem. A* 4: 14194–14203.
41 Santra, P.K. and Kamat, P.V. (2012). Mn-doped quantum dot sensitized solar cells: a strategy to boost efficiency over 5%. *J. Am. Chem. Soc.* 134: 2508–2511.
42 Yang, J., Wang, J., Zhao, K. et al. (2015). CdSeTe/CdS type-I core/shell quantum dot sensitized solar cells with efficiency over 9%. *J. Phys. Chem. C* 119: 28800–28808.
43 Chen, Y., Tao, Q., Fu, W. et al. (2015). Synthesis of PbS/Ni^{2+} doped CdS quantum dots cosensitized solar cells: enhanced power conversion efficiency and durability. *Electrochim. Acta* 173: 812–818.
44 Parsi Benehkohal, N., González-Pedro, V., Boix, P.P. et al. (2012). Colloidal PbS and PbSeS quantum dot sensitized solar cells prepared by electrophoretic deposition. *J. Phys. Chem. C* 116: 16391–16397.
45 Braga, A., Giménez, S., Concina, I. et al. (2011). Panchromatic sensitized solar cells based on metal sulfide quantum dots grown directly on nanostructured TiO_2 electrodes. *J. Phys. Chem. Lett.* 2: 454–460.
46 Chen, Y., Tao, Q., Fu, W. et al. (2014). Enhanced photoelectric performance of PbS/CdS quantum dot co-sensitized solar cells via hydrogenated TiO_2 nanorod arrays. *Chem. Commun.* 50: 9509–9512.
47 Jiao, S., Wang, J., Shen, Q. et al. (2016). Surface engineering of PbS quantum dot sensitized solar cells with a conversion efficiency exceeding 7%. *J. Mater. Chem. A* 4: 7214–7221.
48 Jackson, P., Hariskos, D., Lotter, E. et al. (2011). New world record efficiency for Cu(In,Ga)Se_2 thin-film solar cells beyond 20%. *Prog. Photovolt. Res. Appl.* 19: 894–897.
49 Peng, Z., Liu, Y., Chen, K. et al. (2014). Fabrication of the protonated pentatitanate nanobelts sensitized with $CuInS_2$ quantum dots for photovoltaic applications. *Chem. Eng. J.* 244: 335–342.
50 Pan, Z., Mora-Seró, I., Shen, Q. et al. (2014). High-efficiency "green" quantum dot solar cells. *J. Am. Chem. Soc.* 136: 9203–9210.
51 Du, J., Du, Z., Hu, J.-S. et al. (2016). Zn–Cu–In–Se quantum dot solar cells with a certified power conversion efficiency of 11.6%. *J. Am. Chem. Soc.* 138: 4201–4209.
52 Wang, Z. and Demopoulos, G.P. (2015). Growth of Cu_2ZnSnS_4 nanocrystallites on TiO_2 nanorod arrays as novel extremely thin absorber solar cell structure via the successive-ion-layer-adsorption-reaction method. *ACS Appl. Mater. Interfaces* 7: 22888–22897.
53 Chen, K., Zhou, J., Chen, W. et al. (2016). A green synthesis route for the phase and size tunability of copper antimony sulfide nanocrystals with high yield. *Nanoscale* 8: 5146–5152.
54 Yang, C., Zhou, B., Miao, S. et al. (2013). $Cu_2Ge(S_{3-x}Se_x)$ colloidal nanocrystals: synthesis, characterization, and composition-dependent band gap engineering. *J. Am. Chem. Soc.* 135: 5958–5961.
55 Zaban, A., Mićić, O.I., Gregg, B.A. et al. (1998). Photosensitization of nanoporous TiO_2 electrodes with InP quantum dots. *Langmuir* 14: 3153–3156.

56 Yu, P., Zhu, K., Norman, A.G. et al. (2006). Nanocrystalline TiO_2 solar cells sensitized with InAs quantum dots. *J. Phys. Chem. B* 110: 25451–25454.
57 Zumeta-Dubé, I., Ruiz-Ruiz, V.-F., Díaz, D. et al. (2014). On the TiO_2 sensitization with Bi_2S_3 quantum dots: the inconvenience of sodium ions in the deposition procedure. *J. Phys. Chem. C* 118: 11495–11504.
58 Itzhaik, Y., Niitsoo, O., Page, M. et al. (2009). Sb_2S_3-sensitized nanoporous TiO_2 solar cells. *J. Phys. Chem. C* 113: 4254–4256.
59 Tubtimtae, A., Wu, K.-L., Tung, H.-Y. et al. (2010). Ag_2S quantum dot-sensitized solar cells. *Electrochem. Commun.* 12: 1158–1160.
60 Kamat, P.V. (2012). Boosting the efficiency of quantum dot sensitized solar cells through modulation of interfacial of interfacial charge transfer. *Acc. Chem. Res.* 45: 1906–1915.
61 Hodes, G. (2008). Comparison of dye- and semiconductor-sensitized porous nanocrystalline liquid junction solar cells. *J. Phys. Chem. C* 112: 17778–17787.
62 Liang, C., Li, L., Su, J. et al. (2014). A general method for preparing anatase TiO_2 tree-like-nanoarrays on various metal wires for fiber dye-sensitized solar cells. *Sci. Rep.* 4: 4420.
63 Park, S., Wang, G., Cho, B. et al. (2012). Flexible molecular-scale electronic devices. *Nat. Nanotechnol.* 7: 438–422.
64 Fu, F., Feurer, T., Weiss, T. et al. (2016). High-efficiency inverted semi-transparent planar perovskite solar cells in substrate configuration. *Nat. Energy* 2: 16190.
65 Mali, S., Desai, S., Kalagi, S. et al. (2012). PbS quantum dot sensitized anatase TiO_2 nanocrystals for quantum dot-sensitized solar cell applications. *Dalton Trans.* 41: 6130–6136.
66 Wu, D., Zhang, S., Jiang, S. et al. (2015). Anatase TiO_2 hierarchical structures composed of ultra-thin nano-sheets exposing high percentage {001} facets and their application in quantum-dot sensitized solar cells. *J. Alloys Compd.* 624: 94–99.
67 Weerasinghe, H., Sirimanne, P., Simon, G. et al. (2012). Cold isostatic pressing technique for producing highly efficient flexible dye-sensitized solar cells on plastic substrates. *Prog. Photovolt. Res. Appl.* 20: 321–332.
68 Zhu, G., Cheng, Z., Lv, T. et al. (2010). Zn-doped nanocrystalline TiO_2 films for CdS quantum dot sensitized solar cells. *Nanoscale* 2: 1229–1232.
69 Shu, T., Xiang, P., Zhou, X. et al. (2012). Mesoscopic nitrogen-doped TiO_2 spheres for quantum dot-sensitized solar cells. *Electrochim. Acta* 68: 166–172.
70 Li, L., Xiao, J.Y., Yang, X. et al. (2015). High efficiency CdS quantum-dot-sensitized solar cells with boron and nitrogen co-doped TiO_2 nanofilm as effective photoanode. *Electrochim. Acta* 169: 103–108.
71 Liu, Y., Cheng, Y., Shu, W. et al. (2014). Formation and photovoltaic performance of few-layered graphene-decorated TiO_2 nanocrystals used in dye-sensitized solar cells. *Nanoscale* 6: 6755–6762.
72 Wang, H., Bai, Y., Zhang, H. et al. (2010). CdS quantum dots-sensitized TiO_2 nanorod array on transparent conductive glass photoelectrodes. *J. Phys. Chem. C* 114: 16451–16455.

73 Peng, Z., Liu, Y., Zhao, Y. et al. (2013). Efficiency enhancement of TiO_2 nanodendrite array electrodes in $CuInS_2$ quantum dot sensitized solar cells. *Electrochim. Acta* 111: 755–761.

74 Peng, Z., Liu, Y., Zhao, Y. et al. (2014). Incorporation of the TiO_2 nanowire arrays photoanode and Cu_2S nanorod arrays counter electrode on the photovoltaic performance of quantum dot sensitized solar cells. *Electrochim. Acta* 135: 276–283.

75 Hsiao, P., Liou, Y., and Teng, H. (2011). Electron transport patterns in TiO_2 nanotube arrays based dye-sensitized solar cells under frontside and backside illuminations. *J. Phys. Chem. C* 115: 15018–15024.

76 Wang, C., Jiang, Z., Wei, L. et al. (2012). Photosensitization of TiO_2 nanorods with CdS quantum dots for photovoltaic applications: a wet-chemical approach. *Nano Energy* 1: 440–447.

77 Rao, H., Wu, W., Liu, Y. et al. (2014). CdS/CdSe co-sensitized vertically aligned anatase TiO_2 nanowire arrays for efficient solar cells. *Nano Energy* 8: 1–8.

78 Shen, Q., Yamada, A., Tamura, S., and Toyoda, T. (2010). CdSe quantum dot-sensitized solar cell employing TiO_2 nanotube working-electrode and Cu_2S conter-electrode. *Appl. Phys. Lett.* 97: 123107.

79 Wu, D., He, J., Zhang, S. et al. (2015). Multi-dimensional titanium dioxide with desirable structural qualities for enhanced performance in quantum-dot sensitized solar cells. *J. Power Sources* 282: 202–210.

80 Liu, Y., Cheng, Y., Chen, K. et al. (2014). Enhanced light-harvesting of the conical TiO_2 nanotube arrays used as the photoanodes in flexible dye-sensitized solar cells. *Electrochim. Acta* 146: 838–844.

81 Sun, W., Yu, Y., Pan, H. et al. (2008). CdS quantum dots sensitized TiO_2 nanotube-array photoelectrodes. *J. Am. Chem. Soc.* 139: 1124–1125.

82 Burungale, V., Satale, V., Teli, A. et al. (2016). Surfactant free single step synthesis of TiO_2 3-D microflowers by hydrothermal route and its photoelectrochemical characterizations. *J. Alloys Compd.* 656: 491–499.

83 Xu, X., Wu, W., and Rao, H. (2015). CdS/CdSe co-sensitized TiO_2 nanowire-coated hollow spheres exceeding 6% photovoltaic performance. *Nano Energy* 11: 621–630.

84 Liu, Y., Cheng, Y., Chen, K. et al. (2014). Fabrication of TiO_2 nanotube arrays and their application in flexible dye-sensitized solar cells. *RSC Adv.* 4: 45592–45597.

85 Wang, G., Yang, X., Qian, F. et al. (2010). Double-sided CdS and CdSe quantum dot co-sensitized ZnO nanowire arrays for photoelectrochemical hydrogen generation. *Nano Lett.* 10: 1088–1092.

86 Raj, C., Karthick, S., Park, S. et al. (2014). Improved photovoltaic performance of CdSe/CdS/PbS quantum dot sensitized ZnO nanorod array solar cell. *J. Power Sources* 248: 439–446.

87 Chen, J., Lei, W., Li, C. et al. (2011). Flexible quantum dot sensitized solar cell by electrophoretic deposition of CdSe quantum dots on ZnO nanorods. *Phys. Chem. Chem. Phys.* 13: 13182–13184.

88 Chetia, T., Ansari, M., and Qureshi, M. (2015). Ethyl cellulose and cetrimonium bromide assisted synthesis of mesoporous, hexahon shaped ZnO nanodisks with

exposed ±{0001} polar facet for enhanced photovoltaic performance in quantum dot sensitized solar cells. *ACS Appl. Mater. Interfaces* 7: 13266–13279.
89 Lee, W., Kang, S., Hwang, T. et al. (2015). Facile conversion synthesis of densely-formed branched ZnO-nanowire arrays for quantum-dot-sensitized solar cells. *Electrochim. Acta* 167: 194–200.
90 Chetia, T., Barpuxary, D., and Qureshi, M. (2014). Enhanced photovoltaic performance utilizing effective charge transfers and light scattering effects by the combination of mesoporous, hollow 3D-ZnO along with 1D-ZnO in CdS quantum dot sensitized solar cells. *Phys. Chem. Chem. Phys.* 16: 9625–9633.
91 Tian, J., Uchaker, E., Zhang, Q. et al. (2014). Hierarchically structured ZnO nanorods–nanosheets for improved quantum-dot-sensitized solar cells. *ACS Appl. Mater. Interfaces* 6: 4466–4472.
92 Tian, J., Zhang, Q., Zhang, L. et al. (2013). ZnO/TiO_2 nanocable structured photoelectrodes for CdS/CdSe quantum dot co-sensitized solar cells. *Nanoscale* 5: 936–943.
93 Feng, H., Wu, W., Rao, H. et al. (2015). Three dimensional TiO_2/ZnO hybrid array as a heterostructured anode for efficient quantum-dot-sensitized solar cells. *ACS Appl. Mater. Interfaces* 7: 5199–5205.
94 Huang, Q., Li, F., Gong, Y. et al. (2013). Recombination in SnO_2-based quantum dots sensitized solar cells: the role of surface states. *J. Phys. Chem. C* 117: 10965–10973.
95 Ganapathy, V., Kong, E., Park, Y. et al. (2014). Cauliflower-like SnO_2 hollow microspheres as anode and carbon fiber as cathode for high performance quantum dot and dye-sensitized solar cells. *Nanoscale* 6: 3296–3301.
96 Patil, S., Shinde, D., Bhande, S. et al. (2013). Current density enhancement in ZnO/CdSe photoelectrochemical cells in the presence of a charge separating SnO_2 nanoparticles interfacing-layer. *Dalton Trans.* 42: 13065–13070.
97 Xiao, J., Huang, Q., Xu, J. et al. (2014). CdS/CdSe co-sensitized solar cells based on a new SnO_2 photo-anode with a three-dimensionally interconnected ordered porous structure. *J. Phys. Chem. C* 118: 4007–4015.
98 Zhu, G., Pan, L., Xu, T. et al. (2011). CdS/CdSe-cosensitized TiO_2 photoanode for quantum-dot-sensitized solar cells by a microwave-assisted chemical bath deposition method. *ACS Appl. Mater. Interfaces* 3: 3146–3151.
99 Becker, M.A., Radich, J.G., Bunker, B.A. et al. (2014). How does a SILAR CdSe film grow? Tuning the deposition steps to suppress interfacial charge recombination in solar cells. *J. Phys. Chem. Lett.* 5: 1575–1582.
100 Tao, L., Xiong, Y., Liu, H., and Shen, W. (2014). High performance PbS quantum dot sensitized solar cells via electric field assisted in situ chemical deposition on modulated TiO_2 nanotube arrays. *Nanoscale* 6: 931–938.
101 Xu, X.Y., Liao, J., Qiu, K. et al. (2011). Dynamic study of highly efficient CdS/CdSe quantum dot-sensitized solar cells fabricated by electrodeposition. *ACS Nano* 5: 9494–9500.
102 Lee, Y., Im, S., Lee, J. et al. (2011). Porous CdS-sensitized electrochemical solar cells. *Electrochim. Acta* 56: 2087–2090.
103 Huang, X., Huang, S., Zhang, Q. et al. (2011). A flexible photoelectrode for CdS/CdSe quantum dot-sensitized solar cells (QDSSCs). *Chem. Commun.* 47: 2664–2666.

104 Pernik, D., Tvrdy, K., Radich, J. et al. (2011). Tracking the adsorption and electron injection rates of CdSe quantum dots on TiO_2: linked verous direct attachment. *J. Phys. Chem. C* 115: 13511–13519.

105 Santra, P., Nair, P., Thomas, K. et al. (2013). $CuInS_2$-sensitized quantum dot solar cell. Electrophoretic deposition, excited-state dynamics, and photovoltaic performance. *J. Phys. Chem. Lett.* 4: 722–729.

106 Peng, Z., Liu, Y., Shu, W. et al. (2012). Synthesis of various sized $CuInS_2$ quantum dots and their photovoltaic properties as sensitizers for TiO_2 photo-anodes. *Eur. J. Inorg. Chem.* 32: 5239–5244.

107 Xie, R., Rutherford, M., and Peng, X. (2009). Formation of high-quality I-III-VI semiconductor nanocrystals by tuning relative reactivity of cationic precursors. *J. Am. Chem. Soc.* 131: 5691–5697.

108 Huang, W., Tseng, C., Chang, S. et al. (2012). Solvethermal synthesis of zincblende and wurtzite $CuInS_2$ nanocrystals and their photovoltaic application. *Langmuir* 28: 8496–8501.

109 Zhu, L., An, W.J., Springer, J.W. et al. (2012). Linker-free quantum dot sensitized TiO_2 photoelectrochemical cells. *Int. J. Hydrogen Energy* 37: 6422–6430.

110 Zhou, N., Chen, G., Zhang, X. et al. (2012). Highly efficient PbS/CdS co-sensitized solar cells based on photoanodes with hierarchical pore distribution. *Electrochem. Commun.* 20: 97–100.

111 Yu, X., Lei, B., Kuang, D. et al. (2012). High performance and reduced charge recombination of CdSe/CdS quantum dot-sensitized solar cells. *J. Mater. Chem.* 22: 12058–12063.

112 Peng, Z., Liu, Y., Shu, W. et al. (2013). Efficiency enhancement of $CuInS_2$ quantum dot sensitized TiO_2 photo-anodes for solar cell applications. *Chem. Phys. Lett.* 586: 85–90.

113 Shen, Q., Kobayashi, J., Diguna, L. et al. (2008). Effect of ZnS coating on the photovoltaic properties of CdSe quantum dot-sensitized solar cells. *J. Appl. Phys.* 103 (8): 084304.

114 Shen, C., Fischu, D., and Wang, Q. (2016). Interfacial engineering for quantum-dot-sensitized solar cells. *Chem. Asian J.* 11: 1183–1193.

115 Gopi, C.V., Venkata-Haritha, M., Kim, S. et al. (2015). Improved photovoltaic performance and stability of quantum dot sensitized solar cells using Mn-ZnSe shell structure with enhanced light absorption and recombination control. *Nanoscale* 7 (29): 12552–12563.

116 Li, T., Lee, Y., and Teng, H. (2012). High-performance quantum dot-sensitized solar cells based on sensitization with $CuInS_2$ quantum dots/CdS heterostructure. *Energy Environ. Sci.* 5 (1): 5315–5324.

117 Luo, J., Wei, H., Huang, Q. et al. (2013). Highly efficient core-shell $CuInS_2$-Mn doped CdS quantum dot sensitized solar cells. *Chem. Commun.* 49 (37): 3881–3883.

118 Shalom, M., Dor, S., Ruhle, S. et al. (2009). Core/CdS quantum dot/shell mesoporous solar cells with improved stability and efficiency using an amorphous TiO_2 coating. *J. Phys. Chem. C* 113 (9): 3895–3898.

119 Roelofs, K., Brennan, T., Dominguez, J. et al. (2013). Effect of Al_2O_3 recombination barrier layers deposited by atomic layer deposition in

solid-state CdS quantum dot-sensitized solar cells. *J. Phys. Chem. C* 117 (11): 5584–5592.

120 Liu, Z., Miyauchi, M., Uemura, Y. et al. (2010). Enhancing the performance of quantum dots sensitized solar cell by SiO_2 surface coating. *Appl. Phys. Lett.* 96 (23): 233107.

121 Tachan, Z., Hod, I., Shalom, M. et al. (2013). The importance of the TiO_2/quantum dots interface in the recombination processes of quantum dot sensitized solar cells. *Phys. Chem. Chem. Phys.* 15 (11): 3841–3845.

122 Rühle, S., Greenshtein, M., Chen, S. et al. (2005). Molecular adjustment of the electronic properties of nanoporous electrodes in dye-sensitized solar cells. *J. Phys. Chem. B* 109 (40): 18907–18913.

123 Mora-Seró, I., Giménez, S., Moehl, T. et al. (2008). Factors determining the photovoltaic performance of a CdSe quantum dot sensitized solar cell: the role of the linker molecule and of the counter electrode. *Nanotechnology* 19 (42): 424007.

124 Shalom, M., Ruhle, S., Hod, I. et al. (2009). Energy level alignment in CdS quantum dot sensitized solar cells using molecular dipoles. *J. Am. Chem. Soc.* 131 (29): 9876–9877.

125 Sykora, M., Petruska, M., Alstrum-Acevedo, J. et al. (2006). Photoinduced charge transfer between CdSe nanocrystal quantum dots and Ru-polypyridine complexes. *J. Am. Chem. Soc.* 128 (31): 9984–9985.

126 Mora-Seró, I., Gross, D., Mittereder, T. et al. (2010). Nanoscale interaction between CdSe or CdTe nanocrystals and molecular dyes fostering or hindering directional charge separation. *Small* 6 (2): 221–225.

127 Hod, I., González-Pedro, V., Tachan, Z. et al. (2011). Dye versus quantum dots in sensitized solar cells: participation of quantum dot absorber in the recombination process. *J. Phys. Chem. Lett.* 2 (24): 3032–3035.

128 Shen, C., Wang, X., Jiang, X. et al. (2015). Fast charge separation at semiconductor sensitizer–molecular relay interface leads to significantly enhanced solar cell performance. *J. Phys. Chem. C* 119 (18): 9774–9781.

129 Barea, E.M., Shalom, M., Giménez, S. et al. (2010). Design of injection and recombination in quantum dot sensitized solar cells. *J. Am. Chem. Soc.* 132 (19): 6834–6839.

130 Choi, H., Nicolaescu, R., Paek, S. et al. (2011). Super sensitization of CdS quantum dots with a near-infrared organic dye: toward the design of panchromatic hybrid-sensitized solar cells. *ACS Nano* 5 (11): 9238–9245.

131 Wang, J., Li, Y., Shen, Q. et al. (2016). Mn doped quantum dot sensitized solar cells with power conversion efficiency exceeding 9%. *J. Mater. Chem. A* 4 (3): 877–886.

132 Meng, K., Chen, G., and Thampi, K. (2015). Metal chalcogenides as counter electrode materials in quantum dot sensitized solar cells: a perspective. *J. Mater. Chem. A* 3: 23074–23089.

133 Raj, C., Prabakar, K., Savariraj, A. et al. (2013). Surface reinforced platinum counter electrode for quantum dots sensitized solar cells. *Electrochim. Acta* 103: 231–236.

134 Kiyonaga, T., Akita, T., and Tada, H. (2009). Au nanoparticle electrocatalysis in a photoelectrochemical solar cell using CdS quantum dot-sensitized TiO_2 photoelectrodes. *Chem. Commun.* 45 (15): 2011–2013.

135 Zhu, G., Pan, L., Sun, H. et al. (2012). Electrophoretic deposition of a reduced graphene-Au nanoparticle composite film as counter electrode for CdS quantum dot-sensitized solar cells. *ChemPhysChem* 13 (3): 769–773.
136 Yoon, Y., Kim, J., Kang, S. et al. (2014). Enhanced electrocatalytic activity of the Au-electrodeposited Pt nanoparticles-coated conducting oxide for the quantum dot-sensitized solar cells. *Appl. Phys. Lett.* 105 (8): 083116.
137 Fan, S., Fang, B., Kim, J. et al. (2010). Hierarchical nanostructured spherical carbon with hollow core/mesoporous shell as a highly efficient counter electrode in CdSe quantum-dot-sensitized solar cells. *Appl. Phys. Lett.* 96 (6): 063501.
138 Fan, S., Fang, B., Kim, J. et al. (2010). Ordered multimodal porous carbon as highly efficient counter electrodes in dye-sensitized and quantum-dot solar cells. *Langmuir* 26 (16): 13644–13649.
139 Lee, J., Sohn, K., and Hyeon, T. (2002). Low-cost and facile synthesis of mesocellular carbon foams. *Chem. Commun.* 38 (22): 2674–2675.
140 Dong, J., Jia, S., Chen, J. et al. (2012). Nitrogen-doped hollow carbon nanoparticles as efficient counter electrodes in quantum dot sensitized solar cells. *J. Mater. Chem.* 22 (19): 9745–9750.
141 Giordano, C., Erpen, C., Yao, W. et al. (2008). Synthesis of Mo and W carbide and nitride nanoparticles via a simple 'urea glass' route. *Nano Lett.* 8 (12): 4659–4663.
142 Zeng, X., Xiong, D., Zhang, W. et al. (2013). Spray deposition of water-soluble multiwall carbon nanotube and $Cu_2ZnSnSe_4$ nanoparticle composites as highly efficient counter electrodes in a quantum dot-sensitized solar cell system. *Nanoscale* 5 (15): 6992–6998.
143 Radich, J., Dwyer, R., and Kamat, P. (2011). Cu_2S reduced graphene oxide composite for high-efficiency quantum dot solar cells. Overcoming the redox limitations of $S^{2-}/S_n^{\ 2-}$ at the counter electrode. *J. Phys. Chem. Lett.* 2 (19): 2453–2460.
144 Meng, K., Surolia, P., Byrne, O., and Thampi, K. (2014). Efficient CdS quantum dot sensitized solar cells made using novel Cu_2S counter electrode. *J. Power Sources* 248: 218–223.
145 Jiang, Y., Zhang, X., Ge, Q. et al. (2014). ITO@Cu_2S tunnel junction nanowire arrays as efficient counter electrode for quantum-dot-sensitized solar cells. *Nano Lett.* 14: 365–372.
146 Li, D., Cheng, L., Zhang, Y. et al. (2014). Development of Cu_2S/carbon composite electrode for CdS/CdSe quantum dot sensitized solar cell modules. *Sol. Energy Mater. Sol. Cells* 120: 454–461.
147 Kundu, M., Hasegawa, T., Terabe, K. et al. (2008). Effect of sulfurization conditions on structural and electrical properties of copper sulfide films. *J. Appl. Phys.* 103 (7): 073523.
148 Wang, F., Dong, H., Pan, J. et al. (2014). One-step electrochemical deposition of hierarchical CuS nanostructures on conductive substrates as robust, high-performance counter electrodes for quantum-dot-sensitized solar cells. *J. Phys. Chem. C* 118 (34): 19589–19598.

149 Savariraj, A., Viswanathan, K., and Prabakar, K. (2014). Influence of Cu vacancy on knit coir mat structured CuS as counter electrode for quantum dot sensitized solar cells. *ACS Appl. Mater. Interfaces* 6 (22): 19702–19709.

150 Ye, M., Wen, X., Zhang, N. et al. (2015). In situ growth of CuS and $Cu_{1.8}S$ nanosheet arrays as efficient counter electrodes for quantum dot-sensitized solar cells. *J. Mater. Chem. A* 3 (18): 9595–9600.

151 Zhao, K., Yu, H., Zhang, H. et al. (2014). Electroplating cuprous sulfide counter electrode for high-efficiency long-term stability quantum dot sensitized solar cells. *J. Phys. Chem. C* 118 (11): 5683–5690.

152 Kim, C., Choi, S., and Bang, J. (2014). New insight into copper sulfide electrocatalysts for quantum dot-sensitized solar cells: composition-dependent electrocatalytic activity and stability. *ACS Appl. Mater. Interfaces* 6 (24): 22078–22087.

153 Yang, Z., Chen, C., Liu, C. et al. (2010). Electrocatalytic sulfur electrodes for CdS/CdSe quantum dot-sensitized solar cells. *Chem. Commun.* 46 (30): 5485–5487.

154 Yuan, H., Lu, J., Xu, X. et al. (2013). Electrochemically deposited CoS films as counter electrodes for efficient quantum dot-sensitized solar cells. *J. Electrochem. Soc.* 160 (9): H624–H629.

155 Tachan, Z., Shalom, M., Hod, I. et al. (2011). PbS as a highly catalytic counter electrode for polysulfide-based quantum dot solar cells. *J. Phys. Chem. C* 115 (13): 6162–6166.

156 Parand, P., Samadpour, M., Esfandiar, A. et al. (2014). Graphene/PbS as a novel counter electrode for quantum dot sensitized solar cells. *ACS Photonics* 1 (4): 323–330.

157 Eskandari, M., Ahmadi, V., and Ghahary, R. (2015). Copper sulfide/lead sulfide as a highly catalytic counter electrode for zinc oxide nanorod based quantum dot solar cells. *Electrochim. Acta* 151: 393–398.

158 Song, X., Wang, M., Deng, J. et al. (2014). ZnO/PbS core/shell nanorod arrays as efficient counter electrode for quantum dot-sensitized solar cells. *J. Power Sources* 269: 661–670.

159 Faber, M., Lukowski, M., Ding, Q. et al. (2014). Earth-abundant metal pyrites (FeS_2, CoS_2, NiS_2, and their Alloys) for highly efficient hydrogen evolution and polysulfide reduction electrocatalysis. *J. Phys. Chem. C* 118: 21347–21356.

160 Cao, Y., Xiao, Y., Jung, J. et al. (2013). Highly electrocatalytic $Cu_2ZnSn(S_{1-x}Se_x)_4$ counter electrodes for quantum-dot-sensitized solar cells. *ACS Appl. Mater. Interfaces* 5 (3): 479–484.

161 Choi, H., Ji, I., and Bang, J. (2014). Metal selenides as a new class of electrocatalysts for quantum dot-sensitized solar cells: a tale of $Cu_{1.8}Se$ and PbSe. *ACS Appl. Mater. Interfaces* 6 (4): 2335–2343.

162 Bo, F., Zhang, C., Wang, C. et al. (2014). From red selenium to cuprous selenide: a novel and facile route to a high performance metal selenide cathode for sensitized solar cells. *J. Mater. Chem. A* 2 (35): 14585–14592.

163 Liu, F., Zhu, J., Hu, L. et al. (2015). Low-temperature, solution-deposited metal chalcogenide films as highly efficient counter electrodes for sensitized solar cells. *J. Mater. Chem. A* 3 (12): 6315–6323.

164 Chen, X., Li, Z., Bai, Y. et al. (2015). Room-temperature synthesis of $Cu_{2-x}E$ (E=S, Se) nanotubes with hierarchical architecture as high-performance counter electrodes of quantum-dot-sensitized solar cells. *Chem. Eur. J.* 21 (3): 1055–1063.

165 Yeh, M., Lee, C., Chou, C. et al. (2011). Conducting polymer-based counter electrode for a quantum-dot-sensitized solar cell (QDSSC) with a polysulfide electrolyte. *Electrochim. Acta* 57: 277–284.

13

Flexible Triboelectric Nanogenerators

Fang Yi, Yue Zhang, Qingliang Liao, Zheng Zhang, and Zhuo Kang

University of Science and Technology Beijing, School of Materials Science and Engineering, Department of Material Physics, State Key Laboratory for Advanced Metals and Materials, Beijing Municipal Key Laboratory of New Energy Materials and Technologies, No. 30 Xueyuan Road, Haidian District, Beijing 100083, China

13.1 Introduction

13.1.1 Motivation for the Development of Flexible Triboelectric Nanogenerators

Energy is fundamental to the quality of our lives and is a key ingredient in all sectors of modern economies. In today's world, we are dependent on an uninterrupted supply of energy for living and working. It is predicted that the world-wide energy demand will continue to increase. The limited supply of fossil energy and greenhouse gas emission from burning of fossil fuels has given rise to great concern. Consequently, searching for renewable and sustainable energy technologies is attracting more and more attention all around the world. There are many renewable energy sources that exist in our ambient environment, such as solar energy, thermal energy, wind energy, wave energy, and biomechanical energy. Among the technologies aiming to harvest such renewable energy sources, solar cells absorb sunlight and convert it into electricity, thermoelectric generators generate electricity under a temperature gradient, electromagnetic generators harvest mechanical energy through electromagnetic induction, and piezoelectric nanogenerators (PiENGs) produce electricity upon applying a force [1–5]. All these technologies have greatly promoted the development of sustainable energy. However, they all have undesirable features, for example, solar cells cannot work continuously due to the intermittency of sunlight; thermoelectric generators have relatively limited applications due to their stringent working requirements; electromagnetic generators have cumbersome structures and are difficult for further miniaturization; and PiENGs must use piezoelectric materials and have monotonous working mode as well as low output power density. The development of energy harvesting technologies with high energy conversion efficiency, light weight, low cost, and high stability is still one of the hottest topics among scientists today. Recently, triboelectric nanogenerators (TENGs) have been invented to harvest mechanical energy, which work based on the triboelectric effect and electrostatic induction [6]. With the advantages

Flexible Energy Conversion and Storage Devices, First Edition. Edited by Chunyi Zhi and Liming Dai.

of high output power density and energy conversion efficiency, easy fabrication, low cost, light weight, wide selection of materials, and various working modes, TENGs have attracted intensive attention since their invention [7–19].

In the present day, almost all phases of modern technological society, such as the home, the car, and the workplace, use electronics. More and more types of sensors for various applications such as environmental protection, infrastructure monitoring, security, health monitoring, medical care, and robotics have been developed. The power source for these sensors form an indispensable component to ensure an ordinary operation. Specifically, the Internet of Things, which connects any device with an on and off switch to the internet and includes almost everything such as cellphones, people, software, machines, etc., has greatly changed people's understanding about energy supply. Traditional power sources for electronics such as batteries, supercapacitors, and fuel cells have limited life time, which brings about issues relating to frequent replacement and proper disposal of such power supply devices. Although tremendous efforts have been made to increase both the energy density and the power density of such power sources, new approaches for electronics to operate independently and sustainably and be maintenance-free are still highly demanded. To address this challenge, self-powered sensors and self-charging power units (SCPUs) have emerged and are under rapid development. Self-powered sensors can themselves generate electrical signals as a response to the trigger or change in the environment and thus do not need any external power sources during operation. The development of world electronic technology has managed to largely improve the work efficiency and thereby decrease the power consumption of the electronics, which makes it possible to use the energy harvested from the ambient environment to power electronics. This concept is realized by the SCPU. Since ambient energy harvesters generally generate fluctuated and inconstant outputs according to the uncontrollable ambient environmental conditions, the SCPU combines the energy harvesters with energy storage devices. In this way, the SCPU is able to provide a continuous and sustainable power supply.

Flexible/stretchable electronics that exceed the scope of rigid, brittle, and planar nature of traditional micro/macroelectronics are an emerging class of electronics receiving significant attention. They can conform to complex, curvilinear surfaces while maintaining levels of performance, which enable a wide range of applications such as electronic skins, implantable devices, robotics, smart textiles, and prosthetics [20–25]. The fast development of flexible/stretchable electronics calls for flexible/stretchable power sources. Since TENGs have many advantages such as light weight, wide selection of materials and low cost, they can be mounted on the human body or objects that have curvilinear surfaces and may deform as they move. This makes them a desirable flexible/stretchable power source.

Thus, developing flexible/stretchable TENG is not only an approach to scavenge the wasted ambient energy that helps to promote environmentally sustainable development of economy and technology, but also a promising solution to meet the challenge of providing continuous and sustainable power supply for flexible/stretchable, wearable, and portable electronics.

13.1.2 Basic Working Mechanism and Working Modes of Flexible Triboelectric Nanogenerators

Flexible triboelectric nanogenerators (FTENGs) have the same working mechanism and working modes as traditional TENGs, and can generate electricity from various types of mechanical motion such as touching, patting, sliding, rotation, and vibration.

Because of the triboelectric effect, when two materials come into contact, the surface of one material will become negatively charged while the surface of the other material will become positively charged, with the same amount of charges. In this regard, there are a lot of options when it comes to the selection of materials for fabricating a TENG. The polarity of charge separation when one material touches the other can refer to the triboelectric series [26]. The farther away the two materials are from each other on the series, the greater the charge transferred. Generally, the negatively charged triboelectric material in a TENG is a dielectric material; whereas the positively charged triboelectric material can be conductors, insulators, and semiconductors, in the form of solid, gel, or liquid. The triboelectric charges on the surface of an insulator can remain on the material for a long time, which serves as an electrostatic induction source for the electricity generation process of the TENG. When relative motion occurs between the two triboelectrically charged materials, potential difference will be generated between the two corresponding electrodes in the open-circuit condition, which will drive electrons to flow between the two electrodes (for single-electrode mode, electrons transfer between the working electrode and the ground). Large contact area and high separation distance are preferable for the generation of high electrical outputs.

The working modes of the TENG can be categorized into three types based on the fundamental electrostatic induction process: attached-electrode mode, single-electrode mode, and freestanding mode (Figure 13.1). All the three working modes can harvest the two basic kinds of mechanical motion: vertical contact/separation motion and lateral sliding motion (including rotation and rolling).

The attached-electrode-mode TENGs have two components, and each component has an attached electrode [27, 28]. As shown in Figure 13.1a,b, one component is the negatively charged dielectric layer with an electrode attached on the back side; while the other component is the positively charged layer, which can be a conductive layer that also serves as the electrode or a dielectric layer attached to an electrode. The single-electrode-mode TENGs have just one component (Figure 13.1c,d) [29, 30]. This component can be a conductive layer that also serves as the electrode or a dielectric layer attached to an electrode. The freestanding-mode TENGs have three components: a freestanding component and two stationary components (Figure 13.1e,f) [31, 32]. When the freestanding triboelectric component is a positively charged material, the two stationary components usually are a dielectric material attached to an electrode. When the freestanding component is a negatively charged material, the two stationary components can be a dielectric material attached to an electrode or a conductive material such as metal. The corresponding electrodes of the two stationary

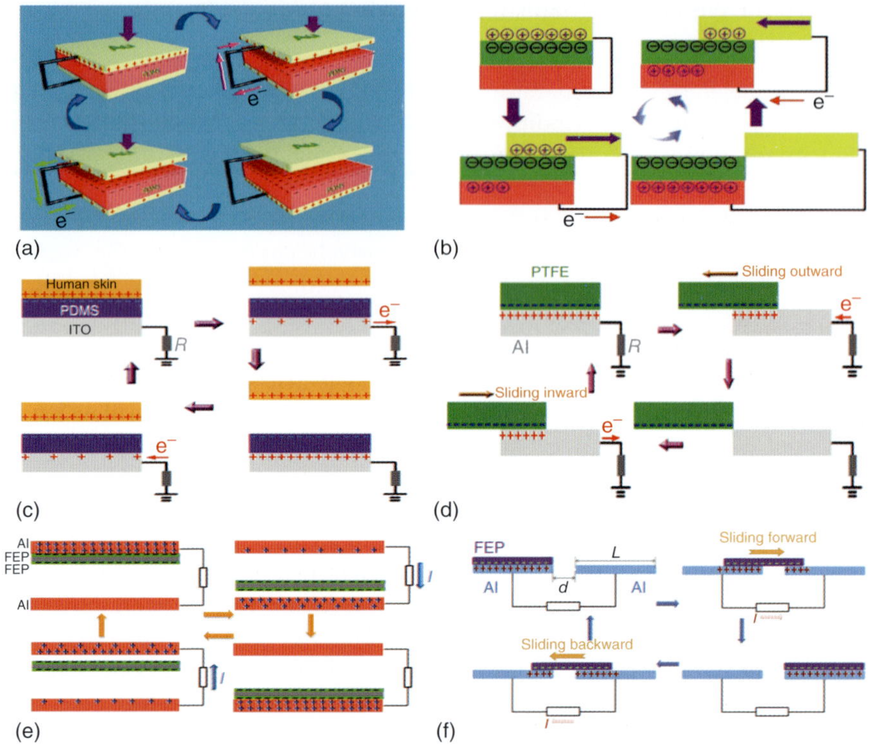

Figure 13.1 Basic working modes of the TENG. The attached-electrode (a) contact/separation and (b) sliding mode. *Source:* (a, b) Zhu et al. 2013 [27, 28]. Reproduced with permission from American Chemical Society. The single-electrode (c) contact/separation and (d) sliding mode. *Source:* (c, d) Yang et al. 2013 [29, 30]. Reproduced with permission from American Chemical Society. The freestanding (e) contact/separation and (f) sliding mode. *Source:* (e, f) Wang et al. 2014 [31, 32]. Reproduced with permission from American Chemical Society.

components are usually designed into interdigital structures so as to enhance the outputs.

Each working mode for the TENG has its own advantages. The single-electrode-mode TENG has the easiest structure and can conveniently harvest energy from an arbitrary, freely moving object because the other triboelectric part can move without restriction. The attached-electrode-mode TENG can provide higher outputs because it is harder to reach air breakdown conditions. The freestanding-mode TENG can efficiently and conveniently scavenge energy from sliding or contact/separation motion with one freely moving triboelectric part. Among these working modes, the TENG working in the single-electrode mode usually has lower outputs because of the negative impact caused by the electrostatic shield effect of the primary electrode.

The theories involving TENGs have been systematically studied by Wang's group and some other groups, which will not be discussed in detail here. Readers who are interested in the theoretical studies of TENGs can have a look at their works [33–42].

13.2 Materials Used for Flexible Triboelectric Nanogenerators

To enable flexibility at the device level, all the materials for fabricating the FTENG should be flexible. Generally, the fabrication of an FTENG with stretchability faces more challenges than that of an FTENG without stretchability.

For the material design of an FTENG without stretchability, the negatively charged triboelectric layer can be flexible dielectric materials such as fluorinated ethylene propylene (FEP), polytetrafluoroethylene (PTFE), polydimethylsiloxane (PDMS), polyethylene (PE), epoxy, polyester, Kapton, polycaprolactone (PCL), and polyimide (PI). The positively charged triboelectric layer can be conductive materials such as aluminum (Al), copper (Cu), gold (Au), silver (Ag), and indium tin oxide (ITO), which is either a flexible foil or a film deposited on flexible substrates such as polyethylene terephthalate (PET). The positively charged triboelectric layer can also be flexible dielectric materials such as nylon and polyaniline (PANi) or nanostructured semiconductors such as zinc oxide and titanium dioxide. The corresponding electrode for the negatively or positively charged dielectric layer can be a metal film deposited on the dielectric layer or a conductive organic film such as poly(3,4-ethylene dioxythiophene) (PEDOT) coated on the dielectric layer. Some researchers have applied graphene as the positively charged triboelectric layer but have observed undesirable performance [43, 44]. Although carbon material such as graphene or carbon nanotube may not be an ideal triboelectric layer, they can be a good material for the electrode of FTENGs owing to their good flexibility.

For the material design of an FTENG with stretchability, namely stretchable TENG, there can be two types of designs. One common material design is using intrinsically stretchable materials to construct the FTENG. The other material design is using intrinsically nonstretchable flexible materials but with specially designed device structures to endow the FTENG with stretchability [45]. The intrinsically stretchable dielectric materials used for stretchable TENGs can be PDMS and silicone rubber. Compared with PDMS, silicone rubbers (Ecoflex rubbers) are a more desirable negatively charged dielectric material that can endow the FTENGs with relatively higher performance because Ecoflex rubbers have a stronger ability to attract electrons that can lead to a higher surface triboelectric charge density and a softer texture that can enable more effective contact between the two triboelectric parts. However, PDMS has higher transparency than Ecoflex silicone rubbers, which makes it more suitable for highly transparent FTENGs. The key challenge for the fabrication of stretchable TENGs is the stretchable electrodes. The stretchable electrodes can be made of nonstretchable conductive materials shaped into wavy, serpentine, or spiral structures; but this kind of stretchable electrodes only has stretchability in limited directions. The intrinsically stretchable electrodes can be conductive solutions, water, conductive hydrogels, network of silver nanowires, liquid metals, liquid alloys, and composites made by mixing silicone rubber or PDMS with conductive powders such as carbon black and carbon nanotube. The resistance of the intrinsically stretchable electrode usually will increase with increasing stretchment, but this increase generally has little impact on the performance because the high inherent impedance of the TENG allows the electrode resistance to vary within a wide range [46, 47].

To enhance the performance of FTENGs, nanostructures are often created on the surface of the triboelectric layers to increase the contact area, with methods such as plasma dry etching, patterned silicone molds and sand papers for dielectric materials, and electrochemical anodization for metal foils [48]. The performance of FTENGs may also be improved by surface modification to increase the triboelectric surface charge density, with methods such as ionized-air injection and molecular surface functionalization [49, 50]. But it should be kept in mind that there is a maximum surface charge density for the triboelectric materials in ambient atmosphere because of the air breakdown limit.

13.3 Flexible Triboelectric Nanogenerators for Harvesting Ambient Energy

As an energy harvester, FTENGs have been applied to generate electricity from various kinds of mechanical energy existing in the ambient environment such as biomechanical energy, wind energy, water energy, and acoustic energy.

13.3.1 Harvesting Biomechanical Energy

The most common applications of FTENGs are for biomechanics. Humans are in constant motion, from the obvious movements of the external body parts to the unnoticeable movements of the internal organs and blood flow. To harvest the external motion, FTENGs can be attached to clothes/shoes, be weaved into fabric, or be made into mats to cover the floor/road that humans walk on; to harvest the internal motion, FTENGs can be implanted into the human body and placed onto the organs like throat and heart.

Human walking/jogging involves the movement of the limbs such as arm and leg swinging, foot rising/falling, corresponding to lateral sliding motion and vertical contact/separation motion that can be converted into electricity by FTENGs with various working modes [51–61]. A multilayered FTENG has been demonstrated by Bai et al. to harvest foot motion during walking (Figure 13.2a) [51]. The FTENG is composed of five layers of units that are fabricated on a zig zag-shaped flexible substrate, which can generate an open-circuit voltage (V_{oc}) of 215 V and a short-circuit current (I_{sc}) of 0.66 mA. Triggered by the pressure exerted by normal human walking, the FTENG attached to a shoe pad was demonstrated to power nine light-emitting diode (LED) bulbs. Kim et al. fabricated a waterproof fabric-structured FTENG by weaving fibers, which has a stable output voltage and current of 40 V and 210 μA respectively (Figure 13.2b) [54]. The fabric-structured TENG has been demonstrated in applications of footstep-driven large-scale power mats during walking and power clothing attached to the elbow. Zhang et al. attached two parts of an FTENG between clothes to harvest body motion energy (Figure 13.2c) [56]. During walking, the maximum output voltage and current density are up to 17 V and 0.02 μA cm^{-2}.

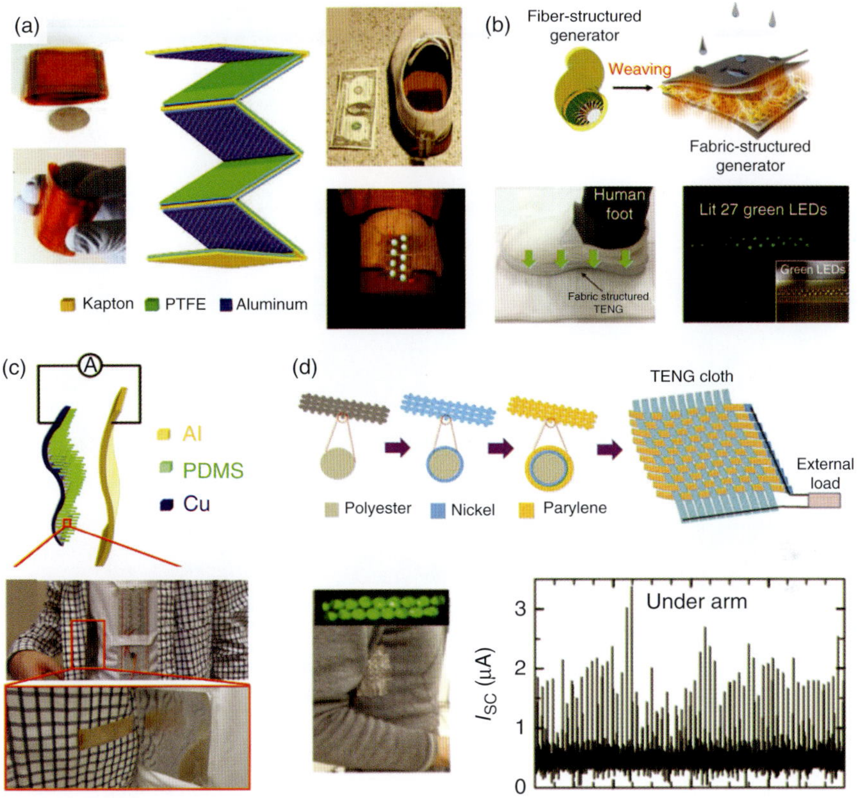

Figure 13.2 FTENGs for harvesting energy from human activities such as walking/jogging. (a) A multilayered FTENG to harvest energy from foot motion during walking. *Source:* Bai et al. 2013 [51]. Reproduced with permission from American Chemical Society. (b) A fabric-structured FTENG and its application of footstep-driven power mats during walking. *Source:* Kim et al. 2015 [54]. Reproduced with permission from American Chemical Society. (c) An FTENG attached to clothes to scavenge energy from body motion. *Source:* Zhang et al. 2013 [56]. Reproduced with permission from Elsevier. (d) An FTENG cloth by weaving two kinds of belts to scavenge energy from body motion. *Source:* Pu et al. 2015 [60]. Reproduced with permission from John Wiley & Sons.

Pu et al. fabricated an FTENG cloth by weaving Ni-cloth belts and parylene-cloth belts into a fabric, which serve as the two triboelectric parts of the FTENG (Figure 13.2d) [60]. The FTENG has a maximum V_{oc} of 50 V and I_{sc} of 4 μA, and has been demonstrated to harvest energy from both foot and arm motion during walking.

FTENGs can also be applied to harvest other kinds of everyday human activities such as hand motion and finger motion while using personal electronics such as cell phones and computers [62–68]. An FTENG with checker-like interdigital electrodes based on freestanding mode was reported by Guo et al., which has a V_{oc} of 210 V and a maximum output power density of 1.9 W m^{-2} (Figure 13.3a) [62]. The FTENG has been demonstrated to harvest the energy from playing computer games. Meng et al. fabricated a single-electrode-mode

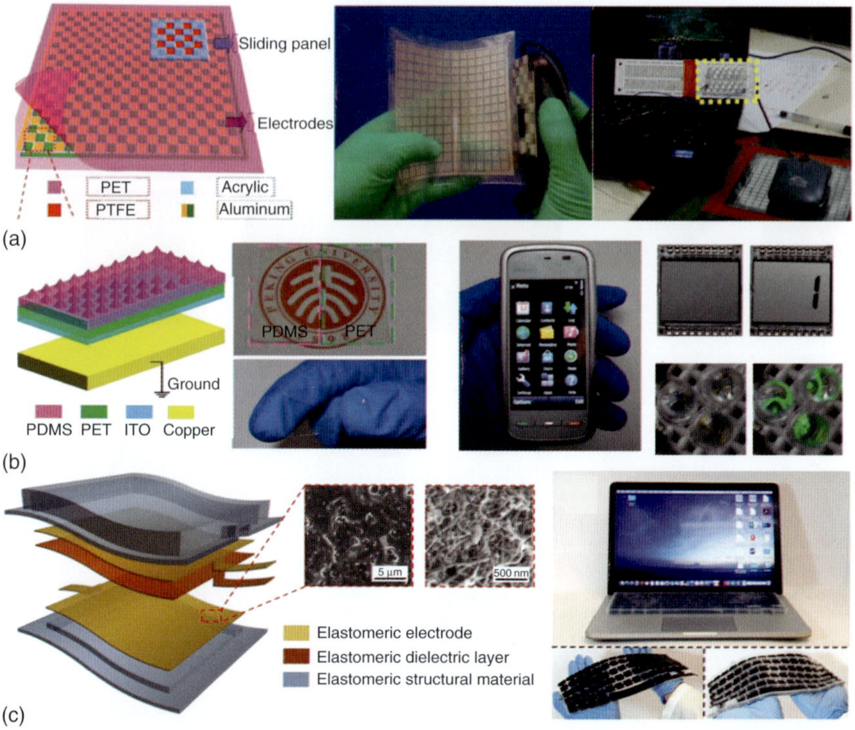

Figure 13.3 FTENGs for harvesting energy from hand or finger motions. (a) An FTENG with checker-like interdigital electrodes to harvest energy from moving the mouse. *Source:* Guo et al. 2015 [62]. Reproduced with permission from John Wiley & Sons. (b) A transparent single-electrode-mode FTENG to scavenge energy from touching the screen of a cell phone. *Source:* Meng et al. 2013 [64]. Reproduced with permission from Royal Society of Chemistry. (c) An FTENG as a keyboard cover to harvest energy from typing motion. *Source:* Li et al. 2016 [65]. Reproduced with permission from American Chemical Society.

FTENG, which has a V_{oc} of 130 V and I_{sc} of 1 μA cm^{-2} (Figure 13.3b) [64]. The FTENG can directly power three LEDs when the phone screen is tapped in daily use. Li et al. constructed an FTENG as a keyboard cover to harvest typing motion, which is based on the attached-electrode contact/separation mode and has a short-circuit transferred charge density of ~140 μC m^{-2} (Figure 13.3c) [65].

Harvesting biomechanical energy in vivo is a possible approach to provide power sources for the operation of medical micro/nanoelectronics. Compared with implantable PiENGs, implantable FTENGs have higher outputs [69–71]. The reported implantable FTENGs are usually working in the attached contact/separation mode, and good sealability is highly demanded in order to prevent leakage. Zheng et al. fabricated a biodegradable FTENG for in vivo biomechanical energy harvesting, which can be degraded and resorbed in an animal body after finishing its working cycle with no adverse long-term effects [70]. The FTENG implanted in the subdermal region of the backs of SD rats can generate a V_{oc} of up to 4 V, triggered with the help of a slight finger tapping on the skin of the implanted region.

13.3.2 Harvesting Wind Energy

Wind energy is plentiful, renewable, clean, widely distributed, and therefore is one of the most promising new energy sources. Conventional wind energy harvesters were generally designed to work under a certain wind direction. However, the blowing direction of most ambient wind is not constant. FTENGs have the advantages of being able to harvest energy from wind blowing in all directions, low cost, and simple fabrication [72, 73]. Zhang et al. reported a transparent FTENG array with vertically freestanding polymer strips, working in the contact/separation mode (Figure 13.4a,b) [72]. The FTENG unit can produce a V_{oc} of up to 98 V and I_{sc} of 16.3 μA, corresponding to a power density of 2.76 W m^{-2}. The wind-harvesting FTENG was demonstrated to be mounted on the roof top of a model house, with 60 units to light up 60 LEDs. Zhao et al. fabricated a freestanding woven flag-shaped FTENG that can harvest wind energy from arbitrary directions (Figure 13.4c,d) [73]. The flag-shaped FTENG was woven by two kinds of belts separated by an air gap, and the contact/separation motion between the two belts driven by wind generates electricity. The FTENG with a unit size of 1.5 × 1.5 cm can have a maximum V_{oc} of 40 V and I_{sc} of 30 μA at a wind speed of 22 m s^{-1}.

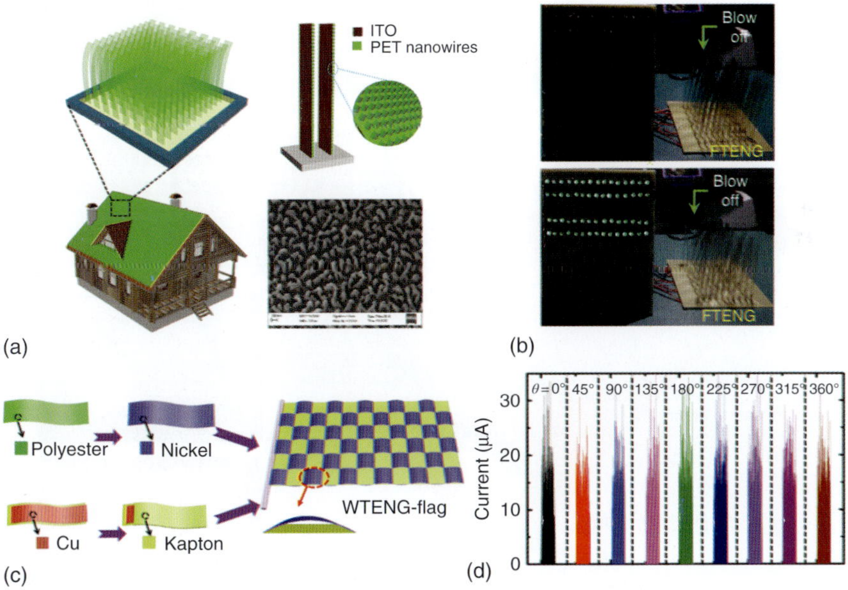

Figure 13.4 FTENGs for harvesting wind energy. A lawn-structured FTENG to harvest wind energy: (a) schematic illustration of the device structure and (b) photographs showing 60 LEDs lighted up by the FTENG by energy harvested from wind blowing. *Source:* (a, b) Zhang et al. 2016 [72]. Reproduced with permission from John Wiley & Sons. A flag-shaped FTENG to scavenge wind energy: (c) schematic illustration of the device structure and (d) output current of the FTENG from wind blowing at various incident angles. *Source:* (c, d) Zhao et al. 2016 [73]. Reproduced with permission from American Chemical Society.

13.3.3 Harvesting Water Energy

Water energy is a universal renewable energy source on earth that is abundant and possesses tremendous mechanical energy. Specifically, water waves such as ocean waves, river waves, and tides are released constantly and continuously, which contain a gigantic reserve of kinetic energy. The reported FTENGs for scavenging water energy are mainly applied to harvest the water wave energy [74, 75]. Su et al. constructed an FTENG with elastic wavy electrodes sandwiched between two nanostructured polymer thin films, which can work in two combined modes (Figure 13.5a,b) [74]. Under water waves propagating at a speed of 0.5 $m\,s^{-1}$, the part working in the contact/separation mode and the part working in the freestanding sliding mode of the FTENG can generate I_{sc} of

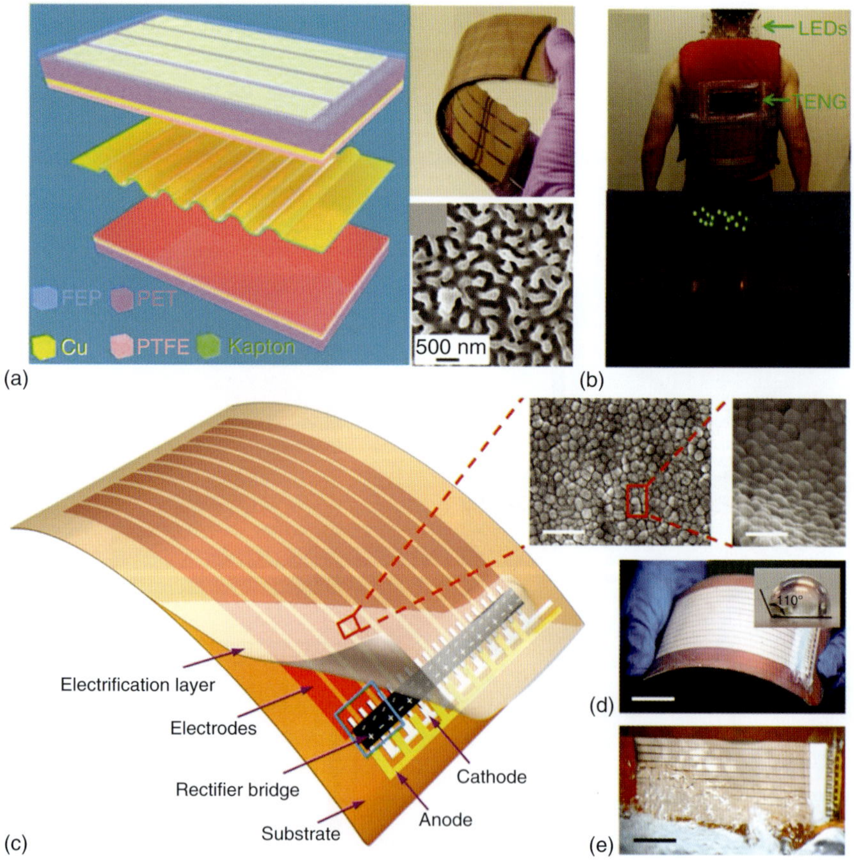

Figure 13.5 FTENGs for harvesting water energy. An FTENG with elastic wavy electrodes to harvest water energy: (a) schematic diagram and photograph of the device and (b) photographs showing the FTENG mounted on a life vest to harvest water wave energy *Source:* (a, b) Su et al. 2014 [74]. Elsevier. An FTENG with multiple parallel electrodes to scavenge water energy: (c) schematic illustration of the device structure; (d) photograph showing the FTENG during bending; and (e) photograph showing the FTENG interacting with water waves. *Source:* (c–e) Zhao et al. 2015 [75]. Reproduced with permission from American Chemical Society.

5.1 and 4.3 μA respectively, which is capable of powering 50 LEDs. Zhao et al. fabricated a freestanding-sliding-mode FTENG with multiple parallel electrodes connected by an array of surface-mounted bridge rectifiers (Figure 13.5c–e) [75]. The induced charge flow between any pair of adjacent electrodes can be rectified and added up to form pulsed direct current. The FTENG can produce a voltage of up to 230 V, a pulsed current of 13 μA, and a peak output power of 1.1 mW.

In addition, FTENGs have been applied to harvest other kinds of energy sources such as acoustic energy. Kim et al. reported an FTENG structured with all organic materials that was prepared by solution casting polymerization on the dielectric polymer films [76]. The FTENG can generate a voltage of 11 V under a sound pressure level of 90–100 dB, which can power five LEDs.

13.4 Flexible Triboelectric Nanogenerators for Self-Powered Sensors

FTENGs can act as self-powered sensors for various kinds of motion because they can produce electrical signals from almost any type of mechanical motion in ambient environment including touching, lateral sliding, rotation, bending, and so on. Besides, FTENGs can be utilized as self-powered sensors to detect humidity or some chemicals since vapor or chemicals such as ethanol and methanol can influence the surface triboelectric charge density of the FTENG, which has a linear relationship with the amplitudes of the generated electrical signals (V_{oc}, short-circuit transferred charge (Q_{sc}), and I_{sc}) when all the other conditions are maintained the same.

13.4.1 Self-Powered Touch/Pressure Sensors

Touch/pressure sensing is a rapidly developing field that has attracted attentive attention due to its vast applications including human–machine interfacing, skin-like electronics, industrial automation, medical procedures, and security systems. Touching motion can cause the FTENG to generate electrical signals, and this basic working principle makes touch sensing the most straightforward application of self-powered sensing for FTENG. Fan et al. reported a transparent FTENG with pyramid microstructures on the dielectric film surface, which gave a voltage of up to 18 V and a current density of 0.13 μA cm^{-2} (Figure 13.6a) [77]. The FTENG can serve as a self-powered touch sensor to detect a water droplet and a falling feather.

A more quantitative understanding and characterization is required when it comes to pressure sensing. The reported FTENGs for quantitatively detecting pressure can be categorized into two types according to the working modes: the single-electrode contact/separation FTENGs and the attached-electrode contact/separation FTENGs. For FTENGs working in the single-electrode contact/separation mode to detect pressure, a dielectric material with strong ability to attract electrons usually acts as the triboelectric surface [29, 78, 79, 81]. Moreover, the dielectric surface should have nano/microstructures, which leads

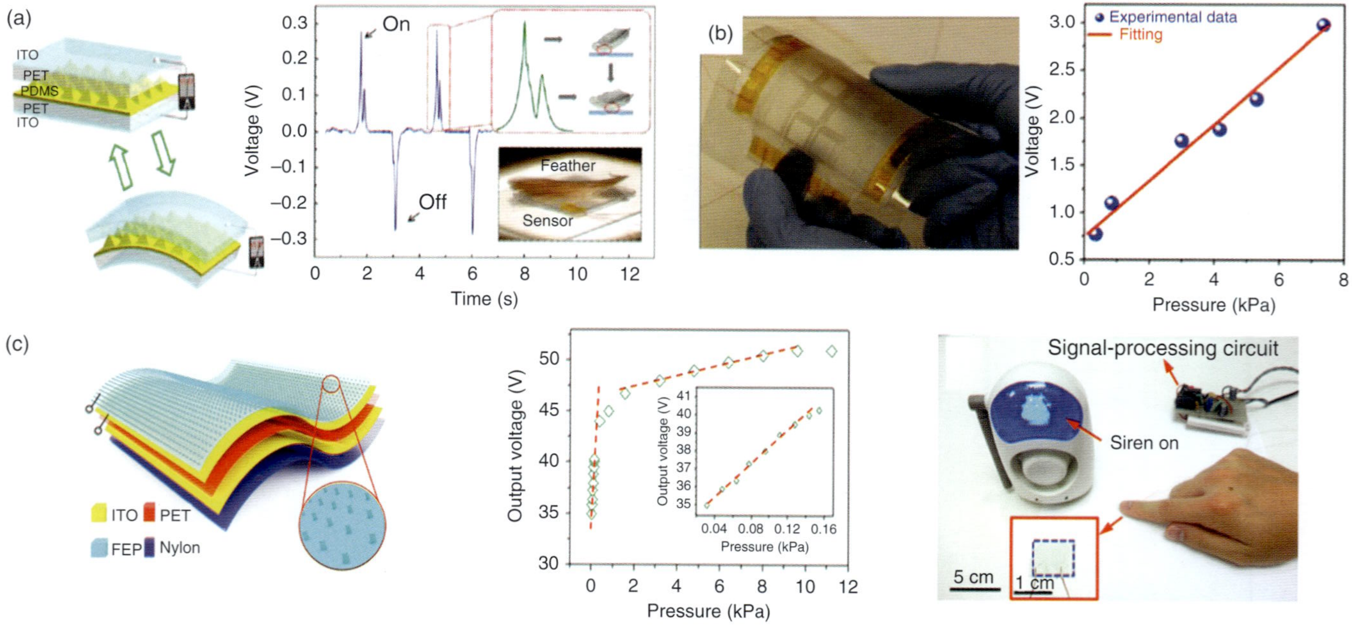

Figure 13.6 FTENGs as self-powered touch/pressure sensors. (a) A transparent FTENG with pyramid microstructures as a touch sensor. The plot shows the output voltage induced by a piece of feather. *Source:* Fan et al. 2012 [77]. Reproduced with permission from American Chemical Society. (b) An FTENG composed of a micropyramid-structured PDMS and an ITO electrode to detect pressure/touch. The plot shows the output voltage as a function of the applied pressure. *Source:* Yang et al. 2013 [29]. Reproduced with permission from American Chemical Society. (c) An FTENG with nanowire-structured FEP film as the triboelectric layer to sense pressure/touch. The plot shows the measured output voltage with increasing pressure and the photograph shows a wireless alarm system triggered by finger tapping on an FTENG. *Source:* Zhu et al. 2014 [78]. Reproduced with permission from American Chemical Society. (d) An FTENG with nanostructured PTFE film as the triboelectric layer to detect pressure/touch. The plot shows the relationship between the forward output current and pressure. *Source:* Chen et al. 2017 [79]. Reproduced with permission from John Wiley & Sons. (e) An integrated FTENG sensor array working in the attached-electrode contact/separation mode for pressure/touch sensing. The plot shows the output voltage as a function of the external pressure. *Source:* Jiang et al. 2016 [80]. Reproduced with permission from American Chemical Society.

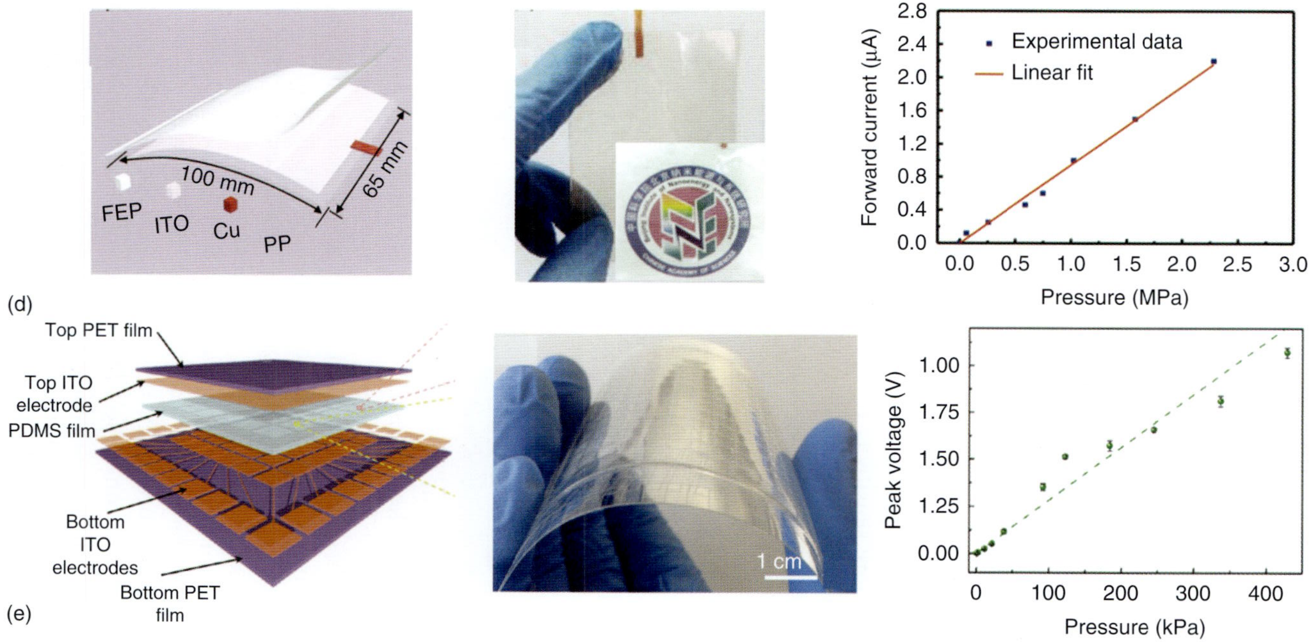

Figure 13.6 (*Continued*)

to a change of contact area corresponding to the different pressure applied on the surface. When other parameters maintain the same, the V_{oc} and Q_{sc} are proportional to the amount of surface triboelectric charges

$$V_{oc}, Q_{sc} \propto Q_{tribo} \tag{13.1}$$

and

$$Q_{tribo} = \sigma S, \tag{13.2}$$

where σ is the surface triboelectric charge density and S is the contact area. Thus the pressure can have a linear relationship with the measured voltage and transferred charge. Yang et al. fabricated an FTENG composed of a PDMS film with a micropyramid-structured surface and a layer of ITO electrode, which produces a voltage of up to 1000 V, a current density of 8 $mA\,cm^{-2}$, and a power density of 500 $mW\,m^{-2}$ (Figure 13.6b) [29]. The pressure sensitivity of the FTENG is ~0.29 ± 0.02 $V\,kPa^{-1}$. Zhu et al. constructed an FTENG composed of thin-film materials that are vertically laminated: FEP/ITO/PET/ITO/nylon (Figure 13.6c) [78]. Vertically aligned polymer nanowires were created on the FEP surface. The FTENG has a pressure sensitivity of 44 $mV\,Pa^{-1}$ (0.09% Pa^{-1}) and a maximum touch sensitivity of 1.1 $V\,Pa^{-1}$ (2.3% Pa^{-1}) in the low pressure region (<0.15 kPa). Chen et al. fabricated an FTENG structured with an Al layer sandwiched between a PTFE layer and a PET layer (Figure 13.6d) [79]. The PTFE surface that acts as the triboelectric surface was etched with nanopores. The pressure sensitivity of the FTENG was ~0.947 $\mu A\,MPa^{-1}$.

For FTENGs working in the attached-electrode contact/separation mode to detect pressure, a specially designed air gap between the two triboelectric layers is required [80, 82]. Since the separation distance between the two triboelectric layers is fixed, the V_{oc} and the applied pressure P can be analyzed by the following equation [83]:

$$\frac{V_{oc,0} - V_{oc}}{V_{oc,0}} = \frac{d_0 - d}{d_0} = \frac{S}{k * d_0} * p, \tag{13.3}$$

where $V_{oc,0}$ is the maximum V_{oc} with no applied pressure, d_0 is the maximum separation distance, S is the effective area of the electrodes and k represents the elastic property of the FTENG. Therefore, the applied pressure can also have a linear relationship with the measured voltage and transferred charge for the FTENG pressure sensors working in the attached-electrode contact/separation mode. Zhong et al. reported an integrated FTENG sensor array for pressure/touch sensing, with each FTENG unit delivering a V_{oc} of up to 1.613 V and I_{sc} of 47.308 $mA\,m^{-2}$ under a pressure of 612.5 kPa (Figure 13.6e) [81]. The pressure sensitivity of the FTENG was ~2.82 ± 0.187 $V\,MPa^{-1}$.

The self-powered pressure/touch sensors based on the two kinds of FTENGs have their own pros and cons. The FTENG pressure/touch sensors working in the single-electrode contact/separation mode have the advantage of simple fabrication; but they are open to the atmosphere and therefore are easily affected by environmental conditions such as humidity. The FTENG pressure/touch sensors working in the attached-electrode contact/separation mode can be waterproof; but their maximum gap distances are affected by deformation, which impacts their performance.

Compared with conventional pressure/touch sensors based on piezoresistive or capacitive effect, FTENG pressure/touch sensors can work without any external power source and have relatively faster response/recover time. However, FTENG pressure/touch sensors have limitations in detecting constant variation of static stimuli because of their instantaneous response to dynamic stimuli.

13.4.2 Self-Powered Motion Sensors

Motion sensing is of significant importance in a broad range of fields such as automation, robotics, prosthetics, medical care, security, sports, and entertainment. FTENGs have been applied to detect motion of both human body and objects with good accuracy and fast response/recover time.

13.4.2.1 Sensing Motion of Human Body

The basic operation mechanism of FTENGs sensing motion of human body is that the human motion triggers deformation in terms of relative displacement between two triboelectric parts in the vertical or lateral direction, which leads to potential difference between the two working electrodes (or the working electrode and the ground) and drives electrons to flow across. The working modes of the reported FTENGs to detect human motion are mainly single-electrode mode and attached-electrode mode [84–92].

Yi et al. reported a stretchable-rubber-based FTENG that consists of a layer of elastic rubber and a layer of Al film (Figure 13.7a) [84]. The FTENG has a unique working principle wherein the electrical signals are generated due to the change of the shape/length and the surface charge density of the rubber, which has a short-circuit current density of 7.5 $\mu A\,m^{-2}$ and output power density of 76.27 $W\,m^{-2}$. This FTENG can accurately detect the rates of diaphragmatic breathing and knee bending, and is capable of distinguishing the bending angle of the knee. Zhao et al. reported a textile-based FTENG that was fabricated by weaving of Cu-coated PET (Cu-PET) warp yarns and PI-coated Cu-PET weft yarns, which works in the attached-electrode contact/separation mode and produces a short-circuit current density of 15.50 $mA\,m^{-2}$ (Figure 13.7b) [85]. This FTENG can monitor human respiratory information including rate and depth. Yang et al. reported an FTENG that was fabricated by assembling serpentine-patterned electrodes and a wavy-structured Kapton film, which works in the attached-electrode contact/separation mode and delivers a output power density of 5 $W\,m^{-2}$ (Figure 13.7c) [86]. The FTENG can be attached onto human skin to monitor motion of joints, muscles, and Adam's apple. Xu et al. fabricated a hydrogel-based FTENG by using physical-crosslinking poly(vinyl alcohol) (PVA) hydrogel as the substrate material, which works in the attached-electrode contact/separation mode and generates a peak output power of 2 mW (Figure 13.7d) [88]. The FTENG can detect the angle and rate of bending the elbow. Lai et al. fabricated a single-thread-based FTENG that was composed of a silicone rubber coated stainless steel thread (Figure 13.7e) [89]. By sewing the serpentine-shaped thread onto an elastic textile, the FTENG produces an output power of 14 mW and can detect finger motion and pulse. Ma et al. reported an implantable FTENG working in the attached-electrode

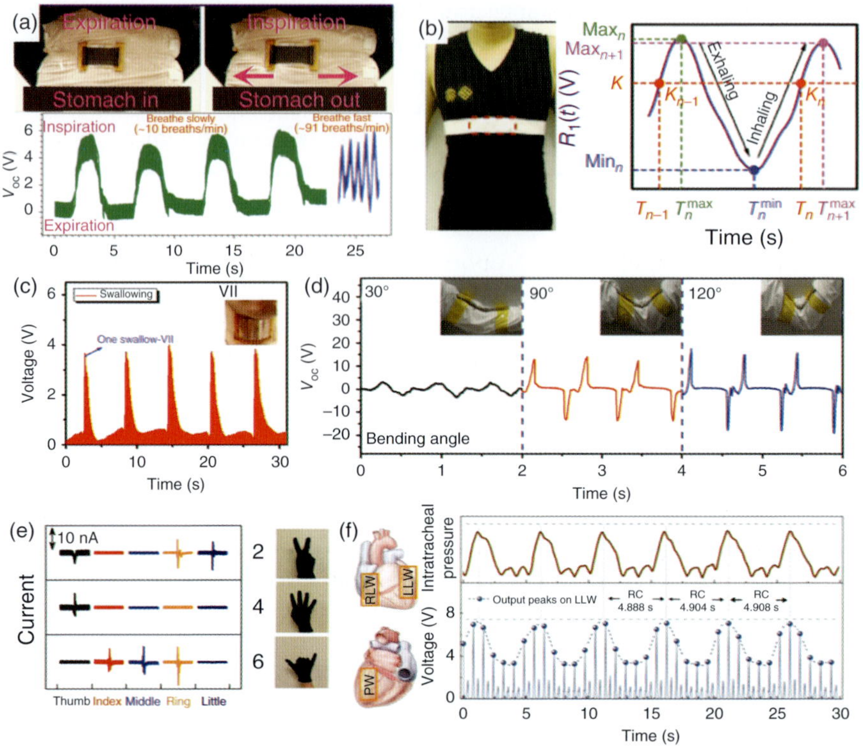

Figure 13.7 FTENGs as self-powered body motion sensors. (a) A stretchable-rubber-based FTENG for body motion sensing. The plot shows the voltage responses to the diaphragmatic breathing. *Source:* Yi et al. 2015 [84]. Reproduced with permission from John Wiley & Sons. (b) A textile-based FTENG to monitor human respiratory information. The plot shows the processed respiratory signal with datum line and other parameters marked for the respiratory data analysis. *Source:* Zhao et al. 2016 [85]. Reproduced with permission from John Wiley & Sons. (c) An FTENG with serpentine-patterned electrodes to detect human motion. The plot shows the voltage response to swallowing motion. *Source:* Yang et al. 2015 [86]. Reproduced with permission from John Wiley & Sons. (d) A hydrogel-based FTENG to sense joint motion. The plot shows the voltage responses when bending the elbow at different angles. *Source:* Xu et al. 2017 [88]. Reproduced with permission from John Wiley & Sons. (e) A single-thread-based FTENG to monitor body motion. The plot shows the current responses to finger gestures. *Source:* Lai et al. 2017 [89]. Reproduced with permission from John Wiley & Sons. (f) An implantable FTENG to sense organ motion. The plot shows the output voltage when the FTENG was anchored over the left lateral wall. *Source:* Ma et al. 2016 [90]. Reproduced with permission from American Chemical Society.

contact/separation mode, which generates a V_{oc} of ~10 V and I_{sc} of ~4 μA triggered by heartbeat and breathing (Figure 13.7f) [90]. This FTENG can monitor the heart and respiration rate, detect life-threatening arrhythmia, and estimate the pressure and velocity of blood flow with the assistance of an arterial pressure catheter.

Generally, the reported FTENG body motion sensors are capable of detecting the frequency of body motion with excellent accuracy and repeatability; but are required to be further developed to obtain a quantitative sensing of the range of body motion such as the distance and direction of a moving joint, with good accuracy and high stability. Moreover, since human body has a curved surface and human motion induces a certain degree of deformation, an FTENG body motion sensor with stretchability is more desirable. Light weight, comfort of wearing, and low cost are also some desirable features that need to be considered when designing an FTENG body motion sensor.

13.4.2.2 Sensing Motion of Objects

When FTENGs are applied to detect motion of objects, they are often designed with arrays of electrodes or assembled into sensor arrays. The reported FTENGs with single device structure to detect motion of objects are mainly based on two kinds of working modes: the freestanding sliding mode and single-electrode sliding mode. For the FTENG object motion sensors working in the freestanding sliding mode, electrical signals will be generated when the mover undergoes relative reciprocation to the stator. From these electrical signals, the displacement, velocity, and acceleration of the object can be acquired. More sets of electrodes are needed to sense the moving direction of the object. Jing et al. reported an FTENG object motion sensor based on the freestanding sliding mode, which generates a voltage of ± 5 V and a peak power density of ≥ 65 mW m^{-2} at a speed of 0.3 m s^{-1} (Figure 13.8a–d) [93]. The FTENG can detect velocity with errors less than 1% and measure accelerations from 0.1 to 0.5 m s^{-2}. For the FTENG object motion sensors working in the single-electrode sliding mode, as an object moving across an arbitrary electrode, an electrical signal with the peak corresponding to the center of the electrode will be generated [95–97]. The trajectory, displacement, velocity, and acceleration of the object can be obtained from the collected electrical signals from all the electrodes. The reported FTENG object motion sensors working in the attached-electrode mode are often assembled into sensor arrays, with each device generating a peak electrical signal as the object moves across. Liang et al. reported a fast soluble, recyclable, and green FTENG, which can generate a power density of 3.8 mW m^{-2} (Figure 13.8e,f) [97]. This FTENG can chemically and physically disappear when triggered by water, and be reproduced with no waste released to the environment. The FTENG can detect vibration and be applied to sense the displacement and velocity of a moving object inside a tube.

13.4.3 Self-Powered Acoustic Sensors

Acoustic waves are a kind of longitudinal waves that have the same direction of vibration as their direction of travel, which can be described by quantities including sound pressure, particle velocity, particle displacement, and sound intensity. The detection of acoustic waves is of great importance for information technology and security. The reported FTENG acoustic sensors work in the single-electrode or attached-electrode contact/separation mode. Yang et al. fabricated a single-electrode-mode FTENG with the structure mimicking the human eardrum, which consists of a thin layer of PET in an oval shape as the bottom supporting substrate, a layer of nylon coated with ITO, and a layer of PTFE tympanic membrane tented outward at the level of the tip of an umbo

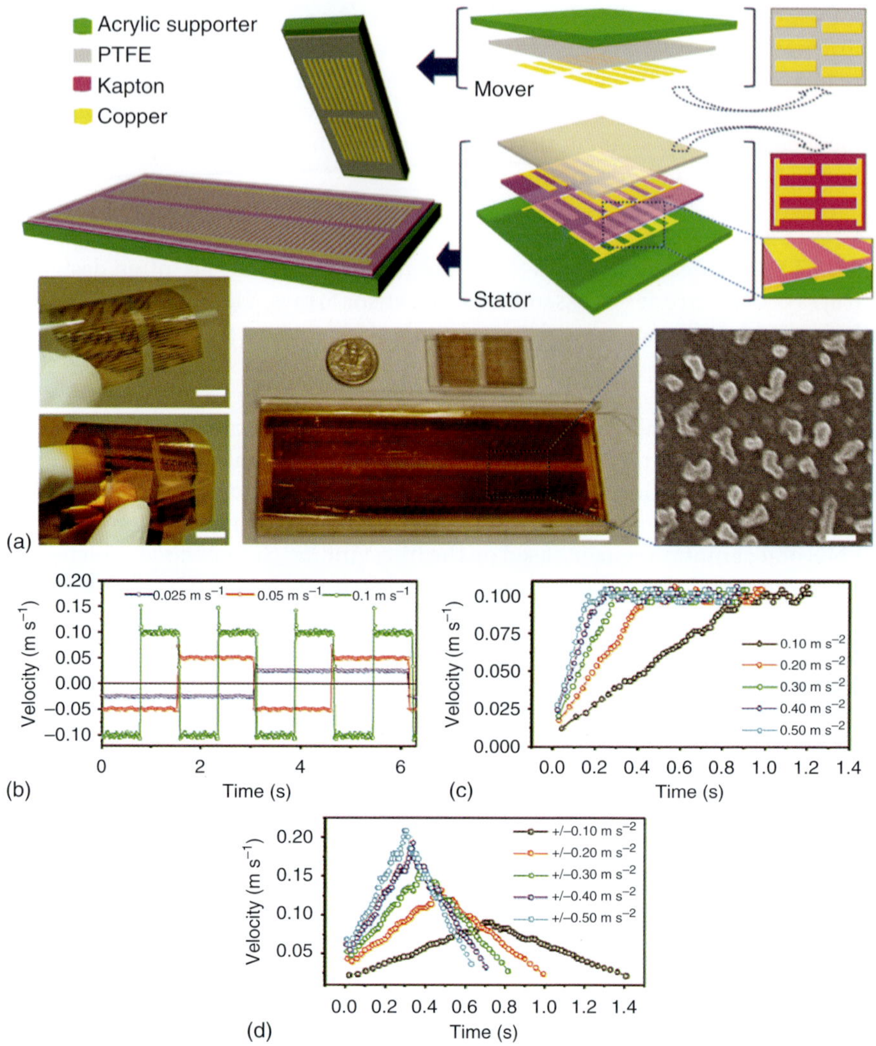

Figure 13.8 FTENGs as self-powered object motion sensors. An FTENG working in the freestanding sliding mode to detect object motion: (a) schematic diagram and photographs of the device; (b) velocity measurement on uniform velocity-reciprocating motion; (c) velocity measurement on uniform acceleration motion; and (d) velocity measurement on uniform acceleration and deceleration motion. *Source:* (a–d) Jing et al. 2015 [93]. Reproduced with permission from Nature Publishing Group. A recyclable FTENG to detect object motion inside a tube: (e) schematic diagram of the device structure and (f) output current response to a moving object inside a tube. *Source:* (e, f) Liang et al. 2016 [97]. Reproduced with permission from John Wiley & Sons.

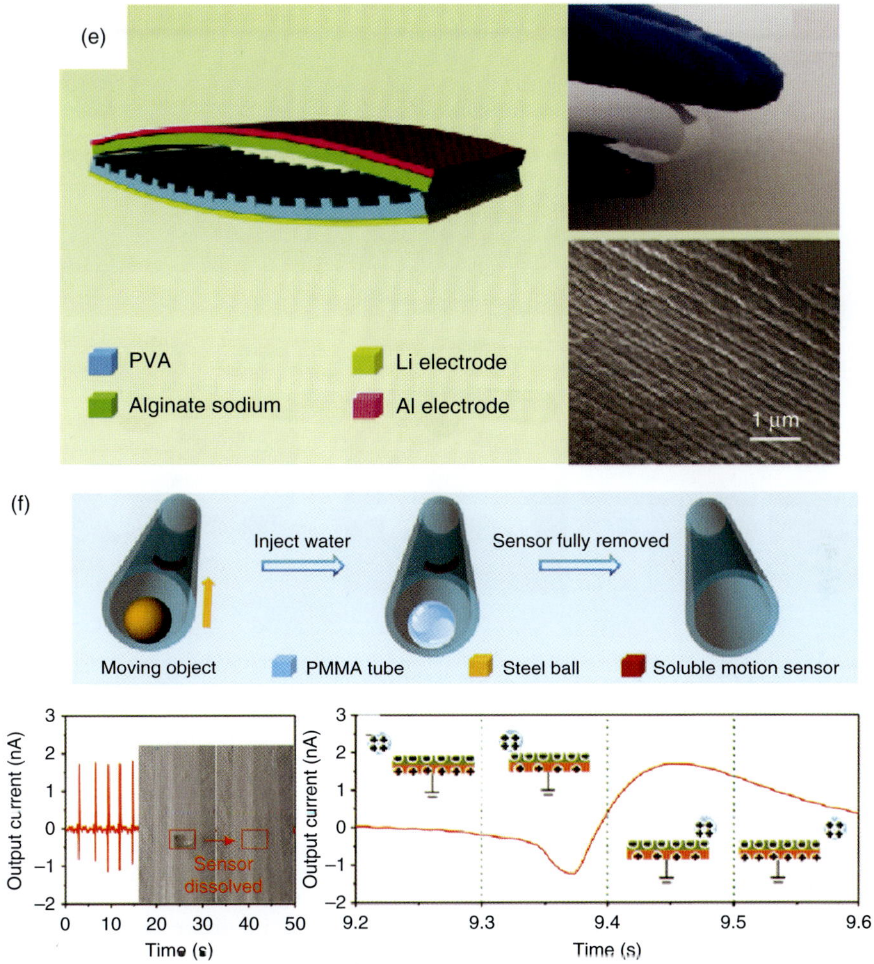

Figure 13.8 (*Continued*)

(Figure 13.9a–c) [98]. The PET acts as the umbo with its two ends tightly attached to the PTFE membrane and the nylon layer. The acoustic waves induce the contact/separation between the PTFE and nylon. This FTENG acoustic sensor has a sensitivity of 51 mV Pa^{-1}, a fast response time less than 6 ms, a frequency range from 0.1 to 3.2 kHz and a pressure detection limit down to 2.5 Pa. The FTENG can pick up and recover human throat voice even in an extremely noisy or windy environment. Fan et al. fabricated an FTENG composed of a layer of Cu-coated paper with multiholes acting as the structural backbone and a layer of PTFE membrane coated with Cu, which works in the attached-electrode contact/separation mode (Figure 13.9d,e) [99]. The acoustic waves trigger the vibration of the PTFE membrane, resulting in contact/separation between the PTFE and the Cu film on the paper. The introduction of holes on the paper can minimize the negative impact caused by air damping but reduce the effective contact area; therefore, the structural parameters of the holes including shape,

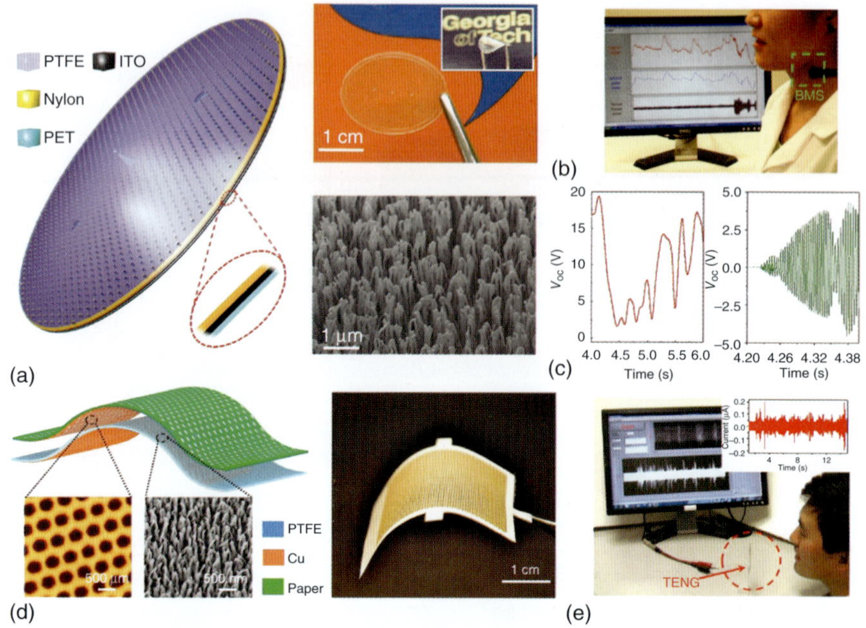

Figure 13.9 FTENGs as self-powered acoustic sensors. A single-electrode-mode FTENG to detect acoustic waves: (a) schematic diagram and photograph of the device; (b) photograph shows the FTENG worn on the neck as a self-powered throat microphone; (c) plots showing the low-frequency component (left) and high-frequency component (right) of the output signals responding to the concrete dynamic change during speaking. *Source:* (a–c) Yang et al. 2015 [98]. Reproduced with permission from John Wiley & Sons. An attached-electrode-mode FTENG to detect acoustic waves: (d) schematic diagram and photograph of the device and (e) photograph showing the FTENG as a self-powered microphone for sound recording. *Source:* (d, e) Fan et al. 2015 [99]. Reproduced with permission from American Chemical Society.

dimensions, and distributions need to be optimized in order to achieve an optimized output for the FTENG. The optimized FTENG delivers a power density of 121 $mW\,m^{-2}$ under a sound pressure of 117 dB_{SPL}. The FTENG is capable of acting as a microphone for sound recording.

13.4.4 Self-Powered Liquid/Gas Flow Sensors

The sensing of liquid/gas flow is important for applications in environmental monitoring and modern analytical science. Ma et al. reported a self-recovering FTENG that is composed of an Al foil inside a PTFE bag, a spring adhered to the PTFE and a polyethylene (polyvinyl chloride, PVC) air inlet (Figure 13.10a–c) [100]. The spring endows the FTENG with self-recovering property. The FTENG works in the single-electrode contact/separation mode, which generates a voltage of 251 V, a current of 56 µA, and a power of 3.1 mW under a helium gas flow rate of 7.5 $l\,min^{-1}$. The FTENG can be applied to detect airflow rate, and other signals such as humidity and force. Chen et al. reported an FTENG composed of a layer of PTFE and a layer of Cu film, which works in the single-electrode sliding mode (Figure 13.10d–f) [101]. The FTENG can detect the flow of both water and air.

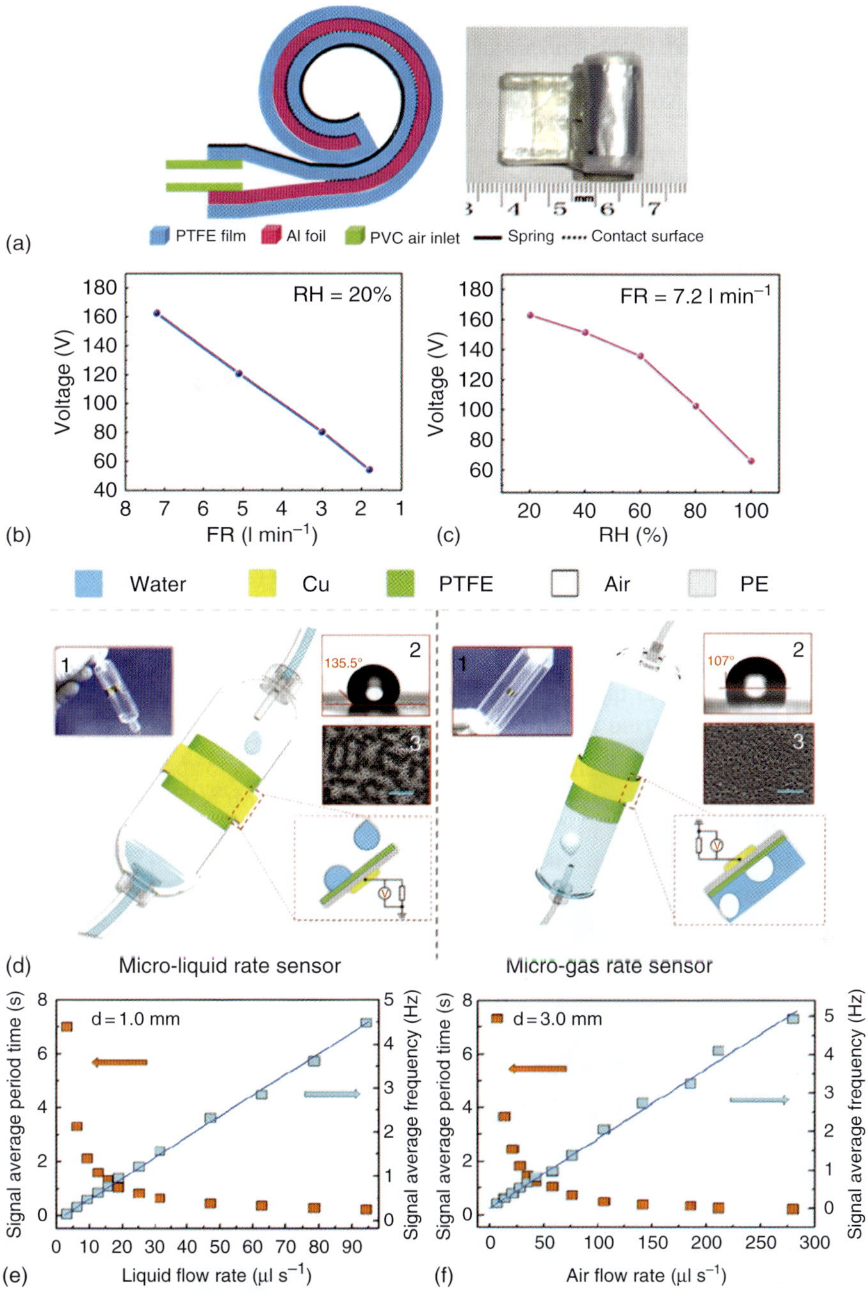

Figure 13.10 FTENGs as self-powered liquid/gas flow sensors. A self-recovering FTENG to detect air flow: (a) schematic diagram and photograph of the device; (b) the relationship between the output voltage and the airflow rate at a certain relative humidity and (c) the influence of humidity on the output voltage at a certain airflow rate. *Source:* (a–c) Ma et al. 2015 [100]. Reproduced with permission from John Wiley & Sons. An FTENG working in the single-electrode sliding mode to detect the flow of both water and air: (d) schematic diagram and photograph of the device; (e) dependence of the signal average period time/frequency on liquid flow rate at 1.0 mm capillarity and (f) dependence of the signal average period time/frequency on gas flow rate at 3.0 mm capillarity. *Source:* (d–f) Chen et al. 2016 [101]. Reproduced with permission from American Chemical Society.

13.5.1 Self-Charging over a Period of Time to Power Electronics

When the energy provided by the FTENG is lower than the energy consumed by the electronic device, the energy storage part of the SCPU needs to be charged by the FTENG for a certain period of time [47, 104–112]. Yi et al. reported a stretchable, waterproof SCPU by seamlessly integrating a stretchable TENG with stretchable supercapacitors, which can be subject to and harvest energy from almost all kinds of large-degree deformation (Figure 13.12a) [47]. The SCPU is made of soft materials and fabricated by casting silicone rubber into molds. The stretchable TENG can generate a V_{oc} of ~400 V and a short-circuit transferred charge density of ~97 μC m^{-2}. The supercapacitors can be charged by the stretchable TENG harvesting human motion and then power an electronic watch. Luo et al. reported a transparent SCPU integrating an FTENG and supercapacitors (Figure 13.12b) [104]. The FTENG can generate a V_{oc} of 31.5 V and I_{sc} of 224 nA by harvesting finger motion. The SCPU can drive a calculator after mechanical energy harvesting and storing. Jung et al. combined FTENGs working in the attached-electrode mode with supercapacitors to form a fabric-based SCPU (Figure 13.12c) [107]. The FTENG delivers a power density of ~0.18 μW cm^{-2} at 1.5 Hz. By mounting the two triboelectric parts of the FTENG in the armpit region and the chest region respectively, energy harvested from human activities such as stretching, walking, running, and sprinting can be charged into the supercapacitors. The SCPU is capable of providing current to a pressure sensor. Hwang et al. reported a transparent stretchable self-powered patchable sensor system by combining an FTENG, a supercapacitor, and a strain sensor, with the stretchable TENG and supercapacitor forming an SCPU to power the strain sensor (Figure 13.12d) [109]. The supercapacitor charged by the TENG provides power for the strain sensor to detect human activities that can induce strains. Seung et al. fabricated a textile-based FTENG for SCPUs (Figure 13.12e) [110]. The FTENG can produce an output voltage of ~120 V and current of ~65 μA. When integrating the FTENG with a capacitor, the capacitor charged by the FTENG can power a liquid crystal display and a keyless vehicle entry system.

13.5.2 Sustainably Powering Electronics

When energy scavenged by the FTENG from ambient environment is higher than the energy consumed by the electronic device, a sustainable operation of the electronic device can be achieved by an SCPU [46, 113–116]. Wang et al. reported a flexible SCPU by combining an FTENG with a Li ion battery (Figure 13.13a) [114]. The SCPU working in the sustainable mode can provide a constant voltage with a direct current of 2 μA for over 40 h, and is capable of continuously powering the operation of an ultraviolet sensor with the FTENG harvesting mechanical energy. Niu et al. reported an SCPU by integrating an FTENG, a capacitor, and a power management circuit (Figure 13.13b) [115]. The power management circuit largely enhances the energy storage efficiency by solving the problem of impedance mismatch between the TENG and capacitor. The 15-layer FTENG can generate a V_{oc} of ~700 V and Q_{sc} of ~2.2 μC. The SCPU can deliver a continuous

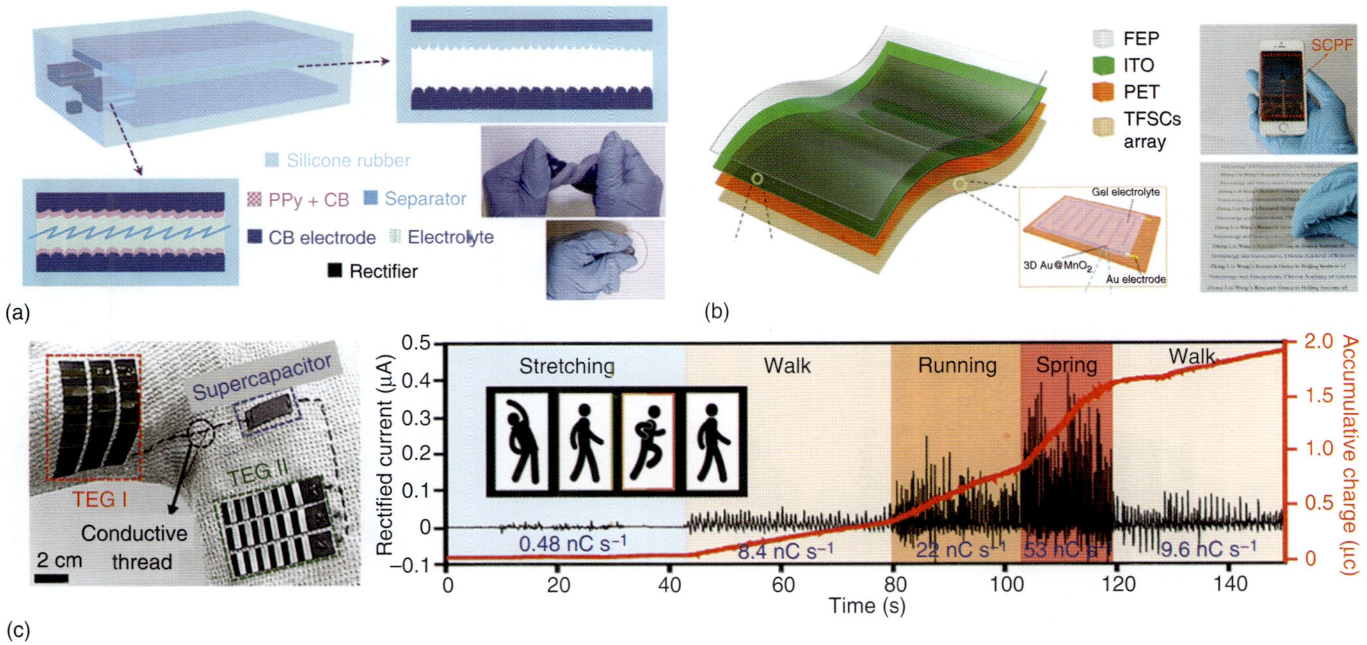

Figure 13.12 FTENGs for SCPUs with self-charging over a period of time to power electronics. (a) A stretchable, waterproof SCPU integrating a stretchable TENG and stretchable supercapacitors to drive an electronic watch. *Source:* Yi et al. 2016 [47]. Reproduced with permission from American Chemical Society. (b) A transparent SCPU integrating an FTENG and supercapacitors to drive a calculator. *Source:* Luo et al. 2016. [104]. American Chemical Society. (c) A fabric-based SCPU integrating FTENGs with supercapacitors to drive a pressure sensor. The plot shows various charging rates of the supercapacitor by the FTENG harvesting energy from different human activities. *Source:* Jung et al. 2014 [107]. Reproduced with permission from John Wiley & Sons. (d) A transparent stretchable sensor system composed of an FTENG, a supercapacitor, and a strain sensor to detect strain. *Source:* Hwang et al. 2015 [109]. Reproduced with permission from American Chemical Society. (e) A textile-based SCPU combining an FTENG and a capacitor to power a liquid crystal display and a remote control. *Source:* Seung et al. 2015 [110]. Reproduced with permission from American Chemical Society.

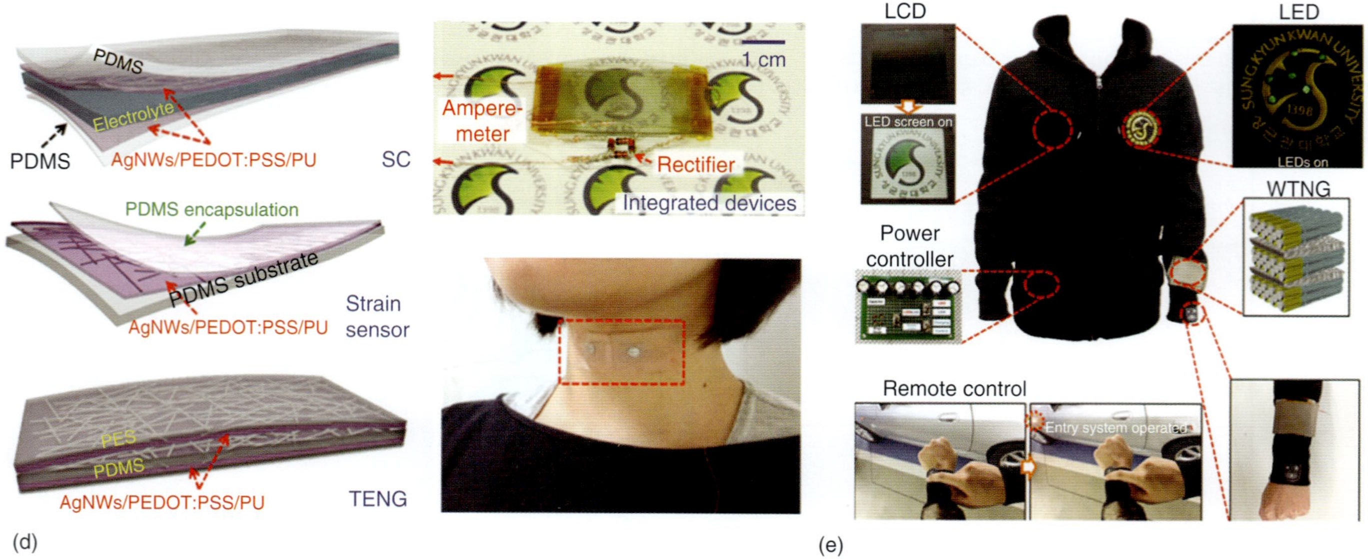

Figure 13.12 (*Continued*)

direct-current electricity of 1.044 mW with the FTENG converting energy from palm tapping, which can sustainably power numerous electronics such as a calculator and a wireless radio-frequency communication system. Yi et al. reported a type of highly shape-adaptive, stretchable TENG based on conductive liquid (Figure 13.13c–f) [46]. With conductive liquid contained in a polymer cover, the FTENG can harvest energy in various working modes and maintain its performance under a strain as large as 300%. The FTENG can deliver a power density of ~200 $mW\,m^{-2}$ at 3 Hz. By integrating two bracelet-like FTENGs with a capacitor and a power management circuit to form an SCPU, the electricity generated by the FTENGs from tapping motion can sustainably power an electronic watch. Wang et al. reported a tube-like stretchable TENG with a rationally designed helix-belt contact structure, working in the attached contact/separation mode (Figure 13.13g,h) [116]. The FTENG can scavenge energy from various deformation and generate a short-circuit transferred charge density of as high as 250 $\mu C\,m^{-2}$. When combined in an SCPU with a Li ion battery, electronic devices such as an electronic watch and a fitness tracker can be sustainably powered with the energy harvested by 40 FTENGs assembled under shoes from walking and jogging.

13.6 Flexible Triboelectric Nanogenerators for Hybrid Energy Cells

Hybrid energy cells aim to achieve maximum harvesting of the existing energy in the ambient environment. The FTENGs can be combined with other mechanical energy harvesters such as PiENGs to improve the total outputs and energy conversion efficiency, or can be integrated with other forms of energy harvesters such as solar cells, pyroelectric nanogenerators (PyENGs), and thermoelectric nanogenerators to simultaneously and individually scavenge available energy such as solar light, heat, and mechanical motion in the ambient environment [117–124]. When combined with an energy harvester that also produces alternative current, the outputs of the FTENG and this energy harvester must be rectified before connected together. The hybrid energy cell can also be combined in an SCPU by integrating with energy storage devices, which can directly or even sustainably power electronic devices.

Wang et al. fabricated a transparent, flexible hybrid energy cell composed of an FTENG, a PiENG and a PyENG (Figure 13.14a,b) [119]. A composite film of poly(vinylidene fluoride) (PVDF) nanowires and PDMS serves as the triboelectric layer, a polarized PVDF film serves as both the piezoelectric and pyroelectric layer, and two layers of ITO film serve as the electrodes. The hybrid energy cell shows much better charging performance for a 10 μF capacitor compared with the TENG-PiENG or the PyENG. Chen et al. reported a microcable power textile by integrating a solar cell with an FTENG to simultaneously harvest energy from ambient sunlight and mechanical movement (Figure 13.14c,d) [120]. The solar cells were fabricated into microcables from lightweight polymer fibers, which were then woven with fiber-based FTENGs to form a smart fabric.

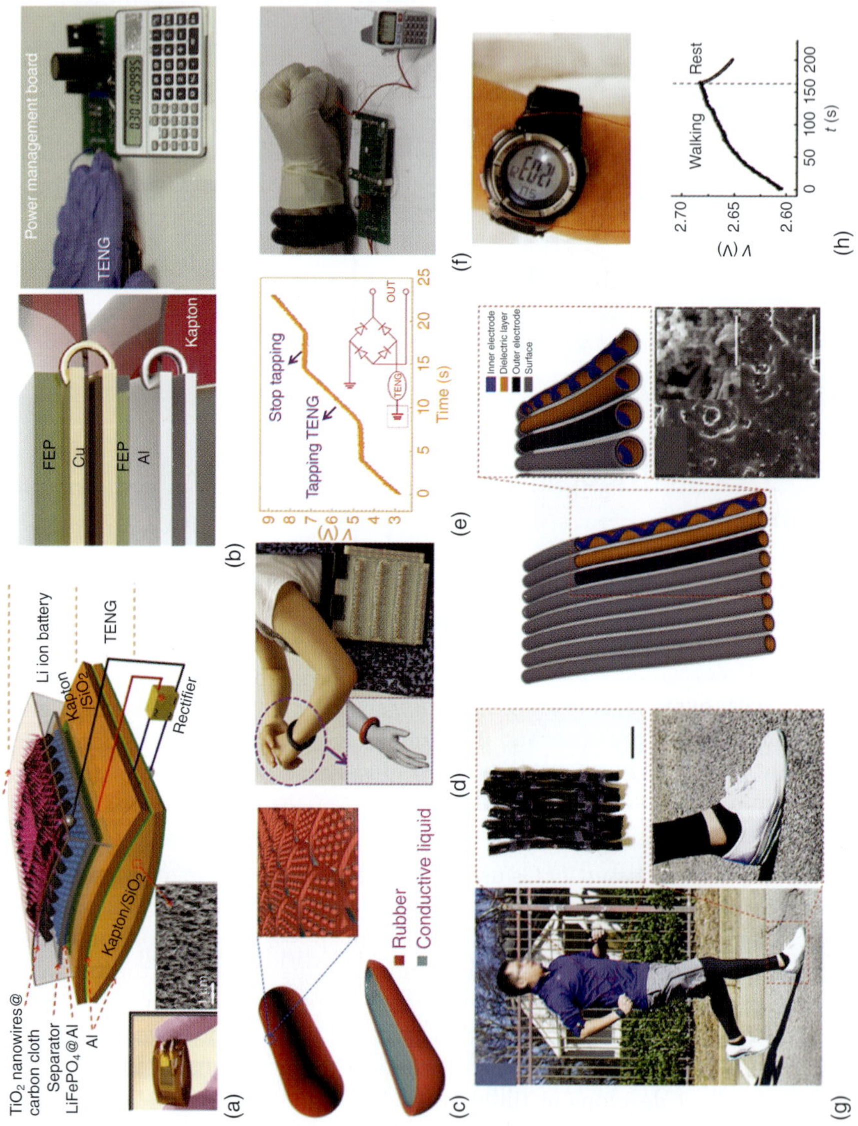
(a)
TiO2 nanowires@carbon cloth
Separator
LiFePO4@Al
Al
Li ion battery
Kapton/SiO2
TENG
Rectifier
(b)
FEP
Cu
FEP
Al
Kapton
TENG
Power management board
(c)
Rubber
Conductive liquid
(d)
(e)
Stop tapping
Tapping TENG
V (V)
Time (s)
TENG
OUT
(f)
Inner electrode
Dielectric layer
Outer electrode
Surface
(g)
(h)
Walking
Rest
V (V)
t (s)

Figure 13.13 FTENGs for SCPUs to sustainably power electronics. (a) A flexible SCPU combining an FTENG with a Li ion battery to sustainably drive an ultraviolet sensor. *Source:* Wang et al. 2013 [114]. Reproduced with permission from American Chemical Society. (b) An SCPU integrating an FTENG, a capacitor, and a power management circuit to sustainably power numerous electronics with the energy harvested from palm tapping. *Source:* Niu et al. 2015 [115]. Reproduced with permission from Nature Publishing Group [115]. A shape-adaptive FTENG combined in an SCPU by integrating with a capacitor and a power management circuit: (c) schematic illustration showing the basic structure of the FTENG; (d) photograph showing a bracelet-like FTENG worn on the wrist; (e) charging a capacitor by an FTENG harvesting tapping motion; (f) demonstration of the SCPU to sustainably power an electronic watch with the FTENG scavenging energy from tapping motion. *Source:* (c–e) Yi et al. 2016 [46]. Reproduced with permission from Science Advances. A tube-like FTENG combined in an SCPU with a battery: (g) photographs and schematic illustration of the FTENG; and (h) demonstration of the SCPU to sustainably power an electronic watch with energy harvested by the FTENG during walking. *Source:* (g, h) Wang et al. 2016 [116]. Reproduced with permission from Nature Publishing Group.

The hybrid energy cell can power an electronic watch, charge a cell phone, and drive water splitting reactions with the energy harvested from sunshine as well as mechanical motion such as human activities and wind blowing. Wen et al. reported a textile-based SCPU by combining a hybrid energy cell with supercapacitors [121]. The hybrid energy cell consists of a fiber-shaped dye-sensitized solar cell and a fiber-shaped FTENG, which can simultaneously scavenge energy from outdoor sunlight and human motion. The fiber-shaped hybrid energy cell and fiber-shaped supercapacitor can be woven into electronic textile to fabricate smart clothes and then power wearable electronic devices with the energy harvested from sunlight and human motion.

13.7 Service Behavior of Triboelectric Nanogenerators

As a new type of multifunctional device, considerable attention was focused on the improvement of conversion efficiency of the TENGs [110–121]. In view of the possible complex application environment, it is essential to explore the service behavior of functional nanodevices [125, 126]. But so far, few reports have focused on the service behavior of TENGs. Actually, addressing the problem of performance deterioration can help enhance their safety, stability, and reliability. The service behavior of TENG was investigated by Zhang group, and the performance and adaptation of TENG were improved [94, 127].

A fast-soluble, recyclable, and green TENG was fabricated, where all the components are water-soluble or can be dissolved by water-triggered cascade reaction [94]. Moreover, it can be reproduced with the generated solution with no waste released to the environment. The developed recyclable and green TENG exhibited excellent performance of scavenging mechanical energy from the environment. In addition, a multifunctional tailorable textile-based TENG (T-TENG) was developed (Figure 13.15a–d) [127]. The fabricated T-TENG was applied as a self-powered human body motion sensor and water energy harvester, and further formed an all textile-based flexible self-charging system by

13.8 Summary and Prospects

With the advantages of light weight, easy fabrication, high output power density and energy conversion efficiency, low cost, wide selection of materials, and various working modes, FTENGs have wide applications in fields such as biomechanics, robotics, security, and medical care. The single-electrode-mode FTENG has the simplest fabrication but suffers from relatively lower outputs due to the electrostatic shield effect. The attached-electrode-mode FTENG is able to prevent environmental contamination but its performance may be impacted by device deformation due to the change of maximum separation gap. For the material design, all the materials made of an FTENG should be flexible. A selection of two materials with great difference in the ability to attract electrons is preferable so as to obtain high surface triboelectric charge density and thereby high performance. The fabrication of FTENGs with stretchable ability faces more challenges than that without stretchability, in terms of the stretchable electrode and packaging. PDMS and Ecoflex silicone rubbers are two common dielectric materials for FTENGs with stretchability. Ecoflex silicone rubbers are favored for producing higher electrical outputs since they are softer and can be more negatively charged; whereas PDMS is preferable for the fabrication of transparent stretchable TENGs due to its higher transparency. The increasing resistance of the stretchable electrode with stretchment usually has little impact on the FTENG's performance due to the high impedance of TENG.

FTENGs have been applied to effectively harvest various kinds of mechanical energy such as biomechanical energy, wind energy, water energy, and acoustic energy, as well as to serve as self-powered sensors to detect diverse stimuli such as body and organ motion, object motion, acoustic waves, liquid/gas flow, liquid volume, and chemicals. Although here the FTENGs are classified according to applications, most FTENGs have multiple functions, which can not only harvest several types of mechanical energy but can also be able to detect several kinds of stimuli. Compared with other kinds of flexible energy harvesters, the FTENG has advantages such as simple fabrication, high output voltage, and high energy conversion efficiency and also disadvantages such as low output current. Compared with other types of sensors, the FTENG has advantages such as self-powered property, fast response/recover time, and high sensitivity; but they also exhibit disadvantages such as inability to sense static status and vulnerability to environmental conditions.

For the purpose of powering electronics, FTENGs usually need to be integrated with energy storage devices to form an SCPU that can provide electronics with direct-current electricity supply. With sustained efforts made to increase FTENG's performance and improve the energy storage efficiency, SCPUs based on FTENGs manage to sustainably drive conventional electronics such as electronic watches, calculators, thermometers, remote controls, and pedometers with FTENG solely harvesting ambient energy. For maximum harvesting of the existing energy in the ambient environment, the FTENG can be combined in a hybrid energy cell with other energy harvesters to enhance the total output and energy conversion efficiency, or to simultaneously and individually scavenge the available energy such as solar light, heat, and mechanical motion.

Alternative output electrical signals should be rectified before connected together and a proper integration design may be required in order to reduce power loss.

To enhance the performance of the FTENG, traditional approaches to increase the contact area by creating nanostructured surfaces and to enlarge the surface triboelectric charge density by surface modification can be applied. Since the surface triboelectric charge density is subject to air breakdown limit, new approaches could be developed to increase the breakdown limit of the surface triboelectric charge, which may be achieved by changing the air medium between the two triboelectric surfaces into vacuum or other kinds of medium that can tolerate higher field strength before electrostatic discharge. But these new approaches may face big challenges with respect to the fabrication process, packaging technology, and cost. Currently, the most promising applications for FTENGs are still revolving around the wearable, portable, and stretchable electronics or sensor systems. To serve as self-powered body or organ motion sensors, further efforts may be needed for FTENGs to acquire a quantitative sensing of the range of body motions such as the distance and direction of a moving joint, with good accuracy and high stability. Sustainably powering personal electronics with daily and normal human activities is of great interest for the market, and future work along this direction could involve further improving the FTENG's performance, developing specially designed multiunit device structures to obtain optimized outputs within a limited volume or weight, and enhancing the total energy storage efficiency of SCPUs based on FTENGs.

References

1 Kraemer, D., Poudel, B., Feng, H.P. et al. (2011). High-performance flat-panel solar thermoelectric generators with high thermal concentration. *Nat. Mater.* 10: 532–538.

2 Vining, C.B. (2009). An inconvenient truth about thermoelectrics. *Nat. Mater.* 8: 83–85.

3 Yang, R., Qin, Y., Dai, L., and Wang, Z.L. (2009). Power generation with laterally packaged piezoelectric fine wires. *Nat. Nanotechnol.* 4: 34–39.

4 Donelan, J.M., Li, Q., Naing, V. et al. (2008). Biomechanical energy harvesting: generating electricity during walking with minimal user effort. *Science* 319: 807–810.

5 Mei, A., Li, X., Liu, L. et al. (2014). A hole-conductor-free, fully printable mesoscopic perovskite solar cell with high stability. *Science* 345: 295–298.

6 Fan, F.-R., Tian, Z.-Q., and Wang, Z.L. (2012). Flexible triboelectric generator. *Nano Energy* 1: 328–334.

7 Zhu, G., Zhou, Y.S., Bai, P. et al. (2014). A shape-adaptive thin-film-based approach for 50% high-efficiency energy generation through micro-grating sliding electrification. *Adv. Mater.* 26: 3788–3796.

8 Wang, S., Lin, L., Xie, Y. et al. (2013). Sliding-triboelectric nanogenerators based on in-plane charge-separation mechanism. *Nano Lett.* 13: 2226–2233.

9 Liang, Q., Zhanga, Z., Yan, X. et al. (2015). Functional triboelectric generator as self-powered vibration sensor with contact mode and non-contact mode. *Nano Energy* 14: 209–216.
10 Xie, Y., Wang, S., Niu, S. et al. (2014). Grating-structured freestanding triboelectric-layer nanogenerator for harvesting mechanical energy at 85% total conversion efficiency. *Adv. Mater.* 26: 6599–6607.
11 Lin, Z.H., Cheng, G., Lee, S. et al. (2014). Harvesting water drop energy by a sequential contact-electrification and electrostatic-induction process. *Adv. Mater.* 26: 4690–4696.
12 Zheng, L., Cheng, G., Chen, J. et al. (2015). A hybridized power panel to simultaneously generate electricity from sunlight, raindrops, and wind around the clock. *Adv. Energy Mater.* 5: 1501152.
13 Wang, Z.L., Chen, J., and Lin, L. (2015). Progress in triboelectric nanogenerators as a new energy technology and self-powered sensors. *Energy Environ. Sci.* 8: 2250–2282.
14 Cheng, G., Lin, Z.H., Lin, L. et al. (2013). Pulsed nanogenerator with huge instantaneous output power density. *ACS Nano* 7: 7383–7391.
15 Zi, Y., Niu, S., Wang, J. et al. (2015). Standards and figure-of-merits for quantifying the performance of triboelectric nanogenerators. *Nat. Commun.* 6: 8376.
16 Wang, X., Niu, S., Yin, Y. et al. (2015). Triboelectric nanogenerator based on fully enclosed rolling spherical structure for harvesting low-frequency water wave energy. *Adv. Energy Mater.* 5: 1501467.
17 Wang, Z.L. (2014). Triboelectric nanogenerators as new energy technology and self-powered sensors – principles, problems and perspectives. *Faraday Discuss.* 176: 447–458.
18 Wang, S., Lin, L., and Wang, Z.L. (2015). Triboelectric nanogenerators as self-powered active sensors. *Nano Energy* 11: 436–462.
19 Zi, Y., Lin, L., Wang, J. et al. (2015). Triboelectric–pyroelectric–piezoelectric hybrid cell for high-efficiency energy-harvesting and self-powered sensing. *Adv. Mater.* 27: 2340–2347.
20 Park, S.I., Brenner, D.S., Shin, G. et al. (2015). Soft, stretchable, fully implantable miniaturized optoelectronic systems for wireless optogenetics. *Nat. Biotechnol.* 33: 1280–1286.
21 Sekitani, T., Nakajima, H., Maeda, H. et al. (2009). Stretchable active-matrix organic light-emitting diode display using printable elastic conductors. *Nat. Mater.* 8: 494–499.
22 Webb, R.C., Bonifas, A.P., Behnaz, A. et al. (2013). Ultrathin conformal devices for precise and continuous thermal characterization of human skin. *Nat. Mater.* 12: 938–944.
23 Tee, B.C., Wang, C., Allen, R., and Bao, Z. (2012). An electrically and mechanically self-healing composite with pressure- and flexion-sensitive properties for electronic skin applications. *Nat. Nanotechnol.* 7: 825–832.
24 Kaltenbrunner, M., Sekitani, T., Reeder, J. et al. (2013). An ultra-lightweight design for imperceptible plastic electronics. *Nature* 499: 458–463.
25 Tee, B.C., Chortos, A., Berndt, A. et al. (2015). A skin-inspired organic digital mechanoreceptor. *Science* 350: 313–316.

26 Wang, Z.L. (2013). Triboelectric nanogenerators as new energy technology for self-powered systems and as active mechanical and chemical sensors. *ACS Nano* 7: 9533–9557.

27 Zhu, G., Lin, Z.H., Jing, Q. et al. (2013). Toward large-scale energy harvesting by a nanoparticle-enhanced triboelectric nanogenerator. *Nano Lett.* 13: 847–853.

28 Zhu, G., Chen, J., Liu, Y. et al. (2013). Linear-grating triboelectric generator based on sliding electrification. *Nano Lett.* 13: 2282–2289.

29 Yang, Y., Zhang, H., Lin, Z.H. et al. (2013). Human skin based triboelectric nanogenerators for harvesting biomechanical energy and as self-powered active tactile sensor system. *ACS Nano* 7: 9213–9222.

30 Yang, Y., Zhang, H., Chen, J. et al. (2013). Single-electrode-based sliding triboelectric nanogenerator for self-powered displacement vector sensor system. *ACS Nano* 7: 7342–7351.

31 Wang, S., Niu, S., Yang, J. et al. (2014). Quantitative measurements of vibration amplitude using a contact-mode freestanding triboelectric nanogenerator. *ACS Nano* 8: 12004–12013.

32 Wang, S., Xie, Y., Niu, S. et al. (2014). Freestanding triboelectric-layer-based nanogenerators for harvesting energy from a moving object or human motion in contact and non-contact modes. *Adv. Mater.* 26: 2818–2824.

33 Vasandani, P., Mao, Z.-H., Jia, W., and Sun, M. (2016). Design of simulation experiments to predict triboelectric generator output using structural parameters. *Simul. Modell. Pract. Theory* 68: 95–107.

34 Yang, B., Zeng, W., Peng, Z.-H. et al. (2016). A fully verified theoretical analysis of contact-mode triboelectric nanogenerators as a wearable power source. *Adv. Energy Mater.* 6: 1600505.

35 Niu, S., Liu, Y., Wang, S. et al. (2014). Theoretical investigation and structural optimization of single-electrode triboelectric nanogenerators. *Adv. Funct. Mater.* 24: 3332–3340.

36 Niu, S., Wang, S., Lin, L. et al. (2013). Theoretical study of contact-mode triboelectric nanogenerators as an effective power source. *Energy Environ. Sci.* 6: 3576.

37 Jiang, T., Chen, X., Han, C.B. et al. (2015). Theoretical study of rotary freestanding triboelectric nanogenerators. *Adv. Funct. Mater.* 25: 2928–2938.

38 Jiang, T., Chen, X., Yang, K. et al. (2016). Theoretical study on rotary-sliding disk triboelectric nanogenerators in contact and non-contact modes. *Nano Res.* 9: 1057–1070.

39 Niu, S. and Wang, Z.L. (2015). Theoretical systems of triboelectric nanogenerators. *Nano Energy* 14: 161–192.

40 Niu, S., Liu, Y., Chen, X. et al. (2015). Theory of freestanding triboelectric-layer-based nanogenerators. *Nano Energy* 12: 760–774.

41 Niu, S., Liu, Y., Wang, S. et al. (2013). Theory of sliding-mode triboelectric nanogenerators. *Adv. Mater.* 25: 6184–6193.

42 Hu, W., Wu, W., and Zhou, H.M. (2016). Wind-blown sand electrification inspired triboelectric energy harvesting based on homogeneous inorganic materials contact: a theoretical study and prediction. *Sci. Rep.* 6: 19912.

43 Kim, S., Gupta, M.K., Lee, K.Y. et al. (2014). Nanogenerators: transparent flexible graphene triboelectric nanogenerators. *Adv. Mater.* 26: 3778–3778.

44 Chandrashekar, B.N., Deng, B., Smitha, A.S. et al. (2015). Roll-to-roll green transfer of CVD graphene onto plastic for a transparent and flexible triboelectric nanogenerator. *Adv. Mater.* 27: 5210–5216.

45 Wu, C., Wang, X., Lin, L. et al. (2016). Paper-based triboelectric nanogenerators made of stretchable interlocking Kirigami patterns. *ACS Nano* 10: 4652–4659.

46 Yi, F., Wang, X., Niu, S. et al. (2016). A highly shape-adaptive, stretchable design based on conductive liquid for energy harvesting and self-powered biomechanical monitoring. *Sci. Adv.* 2: e1501624.

47 Yi, F., Wang, J., Wang, X. et al. (2016). Stretchable and waterproof self-charging power system for harvesting energy from diverse deformation and powering wearable electronics. *ACS Nano* 10: 6519–6525.

48 Jeong, C.K., Baek, K.M., Niu, S. et al. (2014). Topographically-designed triboelectric nanogenerator via block copolymer self-assembly. *Nano Lett.* 14: 7031–7038.

49 Wang, S., Xie, Y., Niu, S. et al. (2014). Maximum surface charge density for triboelectric nanogenerators achieved by ionized-air injection: methodology and theoretical understanding. *Adv. Mater.* 26: 6720–6728.

50 Wang, S., Zi, Y., Zhou, Y.S. et al. (2016). Molecular surface functionalization to enhance the power output of triboelectric nanogenerators. *J. Mater. Chem. A* 4: 3728–3734.

51 Bai, P., Zhu, G., Lin, Z.H. et al. (2013). Integrated multilayered triboelectric nanogenerator for harvesting biomechanical energy from human motions. *ACS Nano* 7: 3713–3719.

52 Chandrasekhar, A., Alluri, N.R., Saravanakumar, B. et al. (2016). Human interactive triboelectric nanogenerator as a self-powered Smart seat. *ACS Appl. Mater. Interfaces* 8: 9692–9699.

53 Zhou, T., Zhang, C., Han, C.B. et al. (2014). Woven structured triboelectric nanogenerator for wearable devices. *ACS Appl. Mater. Interfaces* 6: 14695–14701.

54 Kim, K.N., Chun, J., Kim, J.W. et al. (2015). Highly stretchable 2D fabrics for wearable triboelectric nanogenerator under harsh environments. *ACS Nano* 9: 6394–6400.

55 Meng, X.S., Wang, Z.L., and Zhu, G. (2016). Triboelectric-potential-regulated charge transport through p–n junctions for area-scalable conversion of mechanical energy. *Adv. Mater.* 28: 668–676.

56 Zhang, H., Yang, Y., Hou, T.-C. et al. (2013). Triboelectric nanogenerator built inside clothes for self-powered glucose biosensors. *Nano Energy* 2: 1019–1024.

57 He, X., Zi, Y., Guo, H. et al. (2016). A highly stretchable fiber-based triboelectric nanogenerator for self-powered wearable electronics. *Adv. Funct. Mater.* 1604378.

58 Hou, T.-C., Yang, Y., Zhang, H. et al. (2013). Triboelectric nanogenerator built inside shoe insole for harvesting walking energy. *Nano Energy* 2: 856–862.

59 Lee, S., Ko, W., Oh, Y. et al. (2015). Triboelectric energy harvester based on wearable textile platforms employing various surface morphologies. *Nano Energy* 12: 410–418.

60 Pu, X., Li, L., Song, H. et al. (2015). A self-charging power unit by integration of a textile triboelectric nanogenerator and a flexible lithium-ion battery for wearable electronics. *Adv. Mater.* 27: 2472–2478.

61 Zhu, G., Bai, P., Chen, J., and Wang, Z.L. (2013). Power-generating shoe insole based on triboelectric nanogenerators for self-powered consumer electronics. *Nano Energy* 2: 688–692.

62 Guo, H., Leng, Q., He, X. et al. (2015). A triboelectric generator based on checker-like interdigital electrodes with a sandwiched PET thin film for harvesting sliding energy in all directions. *Adv. Energy Mater.* 5: 1400790.

63 Lai, Y.C., Deng, J., Niu, S. et al. (2016). Electric eel-skin-inspired mechanically durable and super-stretchable nanogenerator for deformable power source and fully autonomous conformable electronic-skin applications. *Adv. Mater.* 28: 10024–10032.

64 Meng, B., Tang, W., Too, Z.-H. et al. (2013). A transparent single-friction-surface triboelectric generator and self-powered touch sensor. *Energy Environ. Sci.* 6: 3235.

65 Li, S., Peng, W., Wang, J. et al. (2016). All-elastomer-based triboelectric nanogenerator as a keyboard cover to harvest typing energy. *ACS Nano* 10: 7973–7981.

66 Leng, Q., Guo, H., He, X. et al. (2014). Flexible interdigital-electrodes-based triboelectric generators for harvesting sliding and rotating mechanical energy. *J. Mater. Chem. A* 2: 19427–19434.

67 Dhakar, L., Pitchappa, P., Tay, F.E.H., and Lee, C. (2016). An intelligent skin based self-powered finger motion sensor integrated with triboelectric nanogenerator. *Nano Energy* 19: 532–540.

68 Meng, B., Tang, W., Zhang, X. et al. (2013). Self-powered flexible printed circuit board with integrated triboelectric generator. *Nano Energy* 2: 1101–1106.

69 Zheng, Q., Shi, B., Fan, F. et al. (2014). In vivo powering of pacemaker by breathing-driven implanted triboelectric nanogenerator. *Adv. Mater.* 26: 5851–5856.

70 Zheng, Q., Zou, Y., Zhang, Y. et al. (2016). Biodegradable triboelectric nanogenerator as a life-time designed implantable power source. *Sci. Adv.* 2: e1501478.

71 Zheng, Q., Zhang, H., Shi, B. et al. (2016). In vivo self-powered wireless cardiac monitoring via implantable triboelectric nanogenerator. *ACS Nano* 10: 6510–6518.

72 Zhang, L., Zhang, B., Chen, J. et al. (2016). Lawn structured triboelectric nanogenerators for scavenging sweeping wind energy on rooftops. *Adv. Mater.* 28: 1650–1656.

73 Zhao, Z., Pu, X., Du, C. et al. (2016). Freestanding flag-type triboelectric nanogenerator for harvesting high-altitude wind energy from arbitrary directions. *ACS Nano* 10: 1780–1787.

74 Su, Y., Wen, X., Zhu, G. et al. (2014). Hybrid triboelectric nanogenerator for harvesting water wave energy and as a self-powered distress signal emitter. *Nano Energy* 9: 186–195.
75 Zhao, X.J., Zhu, G., Fan, Y.J. et al. (2015). Triboelectric charging at the nanostructured solid/liquid interface for area-scalable wave energy conversion and its use in corrosion protection. *ACS Nano* 9: 7671–7677.
76 Kim, Y., Na, J., Park, C. et al. (2015). PEDOT as a flexible organic electrode for a thin film acoustic energy harvester. *ACS Appl. Mater. Interfaces* 7: 16279–16286.
77 Fan, F.R., Lin, L., Zhu, G. et al. (2012). Transparent triboelectric nanogenerators and self-powered pressure sensors based on micropatterned plastic films. *Nano Lett.* 12: 3109–3114.
78 Zhu, G., Yang, W.Q., Zhang, T. et al. (2014). Self-powered, ultrasensitive, flexible tactile sensors based on contact electrification. *Nano Lett.* 14: 3208–3213.
79 Chen, J., Zhu, G., Yang, J. et al. (2015). Personalized keystroke dynamics for self-powered human–machine interfacing. *ACS Nano* 9: 105–116.
80 Chen, S.W., Cao, X., Wang, N. et al. (2017). An ultrathin flexible single-electrode triboelectric-nanogenerator for mechanical energy harvesting and instantaneous force sensing. *Adv. Energy Mater.* 7: 1601255.
81 Zhong, J., Zhu, H., Zhong, Q. et al. (2015). Self-powered human-interactive transparent nanopaper systems. *ACS Nano* 9: 7399–7406.
82 Jiang, X.Z., Sun, Y.J., Fan, Z., and Zhang, T.Y. (2016). Integrated flexible, waterproof, transparent, and self-powered tactile sensing panel. *ACS Nano* 10: 7696–7704.
83 Lin, L., Xie, Y., Wang, S. et al. (2013). Triboelectric active sensor array for self-powered static and dynamic pressure detection and tactile imaging. *ACS Nano* 7: 8266–8274.
84 Yi, F., Lin, L., Niu, S. et al. (2015). Stretchable-rubber-based triboelectric nanogenerator and its application as self-powered body motion sensors. *Adv. Funct. Mater.* 25: 3688–3696.
85 Zhao, Z., Yan, C., Liu, Z. et al. (2016). Machine-washable textile triboelectric nanogenerators for effective human respiratory monitoring through loom weaving of metallic yarns. *Adv. Mater.* 28: 10267–10274.
86 Yang, P.K., Lin, L., Yi, F. et al. (2015). A flexible, stretchable and shape-adaptive approach for versatile energy conversion and self-powered biomedical monitoring. *Adv. Mater.* 27: 3817–3824.
87 Song, W., Gan, B., Jiang, T. et al. (2016). Nanopillar arrayed triboelectric nanogenerator as a self-powered sensitive sensor for a sleep monitoring system. *ACS Nano* 10: 8097–8103.
88 Xu, W., Huang, L.-B., Wong, M.-C. et al. (2017). Environmentally friendly hydrogel-based triboelectric nanogenerators for versatile energy harvesting and self-powered sensors. *Adv. Energy Mater.* 7: 1601529.
89 Lai, Y.-C., Deng, J., Zhang, S.L. et al. (2017). Single-thread-based wearable and highly stretchable triboelectric nanogenerators and their applications in cloth-based self-powered human-interactive and biomedical sensing. *Adv. Funct. Mater.* 27: 1604462.

90 Ma, Y., Zheng, Q., Liu, Y. et al. (2016). Self-powered, one-stop, and multifunctional implantable triboelectric active sensor for real-time biomedical monitoring. *Nano Lett.* 16: 6042–6051.

91 Yang, W., Chen, J., Wen, X. et al. (2014). Triboelectrification based motion sensor for human-machine interfacing. *ACS Appl. Mater. Interfaces* 6: 7479–7484.

92 Cai, F., Yi, C., Liu, S. et al. (2016). Ultrasensitive, passive and wearable sensors for monitoring human muscle motion and physiological signals. *Biosens. Bioelectron.* 77: 907–913.

93 Jing, Q., Xie, Y., Zhu, G. et al. (2015). Self-powered thin-film motion vector sensor. *Nat. Commun.* 6: 8031.

94 Yi, F., Lin, L., Niu, S. et al. (2014). Self-powered trajectory, velocity, and acceleration tracking of a moving object/body using a triboelectric sensor. *Adv. Funct. Mater.* 24: 7488–7494.

95 Cheng, X., Song, Y., Han, M. et al. (2016). A flexible large-area triboelectric generator by low-cost roll-to-roll process for location-based monitoring. *Sensors Actuators A Phys.* 247: 206–214.

96 Su, Y., Zhu, G., Yang, W. et al. (2014). Triboelectric sensor for self-powered tracking of object motion inside tubing. *ACS Nano* 8: 3843–3850.

97 Liang, Q., Zhang, Q., Yan, X. et al. (2016). Recyclable and green triboelectric nanogenerator. *Adv. Mater.* doi: 10.1002/adma.201604961.

98 Yang, J., Chen, J., Su, Y. et al. (2015). Eardrum-inspired active sensors for self-powered cardiovascular system characterization and throat-attached anti-interference voice recognition. *Adv. Mater.* 27: 1316–1326.

99 Fan, X., Chen, J., Yang, J. et al. (2015). Ultrathin, rollable, paper-based triboelectric nanogenerator for acoustic energy harvesting and self-powered sound recording. *ACS Nano* 9: 4236–4243.

100 Ma, M., Liao, Q., Zhang, G. et al. (2015). Self-recovering triboelectric nanogenerator as active multifunctional sensors. *Adv. Funct. Mater.* 25: 6489–6494.

101 Chen, J., Guo, H., Zheng, J. et al. (2016). Self-powered triboelectric micro liquid/gas flow sensor for microfluidics. *ACS Nano* 10: 8104–8112.

102 Wang, H., Xiang, Z., Giorgia, P. et al. (2016). Triboelectric liquid volume sensor for self-powered lab-on-chip applications. *Nano Energy* 23: 80–88.

103 Xue, X., Fu, Y., Wang, Q. et al. (2016). Outputting olfactory bionic electric impulse by PANI/PTFE/PANI sandwich nanostructures and their application as flexible, smelling electronic skin. *Adv. Funct. Mater.* 26: 3128–3138.

104 Luo, J., Tang, W., Fan, F.R. et al. (2016). Transparent and flexible self-charging power film and its application in a sliding unlock system in touchpad technology. *ACS Nano* 10: 8078–8086.

105 Zhong, J., Zhang, Y., Zhong, Q. et al. (2014). Fiber-based generator for wearable electronics and mobile medication. *ACS Nano* 8: 6273–6280.

106 Wang, J., Li, X., Zi, Y. et al. (2015). A flexible fiber-based supercapacitor-triboelectric-nanogenerator power system for wearable electronics. *Adv. Mater.* 27: 4830–4836.

107 Jung, S., Lee, J., Hyeon, T. et al. (2014). Fabric-based integrated energy devices for wearable activity monitors. *Adv. Mater.* 26: 6329–6334.

108 Shi, B., Zheng, Q., Jiang, W. et al. (2016). A packaged self-powered system with universal connectors based on hybridized nanogenerators. *Adv. Mater.* 28: 846–852.

109 Hwang, B.U., Lee, J.H., Trung, T.Q. et al. (2015). Transparent stretchable self-powered patchable sensor platform with ultrasensitive recognition of human activities. *ACS Nano* 9: 8801–8810.

110 Seung, W., Gupta, M.K., Lee, K.Y. et al. (2015). Nanopatterned textile-based wearable triboelectric nanogenerator. *ACS Nano* 9: 3501–3509.

111 Jung, S., Hong, S., Kim, J. et al. (2015). Wearable fall detector using integrated sensors and energy devices. *Sci. Rep.* 5: 17081.

112 Pu, X., Li, L., Liu, M. et al. (2016). Wearable self-charging power textile based on flexible yarn supercapacitors and fabric nanogenerators. *Adv. Mater.* 28: 98–105.

113 Guo, H., Yeh, M.H., Lai, Y.C. et al. (2016). All-in-one shape-adaptive self-charging power package for wearable electronics. *ACS Nano* 10: 10580–10588.

114 Wang, S., Lin, Z.H., Niu, S. et al. (2013). Motion charged battery as sustainable flexible-power-unit. *ACS Nano* 7: 11263–11271.

115 Niu, S., Wang, X., Yi, F. et al. (2015). A universal self-charging system driven by random biomechanical energy for sustainable operation of mobile electronics. *Nat. Commun.* 6: 8975.

116 Wang, J., Li, S., Yi, F. et al. (2016). Sustainably powering wearable electronics solely by biomechanical energy. *Nat. Commun.* 7: 12744.

117 Li, X., Lin, Z.H., Cheng, G. et al. (2014). 3D fiber-based hybrid nanogenerator for energy harvesting and as a self-powered pressure sensor. *ACS Nano* 8: 10674–10681.

118 Wang, G., Xi, Y., Xuan, H. et al. (2015). Hybrid nanogenerators based on triboelectrification of a dielectric composite made of lead-free $ZnSnO_3$ nanocubes. *Nano Energy* 18: 28–36.

119 Wang, S., Wang, Z.L., and Yang, Y. (2016). A one-structure-based hybridized nanogenerator for scavenging mechanical and thermal energies by triboelectric–piezoelectric–pyroelectric effects. *Adv. Mater.* 28: 2881–2887.

120 Chen, J., Huang, Y., Zhang, N. et al. (2016). Micro-cable structured textile for simultaneously harvesting solar and mechanical energy. *Nat. Energy* 1: 16138.

121 Wen, Z., Yeh, M.H., Guo, H. et al. (2016). Self-powered textile for wearable electronics by hybridizing fiber-shaped nanogenerators, solar cells, and supercapacitors. *Sci. Adv.* 2: e1600097.

122 Zhang, H., Zhang, S., Yao, G. et al. (2015). Simultaneously harvesting thermal and mechanical energies based on flexible hybrid nanogenerator for self-powered cathodic protection. *ACS Appl. Mater. Interfaces* 7: 28142–28147.

123 Fang, Y., Tong, J., Zhong, Q. et al. (2015). Solution processed flexible hybrid cell for concurrently scavenging solar and mechanical energies. *Nano Energy* 16: 301–309.

124 Pu, X., Song, W., Liu, M. et al. (2016). Wearable power-textiles by integrating fabric triboelectric nanogenerators and fiber-shaped dye-sensitized solar cells. *Adv. Energy Mater.* 6: 1601048.
125 Zhang, Y., Yan, X.Q., Yang, Y. et al. (2012). Scanning probe study on the piezotronic effect in ZnO nanomaterials and nanodevices. *Adv. Mater.* 24: 4647.
126 Zhang, Y., Yang, Y., Gu, Y.S. et al. (2015). Performance and service behavior in 1-D nanostructured energy conversion devices. *Nano Energy* 14: 30.
127 Zhang, Q., Liang, Q., Liao, Q. et al. (2017). Service behavior of multifunctional triboelectric nanogenerators. *Adv. Mater.* 29: 1606703.

autonomous and flexible wearable technology among other applications such as Internet of Things (IoTs), military, industrial, consumer, and nonconsumer fields is expected to become a highly researched field [18]. Given the growth in wearables and flexible electronics, energy autonomy using flexible devices is of great interest today. Thermoelectric energy harvesting may be able to provide an interesting alternative toward harnessing easily available body heat and is especially suited to wearable sensors and electronics.

14.2 Thermoelectric Energy Conversion Basics

The thermoelectric voltage V generated in a thermoelectric material with Seebeck coefficient S whose ends are maintained at a temperature difference ΔT ($T_H - T_C$) follows the relation (see Figure 14.1):

$$V = -S(T_H - T_C) = -S\Delta T \tag{14.1}$$

where V is measured in volts, ΔT in K, and S has the units of $\mu V\ K^{-1}$. T_H and T_C are the temperature of the hot and cold sides, respectively. Conversely, a potential difference applied across a thermoelectric material would lead to a temperature difference across it, leading to operation in refrigeration mode. In a thermoelectric device, which could be either a generator or a cooler, a number of TE n-type and p-type leg pairs are used to make a thermoelectric module such that each leg is connected electrically in series and thermally in parallel (refer to Figure 14.1).

A higher Seebeck coefficient would lead to a higher generated voltage as can be seen from Equation (14.1). The Seebeck coefficient of common TE materials such as Bi_2Te_3 and its alloys (discussed in detail later) is on the order of 100–200 $\mu V\ K^{-1}$. Power generated by a thermoelectric material, however, depends collectively on its intrinsic material properties, namely electrical and thermal conductivity along with the Seebeck coefficient. These properties are related to each other by a quantity called the *figure of merit* ZT (see Equation (14.2)). A good thermoelectric material should have a higher electrical conductivity and a lower thermal conductivity. The figure of merit ZT is a measure of

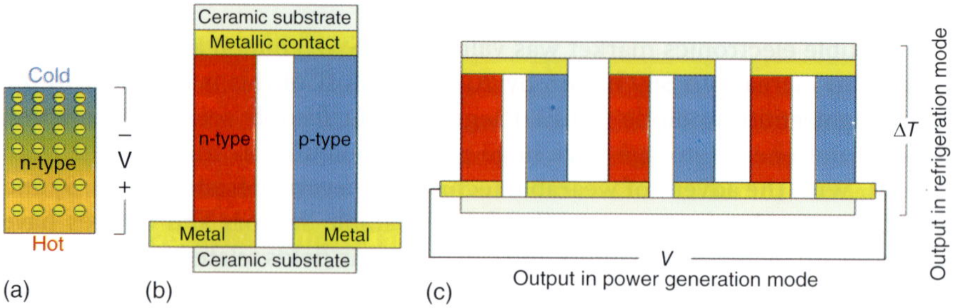

Figure 14.1 Schematic cross-sections of an n-type thermoelectric leg (a), a thermoelectric couple (b), and a thermoelectric module (c) comprising individual thermocouples connected electrically in series and thermally in parallel. With application of temperature difference (potential) the output is a voltage (temperature difference) in the power generation (refrigeration) mode.

the conversion efficiency of a thermoelectric material and higher ZT signifies higher efficiency [1, 2].

$$\mathrm{ZT} = \frac{S^2\sigma}{\kappa}T \tag{14.2}$$

where σ is the electrical conductivity, κ is the thermal conductivity, and T is the absolute temperature. The term $S^2\sigma$ in the numerator is known as the power factor and is a measure of the electrical power produced by a thermoelectric material [19] for a given temperature difference across the material ΔT. Each of the thermoelectric material properties S, σ, κ is temperature-dependent leading to different ZT values at different temperatures. ZT and the efficiency of a thermoelectric material are related by the following equation:

$$\eta = \frac{(\sqrt{\mathrm{ZT}+1}-1)\Delta T}{(\sqrt{\mathrm{ZT}+1}+1)T_{\mathrm{H}} - \Delta T} \tag{14.3}$$

Bismuth telluride and alloys, lead chalcogenides, and silicon germanium, are the state-of-the-art thermoelectric materials that are in use [1]. Bismuth telluride (Bi_2Te_3) is one of the best elemental or simple compounds at ordinary (room) temperatures suitable for thermoelectric refrigeration [1, 20]. Addition of either antimony telluride (Sb_2Te_3) or bismuth selenide (Bi_2Se_3) to Bi_2Te_3 by alloying has been found to be effective in improving ZT due to the decrease in thermal conductivity. Thus, ternary systems $Bi_{(2-x)}Sb_xTe_3$ and $Bi_2Te_{3-y}Se_y$ are among the best p- and n-type thermoelectric materials near room temperature [1]. Bismuth telluride alloys have a ZT of ~1 at room temperature [1]. Lead chalcogenides as thermoelectric materials have been used since 1930s [21, 22]. p- and n-type alloys of lead telluride were used in the RTG in NASA's earliest thermoelectric powered spacecraft [22]. Silicon germanium alloys were studied as early as 1954 and today are almost the only thermoelectric alloys suited for operation between 600 and 1000 °C [2, 23, 24]. Silicon Germanium alloys typically have a ZT of 0.5 at ~600 K [24].

The figure of merit, ZT, of these common thermoelectric materials has varied over the past few decades from <0.5 to >2 obtained by nanostructured materials. Figure 14.2 shows how the ZT of these common thermoelectric materials has varied over the past few decades [30]. The ZT of common materials such as bismuth telluride alloys and lead chalcogenides was around 0.5 in the 1950s and reached ZT values of ~1 around the 1970s [30] (see Figure 14.2). ZT remained stagnant to around unity after the mid-1970s. In the absence of better materials, thermoelectric research has slowed down since then [30]. In 1993, Hicks and Dresselhaus calculated ZT for Bi_2Te_3 quantum wells and quantum wires, showing that low-dimensional materials could achieve much higher ZT than their bulk counterparts [30–32]. This sparked researchers to improve the ZT values of thermoelectric materials by preparing nanostructure material from traditional inorganic materials. Improved ZT values have been reported in nanostructured PbTe, SiGe, and Bi_2Te_3 (see Figure 14.2) [30]. Apart from nanostructured materials, complex bulk materials such as skutterudites, clathrates, and zintl phases have been engineered for high efficiencies [33–36]. Most of the current research in thermoelectric energy harvesting focuses on

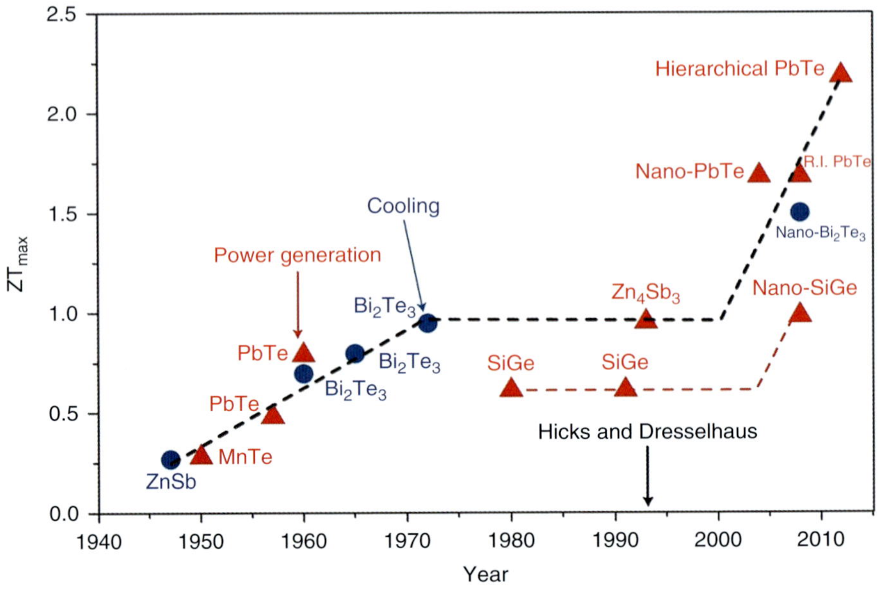

Figure 14.2 Evolution of maximum ZT over time. Blue dots and red triangles represent the cooling and power generation thermoelectric materials, respectively. Bi_2Te_3 and PbTe are the basic constituents of alloys of which ZT is reported. References for Nano SiGe: [25], nano PbTe: [26], RI PbTe (PbTe doped with resonant impurity): [27], nano Bi_2Te_3: [28], and hierarchical PbTe: [29]. *Source:* Figure reprinted from [30].

increasing the ZT and power factors of thermoelectric materials to obtain high power densities through thermoelectric harvesting.

Apart from thermoelectric energy harvesting, vibrations, radio frequency (RF), and photovoltaic energy harvesting are some of the common energy scavenging

Table 14.1 Comparison between different energy scavenging mechanisms.

Harvesting mechanism	Source power	Harvested power
Ambient light		
Indoor	0.1 mW cm^{-2}	10 μW cm^{-2}
Outdoor	100 mW cm^{-2}	10 mW cm^{-2}
Vibration/motion		
Human	0.5 m s^{-2}@1 Hz, 1 m s^{-2}@50 Hz	4 μW cm^{-2}
Industrial	1 m s^{-2}@5 Hz, 10 m s^{-2}@1 kHz	100 μW cm^{-2}
Thermal energy		
Human	20 mW cm^{-2}	30 μW cm^{-2}
Industrial	100 mW cm^{-2}	1–10 mW cm^{-2}
RF		
Cell phone	0.3 μW cm^{-2}	0.1 μW cm^{-2}

Source: Vullers et al. 2009 [37]. Reproduced with permission from Elsevier.

mechanisms which, similarly to thermoelectric energy harvesters are used in both large-scale and micropower harvesting applications. Refer to Table 14.1 for a comparison between different established energy harvesting mechanisms [37]. Among all these energy harvesting mechanisms, thermoelectric energy harvesting can be particularly convenient and useful for wearable devices since human body heat is always available unlike the above mentioned variables.

14.3 Flexible Thermoelectric Materials

The established thermoelectric materials discussed in the previous section are inorganic and brittle. There is growing interest in organic-based thermoelectric materials owing to their inherent properties such as high flexibility, light weight, low cost, and easy processing. Organic-based thermoelectric materials include pure polymer thermoelectric materials, polymer-inorganic TE material hybrids, metal organic polymers along with carbon nanotubes (CNTs), and graphene-based polymer composites as emerging thermoelectric materials [38]. Organic thermoelectric materials are typically based on conducting or pi-conjugated polymers such as polyaniline (PANI), polyacetylene, polypyrrole (PPy), polythiophene, polyphenylene, and poly(3,4-ethylenedioxythiophene) (PEDOT) as well as small molecules such as pentacene, fullerene, tetrathiafulvalene (TTF), and 2,3,5,6-tetrafluoro-7,7,8,8-tetracyanoquinodimethane (F4TCNQ) [39]. Apart from these, high ZT inorganic thermoelectric materials (typically n- and p-type Bi_2Te_3) have been used in the synthesis of inks and pastes for the purpose of screen, inkjet, gravure, roll-to-roll (R2R), and dispenser printing of thermoelectric devices on flexible substrates [19, 40–44]. Some common flexible substrates that have been used in flexible thermoelectric device fabrication are polydimethylsiloxane (PDMS) [45], polyethylene terephthalate (PET), and polyimide (PI) [46] which have also commonly been used in flexible electronics.

Figure 14.3 shows the power factors and ZT values of some emerging thermoelectric materials analyzed through the past decade [38, 47–57]. The power factors of polymer-based thermoelectric materials have varied from less than a hundred to ~600 $\mu W\,m^{-1}\,K^{-2}$ with highest reported ZTs reaching ~0.4, thus far. For a quick comparison, the best inorganic bismuth antimony telluride (BST) alloys have power factor of ~4500 $\mu W\,m^{-1}\,K^{-2}$ with ZT ~1 suitable for low-temperature energy conversion applications [28]. However, these alloys and the processes involved in their fabrication are expensive leading to popularity and preference of comparatively inexpensive polymer-based thermoelectric materials despite their low efficiencies.

As discussed previously, inorganic materials are known to be good candidates for thermoelectric devices. They have high electric conductivity and the highest ZTs in a large temperature range from 300 to 1000 K although their thermal conductivity is typically higher than that of organic materials. Considering the room temperature range, Bi_2Te_3-based materials are most commonly used due to their high thermoelectric performance; for example, nanostructured p-type $Bi_{0.5}Si_{1.5}Te_3$ has a $ZT = 0.8$ at room temperature [28].

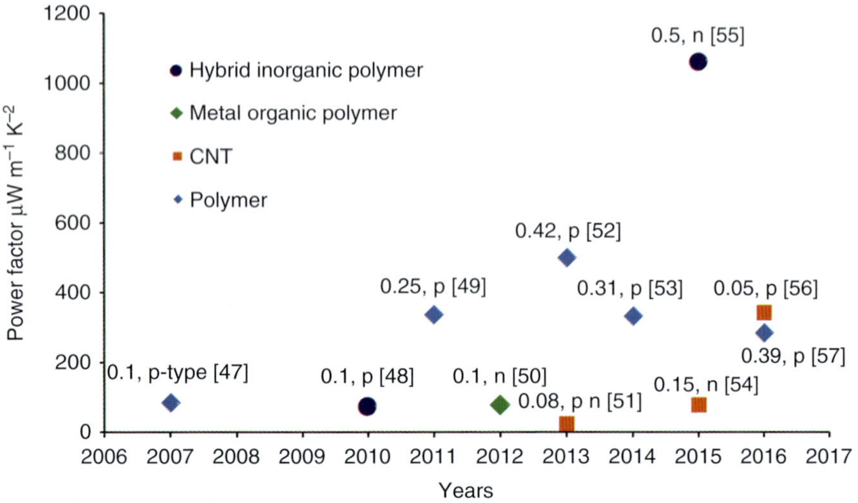

Figure 14.3 Power factors and ZT of emerging polymer and CNT-based thermoelectric materials over the past decades. ZT values are shown above the symbols along with the type of material (n or/and p type) and Refs [38, 47–57].

However, all of these materials are not flexible, difficult to manufacture, and involve high cost. Therefore, researchers have been trying to use different methods to treat the inorganic materials. Kim et al. [58] reported a method that converts the inorganic materials from powder-type to paste-type by mixing them with binders. Later, they use the dispenser-printing method to print them on a flexible polymer substrate. When the ambient temperature is 5 °C, the thermoelectric generator (TEG) made of 20 thermocouples could generate up to 178 nW. Cao et al. [59] mixed $Bi_{1.8}Te_{3.2}$ and Sb_2Te_3 particles with an epoxy binder and used screen printing to print mixture on the substrate. Later, they used the cold isostatic pressing (CIP) technique to lower the resistivity of the thermoelectric materials. The resistivity of Sb_2Te_3 film is lower than the value reported in other reports [15]. The fabricated TEG could generate a maximum power of 40.3 nW at a temperature difference of 20 °C.

Another effective approach to increasing the flexibility as well as maintaining the high conductivity is preparing hybrid nanocomposites by mixing inorganic materials and conducting polymers. Zheng et al. [60] fabricated p-type Sb_2Te_3/PEDOT thermoelectric composites by embedding PEDOT into a Sb_2Te_3 matrix. The thermal conductivity is reduced to 0.14 $Wm^{-1}\,K^{-1}$ compared with Sb_2Te_3 material at the temperature range of 300–523 K. The maximum ZT value of the composites is 1.18 at 523 K, which is almost double of the value of the Sb_2Te_3 compound. Jo et al. mixed p/n-type Bi/Te, multiwalled CNTs and PDMS and then fabricated the composites on the PDMS substrate to prepare TEG devices (S ~150 $\mu V\,K^{-1}$, σ ~6 $S\,cm^{-1}$). It could generate 570 $\mu W\,cm^{-2}$ at a temperature difference of 60 °C. Du et al. [61] prepared Bi_2Te_3 alloy nanosheet/PEDOT:PSS (polystyrene sulfonic acid) composite films by drop-casting and spin-coating technique and achieved power factor of upto 30 μW m^{-1} K^{-2}.

14.3.1 Conducting Polymers

In recent years, researchers have drawn their attention to organic thermoelectric materials (OTEMs) . Compared with traditional inorganic materials, the OTEMs have several unique properties. OTEMs have relatively good average ZT values as well as high flexibility at around room temperature, which makes them potentially good candidates for human body energy harvesting. Also, the thermal conductivities of OTEMs are less by a few orders of magnitude than those of inorganic semiconductors. Besides, most organic materials are solution-processable, which makes them inexpensive and scalable material candidates.

Among organic materials, conjugated polymer-based materials show excellent electrical properties due to their unique π–π conjugated structure. Figure 14.4 shows the molecular structures of some typical organic TE materials. However, since the electrical conductivity of the organic materials is still lower than those of inorganic TE materials, doping of organic materials has been widely explored to enhance their electric conductivities. A good example is PEDOT doped with PSS, which shows good stability and high conductivity. After doping with PSS, PEDOT:PSS will have higher ZT [62], reaching ~0.25 [49]. Presently, it has become a commercially available material used for thermoelectric

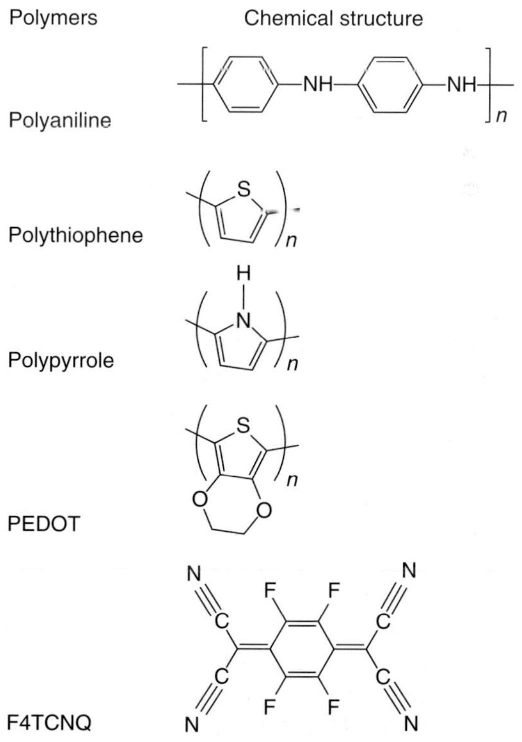

Figure 14.4 Molecular structure of typical organic TE materials.

or photovoltaic devices. Culebras et al. [63] prepared PEDOT doped with bis(trifluoromethylsulfonyl)imide (BTFMSI) and achieved power factor of 150 $\mu W\ m^{-1}\ K^{-2}$ with $S \sim 37\ \mu V\ K^{-1}$ and $\sigma \sim 1100\ S\ cm^{-1}$. Park et al. [64] prepared highly conductive PEDOT films by strictly controlling the oxidation level of the polymer electrochemically during the solution-casting polymerization. A maximum power factor of 1270 $\mu W\ m^{-1}\ K^{-2}$ was obtained. Then the PEDOT film was coated on a PET substrate to fabricate the flexible thermoelectric devices.

There are several other conjugated polymers such as PPy and poly(3-alkylthiophenes) (P3ATs) for thermoelectrics. PPy has attracted great attention recently because it has very low thermal conductivity and good electrical conductivity as well as nontoxicity [65]. Wu et al. [66] prepared different PPy nanotubes and examined the effect of structures and morphologies on the thermoelectric properties and TE performance, achieving power factor $\sim 17\ \mu W\ m^{-1}\ K^{-2}$.

P3AT is another promising TE material. It also has a $\pi-\pi$ conjugated structure between the thiophene rings. Shinohara et al. [67] characterized the effect of the different side chains on the thermoelectric property. The maximum Seebeck coefficient was achieved when the side chain was a dodecyl group, doped with $FeCl_3$. A very large Seebeck coefficient and the maximum ZT value was observed, as the side chain was a hexyl group doped with $FeCl_3$ ($ZT = 2.6 \times 10^{-4}$).

Both chemical and physical properties of organic semiconductors can be improved by the modification of the molecular structure so that both electrical conductivity and Seebeck coefficient can be enhanced due to the dopant materials. Leclerc and coworkers [68] studied thermoelectric poly(2,7-carbazole) derivatives. As a result, $FeCl_3$-doped poly[*N*-9′-heptadecanyl-2,7-carbazole-alt-5,5-(4′,7′-di-2-thienyl-2′,1′,3′-benzothiadiazole)] (PCDTBT) films achieved the highest electrical conductivity of 500 $S\ cm^{-1}$ and the highest power factor was 19 $\mu W\ m^{-1}\ K^{-2}$for PCDTBT.

Schlitz et al. [69] studied an n-type conducting polymer poly{[N,N′-bis(2-octyl-dodecyl)-1,4,5,8-napthalenedicarboximide-2,6-diyl]-alt-5,5′-(2,2′-bithiophene)} (P(NDIOD-T2)) and reported a maximum power factor of 0.6 $\mu W\ m^{-1}\ K^{-2}$ with $S \sim 850\ \mu V\ K^{-1}$ and $\sigma \sim 8 \times 10^{-3}\ S\ cm^{-1}$ when doped with dihydro-1H-ben-zoimidazol-2-yl (N-DMBI). Sun et al. [50] prepared organic thermoelectric devices based on p- and n-type poly(metal 1, 1, 2, 2-ethenetetrathiolate), which have high electric conductivities and reasonable Seebeck coefficients. The n-type material is n-type poly[Nax (Ni-ett)] and n-type poly[Kx (Ni-ett)] and the p-type material is poly[Cux (Cu-ett)]. For n-type material, ZT value of 0.1–0.2 was achieved around 400 K and p-type showed ZT value of 0.01 at the same temperature. Bazan and coworkers [70] studied the effect of counterions and lengths of alky chains on the thermoelectric properties of anionic narrow-bandgap self-doped conjugated polyelectrolytes (CPEs) with π-conjugated cyclopenta-[2,1-b;3,4-b′]-dithiophene-alt-4,7-(2,1,3-benzothiadiazole) backbones. It turns out that the smaller side chain and smaller counterions could increase the doping level of CPEs and therefore increase the Seebeck coefficient. As a result, the CPE-C3-Na achieved a power factor of 0.84 $\mu W\ m^{-1}\ K^{-2}$ with $S \sim 195 \pm 5\ \mu V\ K^{-1}$, $\sigma \sim 0.22 \pm 0.02\ S\ cm^{-1}$. The power factor of a conducting polymer could be increased by increasing the carrier mobility.

Zhu and coworkers [71] reported several polymer semiconductors for TE devices. Two benchmark donor–acceptor (D–A) copolymers poly[(4,4′-bis(2-ethylhexyl)dithieno[3,2-b:2′,3′-d]silole)-2,6-diyl-alt-(2,1,3-benzothiadiazole)-4,7-diyl] (PSBTBT) and poly[{2,5-bis(2-hexyldecyl)-2,3,5,6-tetrahydro-3,6-dioxopyrrolo[3,4-c]pyrrole-1,4-diyl}-alt-{[2,2′:5′,2″-terthiophene]-5,5″-diyl}] (PDPP3T), and the other two donor polymers poly(3-hexylthiophene) (P3HT) and poly(2,5-bis(3-dodecylthiophen-2-yl)thieno[3,2-b]thiophene) (PBTTT-C12) were chemically doped by dipping polymer films into dopant solution. PDPP3T has a maximum power factor of 25 $\mu W\, m^{-1}\, K^{-2}$. Russ et al. [72] synthesized self-doped perylene diimide derivatives with modified side chains as a high-performance n-type organic TE material and achieved a maximum power factor of 1.4 $\mu W\, m^{-1}\, K^{-2}$. Bubnova et al. [73] reported that polymers could be semimetallic, which can have higher Seebeck coefficient and lower thermal conductivity compared with metallic polymers. By creating a bipolaron band, both the electrical conductivity and Seebeck coefficient of PEDOT were enhanced. Recently, Zhang et al. [74] fabricated flexible dual-parameter temperature–pressure sensors based on microstructure-frame-supported organic thermoelectric (MFSOTE) materials. The OTEMs (PEDOT:PSS) were deposited on the deformable microstructure frame polyurethane (PU) to get the active layer. The porous PU layer not only makes the device much more flexible but also adds pressure-sensing features to the organic TE materials. A high Seebeck coefficient of up to 179–200 $\mu V\, K^{-1}$ was achieved. Park et al. [75] came up with a new method to improve the efficiency of the inorganic TE device by using shape-engineerable thermoelectric painting technique (Figure 14.5).

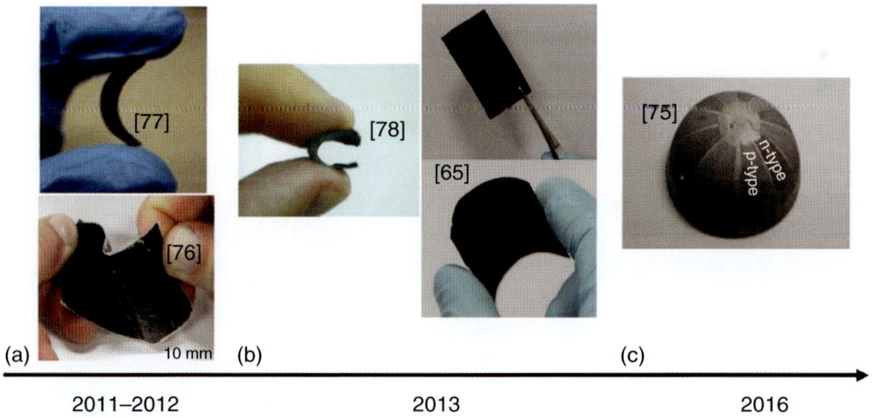

Figure 14.5 Chronological illustration of flexible substrate for TE device (a) from left at bottom 2012 multiple element modules by assembling individual composite films made by multiwalled CNT and PVDF [76]. From left at top 2011 A freestanding flexible nanotube-PEDOT:PSS thin film [77]. (b) 2013 from left: mixed polystyrene and CNTs together and composites printed on a naphthalate film substrate [78]. Right flexible PPy paper-like membrane [65]. (c) 2016 devices directly brush-painted onto curved surfaces by using Bi_2Te_3-based inorganic paints with Sb_2Te_3 chalcogenidometalate as a sintering aid [75].

14.3.2 Graphene and Carbon Nanotube Based TE Materials

Graphene (GN) has attracted a lot of attention due to its excellent physical properties in recent years. It is a two-dimensional (2D) material formed by the carbon atoms with only one atom thick, which makes it exhibit very high electrical conductivity at room temperature because the electrons have high mobility in this $\pi-\pi$ conjugated 2D structure. As mentioned earlier, conducting polymers have lower electric conductivities compared with traditional inorganic materials such as Bi_2Te_3 alloys. Therefore, researchers have studied the thermoelectric properties of conducting polymer–graphene composites. Du et al. [79] reported a method to prepare PANI–graphene composites as a thermoelectric material. The highest power factor (5.6 $\mu W\,m^{-1}\,K^{-2}$) was achieved when the mass percentage of PANI was 50% in the pellet composites. Wang et al. [80] prepared PANI/graphene films with three different types of graphene by a solution-dispersing method. They found that the content of oxygen and structural defects of graphene had a significant effect on the TE performance of the composites. As a result, the composites with lower structural defects and oxygen impurities had a maximum power factor of 19 $\mu W\,m^{-1}\,K^{-2}$. Baitimirova et al. [81] used catalyst the free vapor–solid deposition method to synthesize Bi_2Se_3 nanoplates on graphene substrates. It was found that the non-planar Bi_2Se_3 coating based device was better than planar Bi_2Se_3 coating based device, which generates 1100 μV in a temperature difference of 16 °C.

Using CNTs is another good option for thermoelectric material due to their excellent electrical, thermal, and mechanical properties. Several studies have shown that polymer/CNT composites may be good candidates for further improved OTEMs. Yao et al. [82] prepared SWCNTs (single-walled CNTs)/PANI hybrid nanocomposites (ZT ~0.004) using in situ polymerization. Meng et al. [83] reported a CNT sheet/PANI nanocomposites (S_{max} ~28 $\mu V\,K^{-1}$, maximum power factor ~5 $\mu W\,m^{-1}\,K^{-2}$). Hewitt et al. [76] prepared multiple-element TE modules by assembling individual composite films made of multiwalled CNTs (MWCNTs) and polyvinylidene fluoride (PVDF). Yu and coworkers [84] showed that the thermoelectric properties of CNT–polymer composites could be enhanced by adding PEDOT:PSS as the connecting junctions between CNTs (σ ~400 $S\,cm^{-1}$, S ~25 $\mu V\,K^{-1}$, power factor ~25 $\mu W\,m^{-1}\,K^{-2}$). This electrically conducting PEDOT:PSS significantly enhanced the mobility of electrons in the composites. Later, the same group further optimized the weight ratio of CNT and polymer and increased the TE performance (σ ~1000 $S\,cm^{-1}$, S ~40 $\mu V\,K^{-1}$, power factor ~160 $\mu W\,m^{-1}\,K^{-2}$) with 40 wt% of polymers and 60 wt% of CNTs [84, 85]. Subsequently, they prepared flexible TE materials by functionalizing the surfaces in CNT matrix [85]. Several chemical reductants have been used to treat the CNTs using polyethylenimine (PEI), diethylenetriamine (DETA), and $NaBH_4$. A maximum Seebeck coefficient of 86 $\mu V\,K^{-1}$ and σ ~5200 $S\,cm^{-1}$ for n-type composites were obtained while the weight ratio of PEI to DETA was 67 to 33. Bounioux et al. [86] prepared P3HT: SWCNT composites and P3HT:MWCNT composites with various content of P3HT to enhance the thermoelectric performance of the conducting polymer. For P3HT:SWCNT composites, a maximum power factor of 95 ± 12 μW m^{-1} K^{-2} was obtained when the SWCNT was 42–81 wt%. However, the P3HT:MWCNT could only achieve a maximum power factor of 25 ± 6 μW m^{-1} K^{-2}. The main reason was that SWCNTs had a higher electrical conductivity (>10^3 S cm^{-1}) than the MWCNTs. Screen printing is another technique that could be used to prepare flexible CNT/Polymer TE generators. Suemori et al. [78] mixed polystyrene and CNTs together and then printed the composites on a naphthalate film substrate (σ ~2.1 S cm^{-1}, S ~57 μV K^{-1}).

14.4 Flexible Thermoelectric Energy Harvesters

Studies have shown that heat flow from the skin under indoor sedentary conditions is typically 1–10 mW cm^{-2} and a higher heat flow 10–20 mW cm^{-2} is available at the wrist where the radial artery is located [19, 87, 88]. Temperature of the human skin varies according to the location and ambient conditions with highest temperatures on skin locations at the forehead progressively dropping toward the feet: the average forehead temperature being 31.7 °C in cold (15 °C) and 37 °C in hot (47 °C) ambient conditions [89].

Some of the early research in connection with making use of the heat available from the human body to power wearable devices was carried out by IMEC and Holst Center at Belgium and Netherlands, respectively [19, 87, 90–94] Some of the devices which they first fabricated from 2004 to 2008 can be seen in Figure 14.6. They showed the capability of wearable TEGs to produce power levels of hundreds of microwatts from human body heat [90–94]. One of the first TEGs that used heat flow from a human wrist to generate a power level of 100 μW was fabricated by them in 2004 [91]. They used 48 commercially available Bi_2Te_3 thermopiles, each consisting of 128 thermocouples and demonstrated wireless transmission to a computer at 2.4 GHz [91]. IMEC improved the device in 2005 when they fabricated watch-sized TEGs using thermopile arrangements, which produced output powers of 0.1–0.2 mW at ambient temperatures less than 23 °C when worn by people indoors [90, 92]. A wireless autonomous pulse oximeter was built in 2006, which could operate at 23 °C ambient temperature; it could work at a voltage exceeding 1.2 V, producing a minimal power of 100 μW with the input low power electronics consuming around 62 μW [93, 94].

Later in 2008, IMEC [101] designed an EEG (electroencephalography) system, which was fully powered by human body heat. It could be worn on the forehead and could harvest 2 mW heat at an ambient temperature of 23 °C. The authors used IMEC's ultralow power biopotential electronics chips to record high-quality EEG signals with the complete system consuming only 0.8 mW, which included sampling and transmitting two channels of 12-bit EEG data continuously at 256 Hz. See Figure 14.6 for the EEG system.

All of these TEGs were fabricated using inorganic Bi_2Te_3 alloys. Using microelectromechanicalsystems (MEMSs) and micromachining technologies the size of the TEGs was reduced but using Bi_2Te_3 alloys meant that they were not flexible. To effectively harvest heat from the human skin, a good thermal contact with the human skin is essential. For rigid devices that are hard to bend, it is

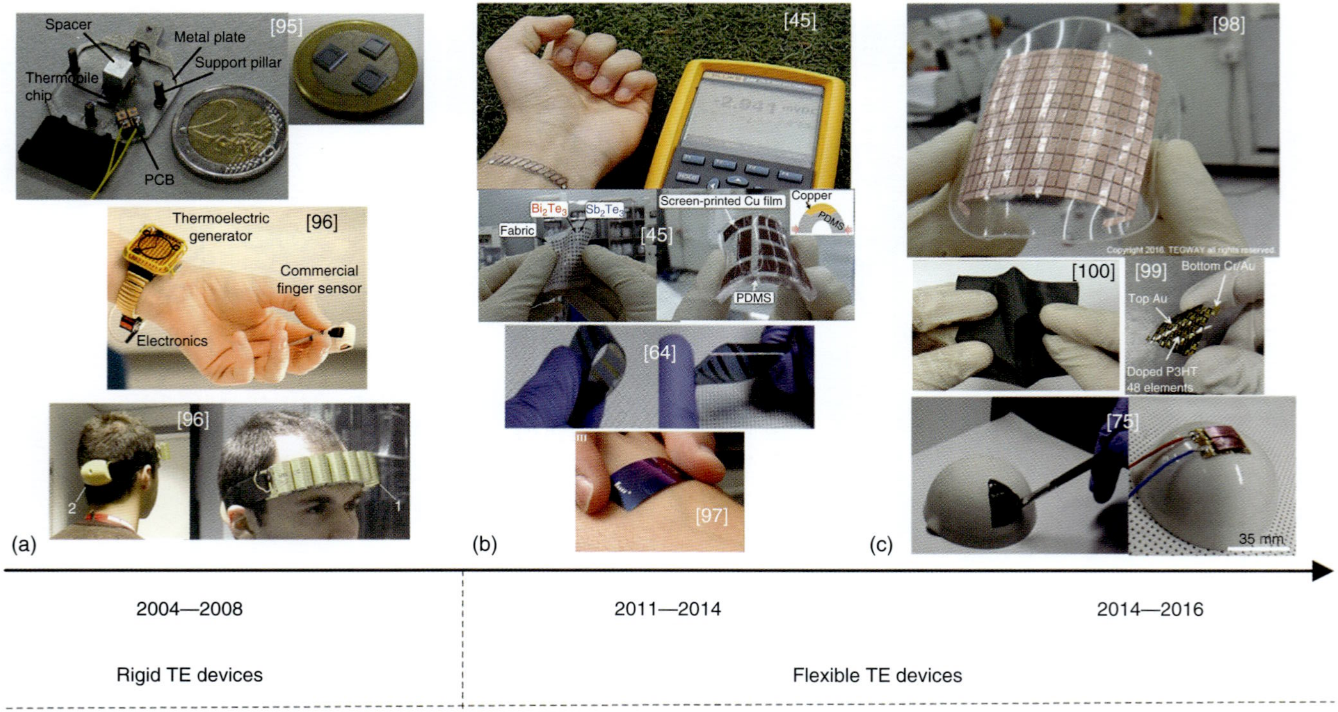

Figure 14.6 Chronological illustration of wearable thermoelectric devices both rigid and flexible . (a) From 2004 to 2008 some of the first-generation wearable TE devices manufactured using Bismuth Telluride as the TE material can be seen. Clockwise from left: assembly of TEG for human body applications, micromachined thermopile chips on a one Euro coin [95], a watch-size TEG used to power a pulse oximeter [96], TEG of the EEG headset stretchable band and its electronic module [96]. (b) From 2011 to 2014: voltage generated by flexible TE device from KAIST, screen printed n- and p-type TE legs on glass fabric, PDMS-embedded copper electrodes for connection of the TE legs 45. PP-PEDOT thermoelectric films [64]. Flexible Si fabric thermoelectric device [97]. (c) 2014–2016 Flexible thermoelectric module from TEGway (*Source:* Copyright 2016. TEGWAY, all rights reserved) [98], a spin-off from KAIST. Flexible uni-leg organic thermoelectric (OTE) module fabricated using a photo etching process from doped P3HT [99]. Flexible thermoelectric fabric composed based on self-assembled tellurium nanorods on PVDF substrate [100], TE painting on substrate, fabricated TE device on hemispherical PET substrate [75].

difficult to have a good thermal contact with the human skin especially when the devices are quite large. This is because the function of the area which is in contact with the skin would be limited due to rigidity and it was one of the reasons for building a segmented EEG system [101]. With a large number of segments it is possible to have a larger contact and more energy could be harvested. Among other drawbacks of these TEGs were the high fabrication cost of Bi_2Te_3 thermopiles and degradation in energy harvested when the devices were made thinner [87].

Polymer films have been used as substrates for flexible electronics previously. Examples of such polymers include PI, PET, polyethylene napthalate (PEN), polycar andbonate (PC), PDMS [102, 103]. Also, conducting polymers such as PANI, PEDOT, and PPy have been used as interconnects and contacts in flexible electronics [102]. Such flexible materials can also be used for fabrication of flexible thermoelectric devices. Some of the first demonstrations of flexible TEGs have been fabricated on PI films, which are also commercially known as Kapton tape. In 1997, Stordeur and Stark showed 20 μW power generation from a temperature difference of 20 K using their compact thermoelectric device microfabricated on flexible polyimide film by sputtering of n- and p-type Bi_2Te_3 as the thermoelectric elements [104]. In 2001, Qu et al. [105] reported a voltage generation of 0.25 V at a temperature difference of 30 K from a TEG fabricated using multiple Sb–Bi thermocouples embedded in 50 μm of epoxy using UV-LIGA (ultraviolet-lithographie, galvanoformung, abformung) microfabrication technique. In 2006, Weber et al. [46] fabricated polymer foil TEG by sputtering antimony and bismuth on polyimide (Kapton) tape and rolling 1.8 m of the strip to form a coin-sized coil. This strip comprised 900 thermocouples and could generate a voltage of 0.8 V at a small temperature difference of 5 °C. The power generated, however was only around 1 μW due to the high electrical resistance of the generator, which was measured to be around 75 kΩ. All these flexible generators were made using flexible polyimide Kapton tape.

PDMS is a versatile elastomer commonly used as substrate in flexible electronics that involve buckling [106] and as a stamp for molding techniques in "soft" lithography [107]. Because of its mechanical flexibility and its low toxicity, microfluidics based on PDMS has been used to study cell biology [108]. PDMS is therefore highly suited for contact with the human skin and is an excellent candidate for use as substrate in flexible wearable devices. In the past decade, PDMS has been extensively used in flexible electronics and sensors [109]. For harvesting human body heat, PDMS has been used extensively by researchers to impart flexibility to their devices [45, 74, 110].

The Cho group at KAIST [45] fabricated a glass fabric based flexible TEG for which they developed n- and p-type Bi_2Te_3 and Sb_2Te_3 alloy pastes and then screen printed them on a glass-based fabric. For electrical connection of the individual n- and p-type layers, they developed PDMS embedded copper films that showed no delamination on bending as opposed to simple adhesion of copper foil to PDMS. Their device was thin (~500 μm) and lightweight (0.13 g cm^{-2}), and their bandage-shaped 11 TE couple device generated 3 μW power when placed on human wrist with the ambient temperature being 15 °C (refer to Figure 14.6). An improved version of their device can be seen on their company's website TEGway, a spin off from KAIST which commercializes this technology [98] (Figure 14.6). Zhang et al. [74] fabricated a flexible dual parameter sensor for both temperature and pressure sensing. Using thermoelectric material PEDOT:PSS and microstructured frame – porous polyurethane – they fabricated what they called MSFOTE (microstructured frame supported organic thermoelectric) material. By using both PU and OTEM PEDOT:PSS their sensor was flexible. Further, they constructed a wearable sensing matrix by using inkjet printing of MSFOTE matrix on a 500 μm thick glove as the microstructured frame and using patterned Ti/Au on PET and PDMS as electrodes. They showed spatially resolved pressure (2–3 kPa) and temperature (0–5 °C) and highlighted the potential of using such an array in intelligent robotics and health monitoring (see Figure 14.6).

In a different approach, bulk silicon (100) has been used by Hussain et al. [111] to transform the rigid, brittle material to flexible Si fabric for fabrication of flexible semitransparent devices. They started with a bulk Si(100) wafer and used a process flow involving photoresist patterning, deep reactive ion etching (DRIE) along with the use of an isotropic etch process. Then they separated the silicon substrate and the Si under their devices that was only 25 μm thick. Their process ensures a lateral etch of 10 μm on either side leading to 20 μm lateral etch (width of the cave) and their devices are present on this 20 μm × 20 μm area. They have fabricated not only thermoelectric devices but also flexible Metal Oxide Semiconductor Field Effect Transistors (MOSFETs), movable MEMS-based actuators for displays, and lithium ion batteries for energy storage [97, 111–114]. Using silicon meant that there was no limitation in charge transport speed, processing could be done at higher temperatures, and lithographic resolution was not limited unlike when polymer substrates were used for conventional flexible devices [111]. They deposited Bi_2Te_3 (n-type alloy) and Sb_2Te_3 (p-type alloy) using a sputtering-based physical vapor deposition (PVD) process followed by flexible Si fabrication processes to obtain a TEG on flexible Si reporting power density of 0.139 mW cm^{-2} from the fabricated device. Even though their flexible Si-based TEGs [97] produced sub microwatts power at temperature difference of 20 K they showcased the possibility of building flexible TEGs on Si substrate. Using bulk materials $Ca_3Co_4O_9$ ceramic and Cu–Ni alloy for hole and electron conduction on flexible mineral band substrate with copper tape for conducting electrical connections, a flexible TEG was realized, which gave an open circuit voltage of 16.52 μV and a short circuit current of 19.4 μA [115].

In another approach, using dispenser printing, Siddique et al. [40] demonstrated a polymer fabric based flexible TEG. They used bismuth telluride based powder with a binder to make a paste and deposited it inside the holes of the polymer fabric with conductive silver thread (hand sewn) in the fabric to connect the n- and p-type legs. Kapton tape was used at one end of their device to insulate human skin from the device and their device showed excellent bending and conformation to the human body. The voltage and power output they obtained from their device was 23.9 mV and 3.107 nW, respectively, at a temperature difference of 22.5 K.

Varghese et al. [42] have very recently fabricated a bismuth telluride based ink and used it for screen printing of flexible thermoelectric devices that showed a high power density of 4.1 mW cm^{-2} with 60 °C temperature difference. The n-type inks show a peak ZT of 0.43, among the highest reported ZT in flexible thermoelectric devices. A number of devices have been fabricated using inkjet printing techniques [41–43] such as flexo, and R2R printing [44]. However, all these devices have produced very low powers, the maximum being a few microwatts. The R2R printed organic thermoelectric device by the Krebs group comprises 18 000 serially connected junctions of flexoprinted silver electrodes and rotary screen printed PEDOT:PSS rolled into a cylinder producing only picowatts of power.

The Adachi group [99] recently fabricated an organic thermoelectric (OTE) module using microfilms of P3HT doped with Fe^{3+}-tos_3·$6H_2O$, which showed a maximum power output of 1.94 nW K^{-2} for a uni-leg module with 48 elements. The microfilms showed power factors of ~30 mW m^{-1} K^{-2} with a conductivity of 55.4 S cm^{-1}. The group states that optimization of the contact resistance of the device would lead to better performance.

Park et al. have very recently published their report on high performance shape engineerable thermoelectric painting that can be directly painted onto curved surfaces producing a high output power of 4 mW cm^{-2} [75]. Their approach makes it possible to utilize heat from curved surfaces or irregular shaped heat sources by simple painting of the n and p-type inks onto the surfaces which is followed by sintering to obtain uniform ~50 μm thin films. They synthesized Bi_2Te_3-based inorganic paints using the molecular Sb_2Te_3 chalcogenidometalate as a sintering aid for thermoelectric particles with ZT values of 0.67 and 1.21 for the n- and p-type ink, respectively (more details on the ink are presented in the materials section). In this work they applied their inks onto a polyimide hemispherical substrate to obtain a planar TEG (Figure 14.6) that generated ~80 mV and ~60 μW of power under a temperature difference of ~50 K. They have shown that this TE paint offers more freedom in the 2D and 3D shape engineering of TE structures and devices at the same time showing performance comparable to the conventional TE devices [75].

14.4.1 Energy Management

The early TEG powered devices manufactured by IMEC used discrete circuit components for power management [93, 94, 101]. For their autonomous pulse oximeter, Torfs et al. [93] used a first rectifier stage consisting of a comparator-driven MOSFET, a second energy storage buffer capacitor stage, and a final high efficiency DC–DC converter stage. Their circuit stepped up the voltage from 0.8 to 2 V, with quiescent power consumption of circuitry being ~12% at input power of 89 μW. The average power consumption of their system was 62 μW at an input power of ~89 μW, out of which maximum power (43%) was used for analog signal processing.

For TEGs that output low voltages, boost converters enabling low voltage startup as low as ~20 mV have been reported [116–118]. Some also feature Maximum Power Point Tracking (MPPT) capabilities to ensure extraction of maximum possible power from the energy harvester. MPPT algorithms are typically processed by digital signal processors that lead to a high power consumption [118, 119] and therefore limit their use in microenergy harvesting devices where energy usage is expensive. MPPT control for TEGs based on impedance matching [120] and control of output voltage to half the open circuit voltage [121] have been implemented. Depending on the application, harvested power, open circuit voltages generated, and storage or charging capabilities required energy management circuits that can feature simple boost conversion mechanisms or more complex MPPT algorithms for efficient energy utilization.

14.4.2 Architecture of Thermoelectric Modules

For conventional longitudinal Π-shaped thermoelectric devices constructed from n- and p-type thermoelements, individual thermocouples are connected electrically in series and thermally in parallel to maximize the generated voltage and the temperature difference. However, the module performance depends not only on the thermoelectric materials but also on the leg geometries and the fill factor. Increasing the leg length or decreasing the cross-section will lead to an increase in ΔT and will also increase the joule heating losses. For a low fill factor, (lower area coverage by TE elements) the ΔT is easier to maintain but the power generation is limited due to the low number of elements.

A thermoelectric energy harvesting device should be both electrically and thermally matched to the external load and the heat exchangers/heat spreaders connected to it, respectively, to convert maximum heat to electricity. Under the thermal matching for maximum power generation condition, the temperature difference across the TEG is approximately one half of that between source and ambient.

For effective thermoelectric operation, it is necessary to obtain high temperature gradients across the thermoelectric legs [19]. When low-cost techniques such as inkjet, gravure, and R2R printing are used for making flexible devices, the thickness of the films is typically in the nanometer or micrometer range and the most effective way to obtain a high temperature gradient across the individual legs is by designing the heat transfer in the in-plane direction of the elements [122] (see Figure 14.7). In this way, by easily controlling the printing of leg widths, the cross-sectional areas of the legs can also be easily controlled. These 2D printed arrays of thin film legs can be easily converted to 3D architectures such that the thermal gradient is parallel to the substrate. Sun et al. used heat shrink PET films to create a corrugated structure from their 2D printed thermoelectric array [123]. Also, as discussed earlier, Weber et al. [46] constructed rolled architecture from their 2D array of printed thermoelectric legs.

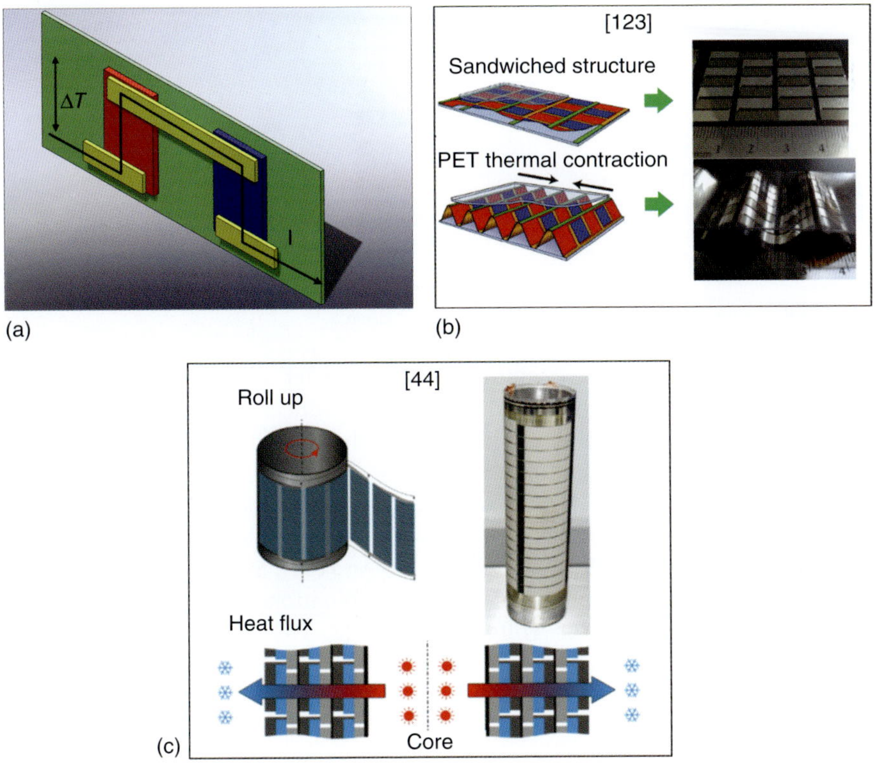

Figure 14.7 Planar TE devices' architecture. (a) A planar TE device printed on a flexible substrate. (b) Printed devices can be used for forming different 3D structures, for example, corrugated TE device. *Source:* Sun et al. 2015 [123]. Reproduced with permission from Elsevier. (c) Rolled architecture *Source:* Søndergaard et al. 2013 [44]. Reproduced with permission from John Wiley & Sons.

14.5 Transverse TE Devices

All of the above discussed TEGs used the longitudinal/Π-shaped thermoelectric devices. Most of the current research in thermoelectric devices uses conventional longitudinal thermoelectric effects for power generation or cooling. Another kind of thermoelectric effect known as the transverse thermoelectric effect also exists, the foundations of which were laid down in the 1960s by Samoilovich [124], and was reported in 1973–1974 by Korolyuk et al. and Babin et al. [1, 125, 126]. The transverse thermoelectric effect or the off-diagonal thermoelectric effect develops in tilted layered materials [127–129]. As suggested by the terminology, the main difference between the transverse and longitudinal thermoelectric devices is that in a transverse device the thermoelectric voltage develops perpendicular to the temperature difference as opposed to a longitudinal device where the thermoelectric voltage develops in parallel in response to temperature difference(see Figure 14.8) [125–131]. Among other characteristics, transverse devices do not require the use of both n- and p-type

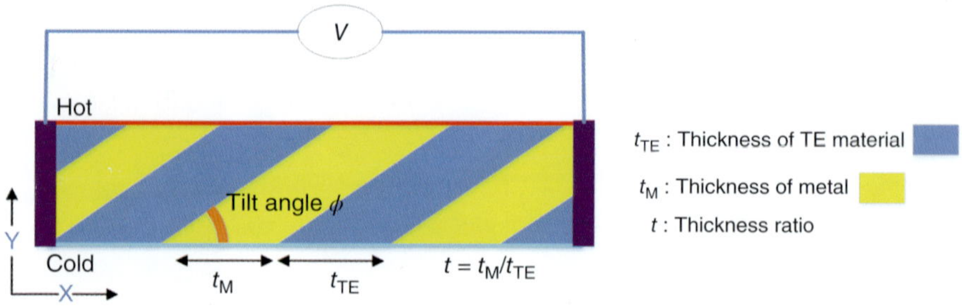

Figure 14.8 Structure of a transverse thermoelectric device.

materials as either of these is sufficient to construct a thermoelectric device [125–133] along with a metal, which potentially reduces material cost substantially. Further, changing the combination and the periodicity of the individual layers in the transverse geometry changes the macroscopic physical properties of the multilayer material [130, 134, 135]. Many anisotropic thermoelectric devices have been fabricated using p-type bismuth telluride–nickel [130], n-type bismuth telluride–lead [129], LSCO ceramic–Ag [131], Si–Al [128], and Bi–Cu [136] as the thermoelectric material–metal combination. Some previous studies have reported unusually high voltages [130, 132, 133, 137] arising from the transverse thermoelectric effect in incline-oriented nanometer-scale thin films and enhanced power factors in artificially tilted multilayer materials [138].

Figure 14.8 shows the structure of a transverse TE device. It comprises one TE material that may be either n- or p-type and a metallic material whose thermal and electrical conductivity is much higher than that of the TE material. The layers are tilted at angle ϕ. Application of a thermal gradient in the Y direction leads to a current flow/potential difference in the X direction.

The figure of merit ZT of a transverse device depends on the cumulative Seebeck coefficient S_{xy}, thermal conductivity in the y direction κ_{yy} (across which ΔT is applied), and electrical resistivity in the x direction ρ_{xx} which is the direction of current flow (refer to Figure 14.8).

$$\mathrm{ZT} = \frac{S_{xy}^2}{\kappa_{yy}\rho_{xx}} T \tag{14.4}$$

With condition: $\rho_{\mathrm{TE}} \gg \rho_M$ and $\kappa_M \gg \kappa_{\mathrm{TE}}$, S, κ, ρ refer to the Seebeck coefficient, thermal conductivity, and electrical resistivity, respectively, and subscripts TE and M indicate thermoelectric and metallic material, respectively.

Electrical power density can be calculated as (d is device thickness or device length parallel to temperature gradient):

$$P_{\mathrm{density}} = \frac{{S_{xy}}^2 \Delta T^2}{4d\rho_{xx}} \tag{14.5}$$

Detailed equations for transverse TEG operation for estimation of the parallel and perpendicular components along with the resulting cumulative Seebeck coefficients, electrical conductivities, and thermal conductivities can be found elsewhere [131].

Yamada group at Panasonic Corporation has recently carried out research to develop TEGs utilizing the transverse thermoelectric effect [130, 134, 135, 138]. In 2012, they fabricated a tubular $Bi_{0.5}Sb_{1.5}Te_3$/Ni composite with cylindrical anisotropy, which showed a high-density thermoelectric power generation of 417 W m^{-2} from a temperature difference of 83 K [134]. Their tubular device measuring around 100 mm in length and 10 mm inner diameter with a thickness of 1.4 mm was designed to harvest energy from fluid heat sources. They reported a power generation of 1.3 W from a hot water flow at 93 °C and 15 l min^{-1} while the cold water flow was at 10 °C and 25 l min^{-1}. Later, they improved the power density from this device to 800 W m^{-2} by improving the fabrication process of the $Bi_{0.5}Sb_{1.5}Te_3$/Ni composite using the spark plasma sintering technique to form the multilayered tube [138] instead of the melt casting technique they used earlier [134]. In 2013, they also reported a power generation of 870 W m^{-2} and a volume power density of 10 kW m^{-3} from harvesting from low-grade heat sources ($\Delta T < 100$ °C) with an efficiency of 0.189% [130]. In this work, water at 90 °C which flowed (at 20 l min^{-1}) inside the BST/Ni transverse TEG was used as the low-grade heat source and cold water at 10 °C flowing outside the tube at 20 l min^{-1} produced a voltage of 0.24 V. Figure 14.9 shows this tubular transverse device manufactured by Panasonic which could have interesting applications in energy harvesting from household hot water pipes.

Very recently, Dressler et al. reported anisotropic thermoelectric energy harvesters consisting of a multilayer stack of $La_{1.97}Sr_{0.03}CuO_4$/Ag showing power density of around 0.4 mW cm^{-2} at a temperature difference of 37.4 K, which could be improved with further optimization [131]. They have used silver paste for connecting the ceramic TE plates in their transverse TEG. The contact resistance between the silver and ceramic layers could have contributed to higher resistance and lower power from this transverse device. In transverse devices from Panasonic, the contact resistance of the BST and Ni layers was lowered and the resulting power increased when they updated their process from cold pressing to spark plasma sintering of the two materials [130, 138]. Dressler et al. reported that only part of the ceramic layer (which was in contact with the silver paste)

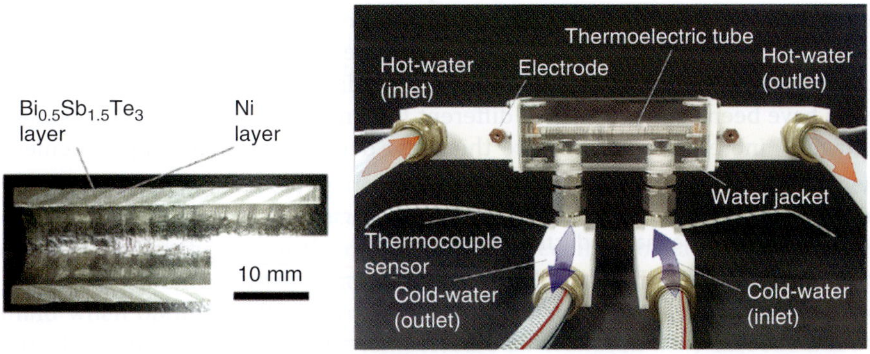

Figure 14.9 BST/Ni tubular-type transverse TEG manufactured by Yamada group with a power density of ~870 W m^{-2} at ΔT ~90 K. *Source:* Takahashi et al. 2013 [130]. Reproduced with permission from Springer Nature.

resistances of the device here are calculated as few milliohms without considering contact resistances between the nickel and the BST layers, which is a good assumption if both the materials are spark plasma sintered to fabricate the device as reported in Ref. [130]. Table 14.2 lists the properties of the materials of the device used in the simulations and calculations. Other parameters such as the tilt angle and ratio of the length of individual layers can be optimized for maximum power generation.

14.6 Thermoelectric Sensors

Thermal sensors used in applications such as noncontact temperature sensing passive infrared (IR) detection are based on thermistor (bolometer), pyroelectric, or thermoelectric (thermopiles) technology [139, 140], where pyroelectric sensors can detect changes in IR radiation thermopiles and bolometers are responsive to constant radiation [141]. Thermopile IR sensors have some inherent advantages over the other two IR-sensing principles in that they do not have any $1/f$ (pink) noise and can generate signals even when either the object or the sensor is moved and therefore do not need choppers to refresh signals [139, 141]. However, they have a larger response time when compared to their pyroelectric counterparts and are more cost-efficient but not as sensitive.

Because of the compatibility of thermopile manufacturing with standard complementary metal-oxide-semiconductor (CMOS) processes, thermopiles made of materials such as polysilicon and aluminum have been integrated in silicon chips in the past and are even today manufactured using MEMS technologies [140–142].

Micromachined thermopiles have been fabricated on rigid silicon substrates, which limit their use in flexible devices. Flexible thermopiles using PI, PET, SU-8, and PDMS substrates have been illustrated in literature [142–146]. Makki et al. [144] showed the possibility of realizing polymer-based flexible thermopiles by fabrication of Ge–Al based thermocouples on PET substrates and the future possibility of using them as IR detectors provided transistors could be integrated on the PET as well. Bergman and Nokolajeff [145] fabricated SU8-based polymer thermopiles with chromel–alumel thermocouples.

For all microfabricated thermopiles, the architecture is in-plane which means that both hot and cold junctions are on one plane. Some three-dimensional thermopiles are also seen, which have a vertical thermocouple arrangement and the hot and cold junctions are on different planes [147–152]. The hot junction is on the top plane and the complete area of the top plane can be utilized for sensing as opposed to in-plane architectures. Lin et al. [147] fabricated flexible 3D thermopiles consisting of 71 Ni–Cu thermocouples on polyimide substrate and reported a measured sensitivity of 0.44 mV K^{-1}. The temperature measured was close to that measured with a digital thermometer thus evincing the potential of using them commercially in the near future.

More recently, in 2012, Yang et al. [153] fabricated a flexible TEG that can be used as a self-powered temperature sensor based on a Te-nanowire/P3HT polymer composite. When used as a self-powered temperature sensor they reported a response time of 17 s and a reset time of 9 s with a detection sensitivity of 0.15 K at room temperature. Sb–Bi based thermopile has also been fabricated on PET substrates for applications of polymer-based MEMS differential scanning calorimeter [146]. A proof of concept flexible thermoelectric based flow sensor was fabricated by Lang and Benecke team from Microsystems center, Bremen, on a flexible 10 μm thick polyimide foil using low-pressure chemical vapor deposition (LPCVD) of silicon, which was used as the thermoelectric material [154]. Their first results showed a change in output voltage with air flow but they concluded that their sensor layout had to be adapted to the properties of the polymer substrate.

14.7 Summary and Outlook

This chapter has discussed the recent research in flexible thermoelectric materials and devices. An introduction to thermoelectric energy conversion was followed by the brief history of the thermoelectric field in the past few decades since the onset of modern thermoelectrics in the 1950s. The conventional inorganic thermoelectric materials with high ZT, which are still very popular for TE energy harvesting, were discussed.

Emerging thermoelectric composites based on conducting polymers, graphene, and CNT as well as bismuth telluride doping were discussed in the materials section. This new research in thermoelectric materials paves a way for flexible thermoelectric devices. Recent research in flexible TE energy harvesting devices was discussed thereafter. A transition from older rigid wearable TE harvesters to newer flexible and wearable TE harvesters was shown. Measurement of TE properties of materials and some recent approaches for characterization of thermal conductivity and subsequently ZT were illustrated. Energy management for TE devices based on the application and TE harvester characteristics itself were briefly discussed.

Although TE devices have been historically based on longitudinal Π-structure comprising p–n leg structure, which is still the most common TE energy harvester structure, there has been research on alternate structures such as the transverse device structure. Researchers have shown that such structures for energy harvesting purposes hold promising results especially because of the lower cost and complexity as compared to conventional structures. However, more work needs to be done for development of flexible transverse TE devices. Simulations for transverse TEG performance for wearable device applications were discussed. Finally, some recent research illustrating flexible TE sensors was discussed.

Although power factors and ZT of emerging, flexible TE materials are lower than that of inorganic TE materials, the cost, ease of manufacture, suitableness to wearable devices, and better possibility to make thermal contact with the human skin make flexible and emerging TE materials an excellent area for research and commercialization. Flexible TE devices for energy harvesting are commercially available [155] and recent research in flexible TE devices has also resulted in commercialization [98] of the technology. Inorganic TE materials still attract great interest for energy harvesting, and have most recently been used by a Bay Area-startup Matrix [156] in their smart watch powered by body heat. As advances are made in materials and manufacturing techniques such as screen printing and inkjet printing, one can expect to see cost-effective high performance flexible TE devices in the market in the very near future.

References

1 Nolas, G.S., Sharp, J., and Goldsmid, J. (2013). *Thermoelectrics: Basic Principles and New Materials Developments*. Springer Science & Business Media.
2 Rowe, D.M. (1995). *CRC Handbook of Thermoelectrics*. CRC Press.
3 Bulman, G., Barletta, P., Lewis, J. et al. (2016). Superlattice-based thin-film thermoelectric modules with high cooling fluxes. *Nat. Commun.* 7: 10302.
4 Fan, X., Zeng, G., LaBounty, C. et al. (2001). SiGeC/Si superlattice micro-coolers. *Appl. Phys. Lett.* 78: 1580–1582.
5 Glover, J., Khalid, A., Stephen, A. et al. (2015). Micro-coolers fabricated as a component in an integrated circuit. *Semicond. Sci. Technol.* 30: 015005.
6 Glover, J., Khalid, A., Cumming, D.R.S. et al. (2014). On wafer thermal characterization of miniature gallium arsenide microcoolers with thermal loading from Dc probes. *Microwave Opt. Technol. Lett.* 56: 2699–2700.
7 IDTechEx (2016). *Printed, Organic & Flexible Electronics Forecasts, Players & Opportunities 2016–2026*. IDTechEx http://www.idtechex.com/research/reports/printed-organic-and-flexible-electronics-forecasts-players-and-opportunities-2016-2026-000457.asp (accessed 3 August 2016.
8 McCann, J. and Bryson, D. (2009). *Smart Clothes and Wearable Technology*. Elsevier.
9 Thorp, E.O. (1998). The invention of the first wearable computer. Digest of Papers. Second International Symposium on Wearable Computers, pp. 4–8.
10 Bonato, P. (2005). Advances in wearable technology and applications in physical medicine and rehabilitation. *J. NeuroEng. Rehabil.* 2: 2.
11 Patel, S., Park, H., Bonato, P. et al. (2012). A review of wearable sensors and systems with application in rehabilitation. *J. NeuroEng. Rehabil.* 9: 21.
12 van Rooij, T. and Marsh, S. (2016). eHealth: past and future perspectives. *Pers. Med.* 13: 57–70.
13 Pantelopoulos, A. and Bourbakis, N.G. (2010). A survey on wearable sensor-based systems for health monitoring and prognosis. *IEEE Trans. Syst. Man Cybern. Part C, Appl. Rev.* 40: 1–12.
14 Desjardins, D. and Hopper, D. (2003). Military display market segment: wearable and portable. Cockpit Displays X, Vol. 5080, pp. 346–360.
15 Lamkin, P. and Contributor(2016). *Wearable Tech Market To Be Worth $34 Billion By 2020*. Paul Lamkin - Forbes.
16 Maleville, C. (2015). Engineered substrates for Moore and more than Moore's law device scaling: entering the substrate era. 2015 IEEE Soi-3D-Subthreshold Microelectronics Technology Unified Conference (S3s).

17 Schlachter, F. (2013). No Moore's Law for batteries. *Proc. Natl. Acad. Sci. U.S.A.* 110: 5273.
18 IDTechEx (2016). *Thermoelectric Energy Harvesting 2016–2026*. IDTechEx http://www.idtechex.com/research/reports/thermoelectric-energy-harvesting-2016-2026-000473.asp (accessed 01 August 2016.
19 Bahk, J., Fang, H., Yazawa, K., and Shakouri, A. (2015). Flexible thermoelectric materials and device optimization for wearable energy harvesting. *J. Mater. Chem. C* 3: 10362–10374.
20 Lange, P.W. (1939). Ein Vergleich zwischen Bi 2 Te 3 und Bi 2 Te 2 S. *Naturwissenschaften* 27: 133–134.
21 Skelton, J.M., Parker, S.C., Togo, A. et al. (2014). Thermal physics of the lead chalcogenides PbS, PbSe, and PbTe from first principles. *Phys. Rev. B* 89: 205203.
22 Pei, Y., Wang, H., and Snyder, G. (2012). Band engineering of thermoelectric materials. *Adv. Mater.* 24: 6125–6135.
23 Wang, X., Lee, H., Lan, Y. et al. (2008). Enhanced thermoelectric figure of merit in nanostructured n-type silicon germanium bulk alloy. *Appl. Phys. Lett.* 93: 193121.
24 Rowe, D.M. and Bhandari, C.M. (1983). *Modern Thermoelectrics*. Prentice Hall.
25 Joshi, G., Lee, H., Lan, Y. et al. (2008). Enhanced thermoelectric figure-of-merit in nanostructured p-type silicon germanium bulk alloys. *Nano Lett.* 8: 4670–4674.
26 Hsu, K.F., Loo, S., Guo, F. et al. (2004). Cubic $AgPb_mSbTe_{2+m}$: bulk thermoelectric materials with high figure of merit. *Science* 303: 818–821.
27 Heremans, J.P., Jovovic, V., Toberer, E.S. et al. (2008). Enhancement of thermoelectric efficiency in PbTe by distortion of the electronic density of states. *Science* 321: 554–557.
28 Poudel, B., Hao, Q., Ma, Y. et al. (2008). High-thermoelectric performance of nanostructured bismuth antimony telluride bulk alloys. *Science* 320: 634–638.
29 Biswas, K., He, J., Blum, I.D. et al. (2012). High-performance bulk thermoelectrics with all-scale hierarchical architectures. *Nature* 489: 414–418.
30 Heremans, J.P., Dresselhaus, M.S., Bell, L.E., and Morelli, D.T. (2013). When thermoelectrics reached the nanoscale. *Nat. Nanotechnol.* 8: 471–473.
31 Hicks, L. and Dresselhaus, M. (1993). Effect of quantum-well structures on the thermoelectric figure of merit. *Phys. Rev. B* 47: 12727.
32 Hicks, L. and Dresselhaus, M. (1993). Thermoelectric figure of merit of a one-dimensional conductor. *Phys. Rev. B* 47: 16631.
33 Snyder, G.J. and Toberer, E.S. (2008). Complex thermoelectric materials. *Nat. Mater.* 7: 105–114.
34 Nolas, G., Morelli, D., and Tritt, T.M. (1999). Skutterudites: a phonon-glass-electron crystal approach to advanced thermoelectric energy conversion applications. *Annu. Rev. Mater. Sci.* 29: 89–116.
35 Nolas, G., Cohn, J., Slack, G., and Schujman, S. (1998). Semiconducting Ge clathrates: promising candidates for thermoelectric applications. *Appl. Phys. Lett.* 73: 178–180.

36 Kauzlarich, S.M., Brown, S.R., and Snyder, G.J. (2007). Zintl phases for thermoelectric devices. *Dalton Trans.* 2099–2107.
37 Vullers, R., van Schaijk, R., Doms, I. et al. (2009). Micropower energy harvesting. *Solid-State Electron.* 53: 684–693.
38 Crispin, X. (2016). Thermoelectrics: carbon nanotubes get high. *Nat. Energy* 1: 16037.
39 Chen, Y., Zhao, Y., and Liang, Z. (2015). Solution processed organic thermoelectrics: towards flexible thermoelectric modules. *Energy Environ. Sci.* 8: 401–422.
40 Siddique, A.R.M., Rabari, R., Mahmud, S., and Van Heyst, B. (2016). Thermal energy harvesting from the human body using flexible thermoelectric generator (FTEG) fabricated by a dispenser printing technique. *Energy* 115: 1081–1091.
41 Lu, Z., Layani, M., Zhao, X. et al. (2014). Fabrication of flexible thermoelectric thin film devices by inkjet printing. *Small* 10: 3551–3554.
42 Varghese, T., Hollar, C., Richardson, J. et al. (2016). High-performance and flexible thermoelectric films by screen printing solution-processed nanoplate crystals. *Sci. Rep.* 6: 33135.
43 Jiao, F., Di, C.A., Sun, Y. et al. (2014). Inkjet-printed flexible organic thin-film thermoelectric devices based on p- and n-type poly(metal 1,1,2,2-ethenetetrathiolate)s/polymer composites through ball-milling. *Philos. Trans. A Math. Phys. Eng. Sci.* 372: 20130008.
44 Søndergaard, R.R., Hösel, M., Espinosa, N. et al. (2013). Practical evaluation of organic polymer thermoelectrics by large-area R2R processing on flexible substrates. *Energy Sci. Eng.* 1: 81–88.
45 Kim, S.J., We, J.H., and Cho, B.J. (2014). A wearable thermoelectric generator fabricated on a glass fabric. *Energy Environ. Sci.* 7: 1959–1965.
46 Weber, J., Potje-Kamloth, K., Haase, F. et al. (2006). Coin-size coiled-up polymer foil thermoelectric power generator for wearable electronics. *Sens. Actuators, A* 132: 325–330.
47 Hiroshige, Y., Ookawa, M., and Toshima, N. (2007). Thermoelectric figure-of-merit of iodine-doped copolymer of phenylenevinylene with dialkoxyphenylenevinylene. *Synth. Met.* 157: 467–474.
48 See, K.C., Feser, J.P., Chen, C.E. et al. (2010). Water-processable polymer–nanocrystal hybrids for thermoelectrics. *Nano Lett.* 10: 4664–4667.
49 Bubnova, O., Khan, Z.U., Malti, A. et al. (2011). Optimization of the thermoelectric figure of merit in the conducting polymer poly(3,4-ethylenedioxythiophene). *Nat. Mater.* 10: 429–433.
50 Sun, Y., Sheng, P., Di, C. et al. (2012). Organic thermoelectric materials and devices based on p- and n-type poly(metal 1,1,2,2-ethenetetrathiolate)s. *Adv. Mater.* 24: 932–937.
51 Nonoguchi, Y., Ohashi, K., Kanazawa, R. et al. (2013). Systematic conversion of single walled carbon nanotubes into n-type thermoelectric materials by molecular dopants. *Sci. Rep.* 3: 3344.
52 Kim, G., Shao, L., Zhang, K., and Pipe, K.P. (2013). Engineered doping of organic semiconductors for enhanced thermoelectric efficiency. *Nat. Mater.* 12: 719–723.

53 Lee, S.H., Park, H., Kim, S. et al. (2014). Transparent and flexible organic semiconductor nanofilms with enhanced thermoelectric efficiency. *J. Mater. Chem. A* 2: 7288–7294.

54 Fukumaru, T., Fujigaya, T., and Nakashima, N. (2015). Development of n-type cobaltocene-encapsulated carbon nanotubes with remarkable thermoelectric property. *Sci. Rep.* 5: 7951.

55 Wang, H., Hsu, J., Yi, S. et al. (2015). Thermally driven large n-type voltage responses from hybrids of carbon nanotubes and poly(3,4-ethylenedioxythiophene) with tetrakis(dimethylamino)ethylene. *Adv. Mater.* 27: 6855–6861.

56 Avery, A.D., Zhou, B.H., Lee, J. et al. (2016). Tailored semiconducting carbon nanotube networks with enhanced thermoelectric properties. *Nat. Energy* 1: 16033.

57 Bae, E.J., Kang, Y.H., Jang, K., and Cho, S.Y. (2016). Enhancement of thermoelectric properties of PEDOT:PSS and tellurium-PEDOT:PSS hybrid composites by simple chemical treatment. *Sci. Rep.* 6: 18805.

58 Kim, M.K., Kim, M.S., Jo, S.E. et al. (2013). Wearable thermoelectric generator for human clothing applications. 2013 Transducers & Eurosensors XXVII: The 17th International Conference on Solid-State Sensors, Actuators and Microsystems (TRANSDUCERS & EUROSENSORS XXVII), Barcelona, Spain, pp. 1376–1379.

59 Cao, Z., Koukharenko, E., Torah, R.N. et al. (2014). Flexible screen printed thick film thermoelectric generator with reduced material resistivity. *J. Phys. Conf. Ser.* 557, 012016.

60 Zheng, W. et al. (2014). Low thermal conductivity and high thermoelectric figure of merit in p-type Sb_2Te_3/poly (3,4-ethylenedioxythiophene) thermoelectric composites. *Appl. Phys. Lett.* 105 (2): 023901.

61 Du, Y., Cai, K., Chen, S. et al. (2014). Facile preparation and thermoelectric properties of Bi_2Te_3 based alloy nanosheet/PEDOT: PSS composite films. *ACS Appl. Mater. Interfaces* 6: 5735–5743.

62 Shi, H., Liu, C., Jiang, Q., and Xu, J. (2015). Effective approaches to improve the electrical conductivity of PEDOT:PSS: a review. *Adv. Electron. Mater.* 1: 1500017.

63 Culebras, M., Gomez, C., and Cantarero, A. (2014). Enhanced thermoelectric performance of PEDOT with different counter-ions optimized by chemical reduction. *J. Mater. Chem. A* 2: 10109–10115.

64 Park, T., Park, C., Kim, B. et al. (2013). Flexible PEDOT electrodes with large thermoelectric power factors to generate electricity by the touch of fingertips. *Energy Environ. Sci.* 6: 788–792.

65 Peng, T., Sun, W., Huang, C. et al. (2013). Self-assembled free-standing polypyrrole nanotube membrane as an efficient FTO- and Pt-free counter electrode for dye-sensitized solar cells. *ACS Appl. Mater. Interfaces* 6: 14–17.

66 Wu, J., Sun, Y., Pei, W. et al. (2014). Polypyrrole nanotube film for flexible thermoelectric application. *Synth. Met.* 196: 173–177.

67 Shinohara, Y., Ohara, K., Nakanishi, H. et al. (2005). Thermoelectric properties of poly(3-alkylthiophenes). In: *Mater. Sci. Forum*, vol. 492–493, 141–144.

68 Aïch, R.B., Blouin, N., Bouchard, A., and Leclerc, M. (2009). Electrical and thermoelectric properties of poly (2,7-carbazole) derivatives. *Chem. Mater.* 21 (4): 751–757.
69 Schlitz, R.A., Brunetti, F.G., Glaudell, A.M. et al. (2014). Solubility-limited extrinsic n-type doping of a high electron mobility polymer for thermoelectric applications. *Adv. Mater.* 26: 2825–2830.
70 Mai, C., Schlitz, R.A., Su, G.M. et al. (2014). Side-chain effects on the conductivity, morphology, and thermoelectric properties of self-doped narrow-band-gap conjugated polyelectrolytes. *J. Am. Chem. Soc.* 136: 13478–13481.
71 Zhang, Q., Sun, Y., Xu, W., and Zhu, D. (2014). What to expect from conducting polymers on the playground of thermoelectricity: lessons learned from four high-mobility polymeric semiconductors. *Macromolecules* 47: 609–615.
72 Russ, B., Robb, M.J., Brunetti, F.G. et al. (2014). Power factor enhancement in solution-processed organic n-type thermoelectrics through molecular design. *Adv. Mater.* 26: 3473–3477.
73 Bubnova, O., Khan, Z.U., Wang, H. et al. (2014). Semi-metallic polymers. *Nat. Mater.* 13: 190–194.
74 Zhang, F., Zang, Y., Huang, D. et al. (2015). Flexible and self-powered temperature–pressure dual-parameter sensors using microstructure-frame-supported organic thermoelectric materials. *Nat. Commun.* 6: 8356.
75 Park, S.H., Jo, S., Kwon, B. et al. (2016). High-performance shape-engineerable thermoelectric painting. *Nat. Commun.* 7: 13403.
76 Hewitt, C.A., Kaiser, A.B., Roth, S. et al. (2012). Multilayered carbon nanotube/polymer composite based thermoelectric fabrics. *Nano Lett.* 12: 1307–1310.
77 Yu, C., Choi, K., Yin, L., and Grunlan, J.C. (2011). Light-weight flexible carbon nanotube based organic composites with large thermoelectric power factors. *ACS Nano* 5: 7885–7892.
78 Suemori, K., Hoshino, S., and Kamata, T. (2013). Flexible and lightweight thermoelectric generators composed of carbon nanotube–polystyrene composites printed on film substrate. *Appl. Phys. Lett.* 103: 153902.
79 Du, Y., Shen, S.Z., Yang, W. et al. (2012). Simultaneous increase in conductivity and seebeck coefficient in a polyaniline/graphene nanosheets thermoelectric nanocomposite. *Synth. Met.* 161: 2688–2692.
80 Wang, L., Yao, Q., Bi, H. et al. (2014). Large thermoelectric power factor in polyaniline/graphene nanocomposite films prepared by solution-assistant dispersing method. *J. Mater. Chem. A* 2: 11107–11113.
81 Baitimirova, M., Andzane, J., Petersons, G. et al. (2016). Vapor–solid synthesis and enhanced thermoelectric properties of non-planar bismuth selenide nanoplates on graphene substrate. *J. Mater. Sci.* 51 (17): 8224–8232.
82 Yao, Q., Chen, L., Zhang, W. et al. (2010). Enhanced thermoelectric performance of single-walled carbon nanotubes/polyaniline hybrid nanocomposites. *ACS Nano* 4: 2445–2451.

83 Meng, C., Liu, C., and Fan, S. (2010). A promising approach to enhanced thermoelectric properties using carbon nanotube networks. *Adv. Mater.* 22: 535–539.

84 Kim, D., Kim, Y., Choi, K. et al. (2009). Improved thermoelectric behavior of nanotube-filled polymer composites with poly(3,4-ethylenedioxythiophene) poly(styrenesulfonate). *ACS Nano* 4: 513–523.

85 Kim, S.L., Choi, K., Tazebay, A., and Yu, C. (2014). Flexible power fabrics made of carbon nanotubes for harvesting thermoelectricity. *ACS Nano* 8: 2377–2386.

86 Bounioux, C., Díaz-Chao, P., Campoy-Quiles, M. et al. (2013). Thermoelectric composites of poly(3-hexylthiophene) and carbon nanotubes with a large power factor. *Energy Environ. Sci.* 6: 918–925.

87 Leonov, V. and Vullers, R. (2007). Thermoelectric generators on living beings. Proceedings of the 5th European Conference on 'Thermoelectrics (ECT), Odessa, Ukraine, pp. 47–52.

88 Leonov, V. (2011). Human machine and thermoelectric energy scavenging for wearable devices. *ISRN Renew. Energy* 2011: 1–11.

89 Suarez, F., Nozariasbmarz, A., Vashaee, D., and Öztürk, M.C. (2016). Designing thermoelectric generators for self-powered wearable electronics. *Energy Environ. Sci.* 9: 2099–2113.

90 Leonov, V. and Vullers, R. (2007). Wireless microsystems powered by homeotherms. Proceedings of the Smart Systems Integrated Conference, Paris, France, pp. 349–358.

91 Leonov, V., Fiorini, P., Sedky, S. et al. (2005). Thermoelectric MEMS generators as a power supply for a body area network. The 13th International Conference on Solid-State Sensors, Actuators and Microsystems. Digest of Technical Papers. TRANSDUCERS'05, Seoul, Korea, pp. 291–294.

92 Leonov, V., Torfs, T., Fiorini, P., and Van Hoof, C. (2007). Thermoelectric converters of human warmth for self-powered wireless sensor nodes. *IEEE Sens. J.* 7: 650–657.

93 Torfs, T., Leonov, V., and Vullers, R. (2007). Pulse oximeter fully powered by human body heat. *Sens. Transducers J.* 80: 1230–1238.

94 Torfs, T., Leonov, V., Van Hoof, C., and Gyselinckx, B. (2006). Body-heat powered autonomous pulse oximeter. 2006 5th IEEE Conference on Sensors, Daegu, Korea, pp. 427–430.

95 Wang, Z., Leonov, V., Fiorini, P., and Van Hoof, C. (2009). Realization of a wearable miniaturized thermoelectric generator for human body applications. *Sens. Actuators, A* 156: 95–102.

96 Leonov, V. and Vullers, R.J. (2009). Wearable electronics self-powered by using human body heat: the state of the art and the perspective. *J. Renewable Sustainable Energy* 1: 062701.

97 Sevilla, G.A.T., Inayat, S.B., Rojas, J.P. et al. (2013). Flexible and semi-transparent thermoelectric energy harvesters from low cost bulk silicon (100). *Small* 9: 3916–3921.

98 TEGWAY Company Overview http://tegway.co/?page_id=39 (accessed 26 January 2017).

99 Hwang, S., Potscavage, W.J. Jr.,, Nakamichi, R., and Adachi, C. (2016). Processing and doping of thick polymer active layers for flexible organic thermoelectric modules. *Org. Electron.* 31: 31–40.
100 Dun, C., Hewitt, C.A., Huang, H. et al. (2015). Flexible thermoelectric fabrics based on self-assembled tellurium nanorods with a large power factor. *Phys. Chem. Chem. Phys.* 17: 8591–8595.
101 Torfs, T., Leonov, V., Yazicioglu, R.F. et al. (2008). Wearable autonomous wireless electro-encephalography system fully powered by human body heat. Sensors, 2008 IEEE, Lecce, Italy, pp. 1269–1272.
102 Wong, W.S. and Salleo, A. (2009). *Flexible Electronics: Materials and Applications.* Springer Science & Business Media.
103 Rogers, J.A., Someya, T., and Huang, Y. (2010). Materials and mechanics for stretchable electronics. *Science* 327: 1603–1607.
104 Stordeur, M. and Stark, I. (1997). Low power thermoelectric generator-self-sufficient energy supply for micro systems. Thermoelectrics, 1997. Proceedings ICT'97. XVI International Conference on, Dresden, Germany, pp. 575–577.
105 Qu, W., Ploetner, M., and Fischer, W. (2001). Microfabrication of thermoelectric generators on flexible foil substrates as a power source for autonomous microsystems. *J. Micromech. Microeng.* 11: 146.
106 Lu, X. and Xia, Y. (2006). Electronic materials: buckling down for flexible electronics. *Nat. Nanotechnol.* 1: 163–164.
107 Xia, Y. and Whitesides, G.M. (1998). Soft lithography. *Annu. Rev. Mater. Sci.* 28: 153–184.
108 Whitesides, G.M. (2006). The origins and the future of microfluidics. *Nature* 442: 368–373.
109 Boland, J.J. (2010). Flexible electronics: within touch of artificial skin. *Nat. Mater.* 9: 790–792.
110 Jo, S., Kim, M., Kim, M., and Kim, Y. (2012). Flexible thermoelectric generator for human body heat energy harvesting. *Electron. Lett.* 48: 1013–1015.
111 Hussain, M., Rojas, J.P., and Sevilla, G.A.T. (2013). Flexible transparent devices from bulk silicon (100). *J Micro/Nano Lithogr.* doi: 10.1117/2.1201311.004863.
112 Fahad, H.M. and Hussain, M.M. (2013). High-performance silicon nanotube tunneling FET for ultralow-power logic applications. *IEEE Trans. Electron. Devices* 60: 1034–1039.
113 Rojas, J.P., Sevilla, G.T., and Hussain, M.M. (2013). Structural and electrical characteristics of high-k/metal gate metal oxide semiconductor capacitors fabricated on flexible, semi-transparent silicon (100) fabric. *Appl. Phys. Lett.* 102: 064102.
114 Rojas, J.P., Syed, A., and Hussain, M.M. (2012). Mechanically flexible optically transparent porous mono-crystalline silicon substrate. 2012 IEEE 25th International Conference on Micro Electro Mechanical Systems (MEMS), Paris, France, pp. 281–284.
115 Geppert, B. and Feldhoff, A. (2016). An approach to a flexible thermoelectric generator fabricated using bulk materials. *Energy Harvest. Syst.* 3: 161–171.
116 Carlson, E.J., Strunz, K., and Otis, B.P. (2010). A 20 mV input boost converter with efficient digital control for thermoelectric energy harvesting. *IEEE J. Solid-State Circuits* 45: 741–750.

117 Goeppert, J. and Manoli, Y. (2015). Fully integrated start-up at 70 mV of boost converters for thermoelectric energy harvesting. European Solid-State Circuits Conference (ESSCIRC), ESSCIRC 2015-41st, Graz, Austria, pp. 233–236.

118 Im, J., Wang, S., Ryu, S., and Cho, G. (2012). A 40 mV transformer-reuse self-startup boost converter with MPPT control for thermoelectric energy harvesting. *IEEE J. Solid-State Circuits* 47: 3055–3067.

119 Kim, R. and Lai, J. (2008). A seamless mode transfer maximum power point tracking controller for thermoelectric generator applications. *IEEE Trans. Power Electron.* 23: 2310–2318.

120 Ramadass, Y.K., Chandrakasan, A.P., and A.P. (2011). A batteryless thermoelectric energy-harvesting interface circuit with 35 mV startup voltage. *IEEE J. Solid-State Circuits* 46 (1): 333–341.

121 Cho, S., Kim, N., Park, S., and Kim, S. (2010). A coreless maximum power point tracking circuit of thermoelectric generators for battery charging systems. Solid State Circuits Conference (A-SSCC), 2010 IEEE Asian, Beijing, China, pp. 1–4.

122 Russ, B., Glaudell, A., Urban, J.J. et al. (2016). Organic thermoelectric materials for energy harvesting and temperature control. *Nat. Rev. Mater.* 1: 16050.

123 Sun, T., Peavey, J.L., Shelby, M.D. et al. (2015). Heat shrink formation of a corrugated thin film thermoelectric generator. *Energy Convers. Manage.* 103: 674–680.

124 Rowe, D.M. (2005). *Thermoelectrics Handbook: Macro to Nano*. CRC Press.

125 Babin, V., Gudkin, T., Dashevskii, Z. et al. (1974). Anisotropic synthetic thermoelements and their maximum capabilities. *Sov. Phys. Semicond.-USSR* 8: 478–481.

126 Korolyuk, S., Pilat, I., Samoilovich, A. et al. (1973). Anisotropic thermoelements. *Sov. Phys. Semicond.-USSR* 7: 502–507.

127 Fischer, K., Stoiber, C., Kyarad, A., and Lengfellner, H. (2004). Anisotropic thermopower in tilted metallic multilayer structures. *Appl. Phys. A* 78: 323–326.

128 Kyarad, A. and Lengfellner, H. (2004). Al–Si multilayers: a synthetic material with large thermoelectric anisotropy. *Appl. Phys. Lett.* 85: 5613.

129 Kyarad, A. and Lengfellner, H. (2006). Transverse Peltier effect in tilted $Pb–Bi_2Te_3$ multilayer structures. *Appl. Phys. Lett.* 89: 2103.

130 Takahashi, K., Kanno, T., Sakai, A. et al. (2013). Bifunctional thermoelectric tube made of tilted multilayer material as an alternative to standard heat exchangers. *Sci. Rep.* 3: 1501.

131 Dreßler, C., Bochmann, A., Schulz, T. et al. (2015). Transversal oxide–metal thermoelectric device for low-power energy harvesting. *Energy Harvest. Syst.* 2: 25–35.

132 Habermeier, H., Li, X., Zhang, P., and Leibold, B. (1999). Anisotropy of thermoelectric properties in $La_{2/3}Ca_{1/3}MnO_3$ thin films studied by laser-induced transient voltages. *Solid State Commun.* 110: 473–478.

133 Lengfellner, H., Kremb, G., Schnellbögl, A. et al. (1992). Giant voltages upon surface heating in normal $YBa_2Cu_3O_{7-\delta}$ films suggesting an atomic layer thermopile. *Appl. Phys. Lett.* 60: 501–503.

134 Kanno, T., Sakai, A., Takahashi, K. et al. (2012). Tailoring effective thermoelectric tensors and high-density power generation in a tubular $Bi_{0.5}Sb_{1.5}Te_3$/Ni composite with cylindrical anisotropy. *Appl. Phys. Lett.* 101: 011906.
135 Sakai, A., Kanno, T., Takahashi, K. et al. (2014). Breaking the trade-off between thermal and electrical conductivities in the thermoelectric material of an artificially tilted multilayer. *Sci. Rep.* 4: 6089.
136 Kanno, T., Yotsuhashi, S., Sakai, A. et al. (2009). Enhancement of transverse thermoelectric power factor in tilted Bi/Cu multilayer. *Appl.Phys.Lett.* 94: 061917.
137 Takahashi, K., Kanno, T., Sakai, A. et al. (2010). Gigantic transverse voltage induced via off-diagonal thermoelectric effect in $CaxCoO_2$ thin films. *Appl. Phys. Lett.* 97: 021906.
138 Sakai, A., Kanno, T., Takahashi, K. et al. (2013). Enhancement in performance of the tubular thermoelectric generator (TTEG). *J. Electron. Mater.* 42: 1612–1616.
139 Socher, E., Bochobza-Degani, O., and Nemirovsky, Y. (2001). A novel spiral CMOS compatible micromachined thermoelectric IR microsensor. *J. Micromech. Microeng.* 11: 574.
140 Van Herwaarden, A. and Sarro, P. (1986). Thermal sensors based on the Seebeck effect. *Sens. Actuators* 10: 321–346.
141 Schieferdecker, J., Quad, R., Holzenkämpfer, E., and Schulze, M. (1995). Infrared thermopile sensors with high sensitivity and very low temperature coefficient. *Sens. Actuators, A* 47: 422–427.
142 Graf, A., Arndt, M., Sauer, M., and Gerlach, G. (2007). Review of micromachined thermopiles for infrared detection. *Meas. Sci. Technol.* 18: R59.
143 Hasebe, S., Ogawa, J., Shiozaki, M. et al. (2004). Polymer based smart flexible thermopile for power generation. Micro Electro Mechanical Systems. 17th IEEE International Conference on (MEMS), Maastricht, The Netherlands, pp. 689–692.
144 Makki, B.S., Moradi, M., Moafi, A. et al. (2004). Fabrication of poly-Ge-based thermopiles on plastic. *IEEE Sens. J.* 4: 743–748.
145 Bergman, D. and Nikolajeff, F. (2003). A novel type of metal-on-polymer micromachined thermopile. TRANSDUCERS, 12th International Conference on Solid-State Sensors, Actuators and Microsystems, pp. 754–757.
146 Jia, Y., Wang, B., Zhang, Z., and Lin, Q. (2015). A polymer-based MEMS differential scanning calorimeter. *Sens. Actuators, A* 231: 1–7.
147 Lin, P., Wu, C., Cheng, Y. et al. (2008). Fabrication and characterization of a three-dimensional flexible thermopile. *Jpn. J. Appl. Phys.* 47: 1787.
148 Huynh, T.P., Zhang, Y., and Yehuda, C. (2015). Fabrication and characterization of a multichannel 3D thermopile for chip calorimeter applications. *Sensors* 15: 3351–3361.
149 Snyder, G.J., Lim, J.R., Huang, C., and Fleurial, J. (2003). Thermoelectric microdevice fabricated by a MEMS-like electrochemical process. *Nat. Mater.* 2: 528–531.
150 Lindeberg, M., Yousef, H., Rödjegård, H. et al. (2008). A PCB-like process for vertically configured thermopiles. *J. Micromech. Microeng.* 18: 065021.

151 Huang, I., Chen, G., and Wu, T. (2013). Development of microthermoelectric generators using integrated suspending bridge-type polysilicon thin-film thermopiles. *J. Micro/Nanolithogr. MEMS and MOEMS* 12: 013001.
152 Allison, S., Smith, R., Howard, D. et al. (2003). A bulk micromachined silicon thermopile with high sensitivity. *Sens. Actuators, A* 104: 32–39.
153 Yang, Y., Lin, Z., Hou, T. et al. (2012). Nanowire-composite based flexible thermoelectric nanogenerators and self-powered temperature sensors. *Nano Res.* 5: 888–895.
154 Buchner, R. et al. (2008). Toward flexible thermoelectric flow sensors: a new technological approach. *J. Microelectromech. Syst.* 17 (5): 1114–1119.
155 OTEGO Thermoelectric Generators (TEG) for Energy Harvesting Applications. http://www.otego.de/en/ (accessed 5 September 2017).
156 Matrix Industries Introducing the world's first smartwatch you never have to recharge. https://www.powerwatch.com/ (accessed 5 September 2017)

15

Carbon-based Electrocatalysts for Water-splitting

Guoqiang Li[1] *and Weijia Zhou*[2]

[1]*South China University of Technology, Engineering Research Center on Solid-State Lighting and its Informationisation of Guangdong Province, State Key Laboratory of Luminescent Materials and Devices, 381 Wushan Road, Guangzhou 510641, China*
[2]*South China University of Technology, New Energy Research Institute, School of Environment and Energy, Guangzhou Higher Education Mega Center, Guangzhou, Guangdong 510006, China*

15.1 Introduction

Development of effective technologies for clean and sustainable hydrogen energy has been attracting great attention lately, as hydrogen is hailed as a promising energy source that can reduce our dependence on fossil fuels and benefit the environment by reducing the emissions of greenhouse and other toxic gases. Toward this end, an effective and promising approach is based on the electrolysis of water for hydrogen production. Electrocatalytic hydrogen evolution reaction (HER) preferably driven by solar energy is a highly attractive means for meeting these requirements [1–3]. Yet, such a promise will be realized only when the production can be carried out in an efficient, low-cost, and environmentally friendly fashion. The acidic electrolytes (e.g. 0.5 M H_2SO_4) are preferred for water electrolysis to produce hydrogen as there are enough H^+ in the solution to adsorbed on the electrode surface to react. In addition, hydroxide-conducting polymeric electrolytes are still in their technological infancy: they are less stable and slightly less conductive than their proton-conducting counterparts and impose greater overpotentials at the hydrogen electrode [4]. Therefore, significant research efforts have been devoted to the design and engineering of acid-stable HER catalysts.

To this day, Pt-based electrocatalysts are known to efficiently catalyze HER, but their widespread use has been limited by their low earth-abundance and high-cost; hence, the development of HER catalysts that are composed of inexpensive and earth-abundant elements and can efficiently catalyze HER has been one of the main targets in renewable energy research in recent years [5–9]. Recently, studies about carbon-based catalysts with a low overpotential for HER have stimulated a great deal of interest. The pristine carbon materials are electrochemically inert. This is the reason why we can use many carbon-based current collectors such as glassy carbon, carbon paper, and carbon cloth in electrochemical experiments. Chemical modification of the carbon surface is

Flexible Energy Conversion and Storage Devices, First Edition. Edited by Chunyi Zhi and Liming Dai.

usually necessary in order to enhance its electrocatalytic activity [5]. Actually, the electronic state density of carbon may be modulated by transition metal nanoparticles [10–12] and nonmetal doping [13–16] such that carbon may serve as active sites for HER. As a rising category of potential candidates for the replacement of Pt-based catalysts, carbon-based materials are considered to have great potential to solve some of the vital issues for HER. Up to now, the carbon-based electrocatalysts have included two kinds: (i) nonmetallic elements doped carbon and (ii) the metal@carbon core–shell structure. Recent rapid increases of carbon-based electrocatalysts for HER have motivated demand for reviewing this research field. To our knowledge, although some investigations or reviews involving transition metal based, especially molybdenum-based, compounds as HER catalysts have been reported [17–24], comprehensive reviews about carbon-based electrocatalysts for HER are still rare [25]. This chapter summarizes the recent achievements in carbon-based electrocatalysts for HER, which will be beneficial for the future development of other novel low-cost catalysts with high activities and long lifetimes for practical applications. In addition, some notable matters and challenges in the electrochemical measurement of HER are discussed in this chapter.

15.2 Nonmetal-doped Carbon for HER

Heteroatoms (e.g. N, S, P, B, or others) as dopants into graphene can modulate the graphene's physical and chemical properties to obtain more reactive sites. More importantly, this process can produce graphene-based materials with improved ability to adsorb the atomic/molecular species undergoing catalytic reactions and without substantially compromising their conductive properties [26]. These types of hetero-doped structures may provide opportunities for further development of low-cost metal-free catalysts with high activities and long lifetimes for practical applications. It has been known that the difference in electronegativity and size between the heteroatoms (N, B, and S) and carbon can polarize adjacent carbon atoms to facilitate the hydrogen evolution process. The HER performances of heteroatom-doped carbon metal-free catalysts are summarized in Table 15.1.

15.2.1 Nitrogen-doped Carbon-based Catalysts for HER

Nitrogen- containing carbons have received increasing interest as they can improve the properties of carbon for various applications, including oxygen reduction reaction (ORR) [33] and photocatalysis [34]. When nitrogen atoms are introduced into the carbon structure, the electrical conductivity, basicity, oxidation stability, and catalytic activity of carbon can be altered. The performance of these materials crucially depends on the amount of nitrogen in the carbon host as well as its local structure. However, the N-doped carbon as efficient catalysts for HER have rarely been reported [27, 35, 36], which can be performed either by direct synthesis or by post-synthetic treatment. Antonietti and coworkers [35] reported that the carbon nitride (C_3N_4) electrodes show

Table 15.1 Summary of heteroatom-doped metal-free catalysts for HER.

Heteroatom-doped carbon	Methods	Doping elements	Onset potential (mV vs RHE)	Tafel slope (mV dec^{-1})	η (V vs RHE) for $J = -10$ mA cm^{-2}	References
g-C_3N_4 nanoribbon-graphene	A one-step hydrothermal melamine	N	80	54	0.207	[27]
Porous C_3N_4@N-graphene-750 film	A simple vacuum filtration ammonia	N	8	49.1	0.08	[13]
B-SuG	A facile wet chemical synthetic BH_3-THF	B	20	99	—	[26]
N,S-doped carbon–500	Chemical vapor deposition pyridine thiophene	N,S	130	80.5	0.28	[14]
N,S-doped carbon	Thermal decomposition peanut root nodules	N,S	27	67.8	—	[15]
N,P-doped graphene	Chemical – doping melamine triph-enylphosphine	N,P	289	91	0.422	[28]
g-C_3N_4@S–Se-pGr	In situ process sulfur and selenium powder	S,Se	92	86	—	[29]
Nitrogen-doped hexagonal carbon	Chemical vapor deposition N_2	N	65	56.7	0.18	[30]
N-doped carbon	A micelle-template cyanamide	N	—	109	0.239	[27]
C_3N_4@N-doped graphene	Anneal dicyandiamide	N	—	51.1	0.24	[31]
N,S-doped carbon-800	Simple pyrolysis human hair	N,S	12	57.4	0.097	[15]
N,P doped graphene-900	A bottom-up approach urea, phosphoric acid	N,P	120	79	0.213	[32]
N,P-doped carbon	Carbonizing ctaphylococcus aureus	N,P	76	58.4	0.204	[32]

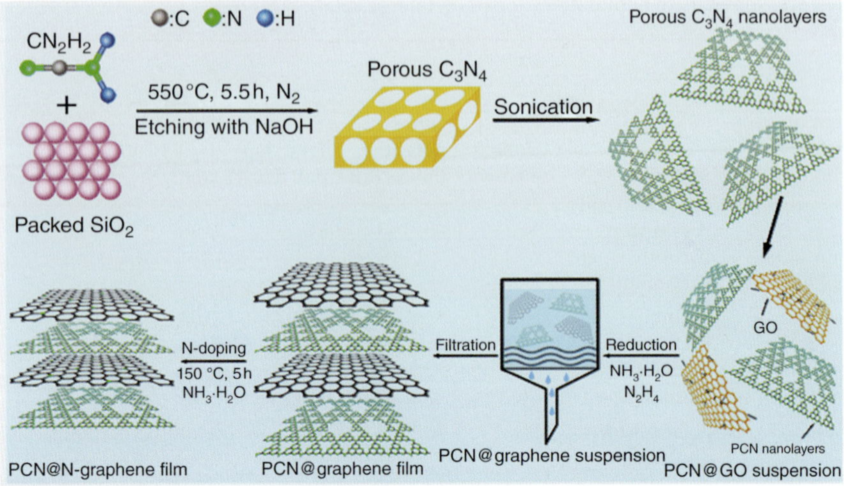

Figure 15.1 Schematic illustration of the preparation process of PCN@N-graphene film. *Source:* Duan et al. 2015 [13]. Copyright 2015. Reprinted with permission from American Chemical Society.

high HER activity with low overpotential and acceptable current densities due to an enhanced H adsorption (Volmer step). Qu's report also suggested that hierarchical architecture of g-C_3N_4 nanoribbon-G provides a large accessible surface area, multielectron transport channel, and short diffusion distance for excellent charge separation and transfer, that effectively accelerates the electrochemical process for HER [30]. As shown in Figure 15.1, Qiao et al. successfully synthesized a flexible three-dimensional (3D) film by integrating porous C_3N_4 nanolayers with nitrogen-doped graphene sheets, which can be directly utilized as HER catalyst electrodes without substrates. This 3D heterostructure has many intriguing properties for favored catalysis, such as highly exposed catalytic centers, hierarchical pores, and strong mechanical flexibility [13]. The electrochemical measurement showed that 3D heterostructured electrodes possess efficient catalytic activity for HER, onset potential (−8 mV vs RHE) close to that of commercial Pt (0 mV vs RHE @ 0.5 mA cm^{-2}), high exchange current density of 0.43 mA cm^{-2}, and remarkable durability (seldom activity loss >5000 cycles).

15.2.2 Other Heteroatom (B, S)-doped Carbon-based Catalysts for HER

Researches involving the electrocatalytic activities of carbon materials doped with other low (e.g. B) [37] or similarly (e.g. S) [26] electronegative atoms have attracted increasing attention. Asefa and coworkers reported a facile wet chemical synthetic method to synthesize electrocatalytically active, B-substituted graphene (B-SuG). The material was synthesized by controlled substitution of the C atoms of defective graphene (DeG) with B atoms using a simple and commercially available borylating agent, borane tetrahydrofuran (BH3-THF). They demonstrated that the B-doped material can serve as an efficient metal-free

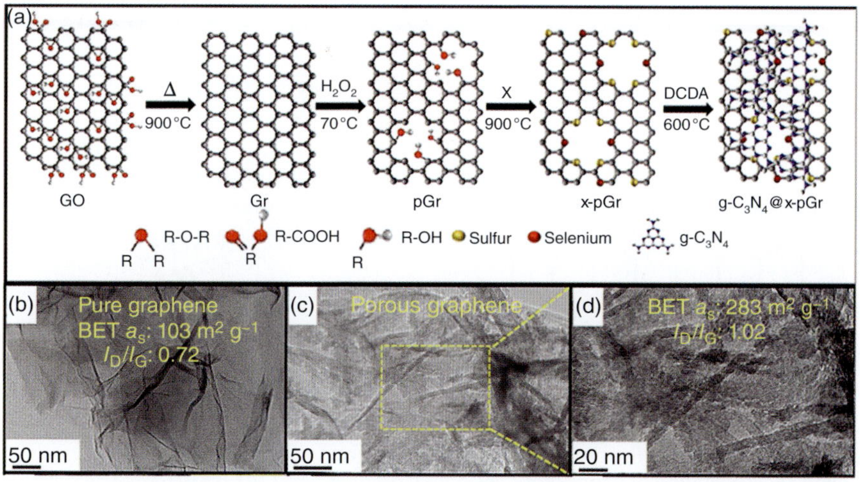

Figure 15.2 (a) Schematic representation of the growth of g-C_3N_4@x-pGr hybrid graphene (x = S, Se, S–Se). TEM images of (b) pure graphene and (c,d) nanoporous graphene (different magnifications). The inset values represent the I_D/I_G ratios and the specific surface areas obtained from Raman spectra and nitrogen adsorption isotherms, respectively. Images (b,c) are the TEM images of pGr with nanopores, and the characteristic features of the graphene surface are clearly visible on this image, while the characteristic features of the pores are not present in the case of pure graphene. *Source:* Shinde et al. 2015 [29]. Copyright 2015. Reprinted with permission from Royal Society of Chemistry.

electrocatalyst for HER [26]. B-SuG shows onset potential of ~0.2 V vs RHE, Tafel slopes of ~99 mV dec^{-1}.

In addition, Amini and coworker [38] succeeded in synthesizing the sulfur-doped graphene nanosheets modified electrode showing good activity for HER. Thus, it is believed that other dopants can effectively tailor the structure of graphene and enhance the HER catalytic activity. Lees's group reported the first successful synthesis of molecularly designed, metal-free catalysts synthesized by coupling a graphitic carbon nitride (g-C_3N_4) with graphene doped by S, Se, or S–Se (Figure 15.2) [29]. In our group, sulfur and nitrogen self-doped carbon nanosheets were prepared as efficient nonmetal catalysts for HER by thermal decomposition of peanut root nodules (an abundant biowaste). Electrochemical measurements show apparent electrocatalytic activity for HER in 0.5 M H_2SO_4, with a small overpotential of only −0.027 V vs RHE, a Tafel slope of 67.8 mV dec^{-1} and good catalytic stability [15]. The catalytic activities increase in the order of nitrogen-doped carbon nanosheets (NC) (onset potential of −0.241 V, vs RHE), nitrogen and sulfur co-doped carbon nanosheets (NSC) (−0.135 V), and sulfur-doped carbon nanosheets (SC) (−0.041 V), which further illustrate that the S doping play a pivotal role in enhancing the HER activity of carbon.

15.2.3 Dual- or Treble-doped Carbons in Metal-free Catalysis

After single heteroatom doping had been confirmed to improve the electrocatalytic performance of carbon materials, development of carbon materials

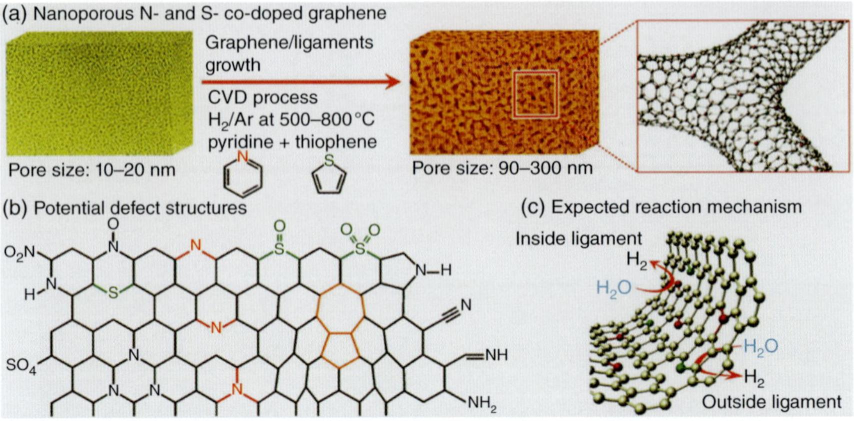

Figure 15.3 Illustration of the preparation process of N and S co-doped nanoporous graphene. (a) Preparation of nanoporous NS-doped graphene by CVD. (b) Potential defect structures in NS-doped nanoporous graphene. (c) Expected reaction mechanism on nanoporous graphene. *Source:* Reprinted with permission from Ref. [14]. Copyright 2014. Wiley-VCH.

doped with dual heteroatoms motivated the curiosity. For example, B/N, S/N, and P/N couples could lead to a unique electron–donor property of carbon by the so-called synergistic coupling effect between two heteroatoms. Chen and coworkers prepared nitrogen and sulfur co-doped nanoporous graphene by nanoporous Ni-based chemical vapor deposition (Figure 15.3) [15]. The obtained nitrogen and sulfur co-doping led to high catalytic activity of nanoporous graphene in HER at low operating potential. N- and S-doped carbon (500 °C) showed the best HER activities with an onset potential of −0.13 V vs RHE, Tafel slope of 80.5 mV dec^{-1}, and operating potential of −0.28 V at 10 mA cm^{-2}. Qiao and coworkers designed and synthesized nitrogen (N) and phosphorus (P) dual-doped graphene by annealing chemically exfoliated graphene oxide powder with a melamine and triphenylphosphine mixture in an Ar atmosphere [28]. The N and P heteroatoms could co-activate the adjacent C atom in the graphene matrix by affecting its valence orbital energy levels to induce a synergistically enhanced reactivity toward HER. Electrochemical measurements (Figure 15.4) showed that N and P dual-doped graphene (overpotential of 420 mV to drive 10 mA cm^{-2}, 91 mV dec^{-1}) was more active than those of the single N-doped graphene (490 mV, 116 mV dec^{-1}) and P-doped graphene (553 mV, 133 mV dec^{-1}).

15.2.4 Metal-doped Carbon for HER

Recently, another kind of metal–carbon hybrid, metal-doped carbon, for efficient HER activity has been developed, such as nanoporous graphene with single-atom nickel dopants [39] and atomic cobalt on nitrogen-doped graphene (Figure 15.5) [40]. Very small amounts of cobalt dispersed as individual atoms on nitrogen-doped graphene possess low overpotentials (30 mV). The catalytically

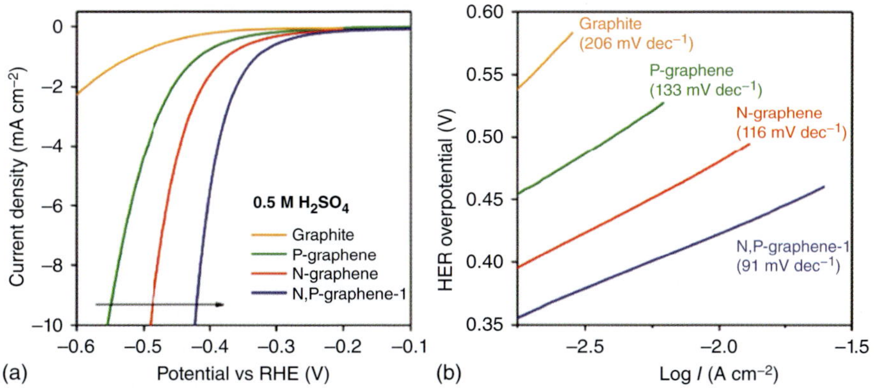

Figure 15.4 (a) HER polarization curves, (b) the corresponding Tafel plots of N- and/or P-doped graphene electrocatalysts in 0.5 M H_2SO_4. *Source:* Zheng et al. 2014 [28]. Copyright 2014. Reprinted with permission from American Chemical Society.

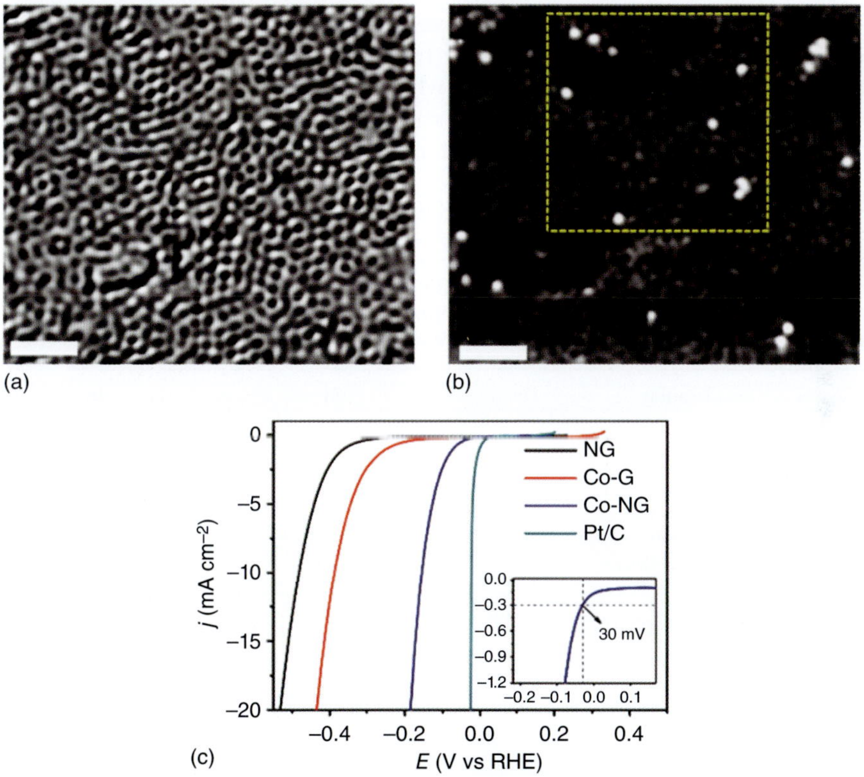

Figure 15.5 (a) Bright-field aberration-corrected STEM image of the Co-NG showing the defective and disordered graphitic carbon structures. Scale bar, 1 nm. (b) HAADF-STEM image of the Co-NG, showing many Co atoms well-dispersed in the carbon matrix. Scale bar, 1 nm. (c) linear sweep voltammetry (LSV) of NG, Co-G, Co-NG and Pt/C in 0.5 M H_2SO_4 at scan rate of 2 mV s^{-1}. The inset shows the enlarged view of the LSV for the Co-NG near the onset region. *Source:* Fei et al. 2015 [40]. Copyright 2015. Reprinted with permission from Nature Publishing Group.

active sites are associated with the metal centers coordinated to nitrogen. In addition, single-atom nickel dopants anchored to 3D nanoporous graphene show superior HER catalysis with a low overpotential of approximately 50 mV and a Tafel slope of 45 mV dec^{-1} in 0.5 M H_2SO_4 solution, together with excellent cycling stability. Experimental and theoretical investigations suggest that the unusual catalytic performance of this catalyst is due to sp–d orbital charge transfer between the Ni dopants and the surrounding carbon atoms.

15.3 Metals Embedded in Carbon for HER

The metal@carbon materials are potentially attractive candidates for HER catalysts. Reports about pure metals of Co, Fe, and Ni used as HER catalysts have been relatively scarce because of their chemical instability in acidic environments. One approach is that of building a thin carbon layer on the surface of metals, which helps to protect the transition metals from dissolution in 0.5 M H_2SO_4 as well as to obtain the prominent catalytic contributions from the entrapped metal nanoparticles due to interfacial charge transfer [6, 10, 11, 15]. Some of these efforts have already led to some efficient metal@carbon HER catalysts (e.g. Fe@carbon, Co@carbon), which possess unique stability in a wide pH range owing to the protection of the carbon shell. The HER performances of metal@carbon catalysts are summarized in Table 15.2.

Table 15.2 Summary of metal@carbon catalysts for HER.

Metal@carbon catalysts	Morphologies	Metal core	Onset potential (mV vs RHE)	Tafel slope (mV dec^{-1})	η (V vs RHE) for $J = -10$ mA cm^{-2}	References
FeCo@N-doped carbon nanotubes-NH_3	Tubular	FeCo	~ −70	63	~ −0.27	[6] b)
Cobalt-embedded nitrogen-rich CNTs	Multiwall carbon nanotubes	Co	−50	69	−0.26	[10] a)
CoNi@N-doped carbon	Nanospheres	CoNi	~0	105	−0.142	[41] b)
Co@Co and nitrogen co-doped carbon	Nanorods	Co	78	~59	0.314	[42] d)
Co-embedded nitrogen-doped CNT/carbon cloth	Nanowire array	Co	—	74	−0.078	[43] a)
Urea-derived carbon nanotubes-900 °C	Fibrous	Co	—	79	−0.27	[44] a)
			—	85	−0.24	[44] c)
			−83	159	−0.34/0.24	[44] e)

continues

Table 15.2 (Continued)

Metal@carbon catalysts	Morphologies	Metal core	Onset potential (mV vs RHE)	Tafel slope (mV dec^{-1})	η (V vs RHE) for $J = -10$ mA cm^{-2}	References
Cobalt–cobalt oxide@ N-doped carbon	Flake-like	CoOx	−85	—	−0.232	[45] c)
Co_3O_4@ graphitic carbon nitride-5%$CoCl_2$ ·$6H_2O$-450 °C	Tubular	Co_3O_4	−30	—	−0.09	[46] a)
CoP-ordered mesoporous carbon	Nanorods	CoP	−77.74	56.7	−0.112	[8] a)
Co@N-doped carbon/N-doped graphene	Wrinkled surface	Co	−49	79.3	−0.2, $J = -13.6$	[47] a)
Co@nitrogen-doped graphene films	Layer alternate	Co	−14	93.9	~ −0.125	[48] a)
Co@N-doped carbon/Ti mesh	Nanorod arrays	Co	−56	78.2	−0.106	[12] a)
Fe@N-doped carbon/rGO	Nanoshells	Fe	—	100	~ −0.39	[49] f)
Ni@N-doped carbon	Nanocoil-like	Ni	—	—	0.19	[50] d)
Ni/carbon quantum dot	Nanoparticles	Ni	~0	98	−0.116	[51] c)
Fe_3C-graphene nanoribbons	Vertically aligned graphene nanoribbons	Fe_3C	−32	46	−0.2, $J = 166.6$	[9] a)
Co_3C-graphene nanoribbons		Co_3C	−41	57	−0.2, $J = 79.6$	
Ni_3C-graphene nanoribbons		Ni_3C	−35	54	−0.2, $J = 116.4$	
Single-shell carbon-encapsulated iron nanoparticle/ single-walled carbon nanotube	Single-shell carbon-encapsulated iron nanoparticles	Fe	~0	40	0.077	[52] a)

a) 0.5 M H_2SO_4.
b) 0.1 M H_2SO_4.
c) 1 M KOH.
d) 0.1 M KOH.
e) Phosphate buffer.
f) 0.5 M $HClO_4$.

15.3.1 Core–Shell Structure for Carbon Nanotube and Nanoparticle

Cobalt-embedded nitrogen-rich carbon nanotubes (CNTs) were synthesized for the first time by Zou et al. (Figure 15.6) [10], which served as highly active electrocatalysts for HER with a small onset potential of −50 mV vs HER and a smaller Tafel slope (69 mV dec^{-1}) at pH 0. The results have guided to synthesize all kinds of metal@carbon electrocatalysts for HER [6, 8, 10–12, 41, 42, 44–46, 49, 50, 52]. The metals as core (Fe, Co, Ni, and alloy) [41, 44, 50, 52], carbon sources to form shell (dicyandiamide [46], urea [44], EDTA [41], cellulose filter [50]), and nanostructures (nanotube [6, 46] and nanoparticles [11]) have been considered to affect the HER activity of metal@carbon. Research progress is summarized below: (i) Up to now, the core metals have been restricted to only a small number of transition metals, such as Fe, Co, and Ni. Extending the core metal to other metals is the important focus of research in the future. (ii) DFT calculations

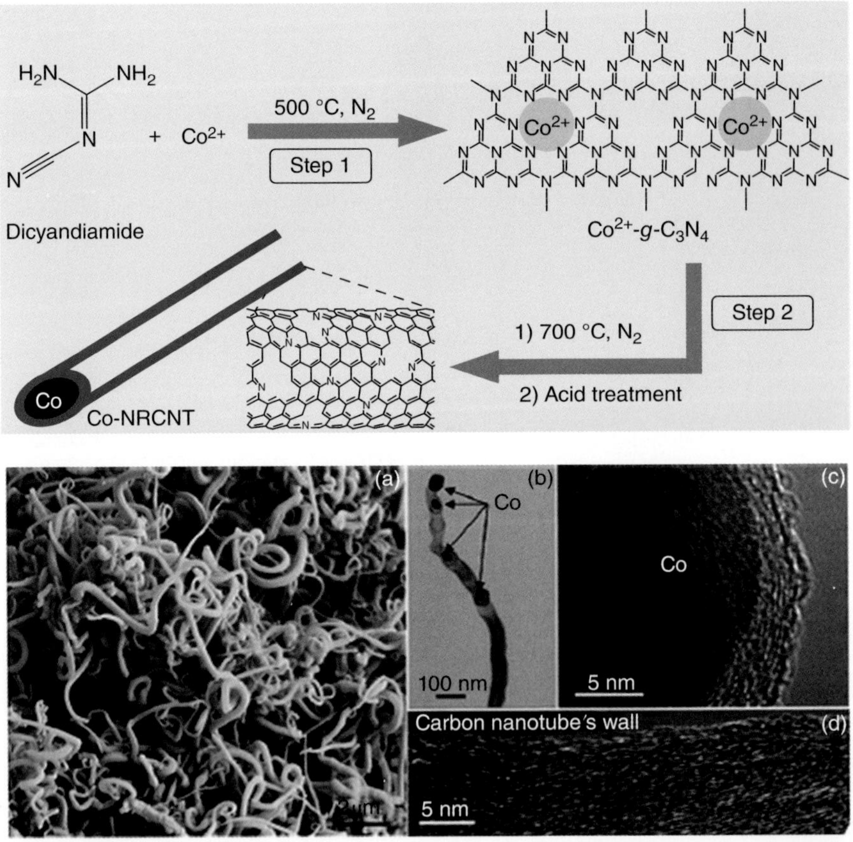

Figure 15.6 Synthesis of Co-NRCNTs: Step 1: thermal treatment at 500 °C in N_2 atmosphere of a mixture of dicyandiamide and $CoCl_2 \cdot 6H_2O$ to form Co^{2+}-*g*-C_3N_4. Step 2: additional thermal treatment at 700 °C in N_2 atmosphere of the Co^{2+}-*g*-C_3N_4, followed by acid treatment of the resulting material to etch away any accessible cobalt species on it. (a) SEM image, (b–d) STEM images of Co-NRCNTs. *Source:* Zou et al. 2014 [10]. Copyright 2014. Reprinted with permission from John Wiley & Sons.

indicate that the ultrathin graphene shells strongly promote electron penetration from the metal core to the graphene surface to enhance the HER activity. Deng et al. [41] designed a hierarchical architecture that consists of ultrathin graphene shells (only 1–3 layers) that encapsulate a uniform CoNi nanoalloy (CoNi@NC), which prepared a bottom-up method by using Co^{2+}, Ni^{2+}, and $EDTA^{4-}$ as precursors to enhance its HER performance in acidic media. The optimized catalyst exhibited high stability and activity with an onset overpotential of almost zero vs RHE and an overpotential of only 142 mV at 10 mA cm^{-2}, which was quite close to that of commercial 40% Pt/C catalysts in 0.1 M H_2SO_4 electrolyte (Figure 15.7). (iii) The doping into carbon shells synergistically increases the electron density on the graphene surface, which results in superior HER activity. Up to now, only N doping into metal@carbon with enhanced HER activity have been reported [10]. The other elements, such as S and P, doped into a carbon shell with high HER activity have been not reported.

According to the above mechanism, the HER catalytic sites were on the surface of metal@carbon, which are only on the end of carbon nanotubes.

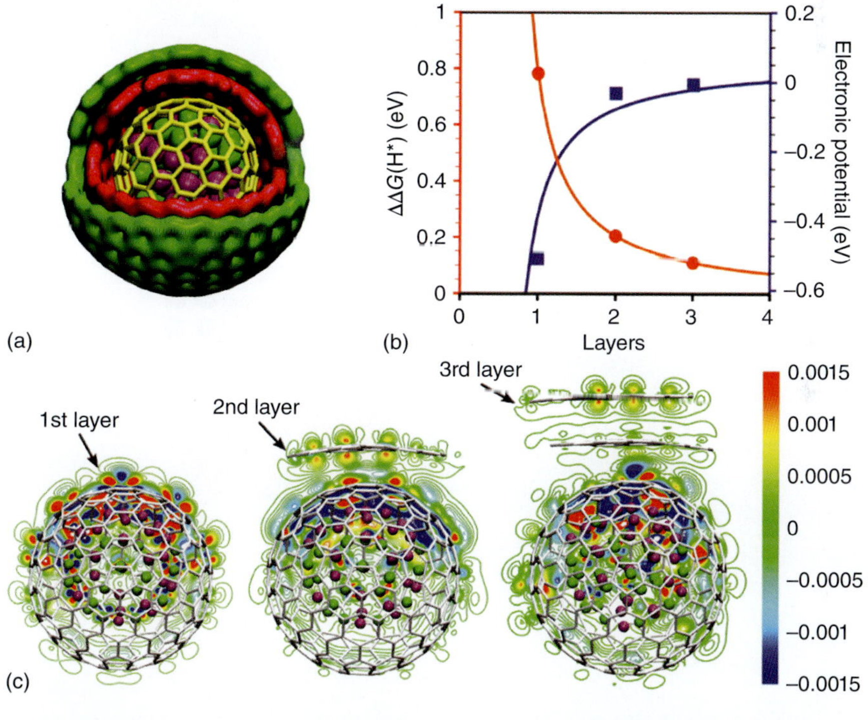

Figure 15.7 (a) Schematic illustration of a Co–Ni alloy encapsulated in three-layer graphene. (b) $\Delta\Delta G(H^*)$ (red line) and electronic potential (blue line) as a function of the number of graphene layers, where $\Delta\Delta G = \Delta G$ (without metal) – ΔG (with metal). (c) Redistribution of the electron densities after the CoNi clusters have been covered by one to three layers of graphene. The differential charge density ($\Delta\rho$) is defined as the difference in the electron density with and without the CoNi cluster. The red and blue regions are regions of increased and decreased electron density, respectively. *Source:* Reprinted with permission from Ref. [41]. Copyright 2014. The Royal Society of Chemistry.

Increasing the interface between metal core and carbon shell in metal@carbon can increase the catalytic sites. So, the metal@carbon structures loaded onto graphene [49], carbon nanotube [10], cellulose filter paper [50], and Ti mesh [12] have been reported. For example, Tavakkoli et al. [52] prepared single-shell carbon-encapsulated iron nanoparticles (SCEINs) decorated on single-walled carbon nanotubes (SWCNTs) as a novel highly active and durable non-noble-metal catalyst for the HER (Figure 15.8). Moreover, the Co@carbon on carbon nanotubes as HER electrocatalysts with a Tafel slope of 40 mV dec^{-1} required a low overpotential of 77 mV to achieve current densities of 10 mA cm^{-2}. Zhou et al. [47] reported that N-doped carbon-wrapped cobalt nanoparticles supported on N-doped graphene nanosheets were prepared

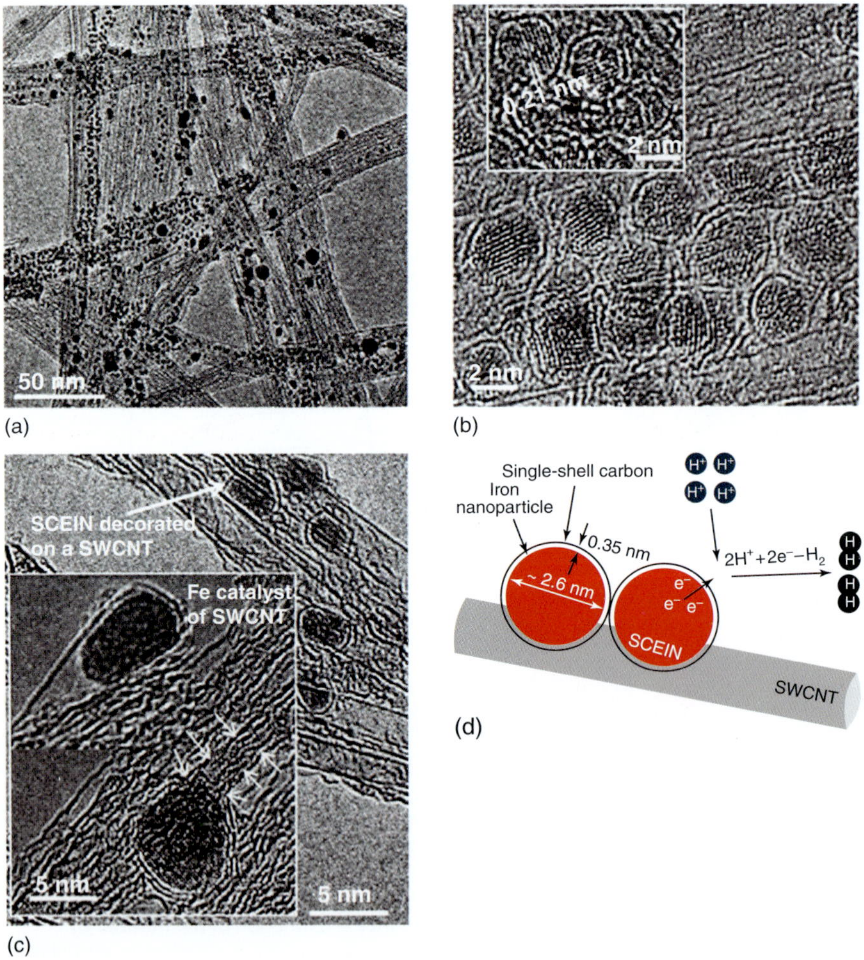

Figure 15.8 (a) TEM image of single-shell carbon-encapsulated iron nanoparticles (SCEINs) supported on SWCNTs showing distribution of the particles on the SWCNTs. (b) HRTEM image of the SCEIN/SWCNT sample, the inset indicates the (110) lattice plane of the Fe particles in SCEINs. (c) HRTEM image of SCEINs decorated on the sidewalls of the SWCNTs, the inset shows Fe catalyst particles for the growth of the SWCNTs (arrows demonstrate the SWCNT). (d) Schematic representation of SCEIN/SWCNT sample simplifying the HRTEM images and HER on SCEINs. *Source:* Reprinted with permission from Ref. [52] Copyright 2014. Wiley-VCH.

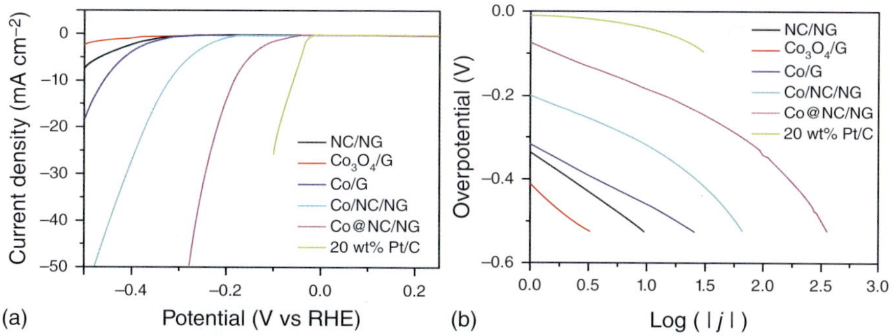

Figure 15.9 (a) Polarization curves for HER in 0.5 M H_2SO_4 at a glassy carbon electrode modified with N-doped carbon/N-doped graphene (NC/NG), Co_3O_4 loaded on graphene (Co_3O_4/G), Co loaded on graphene (Co/G), Co loaded on N-doped carbon/N-doped graphene (Co/NC/NG), Co@N-doped carbon/N-doped graphene (Co@NC/NG) (700 °C), and 20 wt% Pt/C, respectively. Potential sweep rate 5 mV s^{-1}. (b) Corresponding Tafel plots derived from (a). *Source:* Zhou et al. 2015 [47]. Copyright 2015. Reprinted with permission from Royal Society of Chemistry.

by a facile solvothermal procedure and subsequent calcination at controlled temperatures, which were shown a small overpotential of 49 mV with a Tafel slope of 79.3 mV dec^{-1} and prominent electrochemical durability in 0.5 M H_2SO_4. The electrochemical results confirmed that the synergetic effect among N-doped graphene, metal Co, and core–shell structure played an important role in enhancing HER activity (Figure 15.9).

Notably, to develop HER electrodes that are of high efficiency and binder-free, active components are in general directly grown on current-collecting substrates instead of using Nafion or PTFE to immobilize catalysts on electrode surfaces. The 3D electrode entailed a large surface area and abundant exposed active sites. Zhou et al. [12] described the preparation of self-supported N-doped carbon-coated cobalt nanorod arrays supported on a Ti mesh (Co@NC/Ti, Figure 15.10), which exhibited a remarkable HER performance in acid solutions with an onset potential of −56 mV vs RHE, a Tafel slope of 78.2 mV dec^{-1}, and robust stability for 8 h of continuous operation. In addition, Xing et al. [43] described the direct growth of a film consisting of interconnected Co-entrapped, N-doped carbon nanotubes on carbon cloth using chemical vapor deposition from dicyanodiamine using a Co_3O_4 nanowire array as a flexible 3D carbon-based electrode catalyst, which is highly active toward the HER and delivers 10, 20, and 100 mA cm^{-2} at overpotentials of 78, 100, and 155 mV in 0.5 M H_2SO_4, respectively, with a Tafel slope of 97 mV dec^{-1} in 1.0 M phosphate-buffered saline.

15.3.2 Metal Organic Frameworks for HER

Metal organic frameworks (MOFs), built from metal ions and organic ligands by coordination bonds, are a family of crystalline porous solids [53]. Thanks to their particular structure, MOFs possess diverse composition, ultrahigh surface area and ordered and controllable porous structure [54]. Porous carbon and metal oxide/carbon hybrid materials with the morphology of MOFs can be obtained easily by direct pyrolysis of MOFs under an inert atmosphere. For this reason,

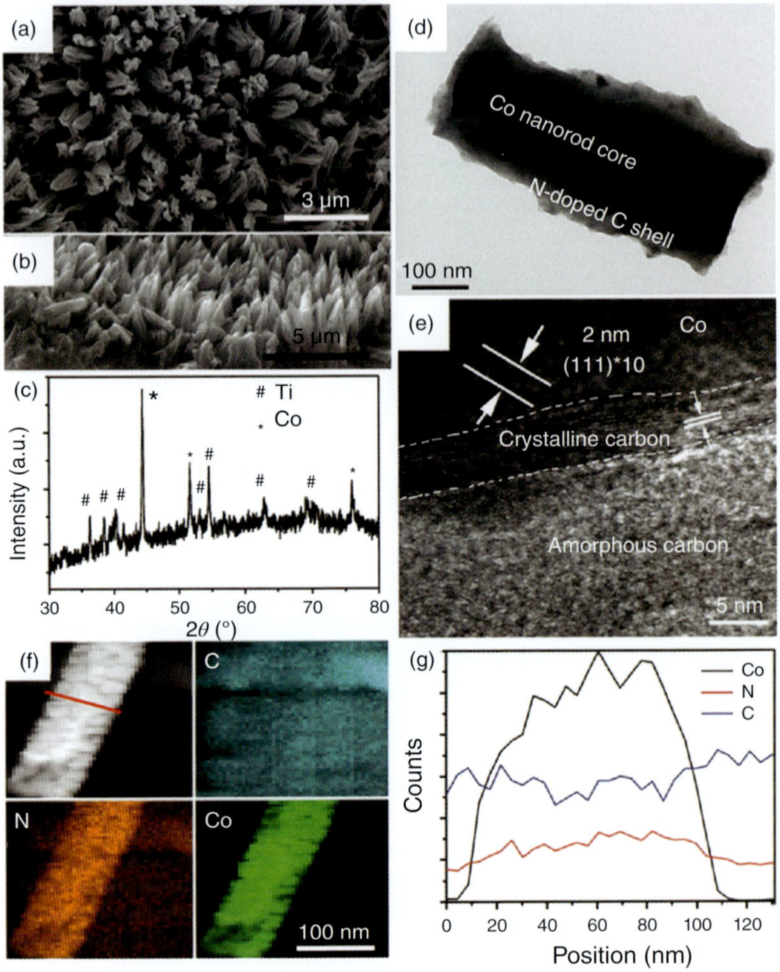

Figure 15.10 (a) Top view and (b) side view SEM images, (c) XRD patterns and (d,e) TEM images of Co@NC/Ti. (f) The corresponding EDX elemental mapping images of C, N, and Co for Co@NC. (g) EDX line scan curves showing C, N, and Co element profiles across the Co@NC indicated by the red line in (f). *Source:* Zhou et al. 2015 [12]. Copyright 2015. Reprinted with permission from Royal Society of Chemistry.

MOFs-derived electrocatalysts have been used in energy conversion and storage applications and exhibit excellent electrochemical activity, like supercapacitors [55, 56], ORR [55, 57, 58], and oxygen evolution reaction (OER) [57, 59]. In addition, MOFs-derived materials are also potential electrocatalysts for HER.

When MOFs were used as electrocatalysts for HER, they can be divided into two groups: (i) MOFs used as electrocatalysts directly without any other procedure. Polyoxometalates (POMs)-based MOFs, sometimes called POMOFs, were widely utilized as electrocatalysts for HER in recent years, for the reason that they have combined the advantages of POMs and MOFs [59]. These POMOFs as HER catalysts exhibit excellent electrocatalytic activity toward HER due to the metal organic ligand structure [60, 61]. Other types of MOFs, like

Cu-MOF [62]and Co-MOF [63] are also applied as electrocatalysts for HER without calcination. However, usually, the catalytic stability of MOFs for HER is not very good. (ii) Metal@carbon materials derived from MOFs calcined at different atmosphere as electrocatalysts for HER. The high catalytic activity is attributed mainly to the following reasons:(i) The porous structure of carbon inherited from the MOFs with a large surface area and (ii) the metal@carbon core–shell structure where electrons can transfer from metal to carbon to improve the catalytic activity.

Co-MOFs and Ni-MOFs can act as good precursors for synthesizing effective HER catalysts [64]. For instance, Hou et al. [64] prepared a nitrogen-doped graphene/cobalt-embedded porous carbon polyhedron (N/Co-doped PCP//NRGO) by calcinating the mixture of Co-MOFs and graphene oxide under N_2 atmosphere (Figure 15.11); the hybrid offering a low onset potential of 58 mV vs RHE and Tafel slope of 126 mV dec^{-1} at acid media. Research about MOFs-derived HER catalysts are not limited only to a handful of elements (Zn, Cu, Co), and other elements (such as Mo) are also widely used. Wu et al. [65] synthesized porous molybdenum carbide nano-octahedrons (MoC_x) by pyrolysis of Mo-based MOFs (Figure 15.12). And MoC_x nano-octahedrons are regarded as good electrocatalysts for HER, offering a low onset potential of −25 and −80 mV in acid and basic electrolyte and a small Tafel slope of 53 and 59 mV dec^{-1}, respectively. However, the carbides M_xC (such as Mo_2C, Fe_3C) catalysts for HER possess different catalytic mechanisms from that of metal@carbon core–shell electrocatalysts.

In addition to the direct carbonation of MOFs, other kinds of treatments to synthesize metal compounds as electrocatalysts for HER have also been reported recently. Direct phosphating Ni-MOFs at a mild condition can transform Ni-MOFs into nickel phosphide, which was proved to be an outstanding catalyst for HER in both acid electrolyte and basic electrolyte. For example, Tian et al. [66] synthesized nickel phosphides (Ni_2P) derived from Ni-MOFs by mixing the Ni-BTC with NaH_2PO_2 and heated at 275 °C in a muffle furnace, and the products showed excellent HER activity in 0.5 M H_2SO_4 with a low onset potential of −75 mV and a small Tafel slope of 62 mV dec^{-1}.

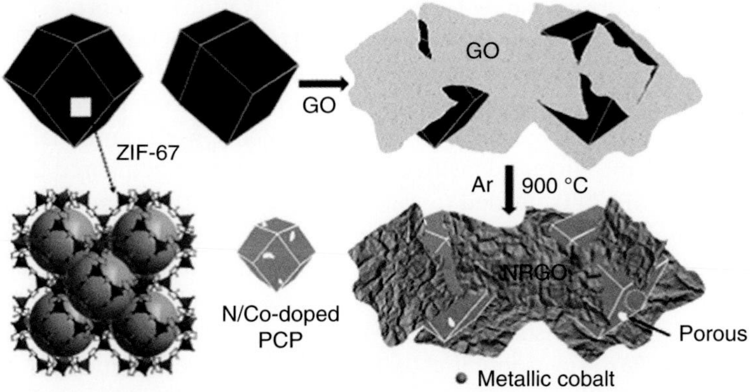

Figure 15.11 Schematic illustration for the synthesis process of N/Co-doped PCP/NRGO. *Source:* Reprinted with permission from Ref. [64]. Copyright 2014. Wiley-VCH.

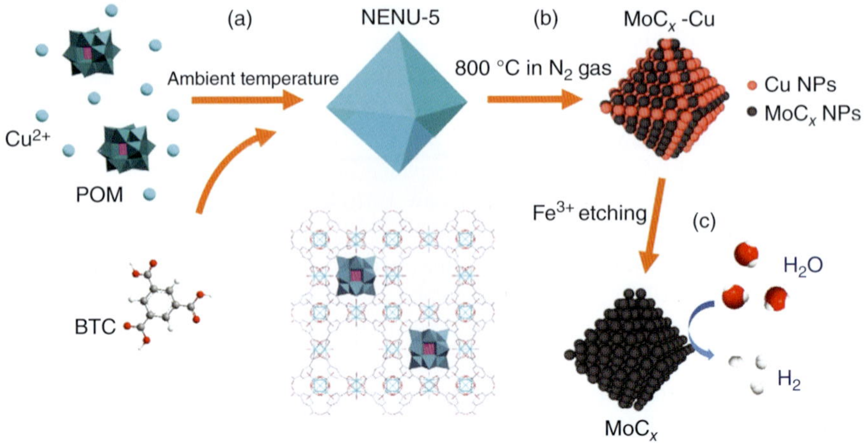

Figure 15.12 Schematic illustration of the synthesis procedure for porous MoC_x nanooctahedrons. (a) Synthesis of NENU-5 nanooctahedrons with Mo-based POMs residing in the pores of HKUST-1 host. (b) Formation of MoC_x-Cu nanooctahedrons after annealing at 800 °C. (c) Removal of metallic Cu nanoparticles by Fe^{3+} etching to produce porous MoC_x nanooctahedrons for electrocatalytic hydrogen production. *Source:* Wu et al. 2015 [65]. Copyright 2015. Reprinted with permission from Nature Publishing Group.

15.4 Electrochemistry

The parameters to evaluate the HER performance of electrocatalysts include overpotentials, current density, Tafel plot, impendence, electrochemical surface areas,stability, etc. However, there are many questions that need to answered with regard to electrochemical measurements, especially for carbon-based electrocatalysts for HER, which should be rigorously evaluated to avoid possible errors in process testing.

15.4.1 Overpotential/Onset Potential and Calibration

To conduct electrochemical water-splitting, voltages above the thermodynamic potential values corresponding to the intrinsic activation barriers present in both half reactions were recognized as overpotential (η) or onset potential [67, 68]. The onset potential is the applied potential with apparent cathodic currents. However, it is difficult to demarcate the precise potential when the circuit start to produce a current density. Yu and coworkers defined η_0 as the potential to achieve 0.5 mA cm^{-2} of the current density. Asefa and coworkers compared the overpotentials of Co-NRCNTs needed to yield a current density of 1 mA cm^{-2}. For the purpose of further quantitative comparison of HER activity, the required overpotential to receive a current density of 10 mA cm^{-2} was used to evaluate the activities of the electrocatalysts in recent reports [64, 69]. In fact, the measured voltage ($E_{electrolysis}$) consists of three parts: $E_{electrolysis} = E_{reversible} + \Delta E_{irreversible} + IR$. The $\Delta E_{irreversible}$ is the overpotential (η) for HER. So, in order to measure the accurate overpotential (η), the potential sweep rate of polarization curve should be measured based on the low sweep rate (such as 2 or 5 mV s^{-1}) to minimize the capacitive current, especially for 3D electrodes with high electrochemical surface area.

In all measurements, the reference electrodes (Ag/AgCl or Hg/Hg_2Cl_2) should be calibrated with respect to a reversible hydrogen electrode (RHE). The revision process should be done at regular intervals due to the possibility of shift of reference electrodes. One method is experimental measurement. The calibration was performed in a high-purity H_2 (99.999%) saturated electrolyte with Pt wires as the working electrode and counter electrode. Cyclic voltammograms (CVs) were collected at a scan rate of 1 $mV\,s^{-1}$, and the average of the two potentials at which the current crossed zero was taken as the thermodynamic potential for the hydrogen electrode reactions. The other method is theoretical calculation by Nernst equation

$$2H^+(aq) + 2e^- \rightleftharpoons H_2\,(g) \qquad H^+ \mid H_2(g) \mid \text{Catalysts}$$

$$E = E^\theta - \frac{RT}{nF}\ln\frac{[H^+]^2}{1} = E^\theta - 0.059\ \text{pH}$$

$$E^\theta(Hg/Hg_2Cl_2,\ \text{saturated KCl}) = 0.242\ \text{V}$$

$$E^\theta(Ag/AgCl,\ \text{saturated KCl}) = 0.197\ \text{V}$$

where, the pH of 0.5 M H_2SO_4 is 0.3. R is the universal gas constant, $R = 8.314\ 472\ \text{J}\,\text{K}^{-1}\,\text{mol}^{-1}$, T is the absolute temperature ($T = 298.15$ K), F is the Faraday constant, the number of coulombs per mole of electrons ($F = 9.648\,533\,99 \times 10^4\ \text{C}\,\text{mol}^{-1}$), n is the number of moles of electrons transferred in the cell reaction or half-reaction ($n = 2$).

15.4.2 Current Density and Electrochemical Surface Area

It is noteworthy that the current density ($mA\,cm^{-2}$) value of catalyst powder and block 3D electrode are achieved from two different testing methods. The first is by loading a certain amount of catalyst onto a glass carbon electrode (GCE), whose area is known as the geometry area. The second is, for block 3D electrode, the geometric area of electrodes is used to calculate the current density. In fact, their mass loading of catalyst and actual electrochemical area are far more than those of powder catalysts loaded onto conductive substrates. In addition, iR-compensation can exclude the influence of the resistance (R_s) from the electrochemical workstation and electrode, which will increase the current density. The obtained results are useful to study the catalytic mechanism, such as the calculation of Tafel plots. However, the current density obtained without iR-compensation can give the integral and actual HER performance of electrodes.

The electrochemical surface area is used as an approximate guide for surface roughness within an order-of-magnitude accuracy [70]. To estimate the effective surface areas, the capacitance of the double layer at the solid–liquid interface was measured. The cyclic voltammograms (CVs) were collected in a certain potential window without faradaic processes [68]. The current density of CV consists of two parts: capacitance current (proportional to r^1, r is the sweep speed) and faradic current (proportional to $r^{1/2}$). Therefore, the sweep speed should be high enough, such as 40, 80, 120, 160, and 200 $mV\,s^{-1}$ to obtain the capacitance current.

15.4.3 Tafel Plot and Exchange Current Density

A Tafel plot is normally used to evaluate the intrinsic property of the catalysts and the efficiency of the catalytic reaction. The polarization curves were replotted as overpotential (η) vs log current ($\log j$) to get Tafel plots. By fitting the linear portion of the Tafel plots to the Tafel equation ($\eta = b \log j + a$), the Tafel slope (b) can be obtained. Note that for hydrogen evolution in acid on metal electrode surfaces, the mechanism typically involves three major reactions [68].

First is the primary discharge step (Volmer reaction):

$$H_3O^+ + e^- \rightarrow H_{ads} + H_2O, b = \frac{2.3RT}{\alpha F} \approx 120 \text{ mV} \tag{15.1}$$

where R is the ideal gas constant, T is the absolute temperature, $R \approx 0.5$ is the symmetry coefficient, and F is the Faraday constant. This step is followed by either an electrochemical desorption step (Heyrovsky reaction),

$$H_{ads} + H_3O^+ + e^- \rightarrow H_2 + H_2O, b = \frac{2.3RT}{(1+\alpha)F} \approx 40 \text{ mV} \tag{15.2}$$

or a recombination step (Tafel reaction),

$$H_{ads} + H_{ads} \rightarrow H_2, b = \frac{2.3RT}{(1+\alpha)F} \approx 30 \text{ mV} \tag{15.3}$$

The Tafel slope is an inherent property of the catalyst that is determined by the rate-limiting step of the HER. For example, HER kinetic models suggest that Tafel slope of about 120, 40, or 30 mV dec^{-1} will be obtained if the Volmer, Heyrovsky, or Tafel reaction is the rate-determining step, respectively [71]. In addition, the logarithmic current density usually deviates from linear dependence at high overpotentials, which is strongly influenced by evolved hydrogen bubbles limiting the available surface area or mass transport. Therefore, Tafel plots are recorded with the linear portions at low overpotential fitted to the Tafel equation [64]. To further evaluate the intrinsic activity of the electrocatalysts, the exchange current density (j_0) is determined by extrapolation of the Tafel plots to j/log(j) axis. For $\eta = b \log j + a$, the constant terms a and b are known from the Tafel plot, we assume that the η is zero, and then the j_0 is determined. The large exchange current density indicates the fast large surface area, electron transfer rate, and favorable HER kinetics.

15.4.4 Electrochemical Impedance

Electrochemical impedance spectroscopy (EIS) is a powerful technique to characterize interface reactions and electrode kinetics in HER. The simple equivalent circuit was usually given, where a constant phase element (CPE) was employed. The series resistance (R_s) represents a combination of ionic resistance of the electrolytes, electronic resistance of the electrode materials, and internal resistance of electrochemical workstation. It was obtained in the high frequency zone, and then sometimes used to correct the polarization curves (IR-corrected). The charge transfer resistance (R_{ct}) is related to the electrocatalytic kinetics and its lower value corresponds to the faster reaction rate, which can be obtained from the semicircle in the low frequency zone. From the Nyquist plots of the EIS response, we can see a depressed semicircle in the high-frequency region (R_{ct})

and a quasi-sloping line in the low-frequency region (corresponding to mass transfer resistance). It can be noted that in the Nyquist plots of the EIS response at various overpotentials, the diameter of the semicircles diminished markedly with increasing overpotential, suggesting decreasing R_{ct} with increasingly negative electrode potentials.

15.4.5 HER Durability and H_2 Production

Durability is another important criterion to evaluate the performance of an electrocatalyst. To probe the durability of the catalyst in acidic environment, long-term potential cycling, and current-time responses at fixed potentials over extended periods were performed. It is noted that after continuous cyclic voltammograms at an accelerated scanning rate of 100 mV s^{-1} for 1000 or 5000 number of cycles, the polarization curves overlay almost exactly with the initial one, with negligible loss of cathodic current [72]. Long-term electrochemical stability was also measured by galvanostatically or chronoamperometry, maintaining for certain hours with a given current density and applied potential [73, 74]. In fact, the i–t testing shows the real reaction activity of hydrogen production, and the measured current densities obtained from the i–t curve are usually less than or equal to that obtained from the polarization curve.

The hydrogen gas production rate was quantified by gas chromatographic measurements. Owing to the increasing pressure that originates during a large hydrogen gas production, the measuring process in the electrolytic cell should proceed carefully. After that, the faradaic efficiency (FE) of the electrocatalytic hydrogen evolution process was calculated by comparing the amount of experimentally quantified hydrogen (gas chromatographic measurements) with the theoretically calculated hydrogen (chronoamperometry). Faraday efficiency for HER is an important factor to confirm complete decomposition of water and other reductive side-reactions.

15.4.6 Activation

Before testing for HER, the electrodes with cured membranes were activated by soaking in electrolyte overnight to improve the contact between catalysts and electrolyte. Another way is that given numbers of cyclic voltammograms were carried out, which can improve the HER activity significantly because of the interactions between active species, especially for the hydrophobic catalysts. In addition, the electrolytes were enhanced, and the contact and the electron transfer at the interface were subsequently dramatically improved. Usually, the activation process can only enhance the current density according to the above activation mechanism. However, sometimes, the onset potentials that were significantly improved by the activation treatment have been reported. Occasionally, there are studies claiming that "metal-free" carbon nanotubes are electrocatalytically active. For example, some groups [75, 76] reported that oxidization of carbon nanotubes introduced abundant carboxylic groups to their surface, which in turn made nanotubes highly active and durable for HER in acid. The electrocatalyst showed an onset potential of −100 mV and required an overpotential of 220 mV to reach 10 mA cm^{-2}.

However, there is a problem that needs special attention. It should be confirmed that the high HER performance was not from the dissolution of the Pt electrode in acid electrolyte. Especially, for the long duration of the activation process with high potential and large current density, the Pt electrode with small physical area can be dissolved from oxidation Pt^0 to Pt^{2+}, which is then electro-reduced onto the working electrode. The other facilitating condition is the existence of chlorine ion in the electrolyte from a reference electrode ($Ag/AgCl$ and Hg/Hg_2Cl_2). The results have been reported by some researchers [77, 78]. When a Pt wire was used for the activation process, the electrochemical catalytic activity of the graphite working electrode was greatly improved to an extent as good as that of commercial Pt/C; however, when a graphite rod was used, the working electrode could not be activated effectively. In our group, we also confirmed the result in earlier studies. The molybdenum oxide loaded on reduced graphene oxide (MoO_x/rGO) was synthesized by hydrothermal method, which was used as electrocatalyst for HER (Figure 15.13). Before activation, the

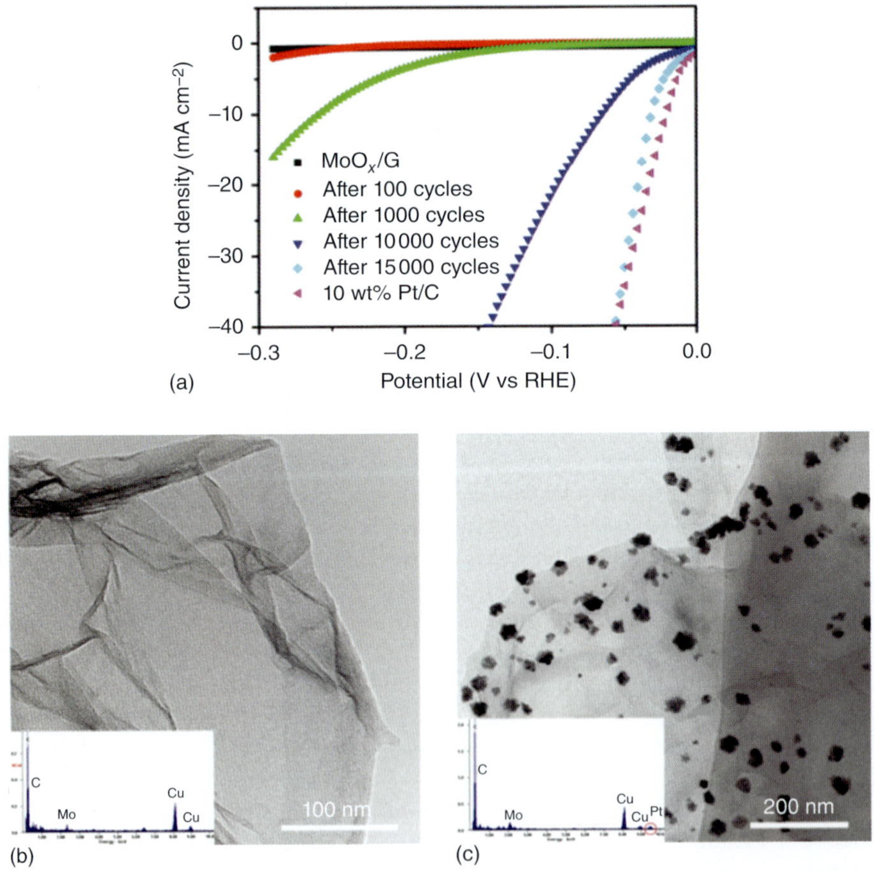

Figure 15.13 (a) Polarization curves for HER in 0.5 M H_2SO_4 at a glassy carbon electrode modified with MoO_x/rGO with different activation time. Potential sweep rate 5 mV s^{-1}. TEM images (b) and EDX (c) of MoO_x/rGO after the activation.

MoO_x/rGO possess bad HER performance. But after activation with different cycles, the HER performance has been improved dramatically. Fortunately, we characterized the MoO_x/rGO after activation by elemental analysis, where apparent peaks for Pt were observed, which confirmed the enhanced HER catalytic performance from the dissolution of the Pt electrode in acid electrolyte.

In order to avoid the dissolution of the Pt electrode in acidic electrolyte, some measures should be employed. (i) The surface area of Pt as counter electrode should be much larger than that of a working electrode; (ii) the applied overpotential should not be too high; (iii) the presence of chloride ions in electrolyte should be avoided; (iv) the porous ceramic membrane between the counter electrode and working electrode should be used. In fact, Pt is not indispensable as a counter electrode. The counter electrode was used to balance the current and voltage from a working electrode. So, materials with stability and conductive properties can be used as counter electrodes, such as Ti mesh, carbon rod, and carbon cloth this study suggests that no Pt should be used in electrochemical measurement as a counter electrode for HER and OER in acidic electrolyte, especially under continuous CV scan and *i*–*t* testing. However, this is not a problem in alkaline electrolyte.

15.5 Outlook and Future Challenges

In recent years, a new family of carbon-based HER catalysts have been developed as alternatives to Pt-based catalysts for HER in acidic medium. Two main types of carbon-based HER catalysts with low overpotentials and good durability have been developed, including nonmetal-doped carbon and metal@carbon. In this chapter, we review the research progress in the development of carbon-based electrocatalysts toward HER during the past few years. Continuing breakthrough about carbon-based electrocatalysts have revealed new catalysts with improved HER performance. However, several unanswered questions about carbon-based electrocatalysts toward HER need to be systematically researched, such as controllable synthesis of metal@carbon, catalytic mechanism, etc.

15.5.1 HER Mechanism for Carbon-based Catalysts

(i) As for nonmetal-doped carbon, the enhanced catalytic activity for HER was interpreted by the same mechanism for ORR in most of the papers. Up to now, the reported results about nonmetal types and doping sites into carbon as HER electrocatalysts were limited, which cannot give the accurate laws to synthesize efficient HER catalysts. For example, N-doped carbons as efficient catalysts were widely used in ORR, although, not all N-doped carbons possess high performance for HER. Usually, there are some possible metal impurities (e.g. Fe, Co, Ni, and Mo) in nonmetal-doped carbons (such as carbon nanotubes with the metal catalyst seeds, activating agent of $MgCl_2$, and template of Ni foam), which will enhance the HER performance of nonmetal-doped carbon. In fact, the single-atom Co and Ni doped into carbon as efficient HER electrocatalysts have been developed [52, 53]. (ii) As for metal@carbon, the core metals have been restricted to only a small number of transition metals, such as Fe, Co, and Ni. Extending the core metal to other metals is the focus of important research in the future, which will give more information for analyzing the catalytic activity caused by different electronic structures. In addition, the controllable thickness and component of shell structures should be established to confirm the catalytic sites from metal core or carbon shell. Lastly, the above mechanism was proposed by DFT calculations. So, more experimental results need to give more direct evidences.

15.5.2 Electrochemistry, Especially for Activation Process

The activation treatment for enhancing HER performance in electrochemical measurement with Pt as counter electrode should be used with caution. If eliminating the dissolution of Pt, the overpotentials for HER were also observed to be markedly improved after the activation; however, the mechanism should be researched, and cannot be simply attributed to the full infiltration between the electrolyte and electrode.

15.5.3 OER in Acidic Electrolyte

The acidic electrolytes (e.g. 0.5 M H_2SO_4) are preferred for water electrolysis to produce hydrogen as there are enough H^+ in the solution to adsorb on the electrode surface to react. Up to now, significant research efforts have been devoted to the design and engineering of acid-stable HER catalysts. However, for the final industrial application, it is also essential to develop the electrolysis of water by two electrodes: anode for OER and cathode for HER. So, going forward, the development of OER catalysts with high performance in acidic electrolyte is more urgently required. Unfortunately, few results about the efficient catalysts for OER in acidic electrolyte have been reported [79, 80].

References

1 Luo, J., Im, J.-H., Mayer, M.T. et al. (2014). *Science* 345: 1593–1596.
2 Popczun, E.J., Mckone, J.R., Read, C.G. et al. (2013). *J. Am. Chem. Soc.* 135: 9267–9270.
3 Sathre, R., Scown, C.D., Morrow, W.R. et al. (2014). *Energy Environ. Sci.* 7: 3264–3278.
4 Varcoe, J.R., Atanassov, P., Dekel, D.R. et al. (2014). *Energy Environ. Sci.* 7: 3135–3191.
5 Zeng, M. and Li, Y. (2015). *J. Mater. Chem. A* 3: 14942–14962.
6 Deng, J., Ren, P., Deng, D. et al. (2014). *Energy Environ. Sci.* 7: 1919.
7 Su, Y., Zhu, Y., Jiang, H. et al. (2014). *Nanoscale* 6: 15080–15089.

8 Walter, M.G., Warren, E.L., Mckone, J.R. et al. (2010). *Chem. Rev.* 110: 6446–6473.

9 Morales-Guio, C.G., Stern, L.-A., and Hu, X. (2014). *Chem. Soc. Rev.* 43: 6555–6569.

10 Zou, X., Huang, X., Goswami, A. et al. (2014). *Angew. Chem. Int. Ed.* 53: 4372–4376.

11 Wang, T., Zhou, Q., Wang, X. et al. (2015). *J. Mater. Chem. A* 3: 16435–16439.

12 Zhou, W., Zhou, Y., Yang, L. et al. (2015). *J. Mater. Chem. A* 3: 1915–1919.

13 Duan, J., Chen, S., Jaroniec, M., and Qiao, S.Z. (2015). *ACS Nano* 9: 931–940.

14 Ito, Y., Cong, W., Fujita, T. et al. (2015). *Angew. Chem. Int. Ed.* 54: 2131–2136.

15 Liu, X., Zhou, W., Yang, L. et al. (2015). *J. Mater. Chem. A* 3: 8840–8846.

16 Zhou, Y., Leng, Y., Zhou, W. et al. (2015). *Nano Energy* 16: 357–366.

17 Zeng, K. and Zhang, D. (2010). *Prog. Energy Combust. Sci.* 36: 307–326.

18 Yan, Y., Xia, B., Xu, Z., and Wang, X. (2014). *ACS Catal.* 4: 1693–1705.

19 Jiao, F. and Frei, H. (2010). *Energy Environ. Sci.* 3: 1018–1027.

20 Yang, J. and Shin, H.S. (2014). *J. Mater. Chem. A* 2: 5979–5985.

21 Merki, D. and Hu, X. (2011). *Energy Environ. Sci.* 4: 3878–3888.

22 Laursen, A.B., Kegnæs, S., Dahl, S., and Chorkendorff, I. (2012). *Energy Environ. Sci.* 5: 5577–5591.

23 Jiao, Y., Zheng, Y., Jaroniec, M., and Qiao, S.Z. (2015). *Chem. Soc. Rev.* 44: 2060–2086.

24 Wang, J., Cui, W., Liu, Q. et al. (2015). *Adv. Mater.* doi: 10.1002/adma.201502696.

25 Zheng, Y., Jiao, Y., and Qiao, S.Z. (2015). *Adv. Mater.* 27: 5372–5378.

26 Sathe, B.R., Zou, X., and Asefa, T. (2014). *Catal. Sci. Technol.* 4: 2023–2030.

27 Huang, X., Zhao, Y., Ao, Z., and Wang, G. (2014). *Sci. Rep.* 4: 7557.

28 Zheng, Y., Jiao, Y., Li, L.H. et al. (2014). *ACS Nano* 8: 5290–5296.

29 Shinde, S.S., Sami, A., and Lee, J.-H. (2015). *J. Mater. Chem. A* 3: 12810–12819.

30 Liu, Y., Yu, H., Quan, X. et al. (2014). *Sci. Rep.* 4: 6843.

31 Zheng, Y., Jiao, Y., Zhu, Y. et al. (2014). *Nat. Commun.* 5: 3783.

32 Wei, L., Karahan, H.E., Goh, K. et al. (2015). *J. Mater. Chem. A* 3: 7210–7214.

33 Liang, J., Du, X., Gibson, C. et al. (2013). *Adv. Mater.* 25: 6226–6231.

34 Liao, L., Zhu, J., Bian, X. et al. (2013). *Adv. Funct. Mater.* 23: 5326–5333.

35 Shalom, M., Gimenez, S., Schipper, F. et al. (2014). *Angew. Chem. Int. Ed.* 53: 3654–3658.

36 Su, D.S., Zhang, J., Frank, B. et al. (2010). *ChemSusChem* 3: 169–180.

37 Chhetri, M., Maitra, S., Chakraborty, H. et al. (2016). *Energy Environ. Sci.* doi: 10.1039/c5ee02521d.

38 Shervedani, R.K. and Amini, A. (2015). *Carbon* 93: 762–773.

39 Qiu, H.J., Ito, Y., Cong, W. et al. (2015). *Angew. Chem. Int. Ed.* 127: 14237–14241.

40 Fei, H., Dong, J., Arellano-Jimenez, M.J. et al. (2015). *Nat. Commun.* 6: 8668.

41 Deng, J., Ren, P., Deng, D., and Bao, X. (2015). *Angew. Chem. Int. Ed.* 54: 2100–2104.

42 Wang, Y., Nie, Y., Ding, W. et al. (2015). *Chem. Commun.* 51: 8942–8945.
43 Xing, Z., Liu, Q., Xing, W. et al. (2015). *ChemSusChem* 8: 1850–1855.
44 Gao, S., Li, G.D., Liu, Y. et al. (2015). *Nanoscale* 7: 2306–2316.
45 Jin, H., Wang, J., Su, D. et al. (2015). *J. Am. Chem. Soc.* 137: 2688–2694.
46 Tahir, M., Mahmood, N., Zhang, X.X. et al. (2015). *Nano Res.* 8: 3725–3736.
47 Zhou, W., Zhou, J., Zhou, Y. et al. (2015). *Chem. Mater.* 27: 2026–2032.
48 Hou, D., Zhou, W., Zhou, K. et al. (2015). *J. Mater. Chem. A* 3: 15962–15968.
49 Wang, J., Wang, G., Miao, S. et al. (2014). *Faraday Discuss.* 176: 135–151.
50 Ren, J.W., Antonietti, M., and Fellinger, T.P. (2015). *Adv. Energy Mater.* 5: doi: 10.1002/aenm.201401660.
51 Yang, Y., Liu, J., Guo, S. et al. (2015). *J. Mater. Chem. A* 3: 18598–18604.
52 Tavakkoli, M., Kallio, T., Reynaud, O. et al. (2015). *Angew. Chem. Int. Ed.* 54: 4535–4538.
53 Zhang, W., Wu, Z.-Y., Jiang, H.-L., and Yu, S.-H. (2014). *J. Am. Chem. Soc.* 136: 14385–14388.
54 Zhao, S., Yin, H., Du, L. et al. (2014). *ACS Nano* 8: 12660–12668.
55 Cao, X., Zheng, B., Shi, W. et al. (2015). *Adv. Mater.* 27: 4695–4701.
56 Chaikittisilp, W., Hu, M., Wang, H. et al. (2012). *Chem. Commun.* 48: 7259–7261.
57 Ma, T.Y., Dai, S., Jaroniec, M., and Qiao, S.Z. (2014). *J. Am. Chem. Soc.* 136: 13925–13931.
58 Jahan, M., Bao, Q., and Loh, K.P. (2012). *J. Am. Chem. Soc.* 134: 6707–6713.
59 Qin, J.-S., Du, D.-Y., Guan, W. et al. (2015). *J. Am. Chem. Soc.* 137: 7169–7177.
60 Rousseau, G., Rodriguez-Albelo, L.M., Salomon, W. et al. (2015). *Cryst. Growth Des.* 15: 449–456.
61 Nohra, B., El Moll, H., Rodriguez Albelo, L.M. et al. (2011). *J. Am. Chem. Soc.* 133: 13363–13374.
62 Jahan, M., Liu, Z., and Loh, K.P. (2013). *Adv. Funct. Mater.* 23: 5363–5372.
63 Gong, Y., Hao, Z., Meng, J. et al. (2014). *ChemPlusChem* 79: 266–277.
64 Hou, Y., Wen, Z., Cui, S. et al. (2015). *Adv. Funct. Mater.* 25: 872–882.
65 Wu, H.B., Xia, B.Y., Yu, L. et al. (2015). *Nat. Commun.* 6: 6512.
66 Tian, T., Ai, L., and Jiang, J. (2015). *RSC Adv.* 5: 10290–10295.
67 Xu, Y.-F., Gao, M.-R., Zheng, Y.-R. et al. (2013). *Angew. Chem. Int. Ed.* 52: 8546–8550.
68 Gao, M.-R., Liang, J.-X., Zheng, Y.-R. et al. (2015). *Nat. Commun.* 6: 5982.
69 Peng, S., Li, L., Han, X. et al. (2014). *Angew. Chem. Int. Ed.* 53: 12594–12599.
70 Mccrory, C.C.L., Jung, S., Peters, J.C., and Jaramillo, T.F. (2013). *J. Am. Chem. Soc.* 135: 16977–16987.
71 Li, Y., Wang, H., Xie, L. et al. (2011). *J. Am. Chem. Soc.* 133: 7296–7299.
72 Yang, Y., Fei, H., Ruan, G. et al. (2014). *Adv. Mater.* 26: 8163–8168.
73 Popczun, E.J., Read, C.G., Roske, C.W. et al. (2014). *Angew. Chem. Int. Ed.* 126: 5531–5534.
74 Callejas, J.F., Read, C.G., Popczun, E.J. et al. (2015). *Chem. Mater.* 27: 3769–3774.
75 Das, R.K., Wang, Y., Vasilyeva, S.V. et al. (2014). *ACS Nano* 8: 8447–8456.
76 Cui, W., Liu, Q., Cheng, N. et al. (2014). *Chem. Commun.* 50: 9340–9342.

77 Dong, G., Fang, M., Wang, H. et al. (2015). *J. Mater. Chem. A* 3: 13080–13086.
78 Zhang, C., Hong, Y., Dai, R. et al. (2015). *ACS Appl. Mater. Interfaces* 7: 11648–11653.
79 Frydendal, R., Paoli, E.A., Chorkendorff, I. et al. (2015). *Adv. Energy Mater.* 5: doi: 10.1002/aenm.201500991.
80 Huynh, M., Bediako, D.K., and Nocera, D.G. (2014). *J. Am. Chem. Soc.* 136: 6002–6010.

Index

d

l